ITALIAN PHYSICAL SOCIETY

PROCEEDINGS

OF THE

INTERNATIONAL SCHOOL OF PHYSICS « ENRICO FERMI »

COURSE XXI

edited by G. CARERI
Director of the Course

VARENNA ON LAKE COMO
VILLA MONASTERO
JULY 3 - JULY 15 1961

Liquid Helium

1963

ACADEMIC PRESS • NEW YORK AND LONDON

SOCIETÀ ITALIANA DI FISICA

RENDICONTI

DELLA

SCUOLA INTERNAZIONALE DI FISICA «ENRICO FERMI»

XXI Corso

a cura di G. Careri
Direttore del Corso

VARENNA SUL LAGO DI COMO
VILLA MONASTERO
3-15 LUGLIO 1961

Elio liquido

1963

ACADEMIC PRESS • NEW YORK AND LONDON

ACADEMIC PRESS INC.
111 Fifth Avenue
New York 3, N. Y.

United Kingdom Edition
Published by
ACADEMIC PRESS INC. (London) Ltd.
Berkeley Square House, London W. 1

Library of Congress Catalog Card Number: 63-14745

PRINTED IN ITALY

INDICE

Lezioni

Seminari

Foreword.

G. Careri
Direttore del corso

Liquid Helium is a 53-year old subject. It started as the strangest liquid, now we realize it as the simplest liquid and a convenient system where new aspects of the many-body problem can be studied.

Therefore we felt that this school should be intended not as a repetition of the glorious past, but should provide the students with the background for the progress which is expected in the future.

The Varenna School is intended mainly for experimentalist research students, and in order to complement their preparation an emphasis has been given to theory at a level which can be familiar to them. Some seminars on contemporary aspects were also included to keep some unity between senior and junior investigators. Needless to say we did not expect to cover all the contemporary work in this field, but we tried to help the new generation to focus its attention on some open questions.

SOCIETÀ ITALIANA DI FISICA

SCUOLA INTERNAZIONALE DI FISICA « E. FERMI »

XXI CORSO - VARENNA SUL LAGO DI COMO - VILLA MONASTERO - 3-15 Luglio 1961

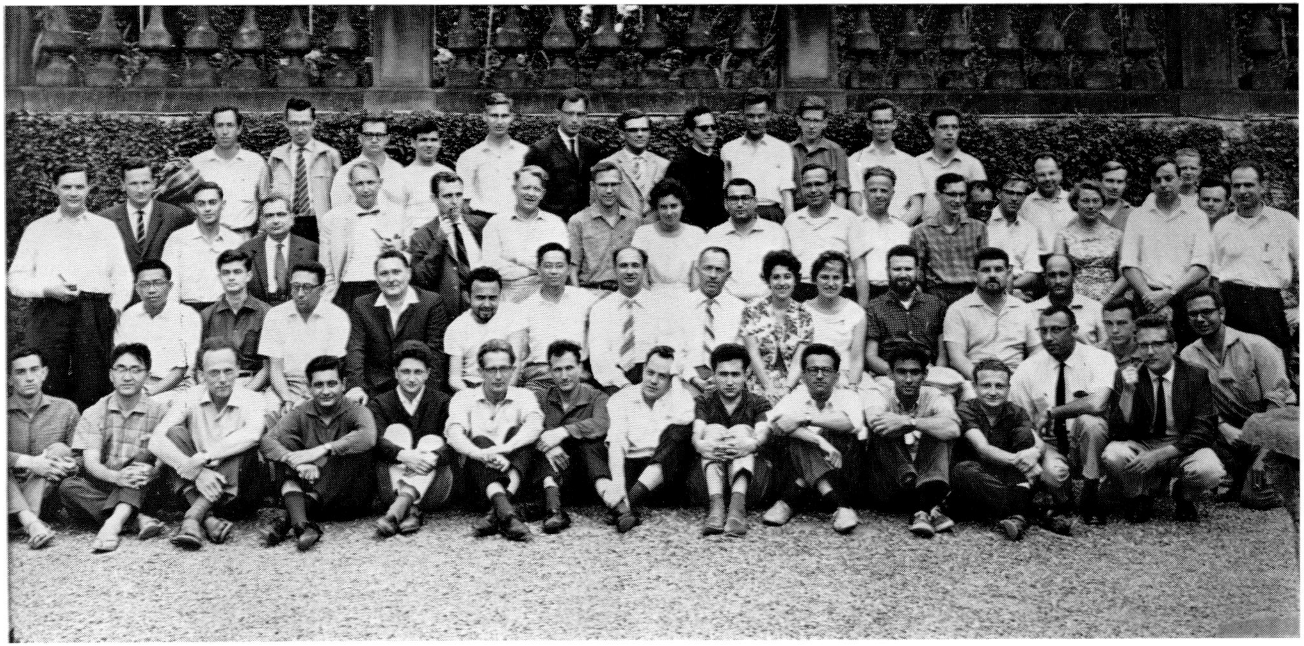

LEZIONI

Excitation Model for Liquid Helium II.

J. DE BOER

Instituut voor Theoretische Fysica, Universiteit van Amsterdam - Amsterdam

I. - Thermodynamic Properties of Liquid Helium.

1. – Discussion of diagram of states on the basis of the principle of corresponding states.

The exceptional position of the substance helium and in particular its condensed phase, can best be demonstrated by comparing its properties with those of the other noble gases and the hydrogen isotopes, which, in first approximation, can also be treated as a monoatomic substance. For such a comparison, in particular if quantum effects are under consideration, the appropriate method is a study on the basis of the principle of corresponding states using molecular units (1948) [1]. From the known values of the molecular parameters ε and σ characterizing the intermolecular potential field

$$\varphi(r) = 4\varepsilon[(\sigma/r)^{12} - (\sigma/r)^6]\,,$$

(r is the intermolecular distance, ε is the depth of the potential field at the minimum occurring at $r_{\min}=2^{\frac{1}{6}}\sigma$ and σ is the « diameter » for which $\varphi(\sigma)=0$, *i.e.* the interaction energy changes sign), one derives « molecular units » for temperature ε/k, for molar volume $N_m\sigma^3$, for molar energy $N_m\varepsilon$ and for pressure ε/σ^3. For helium the values for these molecular units are:

TABLE I. – *Molecular units for* ^{4}He *and* ^{3}He.

$\varepsilon/k = 10.22\ °K$	$N_m\sigma^3 = 10.06\ cm^3/mol$
$\varepsilon/\sigma^3 = 83.4\ atm$	$N_m\varepsilon = 84.9\ J/mol$
$[\varepsilon = 14.11\cdot10^{-16}\ erg$	$\sigma = 2.556\ Å]$

The macroscopic properties for the various substances can now all be expressed in terms of these « molecular units », *e.g.* $T^*=kT/\varepsilon$, $V^*_{\rm m}=V_{\rm m}/N_{\rm m}\sigma^3$, $U^*_{\rm m}=U_{\rm m}/N_{\rm m}\varepsilon$, $P^*=P\sigma^3/\varepsilon$, etc.

In *classical theory* it can now easily be proved from dimensional considerations that, for instance, the reduced equation of state $P^*(V^*, T^*)$ should be the same for all monoatomic substances, provided they are properly described by the Lennard-Jones potential field $\varphi(r)$ and provided that the condition of additivity of intermolecular forces is satisfied. The consequence is that also characteristic quantities like *e.g.* the reduced critical temperatures $T^*_{\rm cr}$ and the reduced triple points temperature $T^*_{\rm tr}$, the reduced critical molar volume $V^*_{\rm m,cr}$ and the reduced molar volume $V^*_{\rm m,0}$ and molar internal energy $U^*_{\rm m,0}$ at the absolute zero-point should be the same for the various substances.

In *quantum theory* this classical principle of corresponding states is no longer valid: again dimensional considerations show that as soon as Planck's constant h plays a role, the reduced equation of state takes the form $P^*=P^*(V^*, P^*, \Lambda^*)$, where $\Lambda^*=\Lambda/\sigma=h/(m\varepsilon)^{\frac{1}{2}}\sigma$ is a dimensionless quantity: the « quantum-mechanical parameter », which is essentially the ratio of a characteristic molecular de Broglie wave length $\Lambda=h/(m\varepsilon)^{\frac{1}{2}}$ to the diameter σ. When this quantum-mechanical parameter is small, quantum effects are negligible, but when Λ^* becomes larger quantum effects become more and more important.

For the heavy noble gases Λ^* is quite small: for Xe: $\Lambda^*=0.064$, for Kr: $\Lambda^*=0.102$ and for A: $\Lambda^*=0.186$ showing that in these cases quantum effects are negligible. In fact these heavy substances satisfy very nicely the classical principle of corresponding states with $T^*_{\rm cr}=1.26$, $T^*_{\rm tr}=0.7$, $V^*_{\rm m,cr}=3.25$, $V^*_{\rm m,0}\approx 0.9$, $U^*_{\rm m,0}\approx -8.6$.

For the lighter substances the quantum-mechanical parameter becomes larger and larger: $\Lambda^*=1.22$ (resp. 1.73) for D_2 (resp. H_2) and reaches the extreme values $\Lambda^*=2.67$ and 3.08 for ^{4}He, and ^{3}He, respectively.

These deviations from classical theory are illustrated in Fig. 1 by plotting the reduced critical temperature as function of the quantum-mechanical parameter Λ^*: one obtains thus a rather smooth curve and in fact this curve was used to predict in 1948 [2] the value of $T_{\rm cr}$ for ^{3}He before this substance was available for macroscopic measurements.

A similar plot of the reduced molar volume at absolute zero shows that the reduced molar volume increases from its classical value 0.95 to $(V^*_{\rm m,0})_{\rm sol}=2.11$ and $(V^*_{\rm m,0})_{\rm liq}=2.75$ for liquid ^{4}He and to the even higher value $(V^*_{\rm m,0})_{\rm liq}=3.86$ for liquid ^{3}He: *as a result of quantum effects* the *light substances are thus « blown up » very much* to volumes 3 to 4 times of the corresponding classical value.

The quantum effect on the reduced internal energy is even more drastic. Here the reduced internal molar energy of the condensed phase, which is

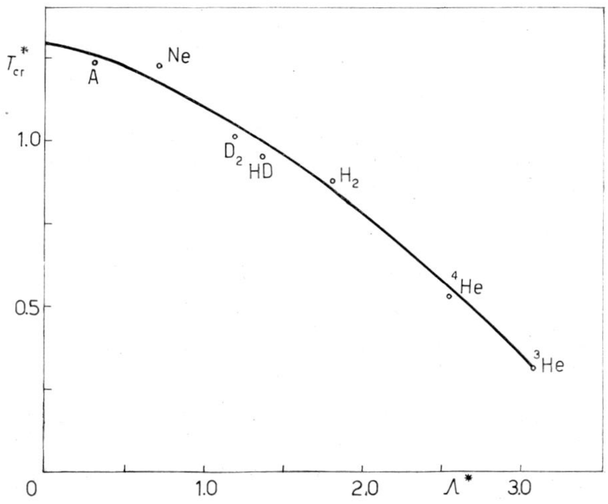

Fig. 1. – The reduced critical temperature plotted as function of the quantum-mechanical parameter Λ^* for various substances, showing the large influence of quantum effects in the case of ^{4}He and ^{3}He.

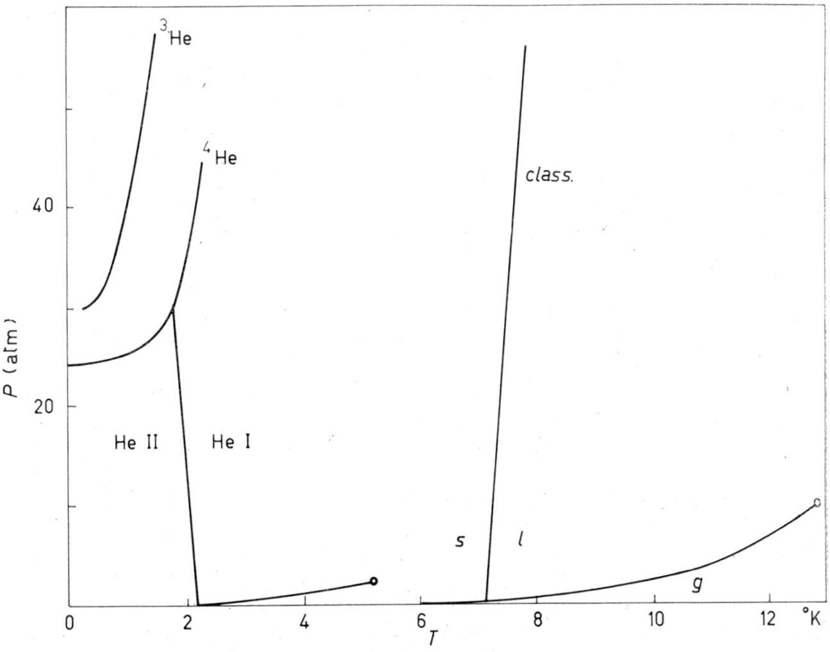

Fig. 2. – The diagram of states of ^{4}He and ^{3}He compared with the corresponding classical diagram of state.

$U^*_{m,0} = -8.6$ for classical substances changes gradually as Λ^* increases and reaches the value $(U^*_{m,0})_{liq} = -0.70$ for liquid ^{4}He and even $(U^*_{m,0})^*_{liq} = -0.25$ for liquid ^{3}He. The *reduced « binding energy » of the condensed phase is thus reduced by a factor larger than* 10 *as a result of the quantum effects.*

An overall picture of the large influence of quantum effects in the case of helium is obtained by considering the diagram of states for ^{4}He and ^{3}He in Fig. 2. In this diagram is also drawn the « classical » diagram of states which would have been the real one if quantum effects would have been absent. Besides a shift of the vapour pressure line describing the gas-liquid equilibrium and the critical point, the really drastic differences result from the big shift of the melting line to lower temperatures and higher pressure which leads to the disappearance of the overcrossing of melting line and vapour pressure line and thus also of the triple point, whith the consequence of the extension of the liquid phase down to the absolute zero. It is well known that in the case of ^{4}He there exists in the liquid phase a phase transition on the λ-line, which separates the two phases liquid He I and liquid He II.

2. – The zeropoint energy. Behaviour at $T = 0$.

A qualitative explanation of the exceptional position of liquid helium was given more than 30 years ago already by SIMON as being due to the zero-point energy. For a solid substance described by the Debye theory the total molar zeropoint energy connected with the zero-point oscillations of all the longitudinal and transversal elastic modes is given by

$$U_{m,z} = 3\int_0^{k_D} \frac{1}{2}\hbar c k \, \mathrm{d}\boldsymbol{k}/8\pi^3 = \frac{9}{8} R\Theta_D \,, \tag{1.1}$$

where $\Theta_D = \hbar c k_D/k$ (c=velocity of sound). The « Debye circular wave number » k_D is defined as usual by

$$N_m = V_m \int_0^{k_D} \mathrm{d}\boldsymbol{k}/8\pi^3 = V_m \cdot \frac{4}{3}\pi k_D^3/8\pi^3 \,. \tag{1.2}$$

As the characteristic Debye temperature increases with decreasing volume, this zero-point energy increases also and consequently exerts an internal pressure which tends to blow up the substance. In classical theory the minimum energy occurs at the reduced volume: $V^*_{m,0} = 0.916$ at which the reduced potential energy of the close-packed structure (calculated with the Lennard-Jones interaction $\varphi(r)$) assumes the value $U^*_{m,0} = -8.610$. It si easy to verify [3] that the above mentioned zero-point pressure causes deviations from those values, which in first approximation are proportional to Λ^*. Also if one describes the solid phase and an Einstein solid, with all particles oscillating independently in the repulsive field of the surrounding particles, one obtains the same result: the zeropoint energy « blows up » the condensed phase and moves away the molecules from the lattice arrangement of minimum potential energy to regions of higher energy, thus causing the very large shift in the internal energy of ^{4}He and ^{3}He, found experimentally.

A very approximate picture for the co-existence of two phases, a solid and a liquid phase, at the absolute zero was given by LONDON (1936) [4]: Although at small molar volume (and higher pressures) the close-packed structure (with co-ordination number 12) has the lowest energy, the large zeropoint energy has the effect that at larger molar volumes and smaller pressure a « structure » with lower co-ordination number (about 6) has a lower energy and thus becomes stable. This then causes the phase transition from the high-pressure close-packed solid phase to the low-pressure liquid phase (at $P=25$ atm for ^{4}He and about 30 atm for ^{3}He) which certainly has a much lower co-ordination number (although this is a somewhat vague concept for quantum liquids like ^{4}He and ^{3}He!).

3. – Caloric properties.

After this rather qualitative description of the situation at the absolute zero of temperature, we now turn to a brief survey of the caloric properties at nonzero temperatures. In Fig. 3 the molar heat capacity for ^{4}He and ^{3}He is drawn as function of temperature. For ^{4}He at low temperatures ($T < 0.6$ °K) $c_v = aT^3$, with a value of $a=(16\pi^5k^4/15h^3c^3)V$ corresponding to the low-temperature limit of a Debye solid with longitudinal waves only ($c=238$ m/s). At temperature $T > 0.6$ °K the heat capacity rises exponentially and shows a phase transition at $T=2.1$ °K. The exact behaviour of the heat-capacity curve in the vi-

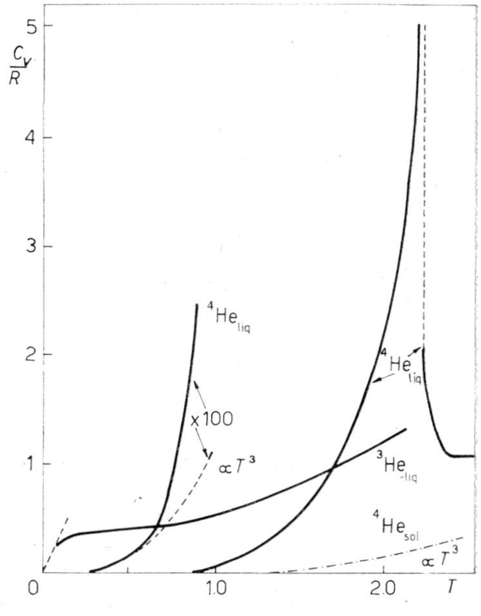

Fig. 3. – The molar heat capacity for liquid ^{4}He and ^{3}He.

cinity of the λ-point has been investigated in very much detail in particular by BUCKINGHAM and FAIRBANK (1957) [5]. There appears to exist a logarithmic infinity on both sides of the λ-point.

For ^{3}He the heat capacity appears to be linear $C_v = bT$ at very low temperatures ($T < 0.1$ °K) as it would be for an ideal gas of Fermi molecules. Using the value $b = (\pi^2 m^*_{\text{eff}} k^2/h^2) N (4\pi V/3N)^{\frac{2}{3}}$ for an ideal gas one finds an effective mass of about $2m_{\text{He}}$. For higher temperatures the curve deviates strongly from this linear curve. So far no singularity in the heat capacity has been found.

4. – Thermal expansion.

Both liquids ^{4}He and ^{3}He have a very remarkable behaviour of the thermal expansion coefficient as a function of temperature.

For ^{4}He the thermal expansion coefficient (see Fig. 4*a*) is proportional to T^3 and positive at low temperatures $T < 0.5$ °K, then the curve changes sign at 1.2 °K and between this temperature and the λ-point the thermal expansion is negative.

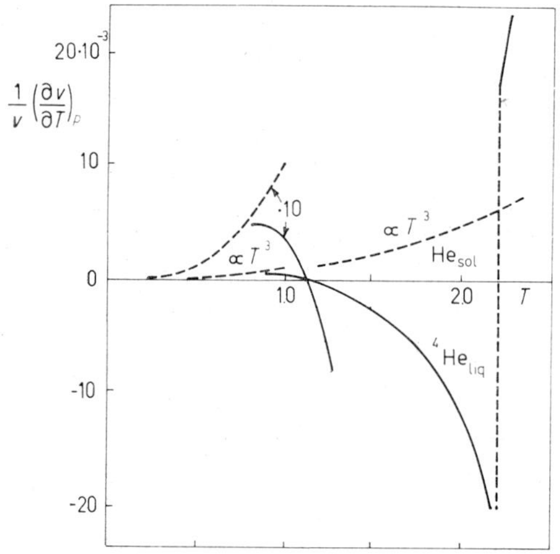

Fig. 4*a*. – The thermal expansion curve for liquid and solid ^{4}He. At temperatures below 1.3 °K the scale has been increased by a factor 10.

For ^{3}He the thermal expansion coefficient (Fig. 4*b*) becomes proportional to T and negative at very low temperatures $T < 0.1$ °K and then changes sign at 0.6 °K and remains positive for higher temperatures.

The behaviour can at least partly be understood using the thermodynamic relation:

$$(1.3)\quad \alpha = \frac{1}{V}\left(\frac{\partial V}{\partial T}\right)_P = \varkappa\left(\frac{\partial P}{\partial T}\right)_V = \varkappa\left(\frac{\partial S}{\partial V}\right)_T,$$

where $\varkappa = -(1/V)(\partial V/\partial P)_T$ is the compressibility.

This gives for ^{4}He at $T < 0.5$ °K:

$$(1.4)\qquad S = \frac{1}{3}aT^3, \qquad \alpha = \frac{1}{3}\varkappa a T^3\left[1 - 3\,\frac{\partial \ln c}{\partial \ln V}\right],$$

and for ^{3}He at $T < 0.1$ °K:

$$S = bT\,, \qquad \alpha = \varkappa bT\left[\frac{2}{3} + \frac{\partial \ln m_{\text{eff}}}{\partial \ln V}\right]. \tag{1.5}$$

In agreement with the experimental facts for ^{4}He that $\partial c/\partial V < 0$, one finds a positive α, proportional to T^3 at low temperatures. At higher temperatures

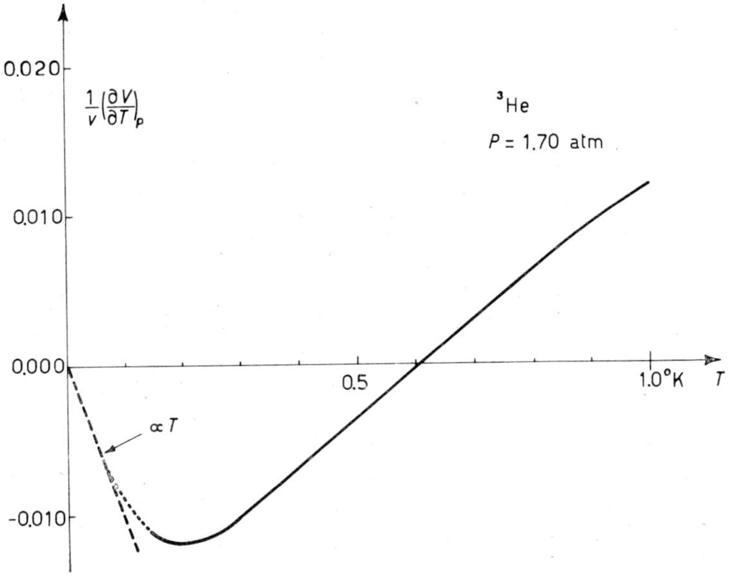

Fig. 4*b*. – The thermal expansion curve for liquid ^{3}He.

the abnormal thermal expansion coefficient shows that there $\partial S/\partial V$ must be negative. On the other hand the fact that experimentally for ^{3}He the thermal expansion coefficient is negative at very low T shows that $\partial \ln m_{\text{eff}}/\partial V < -\frac{2}{3}$, so that the effective mass should become larger on compression.

In the following lecture only the behaviour of ^{4}He will be considered more in detail from the theoretical point of view. The further theoretical discussion of the behaviour of liquid ^{3}He will be given by Prof. SESSLER.

II. - Phenomenological Theory of Two-Fluid Model.

1. – The two-fluid model.

The origin of the two-fluid model for the description of the properties of liquid ^{4}He II came from two different directions. TISZA (1938) [6] starting from the analogy of the situation in liquid helium II with that existing in an ideal B. E. gas below its condensation temperature, suggested the two-fluid model and predicted the possibility of temperature (« second sound ») waves in liquid helium. LANDAU (1941) [7] developed a slightly different version of the two-fluid model, which on the end has proved to give a quite accurate description of the phenomena in liquid helium. We therefore now first present the basic assumptions on which LANDAU based the two-fluid model. For the purpose of describing the many reversible and irreversible transport phenomena occurring in liquid helium II, it is useful to introduce the concept of two interpenetrating fluids: the *superfluid* and the *normal fluid*, which can move with respect to each other practically without friction. The simplest properties to be attached to the two fluids were the following:

a) The *superfluid* moves *without friction* through capillaries and over surfaces as long as the velocity does not become larger than a so-called « critical velocity » the value of which is of the order of magnitude of $(1\div10)$ cm/s for flows through narrow tubes.

b) The *superfluid transports no entropy* as is shown by the well-known mechano-caloric effect: if liquid helium flows through narrow capillaries the heat $Q=TS$ is developed at the entry of the capillary and should be supplied at the exit again if we want to keep the flow isothermal. The same applies to the flow over surfaces.

c) The *normal-fluid density*, *i.e.* the mass density of the fluid which still behaves normal and which shows viscous friction has been determined by the famous experiment of ANDRONIKASHVILI (1946): a pile of discs, at a very small distance, hangs in the liquid attached to a torsion wire. During the oscillations around the axis which the pile of discs can carry out, the superfluid part remains at rest as this is not accelerated with the discs because of lack of friction (the velocities remain smaller than the critical velocities). From the determination of the oscillation period follows the moment of inertia of the whole system which oscillates, and from that one can obtain the normal fluid density as a function of temperature. Later more experiments of this character have been carried out, which confirm the early experiments of Andronikashvili. A direct confirmation of these data came from the superfluid wind tunnel experiments of PELLAM and collaborators which gave directly the superfluid density as a function of temperature. The most accurate data have been obtained, however, in an indirect way from the experiments on second sound (see Section **3** of this chapter). The ratio of the normal fluid density ϱ_n to the total mass density ϱ, plotted as a function of temperature, is given in Fig. 5. At very low temperatures the curve goes to zero proportional to T^4 as is shown on an 10 000 × larger scale for $T < 0.6$ °K. At higher temperatures the curve rises exponentially up to $\varrho_n = \varrho$ at $T = T_\lambda$.

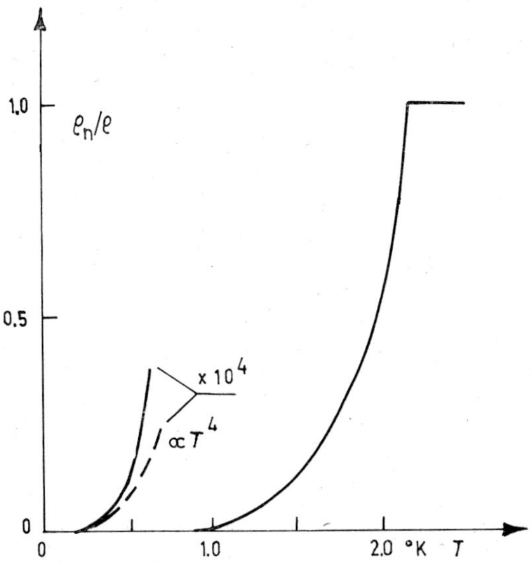

Fig. 5. – The ratio of normal density and total mass density as obtained from oscillating pile and from second sound experiments.

d) The *normal fluid part* shows ordinary liquid viscous behaviour. This follows from the damping of oscillating discs and oscillating cylinders in liquid helium: this damping can be interpreted in the usual way as the damping caused by a fluid of *viscosity* η and of density ϱ_n. The viscosity coefficient appears to show no singular behaviour at the λ-temperature: when the temperature decreases the viscosity decreases also, goes through a minimum of about 12 μP at 1.5 °K and then increases again, the value at 0.8 °K being already about 70 μP.

The heat conduction through liquid helium is a much more complicated phenomenon as this is the result of a flow of superfluid from the low- to the high-temperature region combined with a viscous streaming of normal fluid from the high- to the low-temperature side. This is quite contrary to the usual heat conduction in a liquid where no macroscopic flow occurs. In narrow channel this heat current becomes proportional to the temperature gradient but the size is not determined by the heat-conductivity coefficient but by the inverse of the viscosity.

2. – The equations of Landau.

On the basis of the two-fluid model LANDAU has put forward the equations of motion and the conservation laws which determine the hydrodynamic behaviour of liquid helium. For this purpose LANDAU (1941) introduced the local velocities $\boldsymbol{v}_s$ and $\boldsymbol{v}_n$ of the superfluid and the normal fluid. The total mass current density or momentum density of the liquid is then given by

$$\boldsymbol{j} = \varrho_n \boldsymbol{v}_n + \varrho_s \boldsymbol{v}_s \,. \tag{2.1}$$

As far as the motion of the superfluid is concerned, LANDAU assumed that it is rotation free:

$$\nabla \cdot \boldsymbol{v}_n = 0 \,, \tag{2.2}$$

on the basis of a very interesting attempt to found a real quantum hydrodynamics. A more detailed discussion is, however, outside the scope of these lectures.

The equation put forward by LANDAU (1941) [7] are in linearized form:

$$\partial \boldsymbol{v}_s / \partial t = - \nabla \mu \,, \tag{2.3}$$

$$\partial \widetilde{S} / \partial t = - \nabla \cdot (\widetilde{S} \boldsymbol{v}_n) \,, \tag{2.4}$$

$$\partial \boldsymbol{j} / \partial t = \nabla P + [\eta \nabla^2 \boldsymbol{v}_n + (\phi + \tfrac{1}{3}\eta) \nabla(\nabla \cdot \boldsymbol{v}_n)] \,, \tag{2.5}$$

$$\partial \varrho / \partial t = - \nabla \cdot \boldsymbol{j} \,. \tag{2.6}$$

The first equation is the *acceleration equation for the superfluid*. The argument used by LANDAU for the derivation runs as follows: If one displaces at the *absolute zero* of temperature a mass unit of substance contained in a volume element at $\boldsymbol{r}$ to a volume element at $\boldsymbol{r} + \mathrm{d}\boldsymbol{r}$ the energy change is

$$[-(\partial U / \partial M)_{\boldsymbol{r}} + (\partial U / \partial M)_{\boldsymbol{r}+\mathrm{d}\boldsymbol{r}}] = \nabla(\partial U / \partial M)\, \mathrm{d}\boldsymbol{r} \,.$$

The acting force per mass unit therefore is equal to

$$-\nabla(\partial U / \partial M) = -\nabla(\partial \widetilde{U} / \partial \varrho) = -\nabla \mu \,,$$

where μ is the thermodynamic potential. As at $T=0$ all the fluid is superfluid, this justified eq. (2.3). At higher temperatures the argument has to be refined, but we will not present it here, because we will give a more detailed discussion of a different proof of (2.3) in Chapter IV.

The second equation is *the equation for conservation of entropy*, attached entirely to the normal-fluid flow. $\tilde{S}$ is the entropy per volume and $\tilde{S}\boldsymbol{v}_n$ represents the entropy current density due to the motion of the normal fluid. These two equations together with (2.1) determine the relative motion of the two fluids with respect to each other.

The third equation is *the acceleration equation* for the fluid as a whole, P being the pressure and $\boldsymbol{j}$ the mass current density or momentum density. The terms between square brackets represent the viscous flow resulting from the flow of the normal fluid only. η is the viscosity and ϕ the so-called volume viscosity. These terms which are thus connected with irreversible transport processes will be omitted in Chapters II, III, IV and V.

The fourth equation is *the equation for conservation* of mass or continuity equation. Combining the two acceleration equations (2.3) and (2.5), one may also derive an equation for the time derivative of the *relative* velocity of the two fluids $\boldsymbol{v}=\boldsymbol{v}_n-\boldsymbol{v}_s$, by using $\boldsymbol{j}=\varrho\boldsymbol{v}_s+\varrho_n\boldsymbol{v}$, which gives

$$\partial\varrho_n\boldsymbol{v}/\partial t = -\nabla P + \varrho\nabla\mu = -\tilde{S}\nabla T\,, \tag{2.7}$$

where use has been made of the thermodynamic relation

$$\varrho\,\mathrm{d}\mu = -\tilde{S}\,\mathrm{d}T + \mathrm{d}P\,.$$

This equation (2.7) is very fundamental as it shows that the two fluids are accelerated with respect to each other when there is a temperature gradient: the normal fluid flows in the direction of decreasing temperature and the superfluid in the opposite direction of increasing temperature. We prefer to use this equation (2.7) instead of eq. (2.3) as the fundamental acceleration equation, which determines the relative flow of the two fluids.

We mention yet another version of the two accelerating equations (1949) as used by GORTER [9] and coll., viz.

$$\varrho_s\,\partial\boldsymbol{v}_s/\partial t = -(\varrho_s/\varrho)\nabla P + (\varrho_s/\varrho)\tilde{S}\,\nabla T + [A\varrho_n\varrho_s v^2\boldsymbol{v}]\,, \tag{2.8}$$

$$\varrho_n\,\partial\boldsymbol{v}_n/\partial t = -(\varrho_n/\varrho)\nabla P - (\varrho_s/\varrho)\tilde{S}\,\nabla T + \eta\cdot f(\boldsymbol{v}_n) - [A\varrho_n\varrho_s v^2\boldsymbol{v}]\,. \tag{2.9}$$

Except for the terms between brackets they are equivalent with the two acceleration equations (2.5) and (2.7), but now refer to the two fluids separately. For certain purposes this has advantages, but the physical origin of the terms on the right-hand side is somewhat obscured, because the ∇P accelerating the liquid as a whole and ∇T accelerating the fluids with respect to each other do not occur in both equations. Between brackets we have placed terms proportional to $v^2\boldsymbol{v}$ introduced by GORTER and MELLINK, which appear to describe phenomenologically the mutual friction between the two fluids. We prefer therefore to use the set of eqs. (2.3), (2.4), (2.5) and (2.6).

For later reference it may be useful to quote the form, which these equations take, if we include the hydrodynamic terms [10]:

$$\partial \boldsymbol{v}_s / \partial t = -\nabla(\mu + \tfrac{1}{2}\boldsymbol{v}_s^2)\,, \tag{2.10}$$

$$\partial \tilde{S} / \partial t = -\nabla \cdot (\tilde{S}\boldsymbol{v}_n + \boldsymbol{q}/T) + R\,, \tag{2.11}$$

$$\partial \boldsymbol{j} / \partial t = -\nabla \cdot \mathbf{P}\,, \tag{2.12}$$

$$\partial \varrho / \partial t = -\nabla \cdot \boldsymbol{j}\,, \tag{2.13}$$

where (*)

$$\mathbf{P} = P\mathbf{1} + \varrho_s \boldsymbol{v}_s \boldsymbol{v}_s + \varrho_n \boldsymbol{v}_n \boldsymbol{v}_n - 2\eta \overset{0}{\overline{\overline{\nabla \boldsymbol{v}_n}}} - \phi \nabla \cdot \boldsymbol{v}_n\,, \tag{2.14}$$

$$\boldsymbol{q} = -\lambda \nabla T\,, \tag{2.15}$$

$$R = \lambda (\nabla \ln T)^2 + (2\eta/T)(\overset{0}{\overline{\overline{\nabla \boldsymbol{v}_n}}})^2 + (\phi/T)(\nabla \cdot \boldsymbol{v}_n)^2\,. \tag{2.16}$$

Here $\mathbf{P}$ is the total pressure tensor, including the thermodynamic pressure P, multiplied by the unit tensor $\mathbf{1}$, including two terms of hydrodynamic origin and including the pressure tensor corresponding to the viscous flow. The latter consists of two parts: the pure viscosity part proportional to η and the « second » viscosity part proportional to ϕ. The tensor $\overset{0}{\overline{\overline{\nabla \boldsymbol{v}_n}}}$ represents the symmetrized and trace-free velocity gradient tensor:

$$\begin{pmatrix} \nabla_x v_{nx} - \frac{1}{3}\nabla\cdot\boldsymbol{v}_n & \frac{1}{2}(\nabla_x v_{ny} + \nabla_y v_{nx}) & \frac{1}{2}(\nabla_x v_{nz} + \nabla_z v_{nx}) \\ \frac{1}{2}(\nabla_y v_{ny} + \nabla_x v_{ny}) & \nabla_y v_{ny} - \frac{1}{3}\nabla\cdot\boldsymbol{v}_n & \frac{1}{2}(\nabla_y v_{nz} + \nabla_z v_{ny}) \\ \frac{1}{2}(\nabla_z v_{nx} + \nabla_x v_{nz}) & \frac{1}{2}(\nabla_x v_{ny} + \nabla_y v_{nz}) & \nabla_z v_{nz} - \frac{1}{3}\nabla\cdot\boldsymbol{v}_n \end{pmatrix}. \tag{2.17}$$

(*) *Note regarding vector and tensor notation.*

Vectors are indicated as usual by $\boldsymbol{v}$, $\boldsymbol{p}$, etc. The scalar product is $\boldsymbol{v}\cdot\boldsymbol{p} = \sum_\alpha v_\alpha p_\alpha$; $\nabla\cdot\boldsymbol{v} \equiv \operatorname{div} \boldsymbol{v}$.

Tensors are indicated here by $\mathbf{P}$, the unit tensor as $\mathbf{1}$. Scalar multiplication with a vector is indicated by $\boldsymbol{a}\cdot\mathbf{P}$: this represents a vector with α comp. $\sum_\beta a_\beta P_{\alpha\beta}$. Double scalar multiplication with a tensor is indicated by $\mathbf{P}:\mathbf{Q}$: this represents a scalar: $\sum_{\alpha\beta} P_{\alpha\beta} Q_{\beta\alpha}$. Tensors which can be written in dyadic form, like $\boldsymbol{up}$ are indicated by two vectors without the multiplication dot.

The symmetrized tensor is $(\overline{\overline{Q}})_{\alpha\beta} = \frac{1}{2}(Q_{\alpha\beta} + Q_{\beta\alpha})$.

The trace-free tensor has the diagonal elements $\overset{\circ}{Q}_{\alpha\alpha} = Q_{\alpha\alpha} - \frac{1}{3}\sum_\alpha Q_{\alpha\alpha}$.

If one introduces eq. (2.14) in (2.12) one obtains (2.5), by omitting the hydrodynamic terms, proportional to v^2, and using:

$$2\nabla\cdot\overline{\overline{\nabla\boldsymbol{v}_n}}^{\,0} = \nabla^2\boldsymbol{v}_n + \tfrac{1}{3}\nabla(\nabla\cdot\boldsymbol{v}_n) \equiv f(\boldsymbol{v}_n)\,. \tag{2.18}$$

$\boldsymbol{q}$ is the heat current due to real conduction of heat in the normal fluid and λ the corresponding heat conductivity. R represents the rate of entropy production due to the irreversible transport processes taking place in the normal fluid.

3. – First and second sound.

The set of equations introduced by Landau for the description of liquid helium in terms of the two fluid model has been very succesful in describing the properties of liquid helium. A description of all phenomena goes far beyond the scope of these lectures. We therefore limit ourselves to the description of first and second sound waves, on the basis of the linearized set of eqs. (2.4), (2.5), (2.6) and (2.7) and omitting the irreversible terms leading to damping. Eliminating the quantity $\boldsymbol{j}$ between the eqs. (2.5) and (2.6) leads to

$$\frac{\partial^2\varrho}{\partial t^2} = \nabla^2 P\,. \tag{2.19}$$

Before eliminating $\boldsymbol{v}_n$ and $\boldsymbol{v}$ between eqs. (2.4) and (2.7) we transform eq. (2.7), introducing $\tilde{S}=s\varrho$, using $\partial\varrho/\partial t = -\nabla\cdot\boldsymbol{j}$ and $\boldsymbol{j}=\varrho\boldsymbol{v}_n-\varrho_s\boldsymbol{v}$, and obtain

$$\varrho\frac{\partial s}{\partial t} = -s\frac{\partial\varrho}{\partial t} - \nabla\cdot(s\varrho\boldsymbol{v}_n) = s\nabla\cdot\boldsymbol{j} - s\nabla\cdot(\varrho\boldsymbol{v}_n) = -s\nabla\cdot(\varrho_s\boldsymbol{v})\,.$$

Similarly eq. (2.4) becomes

$$\varrho_n\frac{\partial\boldsymbol{v}}{\partial t} = -\varrho s\nabla T\,.$$

Eliminating $\boldsymbol{v}$ between these two equations one obtains

$$\frac{\partial^2 s}{\partial t^2} = \frac{\varrho_s}{\varrho_n}s^2\nabla^2 T\,. \tag{2.20}$$

Substitution of the solutions:

$$P = P^{(0)} + P^{(1)}\exp[i\omega(t-x/c)]\,,$$

$$T = T^{(0)} + T^{(1)}\exp[i\omega(t-x/c)]\,,$$

leads to the two equations:

$$\left[\left(\frac{\partial \varrho}{\partial P}\right)_T c^2 - 1\right] P^{(1)} + \left(\frac{\partial \varrho}{\partial T}\right)_P c^2\, T^{(1)} = 0\,,$$

$$\left(\frac{\partial s}{\partial P}\right)_T c^2 P^{(1)} + \left[\left(\frac{\partial s}{\partial T}\right)_P c^2 - \frac{\varrho_s}{\varrho_n}\, s^2\right] T^{(1)} = 0\,.$$

This set of equations has the following two solutions:

(2.21) $$c_{\mathrm{I}}^2 = \left(\frac{\partial P}{\partial \varrho}\right)_s, \qquad c_{\mathrm{II}}^2 = \frac{\varrho_s}{\varrho_n}\, s^2 \left(\frac{\partial T}{\partial s}\right)_\varrho,$$

provided that

(2.22) $$\left(\frac{\partial P}{\partial \varrho}\right)_s = \frac{c_P}{c_v}\left(\frac{\partial P}{\partial \varrho}\right)_T \approx \left(\frac{\partial P}{\partial \varrho}\right)_T.$$

The thermal expansion which makes $c_p/c_v \neq 1$ leads to coupling between the two types of waves as shown by LIFSCHITZ (1944), PESHKOV (1948) and DINGLE (1950) [11].

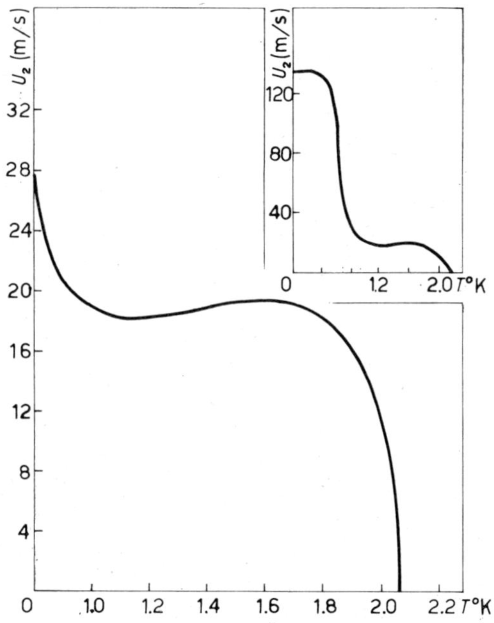

Fig. 6. – Values of second sound velocity $u_2 \equiv c_{\mathrm{II}}$ as function of temperature.

The first solution corresponds to ordinary sound waves in which ϱ, P and $\boldsymbol{j}$ vary periodically in space and time.

The second solution is the velocity for *second sound waves* in which S, T and $\boldsymbol{v}$ vary periodically in space and time. In the first sound waves the relative velocity $\boldsymbol{v}$ remains zero and the two fluids move in the same way in phase. In the second sound waves on the contrary the total mass current density or momentum density remains zero and the two fluids oscillate opposite to each other.

The velocity of first sound does not change very much as function of temperature and approaches the value 238 m/s at the absolute zero. However, the velocity of second sound, which occurs only in Helium II, shows a very remarkable behaviour as may be seen from Fig. 6. We will come back to a further discussion of this curve after having discussed in Chapter V the physical basis of the equations of Landau from which the expression for c_{II} was derived. It should be mentioned here however, that from the values of c_{II} and from the known caloric properties of liquid helium it is possible to derive very accurate values for the ratio $\varrho_s/\varrho_n = (\varrho/\varrho_n) - 1$. In fact the values obtained in this way have been plotted in Fig. 5.

III. - Excitations in Liquid Helium II.

1. – Introduction.

At the absolute zero the liquid is in its quantum-mechanical ground state with entropy $S=0$ and internal energy $U=U_0$. In classical theory this state would correspond to the state of minimum potential energy and kinetic energy zero, but in quantum theory, even at the ground state considerable « zero-point » motion exists.

At nonzero temperatures the system moves into excited states and it appears to be possible to consider these as *localized excitations*, the number of which increases with increasing temperature. It has been the fundamental idea of LANDAU (1941) to *correlate the properties of this system of thermal excitations with those of the « normal fluid »* and in this way to obtain a quantitative explanation of the somewhat mystic concept « normal fluid », which so far could only be considered to be a good working hypothesis in terms of the properties of the system of thermal excitations.

This correlation will be worked out in full detail in Chapter IV. Before doing so, however, one should consider more carefully *which* excitations occur in liquid helium II and what are its properties. This will be done in this chapter.

2. – Basic types of excitations.

In the first article LANDAU (1941) introduced two types of excitations.

1) *phonons*: these were quanta of longitudinal elastic (sound) waves, with energy $\varepsilon=\hbar\omega$ and momentum $p=\hbar k$, where ω and k are the circular frequency $2\pi\nu$ and circular wave number $2\pi/\lambda$ of longitudinal elastic waves. From the dispersion relation $\omega=ck$ (where $c=238$ m/s is the velocity of sound for long longitudinal elastic waves) follows

$$\varepsilon = c\cdot p \; . \tag{3.1}$$

In addition to this spectrum of excitations corresponding to *rotation-free* classical longitudinal elastic waves notations, Landau introduced a second type of excitations, quantized classical vortex motions which were *not rotation free*:

2) *rotons*: these excitations would require a finite excitation energy Δ. From dimensional considerations LANDAU [7] was led to the expression $\Delta \approx m^5/\varrho^2\hbar^2 \approx 1$ °K, but later considerations showed that a more molecular picture [12] of this type of excitations as rotating pairs of neighbouring helium atoms leads to $\Delta = 6(\hbar^2/2m\bar{a}^2) \approx 6$ °K, where $\bar{a} = 3.6$ Å is the average distance between nearest neighbours.

The whole spectrum of excitations would thus consist of two separate branches: the phonon branch of rotation-free motions giving a continuous spectrum starting from the ground state and the roton branch of rotational motions which required a nonzero excitation energy Δ.

In 1947 LANDAU [8] modified this excitation spectrum by proposing for the rotation type of excitations an excitation energy-momentum relation:

$$\varepsilon = \Delta + (p - p_0)^2/2\mu \, , \tag{3.2}$$

where because of the fairly large value of $\Delta \approx 9.6$ °K, which was assumed, only excitations with $p \approx p_0$ can be expected to be excited. Landau came to this modification of the roton spectrum from a study of the caloric equilibrium properties, which follow from these excitations (see Ch. IV). The values for the constants Δ and the effective mass μ estimated by LANDAU (1947) [8] are given in Table II.

After this modification of the spectrum of the roton-type excitations it became more and more clear, that both types of excitations have to be considered as two parts of *one and the same excitation energy-momentum curve* $\varepsilon(p)$, which because of the rather high energies required in the intermediate region can for most practical purposes be considered as two *separate parts*: the phonon region from $p \approx 0$ upwards and the roton region round $p \approx p_0$. The « rotons » then lost completely their « vortex » or rotation character and should instead be considered as short-wavelength longitudinal elastic modes, which are completely degenerated because the wavelength is of the order of magnitude of the average distance between neighbouring atoms. The question whether this excitation spectrum is « complete » is not yet solved. In principle there esists still a possibility, although this is not very likely, that other types of excitations than those presented by (3.1) and (3.2) contribute to the caloric properties, *e.g.* real rotational motions. Although the excitation energy for rotational motions of two atoms round each other is of the same order of magnitude, this motion will be strongly damped, which of course would have a diffusing effect on the excitation energy. We will, however, neglect this possibility in the following, as it appears to be possible to give a reasonable description on the basis of (3.1) and (3.2). Some more remarks about the meaning of this dispersion curve are given in Section **4**.

3. – Determination of the constants of the dispersion curve.

More accurate measurements on the thermodynamic properties of liquid ^{4}He II and of second sound have given the possibility to improve gradually the values of the constants of the excitation energy-momentum curve. Some of the results [13] are collected in Table II.

TABLE II. – *Values of constants of excitation energy-momentum curve.*

source	investigators		Δ/k (°K)	$p_0/\hbar$ (Å^{-1})	μ/m_{He}
thermo-dynamic data, second sound [13]	LANDAU [8]		9.6	1.95	0.77
	KHALATNIKOV [13]		8.9±0.2	1.99±0.05	0.26±0.09
	WIEBES, NIELS-HAKKENBERG, KRAMERS [13]		8.8	1.96	0.23
neutron scattering [14]	YARNELL, ARNOLD, BENDT, KERR [14]	1.1 °K	8.65±0.04	1.92±0.01	0.16±0.01
		1.6 °K	8.43	1.92	0.16
		1.8 °K	8.15	1.92	0.16
	HENSHAW and WOODS [14]	1.1 °K	8.65±0.11	1.91±0.01	0.16
		1.8 °K	8.1	1.91	0.16
	PALEVSKY, OTNES, LARSSON [14]		8.1	—	—

An independent confirmation of this picture of the existence of thermal excitations in liquid helium being described by one excitation energy-momentum curve combining (3.1) and (3.2) came from the experiments of inelastic neutron scattering [14] by liquid helium II.

If monoenergetic neutrons, with energy E_n and momenta p_n are scattered by liquid helium, they loose energy by creating excitations in the liquid. Then the following relations hold:

$$\left\{\begin{aligned} &E_n - E_n' = \varepsilon_{\text{exc}}\,, \\ &\boldsymbol{p}_n - \boldsymbol{p}_n' = \boldsymbol{p}_{\text{exc}}\,, \end{aligned}\right. \tag{3.3}$$

where $E_n = p_n^2/2m_n$ and $E_n' = p_n'^2/2m_n$.

Evidently if we consider the creation of a particular excitation with energy ε_{exc}, this fixes E_n' and consequently p_n'. Then it is only possible for one particular angle of scattering to satisfy also the second (vector) relation, thus leading to a one-one correspondence between the angle of scattering, the momentum $\boldsymbol{p}_{\text{exc}}$ of the excitation created and the energy loss ε_{exc} of the neutrons.

The data obtained from the neutron scattering are given in Fig. 7 and in Table II and there appears to exist now a general consistency with the thermodynamic data.

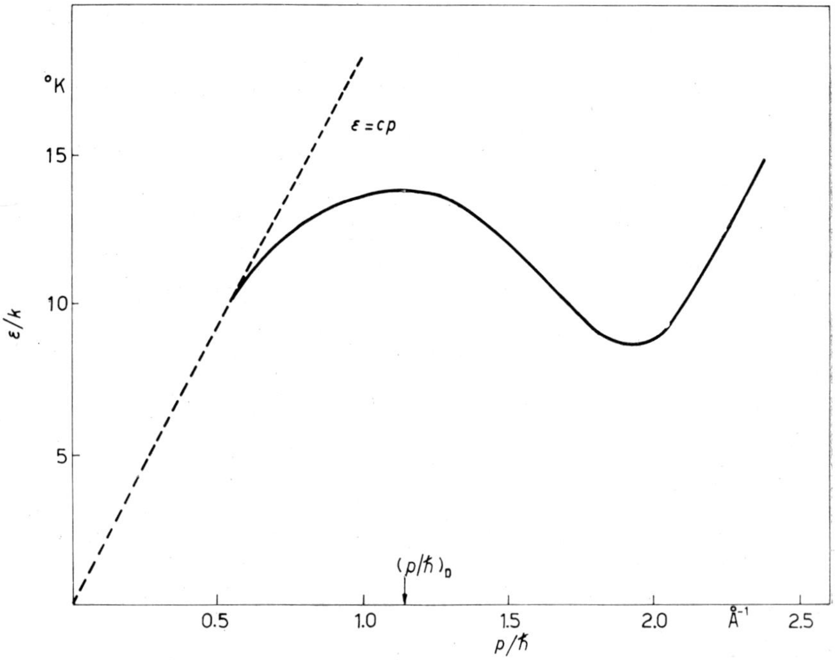

Fig. 7. – Energy-momentum relation for excitations obtained from inelastic neutron scattering: —— YARNELL and coll.

The *density-dependence* of the thermal excitations-momentum curve is not so well known. Indirect knowledge of the change of the constants c, Δ, p_0 and ϱ as a function of temperature has been obtained from consideration of the thermal expansion and the density change of ordinary and second sound. Direct information was obtained from the density-dependence of the neutron scattering. All data are collected in Table III.

TABLE III. – *Density-dependence of* c, Δ, p_0 *and* μ.

	$(\varrho/c)(\partial c/\partial\varrho)$	$(\varrho/\Delta)(\partial\Delta/\partial\varrho)$	$(\varrho/p_0)(\partial p_0/\partial\varrho)$	$(\varrho/\mu)(\partial\mu/\partial\varrho)$
KHALATNIKOV (1952)	+1.8	−0.33	+0.33	0
ATKINS and EDWARDS (1955, 1959)	—	−0.57±0.06	+0.26	−1.8
v. D. MEYDENBERG, TACONIS, DE BRUIN, OUBOTER (1960)	—	−0.9	+0.5	−5.5
HENSHAW and WOODS (1960)	—	−1.0	+0.40	—

YARNELL and collaborators [14] and HENSHAW and WOODS [14] have investigated also the *temperature-dependence* of the excitations. In principle the excitation picture does not allow for such a temperature-dependence: the excitations are quantum-mechanical excited states from the ground state for isolated excitations. However, when the temperature increases, the number of excitations increases, the liquid structure changes and consequently the average value of the thermal excitation energy-momentum curve is changed

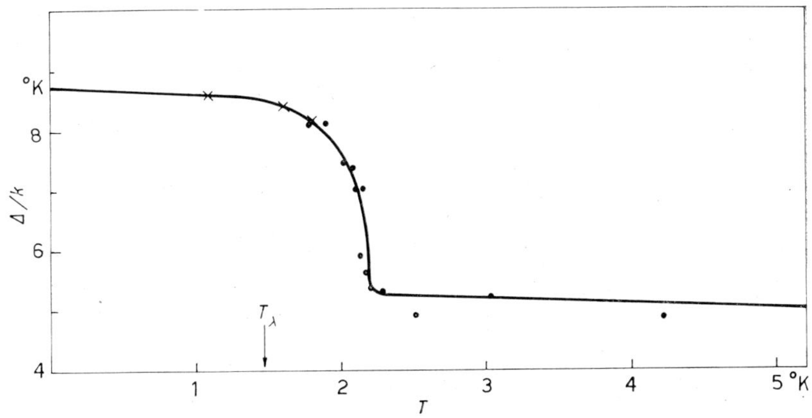

Fig. 8. – Temperature dependence of the excitation energy as obtained (●) by HENSHAW and WOODS [14] and (x) by YARNELL and coll. [14].

also somewhat due to the overall effect of other excitations present in the liquid. The result is that the value of Δ/k appears to decrease more and more quickly above 1.8 °K, reaching a value of about 5 °K at the λ-point as is shown in Fig. 8, obtained from neutron scattering data. This value remains then roughly constant above the transition point in liquid helium I.

Parallel to this shift of the excitation energy-momentum curve due to the average interaction with other excitations goes an increase of the energy broadening when the temperature reaches the transition point. In fact the energy uncertainty in the energy of the excitation $\delta\varepsilon$, which is unnoticeable at the lowest temperatures increases wery quickly above 1.8 °K and reaches a value of about $\delta\varepsilon/k=11.5$ °K at the λ-point. The energy uncertainty or broadening then has become even larger than the excitation energy ε, showing that the whole concept of single excitations moving in the liquid breaks down, because of the fact that so many excitations are present. This finds its counterpart in the average collision times for excitations which can be determined from the transport phenomena (see Chapter VI). At about 0.8 °K the average collision time is 10^{-7} s, corresponding to a quite small energy uncertainty of roughly $(10^{-3}\div10^{-4})$ °K. However, at the λ-point the collision time has decreased to about $5\cdot10^{-12}$ s corresponding to the broadening of about $\delta\varepsilon/k=10$ °K in agreement with the values determined directly from neutron scattering.

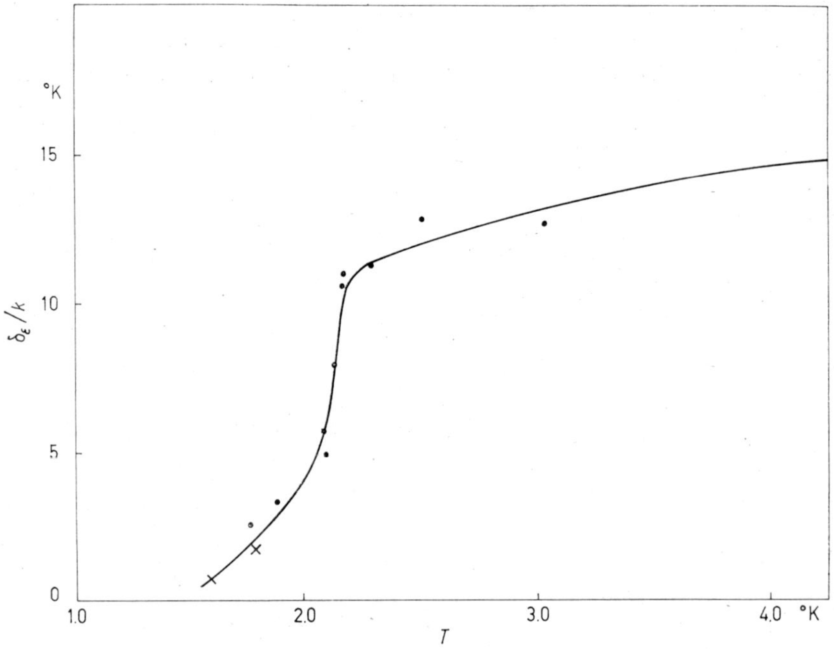

Fig. 9. – Uncertainty in excitation energy obtained from neutron scattering: ● HENSHAW and WOODS, x YARNELL and coll.

The conclusion thus is that the excitation picture is a sound physical interpretation at temperatures not too close to the λ-point, but that in the region of the λ-point the concept looses much of its physical meaning.

4. – Nature of excitations.

Using the relations $\varepsilon=\hbar\omega$ and $\boldsymbol{p}=\hbar\boldsymbol{k}$ the excitation energy-momentum curve can be transformed into a circular frequency-wave number relation which can be considered as the longitudinal sound wave dispersion curve:

$$\text{(3.4)}\qquad \begin{cases} \omega = ck & \text{(small } k) \\ \omega = \Delta/\hbar + \hbar(k-k_0)^2/2\mu & (k\approx k_0)\,, \end{cases}$$

giving the circular frequency ω as a function of the circular wave number k.

Now the interesting point is that this dispersion curve is rather similar to what one could expect to occur in any *condensed substance* [15].

If we consider first a longitudinal elastic wave in an *one-dimensional crystal consisting of atoms* interacting with elastic forces, then the well-known solution is a periodic curve as presented in Fig. 10*a*: for $k=2\pi/\lambda=\pi/a$ (*i.e.* for $\lambda=2a$)

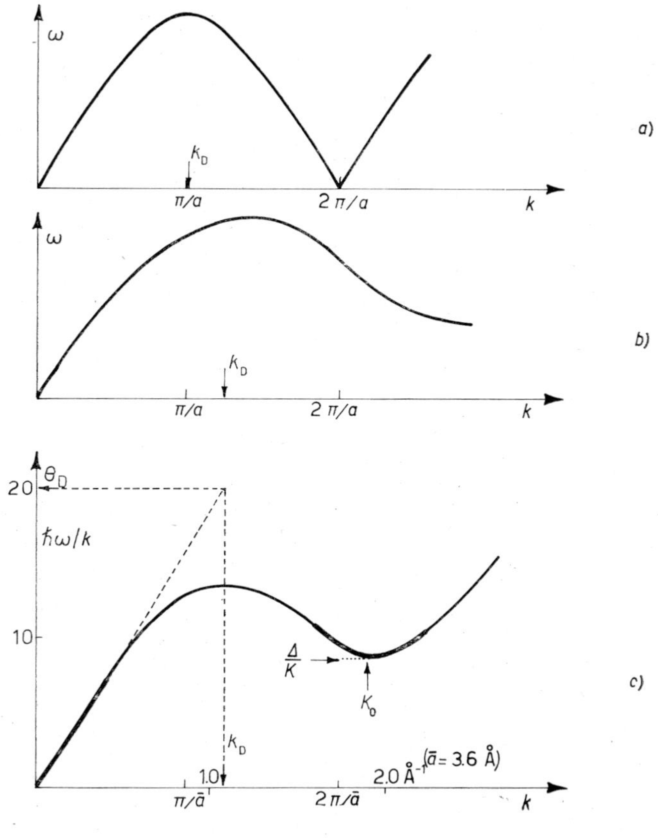

Fig. 10. – *a*) Dispersion curve for linear lattice. *b*) « Average » dispersion curve for cubic lattice (lattice distance *a*). *c*) Dispersion curve for liquid helium, with average intermolecular distance $\bar{a}=3.6$ Å.

the first maximum occurs and the neighbouring molecules move exactly in opposite phase. The range from $k=0$ up to $k=\pi/a$ completes the first Brillouin zone and for this one-dimensional crystal this value of $k=\pi/a$ for the Brillouin zone boundary is equal to the Debye limit $k_D=\pi/a$.

If one continues to higher values of the wave number $k>\pi/a$ (*i.e.* $\lambda<2a$) the mirror picture fo the same curve results and at $k=2\pi/a$ (*i.e.* $\lambda=a$) corresponding to motion in phase of neighbouring atoms, the frequency ω becomes zero again.

This strictly periodic behaviour of the $\omega(k)$ dispersion curve is a well-known direct consequence of the periodic structure of the lattice.

For a *three-dimensional crystal* the situation becomes more complicated, firstly because only in some special directions longitudinal elastic waves in the strict sense of the word can be propagated, and secondly because the dispersion curve depends on the direction in the crystal. If one now takes an average of the dispersion curve over the different directions the result is that the maximum shifts to a somewhat higher value of $k_{\text{max}} \approx 1.45 \times \pi/\bar{a} = 4.5/\bar{a}$, which, however, still corresponds closely to the Debye characteristic circular wave-number $k_{\text{D}} = 2\pi(3N/4\pi V)^{\frac{1}{3}} \approx 3.9/\bar{a}$ (where $\bar{a} = (V/N)^{\frac{1}{3}}$ is the average distance between molecules) (see Fig. 10*b*). The curve now of course no longer repeats itself, but it goes to a flat minimum at higher values of the wave number.

The curve so obtained must be characteristic for the behaviour of all condensed substances in which the constituting particles interact with a short-range repulsive core and a long-range attractive field. In the case that application of quantum mechanics is necessary, nothing essential is changed in the picture. It is therefore not surprising that there is a striking resemblance of this curve with the dispersion curve found for *liquid helium II* (Fig. 10*c*). Also there the first maximum occurs at a wave number corresponding to the Debye limit at which neighbouring molecules move practically in opposite phase. The subsequent minimum occurs at about twice that value of the wave number, where neighbouring molecules would move in phase if the excitation in this region could still be considered to be a longitudinal elastic wave of extremely short wavelength. However, at such short wavelengths the physical picture looses its meaning.

This close relation which exists between the dispersion curve for longitudinal elastic waves and the excitation energy-momentum curve also shows the way to a theoretical calculation of this energy-momentum curve as has been given by FEYNMAN (1954-1956) [16].

If the system of liquid helium could be treated as a solid then the atomic motion could be analysed in terms of a spectrum of elastic waves or vibrations characterized by wave vectors $\boldsymbol{k}$. The Hamiltonian could then be written as a sum of Hamiltonians each one corresponding to one of the wave vectors $\boldsymbol{k}$ and each having exactly the form of a single particle moving in a harmonic potential field. The « co-ordinates » occurring in these Hamiltonians are the so called « normal co-ordinates », being certain linear combinations of the co-ordinates of all constituent particles.

In a liquid system this picture no longer holds exactly, but one still can introduce « collective co-ordinates » $\varrho_{\boldsymbol{k}}$ corresponding to the normal co-ordinates in a solid and defined as the Fourier components $\varrho_{\boldsymbol{k}}$ of the macroscopic number density $\varrho(r)$:

$$\varrho(r) = \int \varrho_{\boldsymbol{k}} \exp[i\boldsymbol{kr}]\, \mathrm{d}\boldsymbol{k}\,, \qquad \varrho_{\boldsymbol{k}} = \int \varrho(r) \exp[-i\boldsymbol{kr}]\, \mathrm{d}\boldsymbol{r}\,. \tag{3.5}$$

The Hamiltonian can now *approximately* be written as a sum of Hamiltonians each one corresponding to one wave number $\boldsymbol{k}$, containing in classical theory the quadratic expressions: $\varrho_{\boldsymbol{k}}\varrho_{\boldsymbol{k}}^*$ and $\dot{\varrho}_{\boldsymbol{k}}\dot{\varrho}_{\boldsymbol{k}}^*$.

By introducing the number density operator

$$\varrho(r)_{\mathrm{op}} = \sum_i \delta(r_i - r)\,,$$

one is able to write

$$(\varrho_k)_{\mathrm{op}} = \int \exp[-i\boldsymbol{kr}]\varrho(r)\, \mathrm{d}\boldsymbol{r} = \sum_i \exp[-i\boldsymbol{kr}_i]\,. \tag{3.6}$$

Now a good variational wave function for the first excited state for any of these wave vectors $\boldsymbol{k}$ can be obtained by multiplying the ground-state wave functions Ψ_0 by the « collective co-ordinate » $\varrho_{\boldsymbol{k}}$, giving the variational trial wave function

$$\Psi_1 = \Psi_0 \varrho_{\boldsymbol{k}} = \Psi_0 \sum_i \exp[-i\boldsymbol{kr}_i]\,, \tag{3.7}$$

in analogy to the fact that the first excited-state wave function of a harmonic oscillator ψ_1, is obtained from the ground-state wave function $\psi_0 = \exp[-\frac{1}{2}\xi^2]$ by multiplication with ξ: $\psi_1 = \psi_0 \xi$. The expression for the excitation energy, *i.e.* the difference between the expection value of the energy calculated with the wave function Ψ_1, given by (3.7) and the ground-state energy corresponding to Ψ_0 could be written in the form

$$\varepsilon(p) = \frac{p^2}{2mS(k)}\,, \tag{3.8}$$

where $S(k)$ is the Fourier transform with wave number k of the molecular pair correlation function. From this the dispersion relation can be obtained as

$$\omega(k) = \hbar k^2 / 2m\, S(k)\,. \tag{3.9}$$

Using the experimental values as obtained from X-ray scattering, FEYNMAN obtained the curve B of Fig. 11 for the excitation energy-momentum relation. Comparison with the experimental curves D and C and with the data in Table II shows that the excitation energy determined in this way, is about a factor 2 too high. Essential is, however, that the curve shows the minimum at about 2.0 Å^{-1}, whereas it can also be shown that (3.8) approaches $c \cdot p$ for small values of p. An improved calculation made by FEYNMAN and COHEN, including the

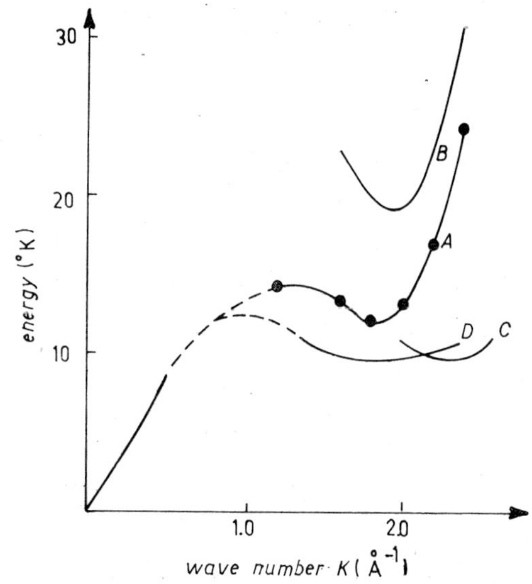

Fig. 11. – Excitation energy-momentum relation obtained by FEYNMAN (1954) (curve B) and FEYNMAN and COHEN (1956) (curve A) compared with the experimental curves: LANDAU (1947) (curve D) and DE KLERCK, HUDSON and PELLAM (1954) (curve C). (Taken from: *Phys. Rev.*, **102**, 1201 (1956).)

so-called « backflow » gives the curve A, which is in better agreement with the experimental curve.

A more detailed discussion of the theoretical calculation of the excitation energy-momentum curve is not in the scope of these lectures, this being the subject of the lectures given by Dr. CHESTER. The present remarks only serve to illustrate the relation between the excitation energy-momentum relation and the dispersion curve for longitudinal elastic sound waves in liquid helium.

IV. - Equilibrium Properties of the Excitation Gas.

1. – Partition function-distribution function.

After having investigated in the previous chapter the character of the dispersion relations, describing the dependence of the energy of a single excitation on its momentum, we now come to the statistical mechanics of a system of excitations, as given in principle by LANDAU [7].

An arbitrary excited state of the whole system may be characterized by the set of numbers $N_{\boldsymbol{p}}, N_{\boldsymbol{p}'}, N_{\boldsymbol{p}''}, \ldots \equiv \{N_{\boldsymbol{p}}\}$ giving the numbers $N_{\boldsymbol{p}}, N_{\boldsymbol{p}'}, \ldots$ of excitations with momenta $\boldsymbol{p}, \boldsymbol{p}', \ldots$ present in the excited state. The energy of this system of excitations with respect to the ground state is then given by

$$E\{N_{\boldsymbol{p}}\} = \sum_{\boldsymbol{p}} N_{\boldsymbol{p}}\, \varepsilon(p) \,. \tag{4.1}$$

The macroscopic properties which result from the equilibrium distribution of excitations of various momenta in the medium can be calculated from a canonical ensemble leading to the canonical partition function:

$$Z(V, \varrho, T) = Z_0 \sum_{\{N_{\boldsymbol{p}}\}} \exp\left[-\beta E\{N_{\boldsymbol{p}}\}\right] , \tag{4.2}$$

where Z_0 is the partition function of the ground state of the medium. Substituting (4.1) one finds

$$Z = Z_0 \prod_{\boldsymbol{p}} \sum_{N_{\boldsymbol{p}}} \exp\left[-\beta N_{\boldsymbol{p}} \varepsilon(p)\right] = Z_0 \prod_{\boldsymbol{p}} \left(1 - \exp\left[-\beta\varepsilon(p)\right]\right)^{-1} . \tag{4.3}$$

The resulting expression for the logarithm of the partition function becomes

$$\ln Z = \ln Z_0 - V \int \ln\left(1 - \exp[-\beta\varepsilon]\right) \mathrm{d}\boldsymbol{p}/h^3 \,, \tag{4.4}$$

where a transition from summation to integration is made, taking the number of quantum states in V and $\mathrm{d}\boldsymbol{p}$ equal to $V\,\mathrm{d}\boldsymbol{p}/h^3$. One thus finds for the Helmholtz free energy of the system

$$F(V, \varrho, T) = F_0 + V \cdot kT \int \ln\left(1 - \exp[-\beta\varepsilon]\right) \mathrm{d}\boldsymbol{p}/h^3 \,. \tag{4.5}$$

The free energy corresponding to the excitations is thus proportional to the total volume of the system and we therefore introduce the free-energy density, *i.e.* the free energy per volume, $\tilde{F}$ which only depends on temperature and on those parameters, like for instance the density of the medium, on which the dispersion curve of the individual excitations $\varepsilon(p)$ depends. Thus

$$\text{(4.6)} \qquad \begin{cases} F = F_0 + V\cdot\tilde{F}_{\text{ex}} , \\ \tilde{F}_{\text{ex}}(\varrho, T) = kT\int \ln\left(1 - \exp[-\beta\varepsilon]\right) \mathrm{d}\boldsymbol{p}/h^3 . \end{cases}$$

From this expression all thermodynamic quantities can be evaluated using the thermodynamic relation:

$$\text{(4.7)} \qquad \mathrm{d}\tilde{F} = -\tilde{S}\,\mathrm{d}T + \mu\,\mathrm{d}\varrho ,$$

where $\tilde{S} = S/V$, and $\mu = \tilde{G}/\varrho$ is the Gibbs function per mass. Thus

$$\text{(4.8)} \qquad \begin{array}{l|l|l} S = V\tilde{S}_{\text{ex}} & S_0 = 0 & \tilde{S}_{\text{ex}} = -\dfrac{\partial \tilde{F}_{\text{ex}}}{\partial T} \\ U = U_0 + V\tilde{U}_{\text{ex}} & U_0 = F_0 & \tilde{U}_{\text{ex}} = \tilde{F}_{\text{ex}} - T\dfrac{\partial \tilde{F}_{\text{ex}}}{\partial T} \\ P = P_0 + P_{\text{ex}} & P_0 = -\dfrac{\partial F_0}{\varrho V} & P_{\text{ex}} = \varrho\dfrac{\partial \tilde{F}_{\text{ex}}}{\partial \varrho} - \tilde{F}_{\text{ex}} \\ \mu = \mu_0 + \mu_{\text{ex}} & \mu_0 = F_0 - V\dfrac{\partial F_0}{VS} & \mu_{\text{ex}} = \dfrac{\partial \tilde{F}_{\text{ex}}}{\partial \varrho} . \end{array}$$

Before evaluating these quantities it is of advantage to introduce the *distribution function* $f_0(p)$ defined in such a way that $f_0(p)V\,\mathrm{d}\boldsymbol{p}$ represents the average number of excitations in V, inside $\mathrm{d}\boldsymbol{p}$.

Obviously this is given by

$$Vf_0(p)\,\mathrm{d}\boldsymbol{p} = \frac{V\,\mathrm{d}\boldsymbol{p}}{h^3}\,\frac{\sum N_{\boldsymbol{p}} \exp[-\beta E\{N_{\boldsymbol{p}}\}]}{\sum \exp[-\beta E\{N_{\boldsymbol{p}}\}]} = \frac{V\,\mathrm{d}\boldsymbol{p}}{h^3}\,\frac{\sum_{N_{\boldsymbol{p}}} N_{\boldsymbol{p}} \exp[-\beta N_{\boldsymbol{p}}\varepsilon(p)]}{\sum_{N_{\boldsymbol{p}}} \exp[-\beta N_{\boldsymbol{p}}\varepsilon(p)]} ,$$

or

$$\text{(4.9)} \qquad f_0(p) = \frac{h^{-3}}{\exp[\beta\varepsilon(p)] - 1} .$$

On introducing this distribution function, one may write

$$\text{(4.10)} \qquad \tilde{F}_{\text{ex}} = -kT\int \ln\left(1 + h^3 f_0(p)\right) \mathrm{d}\boldsymbol{p}/h^3 = -\frac{1}{3}\int f_0\, up\, \mathrm{d}\boldsymbol{p} ,$$

using the isotropy of the distribution function and writing $u=\partial\varepsilon/\partial p$, which is the group velocity.

2. – Expressions for the thermodynamic quantities.

From this expression follows, using the thermodynamic relations (4.8),

$$\text{(4.11)}\quad \begin{cases} \tilde{S}_{\text{ex}} = \dfrac{1}{T}\displaystyle\int f_0\cdot\left(\varepsilon+\frac{1}{3}pu\right)\mathrm{d}\boldsymbol{p} = -\frac{1}{3kT^2}\int f_0' p\varepsilon u\,\mathrm{d}\boldsymbol{p}\,, \\ \tilde{U}_{\text{ex}} = \displaystyle\int f_0\,\varepsilon\,\mathrm{d}\boldsymbol{p}\,, \\ P_{\text{ex}} = \displaystyle\int f_0\cdot\left(\frac{1}{3}pu+\varrho\frac{\partial\varepsilon}{\partial\varrho}\right)\mathrm{d}\boldsymbol{p} = -F_{\text{ex}}+\varrho\mu_{\text{ex}}\,, \\ \mu_{\text{ex}} = \displaystyle\int f_0\frac{\partial\varepsilon}{\partial\varrho}\,\mathrm{d}\boldsymbol{p}\,, \end{cases}$$

where $f_0'=\partial f_0/\partial\beta\varepsilon$ represents a differentiation of f_0 with respect to the argument $\beta\varepsilon$. We finally note that the total number density of excitations, integrated over all momenta,

$$\text{(4.12)}\qquad n_{\text{ex}} = \int f_0\,\mathrm{d}\boldsymbol{p}\,,$$

is not a constant, of course, but depends on the temperature and all parameters, like the density, on which the dispersion curve for the excitations depends.

For the actual calculation of the thermodynamic functions on the basis of the expressions given for the excitations in Chapter III, we will take

$$\text{(4.13')}\qquad \varepsilon=\varepsilon_{\text{ph}}=cp \qquad \text{« phonon region » (small } p)\,,$$

$$\text{(4.13'')}\qquad \varepsilon=\varepsilon_{\text{rot}}=\Delta+\frac{(p-p_0)^2}{2\mu} \qquad \text{« roton region »}\quad (p\approx p_0)\,,$$

because at the actual temperatures occurring in liquid helium the number of excitations is appreciable only in the region of small p, and in the region $p\approx p_0$, close round the minimum in the dispersion curve, this simplification of the excitation curve is a very good approximation to reality. The division has the advantage that we can divide all thermodynamic quantities in two parts, a « phonon part » and a « roton part », which behave rather differently in their thermodynamic behaviour.

The resulting expressions are for the *phonon part*:

$$
(4.14)\qquad
\begin{cases}
\widetilde{F}_{\mathrm{ph}} = -8\pi\zeta_4 \dfrac{k^4T^4}{h^3c^3} & = -n_{\mathrm{ph}} \dfrac{\zeta_4}{\zeta_3} kT\,,\\[2ex]
\widetilde{S}_{\mathrm{ph}} = 32\pi\zeta_4 \dfrac{k^4T^3}{h^3c^3} & = n_{\mathrm{ph}} \dfrac{\zeta_4}{\zeta_3} 4k\,,\\[2ex]
\widetilde{U}_{\mathrm{ph}} = 24\pi\zeta_4 \dfrac{k^4T^4}{h^3c^3} & = n_{\mathrm{ph}} \dfrac{\zeta_4}{\zeta_3} 3kT\,,\\[2ex]
P_{\mathrm{ph}} = 8\pi\zeta_4 \dfrac{k^4T^4}{h^3c^3}\left(1+3\dfrac{\partial \ln c}{\partial \ln \varrho}\right) & = n_{\mathrm{ph}} \dfrac{\zeta_4}{\zeta_3} kT\left(1+3\dfrac{\partial \ln c}{\partial \ln \varrho}\right),\\[2ex]
\mu_{\mathrm{ph}} = 8\pi\zeta_4 \dfrac{k^4T^4}{h^3c^3}\cdot 3\dfrac{\partial \ln c}{\partial \varrho} & = n_{\mathrm{ph}} \dfrac{\zeta_4}{\zeta_3} kT\cdot 3\dfrac{\partial \ln c}{\partial \varrho}\,,
\end{cases}
$$

where

$$
(4.15)\qquad n_{\mathrm{ph}} = 8\pi\zeta_3 \frac{k^3T^3}{h^3c^3}\,,
$$

where $\zeta_4 = \sum 1/n^4 = \pi^4/90 = 1.037$ and $\zeta_3 = \sum 1/n^3 = 1.202$. We see that for the phonon gas the kinetic part of the pressure $P_{\mathrm{ph}} \approx 0.86\, n_{\mathrm{ph}} kT = \frac{1}{3}\widetilde{U}$, whereas for an ordinary gas $P = nkT = \frac{2}{3}\widetilde{U}$.

The thermodynamic expressions for the *roton part are*:

$$
(4.16)\qquad
\begin{cases}
\widetilde{F}_{\mathrm{rot}} = -n_{\mathrm{rot}} kT\,,\\[2ex]
\widetilde{S}_{\mathrm{rot}} = n_{\mathrm{rot}} k\left(\dfrac{3}{2} + \dfrac{\Delta}{kT}\right),\\[2ex]
\widetilde{U}_{\mathrm{rot}} = n_{\mathrm{rot}} kT\left(\dfrac{1}{2} + \dfrac{\Delta}{kT}\right),\\[2ex]
P_{\mathrm{rot}} = n_{\mathrm{rot}}\left(kT + \varrho\dfrac{\partial \Delta}{\partial \varrho}\right),\\[2ex]
\mu_{\mathrm{rot}} = n_{\mathrm{rot}} \dfrac{\partial \Delta}{\partial \varrho}\,,
\end{cases}
$$

where

$$
(4.17)\qquad n_{\mathrm{rot}} = \frac{4\pi p_0^2}{h^2}\left(\frac{2\pi\mu kT}{h^2}\right)^{\frac{1}{2}} \exp[-\Delta/kT]\,,
$$

and where for simplicity only the density-dependence of Δ has been taken into account. The number density of phonons thus increases proportional to T^3 whereas that of the rotons is proportional to $T^{\frac{1}{2}} \exp[-\Delta/kT]$. At very low temperatures $T < 0.5\,^\circ\mathrm{K}$ practically *all* excitations are of the phonon type. At temperatures between 0.5 and 1.0 °K the number of rotons starts to increase very much. Finally at temperatures $T > 1.0$ °K the large majority of the excitations is of the roton type.

These thermodinamic expressions have been used by several investigators to obtain the values of the constants of the excitation energy-momentum relation Δ, p_0 and μ from the thermodynamic data. This method (together with that based on the normal fluid density and second sound to be discussed in Chapter V) has led to the values of the constants as given in the first half of Tables II and III. As has been explained in Chapter III the accuracy of the data obtained from neutron scattering is probably better, so that now the inverse programme, using the neutron scattering data to calculate all thermodinamic properties, becomes important.

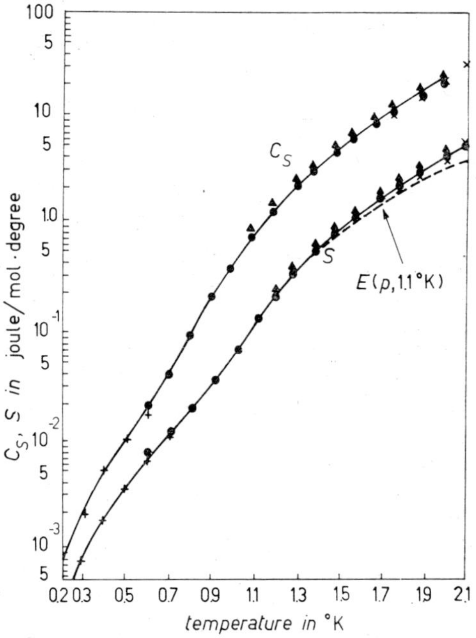

Fig. 12. – Comparison of caloric data of helium II with theoretical expressions (4.11) calculated by Bendt, Cowan and Yarnell, using the energy-momentum relation for the excitations obtained from neutron scattering (taken from: *Phys. Rev.*, **113** 1392 (1959)): —— calculated values using $E(p, T)$; ● Kramers, Wasscher and Gorter; × Hill and Lounasmaa; + Niebes, Wiels-Hakkenberg and Kramers; ▲ Hercus and Wilks.

An interesting calculation has been made by Bendt, Cowan and Yarnell (1959) [18], making use of the expressions (4.11) and using the full dispersion curve without the approximation involved in (4.13). They also made use of the temperature-dependence of the dispersion curve, which in a certain sense accounts for the interaction of the excitations at higher temperatures. Fig. 12 shows that an excellent agreement with the experimental data was found.

3. – The excitation mass density.

For the purpose of an important additional equilibrium property of the system, the excitation mass density or « normal fluid » density, we consider a system of excitations moving with an average velocity in a medium at rest.

According to the principle of statistical mechanics the canonical ensemble, representing such a system, should have a probability distribution in phase space of the form

$$\exp\left[-\beta(E\{N_p\} - \boldsymbol{v}\cdot\boldsymbol{P}\{N_p\})\right], \tag{4.18}$$

where E and $\boldsymbol{P}$ are the total energy and momentum of the excitations:

$$E\{N_p\} = \sum_{p} N_p\,\varepsilon(p) \qquad \boldsymbol{P}\{N_p\} = \sum_{p} N_p\,\boldsymbol{p}\,. \tag{4.19}$$

Consequently the partition function now becomes

$$Z(V, \varrho, T, \boldsymbol{v}) = Z_0 \sum_{\{N_p\}} \exp\left[-\beta\,(E\{N_p\} - \boldsymbol{v}\cdot\boldsymbol{P}\{N_p\})\right], \tag{4.20}$$

giving

$$\ln Z = \ln Z_0 - V\int \ln\left(1 - \exp[-\beta(\varepsilon - \boldsymbol{v}\cdot\boldsymbol{p})]\right) \mathrm{d}\boldsymbol{p}/h^3, \tag{4.21}$$

and

$$\tilde{F}_{\mathrm{ex}}(\varrho, T, \boldsymbol{v}) = kT\int \ln\left(1 - \exp[-\beta(\varepsilon - \boldsymbol{v}\cdot\boldsymbol{p})]\right) \mathrm{d}\boldsymbol{p}/h^3. \tag{4.22}$$

The *distribution function* for the excitations now becomes, following the same reasoning as before,

$$f(p) = \frac{h^{-3}}{\exp[\beta(\varepsilon - \boldsymbol{v}\cdot\boldsymbol{p}) - 1]} = f_0(p) - \boldsymbol{v}\cdot\boldsymbol{p}\,\frac{\partial f_0}{\partial \varepsilon}. \tag{4.23}$$

From this one may also verify that $\boldsymbol{v}$ is the average velocity of the excitations, because

$$\langle \boldsymbol{u} \rangle = \frac{\int f\boldsymbol{u}\,\mathrm{d}\boldsymbol{p}}{\int f\,\mathrm{d}\boldsymbol{p}} = -\frac{\int (\nabla_{\boldsymbol{p}} f_0)\,\boldsymbol{v}\cdot\boldsymbol{p}\,\mathrm{d}\boldsymbol{p}}{\int f_0\,\mathrm{d}\boldsymbol{p}} = \boldsymbol{v}, \tag{4.24}$$

as can be seen directly by partial integration.

An important *new* equilibrium property which occurs in this thermodynamic system, is the average value of the total momentum corresponding to all excitations:

$$\langle \boldsymbol{P} \rangle = \frac{\sum J\{N_p\} \exp[-\beta(E\{N_p\} - \boldsymbol{v}\cdot\boldsymbol{P}\{N_p\})]}{\sum \exp[-\beta(E\{N_p\} - \boldsymbol{v}\cdot\boldsymbol{P}\{N_p\})]} = \frac{\partial \ln Z}{\partial \beta \boldsymbol{v}}. \tag{4.25}$$

Substituting (4.26),

$$j_{ex} = \frac{1}{V}\langle \boldsymbol{P} \rangle = \int f(\boldsymbol{p})\boldsymbol{p}\,\mathrm{d}\boldsymbol{p} = -\frac{1}{3}\boldsymbol{v}\int \frac{\partial f_0}{\partial \varepsilon}\,p^2\,\mathrm{d}\boldsymbol{p}\,. \tag{4.26}$$

One thus arrives at an excitation current mass density $\boldsymbol{j}_{ex}$ proportional to the average velocity $\boldsymbol{v}$. The proportionality constant has to be identified as an

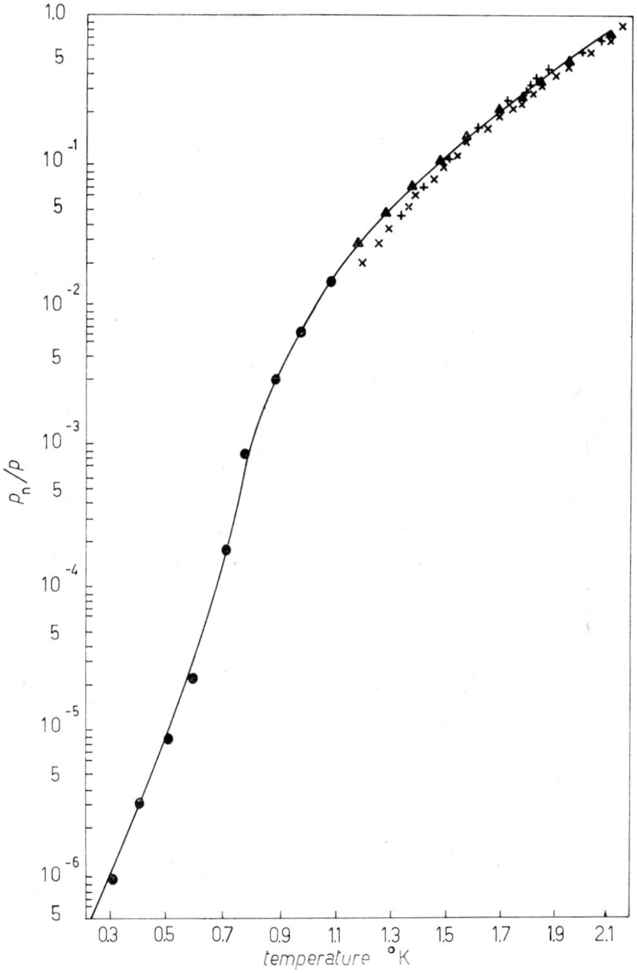

Fig. 13. – Comparison of the experimental data for the excitation mass density (normal-fluid density) with the theoretical curve calculation by Bendt, Cowan and Yarnell, using the energy-momentum relation for excitations obtained from neutron scattering (taken from: *Phys. Rev.*, **113**, 1393 (1959)): —— calculated values; ● v_2 from De Klerk, Hudson and Pellam; ▲ v_2 from Mauer and Herlin, S and c_s for above from Kramers, Wasscher and Gorter, $\varrho_n/\varrho = [(c_s v^2/TS^2)+1]^{-1}$; × Dash and Tailor; + Andronikashvili, torsion pendulum.

excitation mass density, which we will show in the next chapter to be identical with the phenomenological concept of the « *normal-fluid density* » ϱ_n. For briefness sake we will make use of the symbol ϱ_n already now. So we find

$$\text{(4.27)} \qquad \boldsymbol{j}_{ex} = \varrho_n \boldsymbol{v}\,, \qquad \varrho_n = -\frac{1}{3}\int \frac{\partial f_0}{\partial \varepsilon} p^2\, d\boldsymbol{p}\,.$$

The excitation mass density ϱ_n can now be evaluated using the energy-momentum relation as specified by (4.13). The result is that

$$\text{(4.28)} \qquad \varrho_{n\,ph} = \frac{4}{3}\frac{\tilde{U}}{c^2} = 32\pi\zeta_4 \frac{k^4T^4}{h^3c^5} = \frac{4kT}{c^2}\frac{\zeta_4}{\zeta_3} n_{ph}\,,$$

$$\text{(4.29)} \qquad \varrho_{n\,rot} = \frac{4\pi p_0^4}{3h^2kT}\left(\frac{2\pi\mu kT}{h^2}\right)^{\frac{1}{2}} \exp[-\Delta/kT] = \frac{p_0^2}{3kT} n_{rot}\,.$$

These results give a very accurate description of the experimental data for ϱ_n as a function of temperature, obtained by the methods based on the semi-empirical theories presented in Chapter II. Recently BENDT, COWAN, YARNELL (1959) [18], using the full energy-momentum relation calculated numerically ϱ_n as a function of temperature on the basis of the expression (4.27). They found very good agreement with experimental results obtained by various authors (see Fig. 13).

4. – The lambda point.

LANDAU extrapolated the formula up to the λ-point and used the formula as an equation for the λ-temperature by requiring that at the λ-point $\varrho_n = \varrho$, so that

$$\text{(4.30)} \qquad \varrho = 32\pi\zeta_4 \frac{k^4T_\lambda^4}{h^3c^5} + \frac{4\pi p_0^4}{3h^2kT_\lambda}\left(\frac{2\pi\mu kT_\lambda}{h^2}\right)^{\frac{1}{2}} \exp[-\Delta/kT_\lambda]\,.$$

On the first look serious objections can be raised to this procedure. The whole concept of the excitation gas of noninteracting excitations, which is on the background of the calculations in this chapter, becomes worse when we approach the temperatures of the λ-point. One of the most puzzling points therefore seems to be the fact that the present theory for the thermodynamic quantities $\tilde{S}$, $\tilde{U}$ and ϱ_n gives such good results *up to the λ-point*. At the λ-point the number density of rotons is only one sixth of the number density of particles. Knowing that each « roton » considered as a molecular excited state represents a perturbation not only of one particle but probably also of its neighbourhood, one would expect that at the λ-point very large deviations from the single-particle picture occur. On the other hand, from the fact that apparently the calculations give reasonable results even at the λ-point, one might conclude that the interaction between rotons is very much of a δ-character, as was assumed by LANDAU and KHALATNIKOV in their calculations of the transport properties.

5. – The two-fluid model.

So far we have considered the background medium, in which the excitations move, to be at rest and all quantities $\tilde{U}_{ex}$, P_{ex} etc. and also $\boldsymbol{v}$, the average velocity of the excitations, correspond to a co-ordinate system which is at rest with respect to this medium.

We now consider a more general situation in which the background medium, *with* this co-ordinate system, to which the calculations so far referred, moves with constant velocity $\boldsymbol{v}_c$ with respect to a fixed co-ordinate system. For reference we will call this last system the « laboratory system » and we want to derive the expressions for the different quantities in the laboratory system *from* those in the system moving with the background medium. In order to do this we apply the general transformation laws of mechanics:

If a mechanical system with total mass M, energy E and momentum $\boldsymbol{P}$ moves in the laboratory with a velocity $\boldsymbol{v}_s$, the transformed quantities E' and $\boldsymbol{P}'$ in the laboratory system are

$$(4.31)\qquad \begin{cases} E' = E + \boldsymbol{P}\cdot\boldsymbol{v}_s + \frac{1}{2}M\boldsymbol{v}_s^2\,, \\ \boldsymbol{P}' = \boldsymbol{P} + M\boldsymbol{v}_s\,. \end{cases}$$

Applying this to a single excitation on finds, using

$$E = E_0 + \varepsilon \qquad E'_0 \overset{\text{def}}{=} E_0 + \tfrac{1}{2}M\boldsymbol{v}_s^2\,,$$

$$\boldsymbol{P} = \boldsymbol{p} \qquad \boldsymbol{P}'_0 \overset{\text{def}}{=} M\boldsymbol{v}_s\,,$$

that for the excitation the following transformation rules hold:

$$(4.32)\qquad \begin{cases} \varepsilon' = \varepsilon + \boldsymbol{p}\cdot\boldsymbol{v}_s\,, \\ \boldsymbol{p}' = \boldsymbol{p}\,. \end{cases}$$

Application to the system of liquid helium gives:

— total energy density: $\tilde{U}' = \tilde{U} + \boldsymbol{j}_{ex}\cdot\boldsymbol{v}_s + \frac{1}{2}\varrho\boldsymbol{v}_s^2$;

— total momentum density: $\boldsymbol{j}' = \boldsymbol{j}_{ex} + \varrho\boldsymbol{v}_s$;

— total pressure tensor: $\mathbf{P}' = \mathbf{P} + \boldsymbol{j}_{ex}\boldsymbol{v}_s + \boldsymbol{v}_s\boldsymbol{j}_{ex} + \varrho\boldsymbol{v}_s\boldsymbol{v}_s$;

which gives

$$U_0' = U_0 + \tfrac{1}{2}\varrho v_s^2 \qquad U_{ex}' = U_{ex} + \boldsymbol{j}_{ex}\cdot\boldsymbol{v}_s\,,$$

$$\mathbf{P}_0' = P_0\mathbf{1} + \varrho\boldsymbol{v}_s\boldsymbol{v}_s \qquad \mathbf{P}_{ex}' = P_{ex}\mathbf{1} + \boldsymbol{j}_{ex}\boldsymbol{v}_s + \boldsymbol{v}_s\boldsymbol{j}_{ex}\,,$$

thus except for terms *quadratic* in v or v_s the thermodinamic quantities are unchanged by this transformation.

We thus have to distinguish between two systems:

a) the *system of excitations* having the average velocity $\boldsymbol{v}_n = \boldsymbol{v} + \boldsymbol{v}_s$ with respect to the laboratory system. The expressions for the thermodynamic quantities of this gas of excitations, $\tilde{S}_{ex}$, $\tilde{U}_{ex}$ etc. are, disregarding correction terms quadratic in $\boldsymbol{v}$ and $\boldsymbol{v}_s$, again given by the expressions (4.11). To this gas of excitations corresponds a momentum density $\varrho_n\boldsymbol{v}$ and an excitation mass density ϱ_n as given by (4.27);

b) the *background medium* having the average velocity $\boldsymbol{v}_s$ in the laboratory system. To this background medium we have to attach the thermodynamic quantities of the ground state: the energy $U_0 = F_0$, the pressure P_0 and the entropy $S_0 = 0$, as given by eq. (4.8).

These two systems, in terms of which liquid helium is described, *have now to be identified with the two fluids of the two-fluid model.*

a) The *normal fluid* has to be identified with the *system of excitations*: the normal fluid velocity becomes the average velocity $\boldsymbol{v}_n$ of the excitation gas in the laboratory system. The thermodynamic quantities to be attached to this normal fluid are, apart from terms quadratic in the velocities, energy density $\tilde{U}_{ex}$, the pressure P_{ex}, the entropy density $\tilde{S}_{ex}$. The excitation mass density ϱ_n now obtains the name « *normal-fluid density* ».

b) The *superfluid* has to be identified with the background medium: v_s becomes the « *superfluid velocity* ». The thermodynamic quantities to be attached to the superfluid are the ground-state energy U_0 and pressure P_0 and consequently the entropy S_0 of the superfluid is equal to zero. It is sometimes useful for symmetry reasons to introduce also the « *superfluid density* », defined by $\varrho_s = \varrho - \varrho_n$, so that the total density many thus be written as the sum $\varrho = \varrho_n + \varrho_s$ of a normal and a superfluid density. It is then also possible to transform the total momentum density $\boldsymbol{j}'(=\boldsymbol{j}) = \boldsymbol{j}_{ex} + \varrho\boldsymbol{v}_s = \varrho_n(\boldsymbol{v}_n - \boldsymbol{v}_s) + \varrho\boldsymbol{v}_s$

into $\boldsymbol{j}=\varrho_n\boldsymbol{v}_n+\varrho_s\boldsymbol{v}_s$: We will, however, not use this concept, because it brings no simplification in the physical picture.

Of course the relation between the excitation model and the two-fluid model presented in this section needs further support. This will come in the next sections, where we will show that the reversible and irreversible transport phenomena in the two-fluid model, which have empirically been found and which have been explained in Chapter II, follow from this identification.

V. - Reversible Equations of Transport.

1. – Local equilibrium.

In the previous chapter we considered the system of liquid helium in equilibrium and the method of calculating the macroscopic properties of the system. This was done on the basis of the ensemble method of statistical mechanics leading to the partition function. This partition function depends on certain parameters T and $\boldsymbol{v}$ which were assumed to be constant.

We now consider liquid helium in a state which is *non*equilibrium state: The parameters T and $\boldsymbol{v}$ may vary gradually both in time and space. It is assumed, however, that these parameters vary gradually such that *locally* it is still meaningful to use such macroscopic concepts as the *local temperature*, T and the *local average velocity* of the excitations with respect to the medium, $\boldsymbol{v}$, and so that the local values of the thermodynamic quantities are defined by these local values of T and $\boldsymbol{v}$.

In such a system the partition functions method of statistical mechanics no longer applies in its original form. Although it might be possible to make use of methods of statistical mechanics which are more adapted to the present situation, we will follow a somewhat simpler way [17], by using again the concept of *local distribution function*. We thus now consider the distribution function

$$f(\boldsymbol{p})=\frac{h^{-3}}{\exp\left[\beta(\varepsilon-\boldsymbol{v}\cdot\boldsymbol{p})-1\right]}=f_0(p)-\boldsymbol{v}\cdot\boldsymbol{p}\frac{\partial f_0}{\partial\varepsilon}, \tag{5.1}$$

in which T, and $\boldsymbol{v}$ have to be considered as slowly varying functions of $\boldsymbol{r}$ and t. The system of excitations may be considered as an excitation gas of localized excitations, wave packets, each of which is moving with its group velocity $u=\partial\varepsilon(p)/\partial p$ corresponding to its momentum p. The distribution function is then defined by requiring that $f\,\mathrm{d}\boldsymbol{p}\,\mathrm{d}\boldsymbol{r}$ is equal to the number of excitations in $\mathrm{d}\boldsymbol{p}$ and $\mathrm{d}\boldsymbol{r}$ at $\boldsymbol{p}$ and $\boldsymbol{r}$ at time t.

The next step we make is that not only the quantities T and $\boldsymbol{v}$ characterizing the local thermodynamic properties vary in space and time, but also the *local density* ϱ and the *local velocity* $\boldsymbol{v}_s$ of the medium in which the excitations move, are assumed to vary in space and time.

The effect of the density on ε has been considered in Chapter III: the density effects c, Δ, p_0 and μ, all parameters which determine the dispersion relation $\varepsilon(p)$.

If, however, an excitation with excitation energy $\varepsilon(p)$ in the background medium at rest is considered in a co-ordinate (laboratory) system, with respect to which the medium moves with a velocity $\boldsymbol{v}_s$, ordinary transformation laws of mechanics show that then in the laboratory system the energy of the excitation is given by

$$\varepsilon'(p) = \varepsilon(p) + \boldsymbol{v}_s \cdot \boldsymbol{p} \,. \tag{5.2}$$

This relation can also be considered as an ordinary Doppler effect when we write this in the form $\omega'(k)=\omega(k)+\boldsymbol{v}_s\cdot\boldsymbol{k}$. In the laboratory system (at rest) the excitations, whose frequency measured in the co-ordinate system moving with the medium is equal to $\omega(k)$, will be *in*creased to $\omega'(k)$ with an amount $\boldsymbol{v}_s\cdot\boldsymbol{k}$. This expression $\varepsilon'(p)$ may be seen as an Hamiltonian

$$H(\boldsymbol{p}, \boldsymbol{r}) = \varepsilon\big(\boldsymbol{p}; \varrho(\boldsymbol{r})\big) + \boldsymbol{v}_s(\boldsymbol{r})\cdot\boldsymbol{p} \,,$$

which depends on $\boldsymbol{r}$ implicitely through the quantities ϱ and $\boldsymbol{v}_s$. The two Hamiltonian equations then read

$$\dot{r} = u' = \nabla_p H = \nabla_p \varepsilon + \boldsymbol{v}_s = \boldsymbol{u} + \boldsymbol{v}_s \,, \tag{5.3}$$

$$\dot{p} = -\nabla_r H = -\frac{\partial \varepsilon}{\partial \varrho} \nabla \varrho - \nabla \boldsymbol{v}_s \cdot \boldsymbol{p} \,, \tag{5.4}$$

(dyadic notation is used in $\nabla\boldsymbol{v}_s\cdot\boldsymbol{p}=\nabla\sum_i \boldsymbol{v}_{si}\, p_i$). The first relation is obvious as it represents the addition theorem of velocities and the second relation is a direct consequence of the fact that the energy depends on $\boldsymbol{v}_s$ and ϱ: an excitation travelling against a gradient of ϱ or $\boldsymbol{v}_s$ is accelerated or decelerated depending on the direction of the gradient.

2. – The Boltzmann equation.

The equation which determines the time change of the distribution function is the Boltzmann equation:

$$\frac{\partial f}{\partial t} + \boldsymbol{u}\cdot\nabla_r f + \dot{\boldsymbol{p}}\cdot\nabla_p f = \left(\frac{\partial f}{\partial t}\right)_{\text{coll}} . \tag{5.5}$$

During collisions of excitations, momentum and energy are conserved, but the number of excitations may change: they may be annihilated or created during collisions. For the following it is not really necessary to specify further the right-hand side and therefore we will omit a more detailed discussion of the collision process of excitations. It is, however, important to realize that if we substitute the equilibrium distribution function f_0 in the expression $(\partial f/\partial t)_{\text{coll}}$, one should certainly have $(\partial f_0/\partial t)_{\text{coll}} = 0$, this being the *condition* that f_0 is the equilibrium distribution function.

3. – Derivation of the irreversible equations of transport.

In order to derive the reversible equations of transport for momentum and energy, the Boltzmann equation is multiplied by a quantity ψ (which is *conserved during collisions*, *i.e.* either the momentum $\boldsymbol{p}$ or the energy ε') and integration is carried out over all momenta. This leads, after some integration by parts to

$$\frac{\partial}{\partial t}\int f\psi\,\mathrm{d}\boldsymbol{p} + \nabla_r\int f\psi\boldsymbol{u}\,\mathrm{d}\boldsymbol{p} - \int f(\dot{\psi} + \boldsymbol{u}\cdot\nabla_r\psi + \dot{\boldsymbol{p}}\cdot\nabla_p\psi)\,\mathrm{d}\boldsymbol{p} = 0\,. \tag{5.6}$$

Use has been made of the fact that the ψ is assumed to be a quantity which is conserved by collisions. The consequence is the well-known fact that then

$$\int\left(\frac{\partial f}{\partial t}\right)_{\text{coll}}\psi\,\mathrm{d}\boldsymbol{p} = 0\,. \tag{5.7}$$

First case $\psi=\boldsymbol{p}$:

$$\frac{\partial}{\partial t}\int f\boldsymbol{p}\,\mathrm{d}\boldsymbol{p} - \nabla\cdot\int f\boldsymbol{u}\boldsymbol{p}\,\mathrm{d}\boldsymbol{p} + \int f\left(\frac{\partial\varepsilon}{\partial\varrho}\nabla\varrho + \nabla\boldsymbol{v}_s\cdot\boldsymbol{p}\right)\mathrm{d}\boldsymbol{p} = 0\,,$$

where use has been made of (5.4). If we introduce (5.1) and linearize the equation, *i.e.* we drop all terms of the second degree in $\boldsymbol{v}_s$ or $\boldsymbol{v}$, we obtain, using the definition (4.11) for the thermodynamic quantities and (4.27) for the excitation mass density,

$$\frac{\partial}{\partial t}(\varrho_{\mathrm{n}}\boldsymbol{v}) = \nabla\tilde{F}_{\mathrm{ex}} - \mu_{\mathrm{ex}}\nabla\varrho = -\nabla P_{\mathrm{ex}} + \varrho\nabla\mu_{\mathrm{ex}}\,. \tag{5.8}$$

Second case $\psi=\varepsilon$:

$$\frac{\partial}{\partial t}\int f\varepsilon\,\mathrm{d}\boldsymbol{p} + \nabla\cdot\int f\boldsymbol{u}\varepsilon\,\mathrm{d}\boldsymbol{p} - \int f\left(\frac{\partial\varepsilon}{\partial\varrho}\dot{\varrho} + \boldsymbol{p}\cdot\dot{\boldsymbol{v}}_s\right)\mathrm{d}\boldsymbol{p} = 0\,,$$

where use has been made of $\boldsymbol{u}\cdot\nabla_r\,\varepsilon'+\dot{\boldsymbol{p}}\cdot\nabla_p\,\varepsilon'=0$ because of (5.3) and (5.4). Linearizing again, one obtains

$$\frac{\partial}{\partial t}\,\widetilde{U}-\mu_{\text{ex}}\frac{\partial\varrho}{\partial t}=-\nabla\cdot(\boldsymbol{v}_{\text{n}}T\widetilde{S})\,. \tag{5.9}$$

The two eqs. (5.8) and (5.9) are the two equations expressing the conservation of momentum and energy, respectively. The first (5.7) is really the equation of motion in the excitation gas: $\varrho_{\text{n}}\boldsymbol{v}=\boldsymbol{j}_{\text{ex}}$ is the excitation mass current density and $\partial\boldsymbol{j}_{\text{ex}}/\partial t=-\nabla P_{\text{ex}}$ would be the equivalent of Euler's equation in ordinary hydrodynamics. The term $\varrho\nabla\mu_{\text{ex}}$ is added and is related to density changes in the medium, which also accelerate the excitations. The second equation is the equation of conservation of energy. $(\boldsymbol{v}_{\text{n}}T\widetilde{S})$ represents the energy current density of the excitation gas.

Using the thermodynamic relations:

$$\mathrm{d}P_{\text{ex}}=\widetilde{S}_{\text{ex}}\,\mathrm{d}T+\varrho\,\mathrm{d}\mu_{\text{ex}}\qquad \nabla P_{\text{ex}}=\widetilde{S}_{\text{ex}}\nabla T+\varrho\nabla\mu_{\text{ex}}\,, \tag{5.10}$$

$$\mathrm{d}\widetilde{F}_{\text{ex}}=-\widetilde{S}_{\text{ex}}\,\mathrm{d}T+\mu_{\text{ex}}\,\mathrm{d}\varrho\qquad \nabla\widetilde{F}_{\text{ex}}=-\widetilde{S}_{\text{ex}}\nabla T+\mu_{\text{ex}}\nabla\varrho\,, \tag{5.11}$$

$$\mathrm{d}\widetilde{U}_{\text{ex}}=\widetilde{T}\,\mathrm{d}S_{\text{ex}}+\mu_{\text{ex}}\,\mathrm{d}\varrho\,,\qquad \frac{\partial U_{\text{ex}}}{\partial t}=T\frac{\partial S_{\text{ex}}}{\partial t}+\mu_{\text{ex}}\frac{\partial\varrho}{\partial t}\,, \tag{5.12}$$

and linearizing again, one finds

$$\varrho_{\text{n}}\frac{\partial\boldsymbol{v}}{\partial t}=-\widetilde{S}_{\text{ex}}\nabla T\,, \tag{5.13}$$

$$\frac{\partial\widetilde{S}_{\text{ex}}}{\partial t}=-\nabla(\widetilde{S}_{\text{ex}}\boldsymbol{v}_{\text{n}})\,. \tag{5.14}$$

The second eq. (5.14) is identical with the eq. (2.4), which was the equation postulated by LANDAU and which expressed the *conservation of entropy* through the flow of the normal fluid.

The first eq. (5.13) represents the equation of motion (2.7) of superfluid and normal fluid with respect to each other. The accelerating « force » is not the pressure gradient but the *temperature* gradient. If one abstracts this from the equation of Euler

$$\frac{\partial\boldsymbol{j}}{\partial t}=-\nabla P=-\nabla P_{\text{ex}}-\nabla P_0\,, \tag{5.15}$$

one finds, using $\boldsymbol{j}=\varrho_{\text{n}}\boldsymbol{v}+\varrho\boldsymbol{v}_{\text{s}}$,

$$\varrho\frac{\partial\boldsymbol{v}_{\text{s}}}{\partial t}=-\nabla P_{\text{ex}}+S_{\text{ex}}\nabla T-\nabla P_0=-\varrho\nabla\mu_{\text{ex}}-\varrho\nabla\mu_0=-\varrho\nabla\mu\,, \tag{5.16}$$

(using (5.10) and (4.8′)) which is the fundamental equation taken by LANDAU as the basis of the two-fluid model.

4. – Second sound.

All the consequences derived in Chapter II on the basis of the two-fluid model thus automatically follow also from the excitation gas model, in particular also the expression for second sound, eq. (2.21),

$$c_{\mathrm{II}}^2 = \frac{\varrho_s}{\varrho_n} s^2 \left(\frac{\partial T}{\partial s}\right)_\varrho . \tag{5.17}$$

Substituting in this expression the theoretical expressions for s and for ϱ_n, one obtains in the temperature region $T > 1$ °K, where the *roton contribution* is predominant, using (4.16) and (4.29) for s_{rot} and $\varrho_{n,\mathrm{rot}}$, respectively,

$$c_{\mathrm{II}}^2 = \frac{\varrho_s}{\varrho} \frac{3k^2T^2}{p_0^2} \frac{(\Delta/kT)^2 + 3(\Delta/kT) + \frac{9}{4}}{(\Delta/kT)^2 + (\Delta/kT) + \frac{3}{4}} . \tag{5.18}$$

This expression does not change very much with temperature, but goes to zero rapidly near the λ-point due to the factor ϱ_s/ϱ. In the temperature region below 0.5 °K, where only the *phonon contribution* is important, one obtains, using (4.14) and (4.28) for s_{ph} and $\varrho_{n,\mathrm{ph}}$,

$$c_{\mathrm{II}}^2 = \tfrac{1}{3} c^2 (= \tfrac{1}{3} c_{\mathrm{I}}^2) . \tag{5.19}$$

The square of the second sound velocity is thus equal to one third of the velocity of first sound. If one expresses this in terms of the excitation gas one should say: the compression waves in the excitation gas have a velocity of propagation c_{II} which is $1/\sqrt{3}$ times the velocity of the individual excitations (which is $c = c_{\mathrm{I}}$). Similarly the compression waves in a real ideal gas of molecules propagate with a velocity c_{I} which is equal to $1/\sqrt{3}$ times the r.m.s. average speed $\bar{u}$ of the molecules $\left(c_{\mathrm{I}} = \sqrt{(p/\varrho)} = \sqrt{(kT/m)};\;\; \bar{u} = \sqrt{(3kT/m)}\right)$. This relationship between the compression waves in the excitation gas (second sound) and the compression waves in the real ideal gas (« first » sound or ordinary sound) explains the name « second sound ». As compression waves in the excitation gas, they only lead, however, to local increase of the « thermal pressure » P_{ex} corresponding to local increased temperatures, and not to a change of the density ϱ (provided the conditions (2.22) are satisfied).

An interesting experimental discovery supported this picture: At lower temperatures the free path of the phonon type of excitations becomes larger and larger (as we will see in the next chapter) so that in fact in the phonon region the free path becomes of the order of magnitude of the wavelength of the second sound waves. At still lower temperatures the free path becomes even very much larger than this wavelength, so that the whole picture of compression waves in the excitation gas breaks down: perturbations start to propagate with the velocity of the individual excitations, *i.e.* with the velocity of first sound c. Indeed one finds experimentally that the velocity with which second sound pulses propagate through the liquid rises above the low-temperature limit (5.19) and approaches c_{I}.

One final remark should be made to close this chapter on the *reversible* equations of transport: The eq. (5.14), which we derived, is only the true conservation equation for the entropy provided that irreversible effects leading to the terms proportional to the coefficients of heat conductivity and the viscosity occuring in the exact expression (2.11) are neglected. This (not very important) defect of the derivation comes from the fact that the « local distribution function » (5.1), which was on the basis of our derivations, is only an approximation (the so called « zeroth » approximation to the true local distribution function). In the next chapter we will investigate the influence of the irreversible effects.

VI. - Irreversible Transport Phenomena.

1. – Solution of the Boltzmann equation.

As in the previous chapter we consider again liquid helium in the *non-equilibrium state* in which the temperature T and the average velocity of excitations with respect to the background fluid $\boldsymbol{v}$ and also the density ϱ and the velocity $\boldsymbol{v}_s$ of the background fluid vary gradually both in space and time.

In the previous chapter we have introduced the local equilibrium distribution function:

$$f^{(0)}(\boldsymbol{p}) = f_0(p) - \boldsymbol{v}\cdot\boldsymbol{p}\,\frac{\partial f_0}{\partial \varepsilon}\,, \tag{6.1}$$

(which we will indicate *here* with the subscript (0)), which is the equilibrium distribution function corresponding to the local values of T, $\boldsymbol{v}$, ϱ and $\boldsymbol{v}_s$.

The *real* local distribution function which exists in fact in the liquid will differ from this function $f^{(0)}(\boldsymbol{p})$, because $f^{(0)}(\boldsymbol{p})$ is only the correct distribution function provided that the local values T, $\boldsymbol{v}$, ϱ and $\boldsymbol{v}_s$ are the same in the whole system. We therefore write for the true local distribution function:

$$f(\boldsymbol{p}) = f^{(0)}(\boldsymbol{p}) + f^{(1)}(\boldsymbol{p}) . \tag{6.2}$$

We require, however, that $f(\boldsymbol{p})$ satisfies the condition that the quantities T and $\boldsymbol{v}$ occurring in $f(\boldsymbol{p})$ keep their physical meaning, *i.e.* that the correction terms $f^{(1)}(\boldsymbol{p})$ do not affect the local values for $\varrho_n\boldsymbol{v}$ and U obtained in the previous chapter:

$$\int f(\boldsymbol{p})\boldsymbol{p}\,\mathrm{d}\boldsymbol{p} = \int f^{(0)}(\boldsymbol{p})\boldsymbol{p}\,\mathrm{d}\boldsymbol{p} = \varrho_n(T)\boldsymbol{v} , \tag{6.3}$$

$$\int f(\boldsymbol{p})\,\varepsilon\,\mathrm{d}\boldsymbol{p} = \int f^{(0)}(\boldsymbol{p})\varepsilon\,\mathrm{d}\boldsymbol{p} = \tilde{U}(T) , \tag{6.4}$$

i.e. that

$$\int f^{(1)}(\boldsymbol{p})\boldsymbol{p}\,\mathrm{d}\boldsymbol{p} = \int f^{(1)}(\boldsymbol{p})\varepsilon\,\mathrm{d}\boldsymbol{p} = 0 . \tag{6.5}$$

The equation from which the distribution function $f(\boldsymbol{p})$ can be obtained, is again the Boltzmann equation:

$$\mathscr{D}f \equiv \frac{\partial f}{\partial t} + \boldsymbol{u}\cdot\nabla_r f + \dot{\boldsymbol{p}}\cdot\nabla_p f = \left(\frac{\partial}{\partial t}\right)_{\text{coll}} f . \tag{6.6}$$

If we substitute (6.2) in this equation, we first remark that $(\partial/\partial t)_{\text{coll}} f^{(0)} = 0$, because of the definition of $f^{(0)}(\boldsymbol{p})$ being the local *equilibrium* distribution function. The right-hand side thus obtains only a contribution from $f^{(1)}$. For simplicity we give a simplified version of a more elaborate theoretical treatment which leads essentially to the same result: we make the simplifying assumption that $(\partial/\partial t)_{\text{coll}} f^{(1)}$ is proportional to the deviation of equilibrium distribution function, *i.e.* that

$$\left(\frac{\partial}{\partial t}\right)_{\text{coll.}} f^{(1)}(\boldsymbol{p}) = -\frac{f^{(1)}(\boldsymbol{p})}{\tau} , \tag{6.7}$$

where τ is a relaxation time. In fact this is a very useful simplifying assumption which also in the kinetic theory of gases leads to a good approximation.

On the left side of eq. (6.6) substitution of $f^{(0)}$ does not give zero and in agreement with the systematic approximation procedure of Chapman-Enskog we neglect $f^{(1)}$ on the left side in first approximation. One thus obtains the equation:

$$\mathscr{D}f^{(0)} \equiv \frac{\partial f^{(0)}}{\partial t} + \boldsymbol{u}\cdot\nabla_r f^{(0)} + \dot{\boldsymbol{p}}\cdot\nabla_p f^{(0)} = -\frac{f^{(1)}}{\tau} . \tag{6.8}$$

Substitution of $f^{(0)}$ in the left side gives

$$(6.9)\qquad \mathscr{D}f^{(0)} = \beta f^{(0)}(1+h^3 f^{(0)})\Big\{(\varepsilon - \boldsymbol{p}\cdot\boldsymbol{v})\,\frac{\partial \ln T}{\partial t} - \frac{\partial\varepsilon}{\partial\varrho}\frac{\partial\varrho}{\partial t} + \boldsymbol{p}\cdot\frac{\partial \boldsymbol{v}}{\partial t} + (\boldsymbol{u}+\boldsymbol{v}_{\mathrm{s}})(\varepsilon-\boldsymbol{p}\cdot\boldsymbol{v})\,\nabla \ln T - \frac{\partial\varepsilon}{\partial\varrho}(\boldsymbol{v}_{\mathrm{n}}\cdot\nabla\varrho) + \big((\boldsymbol{u}+\boldsymbol{v}_{\mathrm{s}})\cdot\nabla\big)(\boldsymbol{v}_{\mathrm{n}}\cdot\boldsymbol{p}) - (\boldsymbol{v}_{\mathrm{n}}\cdot\nabla)(\boldsymbol{v}_{\mathrm{s}}\cdot\boldsymbol{p})\Big\}\,.$$

The time derivatives of T, ϱ and $\boldsymbol{v}$ are transformed into space derivatives by using the conservation equations. Following the procedure of Chapman and Enskog we use here in a consequent manner also the zero-order conservation eqs. (5.8) and (5.9) derived in the previous chapter from the Boltzmann equation $\mathscr{D}f(\boldsymbol{p})=0$ by substituting $f(\boldsymbol{p})=f^{(0)}(\boldsymbol{p})$. The terms resulting from $f^{(1)}$ would have given rise to the irreversible contributions proportional to η and λ resulting from irreversible viscous streaming and heat conduction, which were included in the *complete* conservation eqs. (2.10)-(2.13), given in Chapter II.

We thus substitute

$$(6.10)\qquad \frac{\partial\varrho}{\partial t} = -\nabla\cdot\boldsymbol{j}\,,$$

$$(6.11)\qquad \varrho_{\mathrm{n}}\frac{\partial\boldsymbol{v}}{\partial t} = -\tilde{S}\nabla T\,,$$

$$(6.12)\qquad \frac{\partial \ln T}{\partial t} = \left(\frac{\partial \ln T}{\partial\varrho}\right)_{\tilde{S}}\frac{\partial\varrho}{\partial t} + \frac{\partial \ln T}{\partial\tilde{S}}\frac{\partial\tilde{S}}{\partial t} = -\left(\frac{\partial \ln T}{\partial\varrho}\right)_{\tilde{S}}(\nabla\cdot\boldsymbol{j}) - \left(\frac{\partial \ln T}{\partial \ln \tilde{S}}\right)_{\varrho}(\nabla\cdot\boldsymbol{v}_{\mathrm{n}})\,,$$

linearize the equations, so obtained, and have

$$\mathscr{D}f^{(0)} = \frac{\partial f^{(0)}}{\partial\varepsilon}\Big\{\Big[\frac{ST}{\varrho_n}\boldsymbol{p} - \varepsilon\boldsymbol{u}\Big]\cdot\nabla \ln T - \overset{0}{\boldsymbol{p}\boldsymbol{u}} : \overset{0}{\overline{\overline{\nabla\boldsymbol{v}_{\mathrm{n}}}}} + \Big[\varepsilon\left(\frac{\partial \ln T}{\partial\varrho}\right)_{\tilde{S}} - \frac{\partial\varepsilon}{\partial\varrho}\Big](\nabla\cdot\boldsymbol{j}) + \Big[\varepsilon\left(\frac{\partial \ln T}{\partial \ln \tilde{S}}\right)_{\varrho} - \frac{1}{3}pu\Big](\nabla\cdot\boldsymbol{v}_{\mathrm{n}})\Big\}\,,$$

where we have written

$$(6.13)\qquad \partial f^{(0)}/\partial\varepsilon = -\beta f^{(0)}[1+h^3 f^{(0)}]\ (^*)\,.$$

(*) See for the dyadic notation note on p. 12.

The Boltzmann eq. (6.8), using (6.13) for the left-hand side, then gives the following solutions for the three cases:

a) Temperature gradient (heat conduction): In the case that a temperature gradient exists only, the solution becomes

$$f_\lambda^{(1)}(\boldsymbol{p}) = -\tau_\lambda \frac{\partial f^{(0)}}{\partial \varepsilon}\left[\frac{\widetilde{S}T}{\varrho_n}\boldsymbol{p} - \varepsilon\boldsymbol{u}\right]\cdot\nabla \ln T\,. \tag{6.14}$$

The velocity distribution $f(\boldsymbol{p})$ is thus modified by a term $f^{(1)}(\boldsymbol{p})$ which is proportional to $\nabla \ln T$. It is clear that the condition $\int f_\lambda^{(1)}\varepsilon\, \mathrm{d}\boldsymbol{p}=0$ is satisfied, because of symmetry reasons. The second condition $\int f_\lambda^{(1)}\boldsymbol{p}\,\mathrm{d}\boldsymbol{p}=0$ is satisfied also because of the expression (4.27) for ϱ_n and (4.11) for $\widetilde{S}$. It is also clear that $\int f_\lambda^{(1)}\,\mathrm{d}\boldsymbol{p}=0$, however that $\int f^{(1)}\boldsymbol{u}\,\mathrm{d}\boldsymbol{p}\neq 0$: there is an average stream of excitations in the direction of $\nabla \ln T$ carrying the heat current:

$$\boldsymbol{q} = \int f\varepsilon\boldsymbol{u}\,\mathrm{d}\boldsymbol{p} = \boldsymbol{q}_\mathrm{ex}^{(0)} + \boldsymbol{q}_\lambda^{(1)}\,, \tag{6.15}$$

$$\boldsymbol{q}_\mathrm{ex}^{(0)} = \int f^{(0)}\varepsilon\boldsymbol{u}\,\mathrm{d}\boldsymbol{p} = \boldsymbol{v}_n\cdot T\widetilde{S}_\mathrm{ex}\,, \tag{6.16}$$

$$\left\{\begin{aligned} q_\lambda^{(1)} &= -\lambda\nabla T\,,\\ \lambda &= \frac{\tau_\lambda}{3kT}\int\frac{\partial f^{(0)}}{\partial\varepsilon}\left(\frac{\widetilde{S}T}{\varrho_n}\boldsymbol{p}-\varepsilon\boldsymbol{u}\right)\cdot\varepsilon\boldsymbol{u}\,\mathrm{d}\boldsymbol{p}\,. \end{aligned}\right. \tag{6.17}$$

There is only a *formal* expression for λ because it expresses λ in terms of the (unknown) « relaxation time » for thermal conductivity τ_λ. It will prove, however, to be quite useful.

b) Velocity gradient (viscous streaming): In case there exists only a gradient in the velocity $\boldsymbol{v}_\mathrm{n}$ the solution takes the form

$$f_\eta^{(1)}(\boldsymbol{p}) = \tau_\eta \frac{\partial f^{(0)}}{\partial\varepsilon}\overset{0}{\boldsymbol{p}\boldsymbol{u}}:\overset{0}{\overline{\overline{\nabla\boldsymbol{v}_\mathrm{n}}}}\,. \tag{6.18}$$

From symmetry reasons it is now clear that

$$\int f_\eta^{(1)}\boldsymbol{p}\,\mathrm{d}\boldsymbol{p} = \int f_\eta^{(1)}\varepsilon\,\mathrm{d}\boldsymbol{p} = 0 \qquad \text{again} \qquad \int f_\eta^{(1)}\,\mathrm{d}\boldsymbol{p} = 0\,.$$

$$\mathsf{P}_\mathrm{ex} = \int f_\eta^{(1)}\left(\boldsymbol{p}\boldsymbol{u} + \varrho\frac{\partial\varepsilon}{\partial\varrho}\mathbf{1}\right)\mathrm{d}\boldsymbol{p} = \mathsf{P}_\mathrm{ex}^{(0)} + \mathsf{P}_\eta^{(1)}\,, \tag{6.19}$$

$$\mathbf{P}_{\text{ex}}^{(0)} = \mathbf{1}\int f^{(0)}\left(\frac{1}{3}\,pu + \varrho\,\frac{\partial\varepsilon}{\partial\varrho}\right)\mathrm{d}\boldsymbol{p} = \mathbf{1}\,P_{\text{ex}}^{(0)}\,, \tag{6.20}$$

$$\begin{cases} \mathbf{P}_{\text{ex}}^{(1)} = -\,2\eta\,\overset{0}{\overline{\overline{\nabla\boldsymbol{v}_{\text{n}}}}}\,, \\ \eta \;\;= -\dfrac{1}{15}\,\tau_\eta\displaystyle\int \frac{\partial f^{(0)}}{\partial\varepsilon}\,u^2p^2\,\mathrm{d}\boldsymbol{p}\,. \end{cases} \tag{6.21}$$

One thus obtains an expression for η which expresses it in terms of the « relaxation time » for viscosity.

c) Divergent motion (second viscosity): The third contribution to $f_\phi^{(1)}$ is proportional to $\nabla\cdot\boldsymbol{j}$ and $\nabla\cdot\boldsymbol{v}_{\text{n}}$ and is of the form

$$f_\phi^{(1)} = -\,\tau_\phi\,\frac{\partial f^{(0)}}{\partial\varepsilon}\left\{\left[\varepsilon\,\frac{\partial\ln T}{\partial\varrho} - \frac{\partial\varepsilon}{\partial\varrho}\right](\nabla\cdot\boldsymbol{j}) + \left[\varepsilon\,\frac{\partial\ln T}{\partial\ln S} - \frac{1}{3}\,pu\right](\nabla\cdot\boldsymbol{v}_{\text{n}})\right\}. \tag{6.22}$$

Evaluation of $\int f_\phi^{(1)}\,\mathrm{d}\boldsymbol{p}$ now gives a *non*zero value:

$$n_\phi^{(1)} = \tau_\phi\left\{\left(\frac{\partial n^{(0)}}{\partial\varrho}\right)_s(\nabla\cdot\boldsymbol{j}) + \left(\frac{\partial n^{(0)}}{\partial T}\right)_\varrho\left(\frac{\partial T}{\partial\ln S}\right)_\varrho(\nabla\cdot\boldsymbol{v}_{\text{n}})\right\}. \tag{6.23}$$

The total number density of excitations is no longer equal to the equilibrium value, but there is a deviation, proportional to $\nabla\cdot\boldsymbol{j}$ and to $\nabla\cdot\boldsymbol{v}_{\text{n}}$.

Such type of terms in the nonequilibrium distribution function find their analogy in the case of the transport phenomena for gases only in the case that molecules possess internal degrees of freedom. In that case one finds that when the internal degrees of freedom are not in equilibrium with the (local) translational degrees of freedom, additional terms in the local nonequilibrium distribution function occur, which would show an effect on the pressure in the case of divergent streaming proportional to the so called second viscosity ϕ. As such a divergent streaming cannot easily be realized and used for measurements, the measurement of the second viscosity is based always on its effect on the attenuation of sound propagation.

The fact that also here such a second viscosity appears results from the deviations in « internal » equilibrium which are possible in helium II. There will appear to exist four second viscosity coefficients, three of which are independent. They can be measured experimentally in this case from the attenuation of first and second sound (see Section **3**).

The contribution $f_\phi^{(1)}$ to the local nonequilibrium distribution function gives rise to the following contributions to the *pressure tensor* and the *thermodynamic potential*:

$$\mathbf{P}_\phi^{(1)} = \int f_\phi^{(1)}\left(\boldsymbol{pu} + \varrho\,\frac{\partial\varepsilon}{\partial\varrho}\,\mathbf{1}\right)\mathrm{d}\boldsymbol{p} = -\,[\phi_1(\nabla\cdot\boldsymbol{j}) + \phi_2(\nabla\cdot\boldsymbol{v}_n)]\mathbf{1}\,, \tag{6.24}$$

and

$$\mu_\phi^{(1)} = \int f_\phi^{(1)} \frac{\partial \varepsilon}{\partial \varrho} \, \mathrm{d}\boldsymbol{p} = -[\phi_3(\nabla \cdot \boldsymbol{j}) + \phi_4(\nabla \cdot \boldsymbol{v}_n)] \,, \tag{6.25}$$

with the following definitions for ϕ (*)

$$\left\{\begin{aligned} \phi_1 &= \tau_\phi \int \frac{\partial f^{(0)}}{\partial \varepsilon} \left(\frac{1}{3} pu + \varrho \frac{\partial \varepsilon}{\partial \varrho} \right) \left(\varepsilon \frac{\partial \ln T}{\partial \varrho} - \frac{\partial \varepsilon}{\partial \varrho} \right) \mathrm{d}\boldsymbol{p} \,, \\ \phi_2 &= \tau_\phi \int \frac{\partial f^{(0)}}{\partial \varepsilon} \left(\frac{1}{3} pu + \varrho \frac{\partial \varepsilon}{\partial \varrho} \right) \left(\varepsilon \frac{\partial \ln T}{\partial \ln S} - \frac{1}{3} pu \right) \mathrm{d}\boldsymbol{p} \,, \\ \phi_3 &= \tau_\phi \int \frac{\partial f^{(0)}}{\partial \varepsilon} \frac{\partial \varepsilon}{\partial \varrho} \left(\varepsilon \frac{\partial \ln T}{\partial \varrho} - \frac{\partial \varepsilon}{\partial \varrho} \right) \mathrm{d}\boldsymbol{p} \,, \\ \phi_4 &= \tau_\phi \int \frac{\partial f^{(0)}}{\partial \varepsilon} \frac{\partial \varepsilon}{\partial \varrho} \left(\varepsilon \frac{\partial \ln T}{\partial \ln S} - \frac{1}{3} pu \right) \mathrm{d}\boldsymbol{p} \,. \end{aligned}\right. \tag{6.26}$$

2. – Confrontation with the experimental data.

The viscosity of liquid helium II which according to the discussion given above is a matter entirely related to the excitation gas, *i.e.* to the *normal fluid* shows according to the experimental data no very strange behaviour below the λ-point, as is shown in Fig. 14.

The *heat conductivity* λ for liquid helium II cannot be measured directly, but the damping of second sound has provided us with indirect data for the heat conductivity, which are given in Fig. 15.

The knowledge of the experimental values for η and λ on one hand and the possibility to make a theoretical calculation of the integral expressions occurring in (6.17) and (6.18), using the neutron scattering excitation energy-momentum relation, on the other hand, gives us the interesting possibility to obtain empirically the values of the relaxation times τ_η and τ_λ. This possibility has been used by BENDT, COWAN and YARNELL (1959). The result of this calculation is given in Fig. 16.

It is the task of the *microscopic theory* to give, starting from the interaction between excitations, a complete *a priori* calculation of the relaxation times τ_η and τ_λ. Very much has been accomplished here by the very elaborate

---

(*) The definition for the four coefficients ϕ differs slightly from those of Khalatnikov:

$$\zeta_1 = \phi_1 ; \qquad \zeta_2 = \phi_2 + \varrho\phi_1 ; \qquad \zeta_3 = \phi_3 ; \qquad \zeta_4 = \phi_4 + \varrho\phi_3 \,.$$

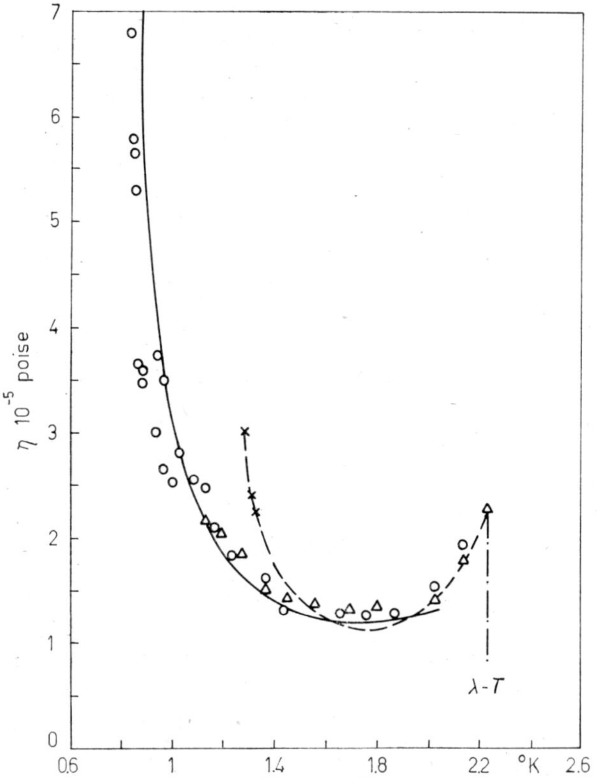

Fig. 14. – Experimental and theoretical curve for the viscosity of the normal fluid of liquid helium II (taken from: KHALATNIKOV [10]).

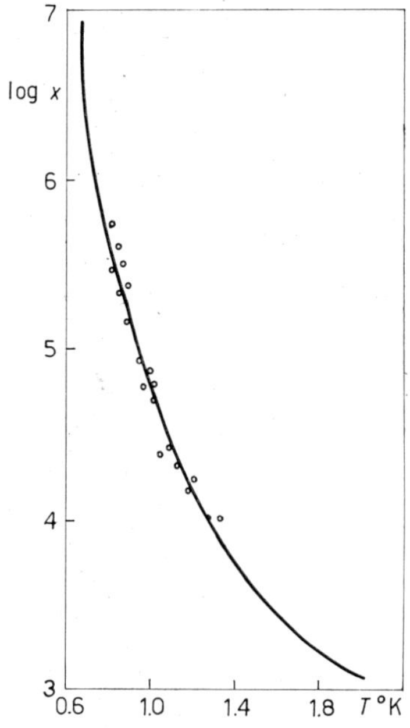

and detailed calculations of L. D. LANDAU and I. M. KHALATNIKOV. It is not our intention to discuss this in these lectures: a very complete and detailed review article of I. M. KHALATNIKOV gives all the necessary information. The nice agreement between the experimental data and the curves drawn in Fig. 14 and Fig. 15 which are theoretical curves from their theory, is very satisfactory. It must be stressed, however, that because of lack of knowledge of the « structure »

Fig. 15. – Indirectly obtained experimental data and the theoretical curve for the true heat conductivity of the normal fluid of liquid helium II (taken from: KHALATNIKOV [10]).

of the excitations (in particular in the roton region), the theoretical discussion of Landau and Khalatnikov necessaryly needed the introduction of arbitrary and adjustable constants, the justification of which came from the nice agreement with the experimental data.

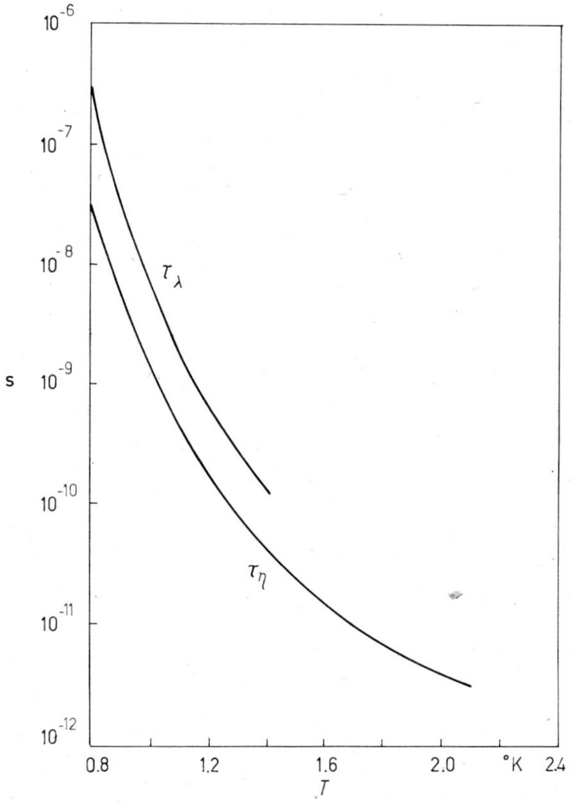

Fig. 16. – Relaxation times τ_λ and τ_η calculated by BENDT, COWAN and YARNELL from (6.17) and (6.21) using the experimental values for λ and η and evaluating the integral expressions on the basis of the excitation energy-momentum relation obtained from neutron scattering.

We only add here briefly one remark about the microscopic theory in order to make clear the general behaviour of the relaxation times τ_λ and τ_η as a function of temperature. It appears to be so that roton as well as phonon type of excitations practically always end their free path in a collision with a roton, which has the effect that the relaxation times should be inversely proportional to the roton number density. As the roton number density depends on temperature through a factor $\exp[-\Delta/kT]$, this is a dominating factor in the temperature dependence of the relaxation times. In Fig. 17 the inverse of the roton number density is plotted on the same logarithmic scale as used in

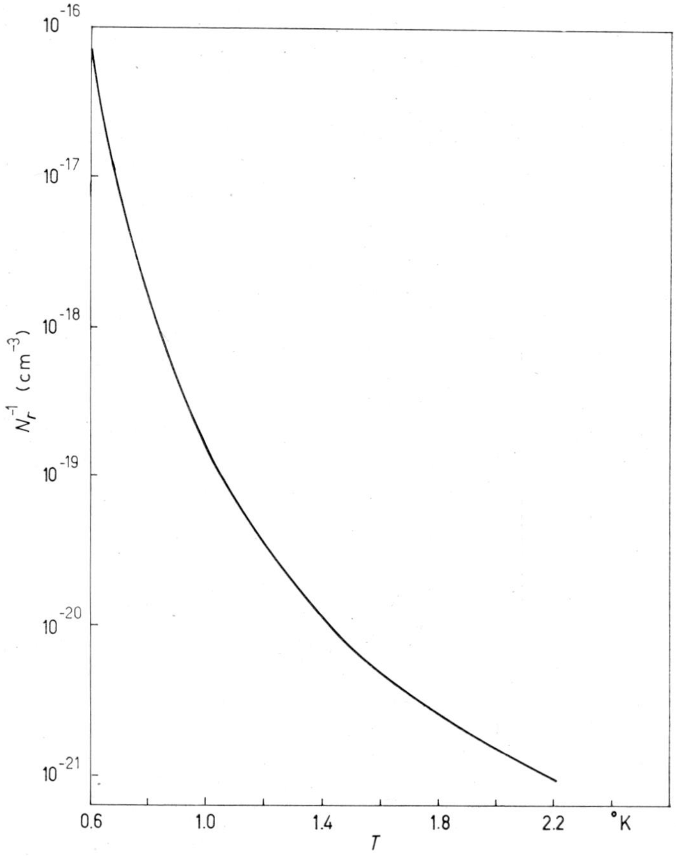

Fig. 17. – Values of the reciprocal of the roton density as function of temperature.

Fig. 16. The fact that both curves go rather parallel demonstrates clearly the dominant effect of the roton number density on the relaxation time.

3. – Effects of second viscosity.

The second viscosity coefficients ϕ could only be determined experimentally in an indirect way from damping of first and second sound.

The attenuation coefficient of first sound, defined by the equation for the density wave

$$\varrho(x, t) = \varrho_0 + \varrho' \exp\left[-\alpha_{\mathrm{I}} x\right] \exp\left[i\omega(t - x/c_{\mathrm{I}})\right],$$

is given by the following expression:

$$\alpha_{\mathrm{I}} = \frac{\omega^L}{2\varrho c_{\mathrm{I}}^2}\left\{\frac{4\eta}{3} + \left(\frac{c_p}{c_v} - 1\right)\frac{\lambda}{T}\left(\frac{\partial T}{\partial S}\right)_\varrho + \phi_2 + \varrho\phi_1\right\}. \tag{6.27}$$

Measurements of α_{I} in the liquid helium I region showed nice agreement with the theoretical value derived from the first viscosity contribution alone. The second term does not contribute very much and the third is not present. At the λ-point the curve goes through a minimum and rises again below the λ-point. A maximum is reached at about 1 °K where the value of α_{I} is about 10 times larger than the « classical value » obtained from the first two terms alone. This additional attenuation could be explained by Khalatnikov quantitatively on the basis of the expressions given for ϕ_1 and ϕ_2.

The *attenuation coefficient for second sound* is given by

$$\alpha_{\text{II}} = \frac{\omega^L}{2\varrho c_{\text{II}}^2}\left\{\lambda\left(\frac{\partial \ln T}{\partial S}\right) + \frac{\varrho_s}{\varrho_n}\left[\frac{4\eta}{\varrho} + (\phi_2 - \varrho\phi_4)\right]\right\}. \tag{6.28}$$

The values of $\alpha_{\text{II}} \gg \alpha_{\text{I}}$ because of the ratio ϱ_s/ϱ_n and the fact that c_{II}^2 occurs instead of c_{I}^2 in the denominator. The second term could now be evaluated using the information obtained from α_{I} and as its contribution is about 50% of that of the first term, one obtains in this manner rather satisfactory information about the heat conductivity and its dependence on temperature.

The *relaxation time* τ_ϕ appears to be of the order of 10^{-10} s at 2 °K and increases to about 10^{-7} s at 1 °K, thus being approximately about 10 times larger than the relaxation times τ_ϕ and τ_η.

For a more detailed review of the various theoretical calculations made in this field we refer again to the original papers.

REFERENCES

[1] J. DE BOER: *Physica*, **14**, 139 (1948).
[2] J. DE BOER and R. J. LUNBECK: *Physica*, **14**, 318, 509 (1948).
[3] J. DE BOER and B. S. BLAISSE: *Physica*, **14**, 149 (1948); R. LUNBECK: *Diss. Amsterdam* (1951), chap. IV.
[4] F. LONDON: *Proc. Roy. Soc. London*, A **153**, 576 (1936). See also F. LONDON: *Superfluids*, vol. **2**, chap. 4 and 5; J. DE BOER: *Progr. Low. Temp. Phys.*, vol. **2**, chap. I, ed. by C. J. GORTER.
[5] See M. J. BUCKINGHAM and W. M. FAIRBANK: *Progr. Low. Temp. Phys.*, vol. **3**, chap. III, ed. by C. J. GORTER.
[6] L. TISZA: *Nature*, **141**, 913 (1938); *Journ. Phys. et Rad.*, **1**, 164, 350 (1940).
[7] L. LANDAU: *Žurn. Ėksp. Teor. Fiz.*, **5**, 71 (1941).
[8] L. LANDAU: *Žurn. Ėksp. Teor. Fiz.*, **11**, 91 (1947).
[9] C. J. GORTER and J. H. MELLINK: *Physica*, **15**, 285 (1949); **16**, 113 (1950); see C. J. GORTER: *Progr. Low. Temp. Phys.*, vol. **1**, chap. I, ed. by C. J. GORTER.
[10] See I. M. KHALATNIKOV: *Fortschr. d. Phys.*, **5**, 211 (1957); **5**, 286 (1957).
[11] See R. B. DINGLE: *Proc. Phys. Soc.*, A **63**, 638 (1950); J. DE BOER and E. G. D. COHEN: *Physica*, **21**, 79 (1955).

[12] See *e.g.* J. DE BOER: *Progr. Low Temp. Phys.*, vol. **2**, chap. I, p. 45; ed. by C. J. GORTER.

[13] L. LANDAU: *Žurn. Ėksp. Teor. Fiz.*, **11**, 91 (1947); I. M. KHALATNIKOV: *Dissertation* (1952): see *Fortschr. d. Phys.*, **5**, 211 (1957); J. WIEBES, C. G. NIELS-HAKKENBERG and H. C. KRAMERS: *Physica*, **23**, 625 (1957).

[14] J. L. YARNELL, G. P. ARNOLD, P. J. BENDT and E. C. KERR: *Phys. Rev.*, **113**, 1379 (1959); D. G. HENSHAW and A. D. B. WOODS: *Phys. Rev.*, **121**, 1266 (1961); H. PALEVSKY, K. OTNES and K. E. LARSSON: *Phys. Rev.*, **112**, 11 (1958).

[15] J. DE BOER: *Suppl. Nuovo Cimento*, **9**, 25 (1958).

[16] R. P. FEYNMAN: *Phys. Rev.*, **94**, 262 (1954); R. P. FEYNMAN and M. COHEN: *Phys. Rev.*, **120**, 189 (1956).

[17] H. A. KRAMERS: *Physica*, **18**, 653 (1952); R. KRONIG: *Physica*, **19**, 535 (1953). See also J. DE BOER: *Suppl. Nuovo Cimento*, **9**, 1 (1958).

[18] P. J. BENDT, R. D. COWAN and J. L. YARNELL: *Phys. Rev.*, **113**, 1386 (1959).

Superfluidity.

G. V. Chester

University of Birmingham - Birmingnam

I. - The Spectrum of Elementary Excitations and the Motion of Impurities in Liquid Helium.

1. - Introduction.

In these two Sections I want to discuss some of the properties *a*) of the elementary excitations that we know exist in liquid helium and *b*) of the motion of the impurities in liquid helium. The excitation spectrum Fig. 1 has, in the last few years, been accurately measured by scattering low-energy neutrons from the liquid [1]. At the moment we know experimentally *a*) the detailed shape of the excitation spectrum, *b*) something about the lifetimes of the excitations from the line widths observed in the neutron scattering. The problem I want to discuss is what are the nature of the excited states of liquid helium that have this rather peculiar energy spectrum? Since we have very little experimental information on this point we shall have to approach the problem theoretically. I shall do this using the method invented by FEYNMAN and COHEN [2]. This method is a variational method and it may therefore be very misleading to discuss the detailed properties of the variational wave functions. However for my purpose it is the only method available and I think that we can learn something from it. We shall see that in this method there is a close connection between the excitation spectrum and the motion of impurities in the liquid. For theis reason we group these topics together. We begin by outlining the basic theory of the method.

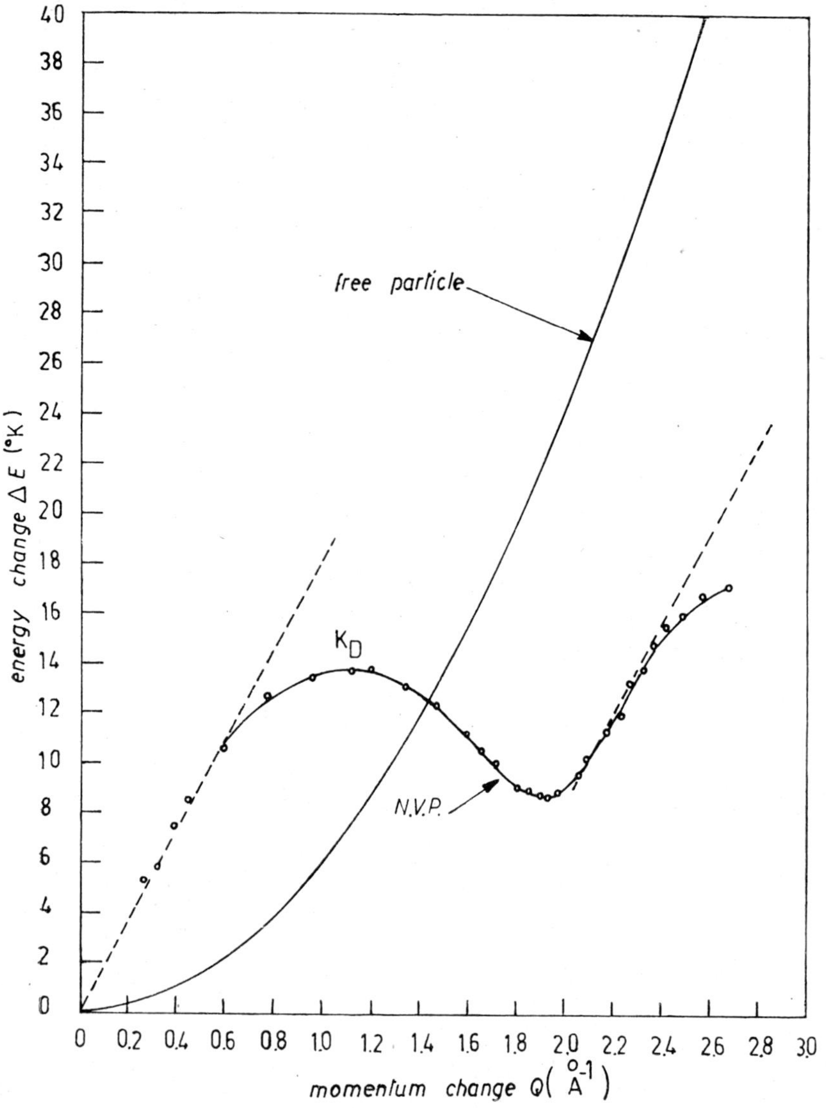

Fig. 1. – The energy spectrum of elementary excitations in liquid helium. HENSHAW [1]. Temperature 1.12 °K, neutron wavelenght 4.04 Å.

2. – Basic theory.

If we consider a given volume of liquid helium then we can label any of the excited states of the system by the linear momentum associated with it. This is of course only permissible if the momentum is a constant of the motion. To achieve this we apply periodic boundary conditions to our system — this is a mathematical trick; it will make a negligible difference to our final answers. Now if we have any physical quantity A which is a constant of the motion, then the variational principle (SCHIFF [3]) tells us that if $E(\alpha)$ is the lowest energy eigenvalue of the system corresponding to the eigenvalue α of the constant of the motion A then

(1) $$E(\alpha) \leqslant (\psi_T(\alpha), H\psi_T(\alpha))/(\psi_T(\alpha), \psi_T(\alpha)) .$$

In this equation H is the Hamiltonian of the system, and $\psi_T(\alpha)$ is any trial function which is an exact eigenfunction of the operator A with eigenvalue α. In our case we are interested in labelling the states with the momentum associated with them, so we can write

(2) $$E(\boldsymbol{p}) \leqslant (\psi_T(\boldsymbol{p}), H\psi_T(\boldsymbol{p}))/(\psi_T(\boldsymbol{p}), \psi_T(\boldsymbol{p})) ,$$

where $E(\boldsymbol{p})$ is the lowest energy eigenvalue with momentum $\boldsymbol{p}$. This much is standard theory. The first important step in the application of this method is to notice that if

(3) $$\psi_T(\boldsymbol{p}) = F(\boldsymbol{p})\varphi_0 ,$$

where φ_0 is the exact *ground state* of the system (with zero momentum) and $F(\boldsymbol{p})$ is an eigenfunction of the momentum $\boldsymbol{p}$ then we can obtain a very useful inequality, namely

(4) $$E(p) - E_0 \leqslant \frac{\hbar^2}{2M} \sum_i (\varphi_0, |\nabla_i F(p)|^2 \varphi_0)/(\varphi_0, |F|^2 \varphi_0) .$$

This inequality is useful because it is an inequality for the *difference* between the exact excited state energy $E(p)$ and the *exact* ground state energy E_0. Consequently if we choose functions $F(p)$ that correspond to excited states with a single excitation present we can write

(5) $$\varepsilon(p) = E(p) - E_0 \leqslant \frac{\hbar^2}{2M} \sum_i (\varphi_0, |\nabla_i F|^2 \varphi_0)/(\varphi_0, |F|^2 \varphi_0) ,$$

where $\varepsilon(p)$ is the exact energy of single excitation. Since the inequality is true for all values of p, we have obtained an expression for an upper bound to the exact spectrum. We have assumed only two things to get this result:

a) That the Hamiltonian can be written in the form

$$H = T + V ,$$

where T is entire kinetic energy operator for the helium, and V is the total potential energy, which we assume is a function of the co-ordinates of the system and does not depend on the momenta. We do not need to assume anything at all about the detailed form of V—except that it is independent of the momenta of the particles. We can even assume that it is the entire potential energy of a collection of α-particles and electrons; we can then take for T the kinetic energy operator for the α-particles and electrons. The inequality (5) is still true for this « exact » Hamiltonian provided $F(\boldsymbol{p})$ is taken to be a symmetric function depending only on the co-ordinates of the α-particles.

b) We have also « assumed » that φ_0 is real. This has been proved to be so by ONSAGER and PENROSE [4].

3. – The trial wave function.

This is then the theoretical basis — we merely have to think up suitable trial functions $F(\boldsymbol{p})$ and calculate. It is here, of course, that we must introduce some physical ideas in order to get an accurate upper bound to $G(\boldsymbol{p})$. But we must also not make $F(\boldsymbol{p})$ too complicated or we shall not be able to work out the integrals! FEYNMAN [5] and FEYNMAN and COHEN [2] have given extensive physical arguments as to why we should choose certain particular form for $F(\boldsymbol{p})$. I do not intend to go through these here; I shall summarise the main points:

a) The excited state function $F(\boldsymbol{p})\,\varphi_0$ must be orthogonal to φ_0; but φ_0 is real and always positive (ONSAGER and PENROSE [4]). Consequently $F(\boldsymbol{p})$ must be positive for about one half of the possible configurations of the atoms, and negative for the other half.

b) φ_0 is completely symmetric; $F(\boldsymbol{p})\,\varphi_0$ must also be as it is the wave function for a Bose system.

c) Any state of the system which can be reached from φ_0 by permutations of the particles cannot be a new state — and in particular it cannot be orthogonal to φ_0.

Detailed considerations of these points lead Feynman to suggest that at least for small momenta one should choose

$$F(\boldsymbol{p}) = \sum_{\alpha} \exp\,[i\boldsymbol{p}\cdot\boldsymbol{r}_\alpha]\,. \tag{6}$$

This is an exact eigenfunction of the momentum operator with momentum $\boldsymbol{p}$. One can arrive at the same $F(\boldsymbol{p})$ by more theoretical arguments. Since $F(\boldsymbol{p})$ must be completely symmetrical in the coordinates and must be an eigen function of the momentum operator *the simplest* $F(\boldsymbol{p})$ is either of the form

$$\prod_{\alpha} f(\boldsymbol{r}_{\alpha}) , \tag{7a}$$

or

$$\sum_{\alpha} f(\boldsymbol{r}_{\alpha}) , \tag{7b}$$

where $f(r_{\alpha})$ is an eigenfunction of the momentum operator for the α-th particle with eigenvalue $\boldsymbol{p}$

$$\boldsymbol{p}_{\alpha} f(\boldsymbol{r}_{\alpha}) = \boldsymbol{p} f(\boldsymbol{r}_{\alpha}) . \tag{8}$$

The solution of this equation is

$$f(\boldsymbol{r}_{\alpha}) = \exp[i\boldsymbol{p}\cdot\boldsymbol{r}_{\alpha}] , \tag{9}$$

where we have chosen our units so that $\hbar=1$.

Now the first form for $F(p)$, eq. (1a), becomes

$$\exp[i\boldsymbol{p}\cdot\boldsymbol{R}]\varphi_0 , \tag{10}$$

where $\boldsymbol{R}$ is the centre-of-mass co-ordinate for the system. This corresponds to motion of the system as a whole, with velocity P/M. This function, therefore, does not correspond to an internal excitation of the system. It should also be noticed that the energy corresponding to this state is very large if we apply periodic boundary conditions, (FEYNMAN [5]). The second form for $F(\boldsymbol{p})$ is identical with that given by eq. (6).

This argument merely tells one how to construct a simple trial, $F(p)$. It does not tell one how good it might be. Let us compare the calculated spectrum with the experimental one. We find

$$\varepsilon(\boldsymbol{p}) \leqslant \frac{p^2}{2MS_0(\boldsymbol{p})} , \tag{11}$$

where

$$S_0(\boldsymbol{p}) = \int \exp[i\boldsymbol{p}\cdot\boldsymbol{v}] P_0(\boldsymbol{r})\, \mathrm{d}^3\boldsymbol{r}, \tag{12}$$

where $P_0(r)$ is the pair correlation function for the ground state. Here we come to a very important point. Since we have assumed that $\psi_r(\boldsymbol{p}) = F(\boldsymbol{p})\varphi_0$ we find that the calculated spectrum depends on φ_0; namely through the Fourier transform $S_0(\boldsymbol{p})$ of the pair correlation function. To proceed further we must either *a*) calculate $P_0(\boldsymbol{r})$ by some means or *b*) take it from experimental data. Now what is measured in any experiment on the pair correlation function is not $S_0(\boldsymbol{p})$ but the structure function $S_T(\boldsymbol{p})$ at a finite temperature T. The difference between these functions is important for small $\boldsymbol{p}$. FEYNMAN and COHEN have shown on a plausible argument that $S_0(\boldsymbol{p})=c|\boldsymbol{p}|$ for small $|\boldsymbol{p}|$. This is quite different from the behaviour of $S_T(\boldsymbol{p})$ for $|\boldsymbol{p}|$; $S_T(p)$ goes to a constant as $|\boldsymbol{p}|\to 0$. The reasons for this are discussed in detail in the paper by FEYNMAN and COHEN [2]. It is interesting to calculate $S_0(\boldsymbol{p})$ by the method used by FEYNMAN and COHEN [2]. The essential idea is that if we want to calculate $S_0(\boldsymbol{p})$ for very small $|\boldsymbol{p}|$ then we can use a continuum approximation to the fluid. This is because very small momenta are equivalent to very long wave lengths and for such wave lengths the atomistic structure should be unimportant.

If we write

$$\varrho_p = \sum_\alpha \exp[i\boldsymbol{p}\cdot\boldsymbol{r}_\alpha]\,, \tag{13}$$

then

$$S_0(\boldsymbol{p}) = (\varphi_0, \varrho_{\boldsymbol{p}}\varrho_{-\boldsymbol{p}}\varphi_0)\,. \tag{14}$$

Now $\varrho_{\boldsymbol{p}}$ is just the Fourier transform of the density operator of the system. Consequently $S_0(\boldsymbol{p})$ is proportional to the Fourier transform of the density fluctuation in the system. In a continuum approximation this is related to the normal mode frequencies.

In a continuum approximation the Hamiltonian is given by [6],

$$H=\int d^3\boldsymbol{r}\left(\frac{1}{2}\varrho\boldsymbol{v}^2 + U(\varrho)\right), \tag{15}$$

where ϱ is the local density operator, $\boldsymbol{v}$ the local velocity operator and $U(\varrho)$ is the internal energy of the system considered to be a function of ϱ alone. If we now consider states which represents only small deviations from uniform density, then we can approximate H by the expression

$$\int d^3\boldsymbol{r}\left(\frac{1}{2}\varrho_0\boldsymbol{v}^2 + \frac{1}{2}(\varrho-\varrho_0)^2\frac{\partial^2 u}{\partial\varrho_0^2}\right), \tag{16}$$

where ϱ_0 is the mean density and we have assumed that the external pressure on the system is zero; $\partial u/\partial\varrho_0=0$. This Hamiltonian can easily be written in the form

$$H = \tfrac{1}{2}\varrho_0\sum_{\boldsymbol{p}} v_{\boldsymbol{p}}v_{-\boldsymbol{p}} + \tfrac{1}{2}c^2\sum_{\boldsymbol{p}}\varrho_{\boldsymbol{p}}\varrho_{-\boldsymbol{p}}\,, \tag{17}$$

where we have used the relation

$$c^2 = \frac{\partial^2 u}{\partial \varrho_0^2}. \tag{18}$$

The continuity equation for the system is

$$\dot{\varrho}_{\boldsymbol{p}} + \varrho_0 i\boldsymbol{p}\cdot\boldsymbol{v}_{\boldsymbol{p}} = 0. \tag{19}$$

So H becomes

$$\frac{1}{2}\varrho_0 \sum_{\boldsymbol{p}} \frac{\dot{\varrho}_{\boldsymbol{p}}\dot{\varrho}_{-\boldsymbol{p}}}{p^2} + \frac{1}{2}c^2 \sum_{\boldsymbol{p}} \varrho_{\boldsymbol{p}}\varrho_{-\boldsymbol{p}}. \tag{20}$$

This is an harmonic oscillator Hamiltonian and the expectation value of the potential energy of any mode $\boldsymbol{p}$ is equal to one half the zero-point energy of that mode. If $\frac{1}{2}\,\omega(p)$ is the zero-point energy then we have

$$\tfrac{1}{2}c^2(\varphi_0, \varrho_{\boldsymbol{p}}\varrho_{-\boldsymbol{p}}\varphi_0) = \omega(\boldsymbol{p}). \tag{21}$$

Or using eq. (14),

$$S_0(\boldsymbol{p}) = \frac{1}{2}\frac{\omega(\boldsymbol{p})}{c^2}. \tag{22}$$

Finally in the continuum approximation

$$\omega(\boldsymbol{p}) = c|\boldsymbol{p}|, \tag{23}$$

where c is the velocity of sound. We therefore obtain,

$$S_0(\boldsymbol{p}) = \frac{|\boldsymbol{p}|}{2Mc}. \tag{24}$$

If we substitute this result into eq. (11), then we find that for small $|\boldsymbol{p}|$

$$\varepsilon(\boldsymbol{p}) \leqslant c|\boldsymbol{p}|. \tag{25}$$

Experimentally we now know, (HEENSHAW [1]), that the excitation spectrum is exactly given by the phonon formula

$$\varepsilon(p) = c|\boldsymbol{p}|, \tag{26}$$

for small $|\boldsymbol{p}|$. This suggests *a*) that the variational wave functions given by eqs. (3) and (6) are very accurate for small $|\boldsymbol{p}|$ and that the continuum approximation to $S_0(\boldsymbol{p})$ is also reliable.

For $|\boldsymbol{p}| \sim 2 Å^{-1}$ the theoretical spectrum, with $S_0(\boldsymbol{p}) = S_T(\boldsymbol{p})$, and $S_T(\boldsymbol{p})$ taken from experiment, (HENSHAW [1]), is very much too high in energy. The experimental spectrum has an energy of about 8.7 °K for $|\boldsymbol{p}| \sim 1.8 Å^{-1}$; eq. (11) gives an energy of 19 °K at the same $|\boldsymbol{p}|$ value. A comparison of the spectra is given in Fig. 2. We can summarize the situation as follows:

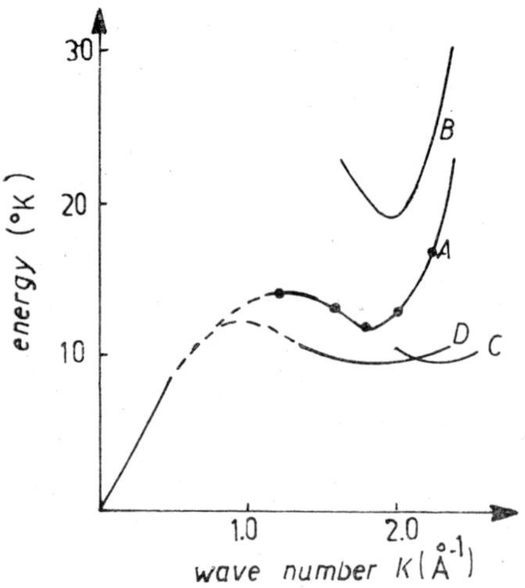

Fig. 2. – Comparison of the theoretical spectrum with the experimental one. Curve A: Feynman-Cohen theory, data of Reekie and Hutchison; curve B: Feynman theory with no backflow; curves D-C: possible experimental curves to fit thermodynamic data.

a) The spectrum given by (11) is in exact agreement with experiment for small $|\boldsymbol{p}|$ if we calculate $S_0(p)$ by a continuum approximation.

b) If we set $S_0(\boldsymbol{p}) = S_T(p)$ for $|\boldsymbol{p}| \geqslant \frac{1}{2} Å^{-1}$ then we have a relation between the theoretical spectrum $\varepsilon(p)$ and an experimentally measurable quantity $S_T(\boldsymbol{p})$.

c) The experimental data for $S_T(\boldsymbol{p})$ show that eq. (11) gives a rather poor bound to the spectrum for $|\boldsymbol{p}| \sim 2 Å^{-1}$.

We conclude that a very much improved wave function is required for higher momentum values. The improved wave function should, however, become identical with those given by eqs. (3) and (6) for small momenta.

4. – The backflow problem.

FEYNMAN and COHEN [2] have suggested a much better form for $F(\boldsymbol{p})$ which greatly improves the spectrum for larger $|\boldsymbol{p}|$. They pointed out that if one makes a wave packet from the simple function we have used, then the local current is not conserved. They, therefore, set about constructing a function which would conserve the current.

They chose

$$F(\boldsymbol{p}) = \sum_\alpha \exp[i\boldsymbol{p}\cdot\boldsymbol{r}_\alpha] \exp\left[i \sum_\beta g(\boldsymbol{r}_\beta - \boldsymbol{r}_\alpha)\right], \tag{27}$$

with

$$g(\boldsymbol{r}) = A\boldsymbol{p}\cdot\boldsymbol{r}/r^3, \tag{28}$$

where A is a constant.

This is an eigenfunction of the momentum and conserves the local current (FEYNMAN and COHEN [2]). Unfortunately this function is too complex to be used as it stands. In order to « do » the necessary integrals we have to approximate,

$$\exp[ig] \simeq 1 + ig\,. \tag{29}$$

For the interesting values of $|\boldsymbol{p}| \sim 2\text{Å}^{-1}$, the magnitude of g is about $\frac{1}{2}$; so this approximation should be good to about 10 %. In any case it does not spoil our variational method. The calculation is still rather complicated. The spectrum can be expressed in terms of about half a dozen integrals most of which depend on $S_0(\boldsymbol{p})$; for, the $\boldsymbol{p}$-values we are interested in, we can set $S_0(\boldsymbol{p}) = S_T(\boldsymbol{p})$.

FEYNMAN and COHEN used the experimental data of REEKIE and HUTCHINSON [8]. They found a very much improved spectrum, Fig. 3; for

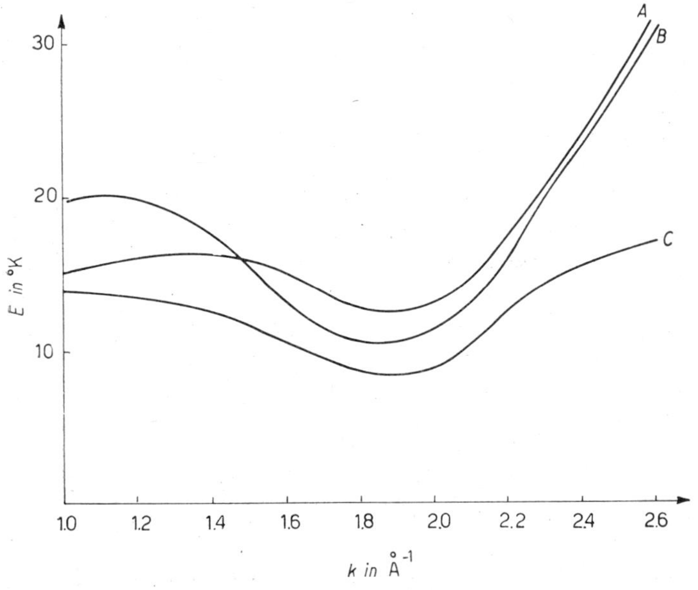

Fig. 3. – Comparison of experimental spectrum, HENSHAW [1], with theoretical calculations. Curve A: Feynman-Cohen theory using data of Reekie and Hutchison; curve B: Feynman-Cohen theory using Henshaw data; curve C: experimental spectrum (HENSHAW).

$|\boldsymbol{p}| \sim 2\,\text{Å}^{-1}$ there was a minimum of about 11.5° K. Recent calculations at BIRMINGHAM [9] using more accurate data (due to HENSHAW [1]) for $S_T(\boldsymbol{p})$ have indicated that the minimum may be as low as 10.5° K. Experiment gives 8.7 °K. The general « shape » of the spectrum is also good. For small $|\boldsymbol{p}|$ the spectrum is again a phonon spectrum. For larger $|\boldsymbol{p}|$ it behaves like $p^2/2M^*$, where M^* is about $1.55M_4$.

5. – Trial functions for impurities.

FEYNMAN and COHEN chose this form for $F(p)$ by considering a simpler problem. This is the problem of the energy spectrum of an impurity moving through liquid helium. The impurity could be helium three atom, or a positive or negative ion. Since it is a different particle from the atoms in the liquid it is distinguishable from them and it is, therefore, meaningful to discuss states of the system in which it moves through the fluid with momentum $\boldsymbol{p}$. The simplest situation is when it moves through the fluid and no excitation are present. This problem can also be tackled by the method we have described. We now write

$$\varphi_T(\boldsymbol{p}) = F(\boldsymbol{p})\varphi_0^{(i)}\,, \tag{30}$$

where $\varphi_0^{(i)}$ is the ground state of the helium with impurity atom present, and $F(\boldsymbol{p})$ is a trial modulating factor. This factor must again be an eigenfunction of the total momentum operator but need only be symmetrical with respect to the coordinates of the helium atoms.

The simplest trial function is now,

$$\psi_T(\boldsymbol{p}) = \exp[i\boldsymbol{p}\cdot\boldsymbol{r}]\varphi_0^{(i)}\,, \tag{31}$$

where $\boldsymbol{r}$ is the co-ordinate of the impurity atom. This leads to an energy spectrum

$$\varepsilon_i(\boldsymbol{p}) \leqslant p^2/2M_i\,. \tag{32}$$

This is an interesting result for it tells us that $\varepsilon_i(\boldsymbol{p})$ is always less than the free particle energy $p^2/2M_i$. This strongly suggests that $\varepsilon_i(\boldsymbol{p})$ is probably always given by

$$\varepsilon_i(\boldsymbol{p}) = p^2/2M_i^*\,, \tag{33}$$

with M_i^* an effective mass greater than M_i. Notice we have not proved that $\varepsilon_i(\boldsymbol{p})$ has this form; but for small $|\boldsymbol{p}|$, $\varepsilon_i(\boldsymbol{p})$ would have to behave either like this or like $|\boldsymbol{p}|^n$, $n > 2$; this latter dependence on $|\boldsymbol{p}|$ is most unplausible.

6. – The backflow round an impurity.

This simple wave function really corresponds to the impurity moving through the liquid and not disturbing the liquid at all; that is why the energy comes out to be the free-particle energy. But we know that when an object moves through a fluid, the fluid streams round it and an appreciable disturbance is set up. In particular there is a large « backflow » round the object. This situation is familiar from classical hydrodynamics. By considering the classical situation FEYNMAN and COHEN were lead to suggest that the simple wave function given by eq. (31) should be modified to include this effect. Classically the streaming of the fluid round a sphere can be represented by a velocity potential

$$g_i(\boldsymbol{r}) = \frac{A\boldsymbol{p}\cdot\boldsymbol{v}}{r^3}\,. \tag{34}$$

The velocity in the fluid is given by the equation

$$v(\boldsymbol{r}) = -\nabla g_i(\boldsymbol{r})\,. \tag{35}$$

Here $\boldsymbol{p}/M$ is the velocity of the body, and A is a constant. We can include this effect in the wave function by noticing that the velocity in the fluid is directly related to the gradient of the phase of the wave function. But since this velocity is associated with the helium atoms, the phase will have to depend on the coordinates of these; and in a symmetrical manner. These requirements will be satisfied if we write

$$\psi_T(p) = \exp\left[i\boldsymbol{p}\cdot\boldsymbol{r}\right]\exp\left[i\sum_l g(r-r_l)\right]\varphi_0^{(i)}\,, \tag{36}$$

where g is the classical velocity potential. The velocity $v(\boldsymbol{r})$ in the fluid as seen from the impurity atom is now given by

$$v(\boldsymbol{r}) = -(1/M)\,p_2(\boldsymbol{r})\,\nabla g(\boldsymbol{r})\,, \tag{37}$$

where $p_2(\boldsymbol{r})$ is the pair correlation function between the impurity atom and the fluid atoms. Since $p_2(\boldsymbol{r})\simeq 1$ for large r we see that $\boldsymbol{v}(\boldsymbol{r})$ is exactly the classical velocity field at large distances. So far the argument has been based on a classical analogy. We can do two things to generalise it.

a) We can simply take the functional form suggested by the classical argument; and see if we can determine the function $g(\boldsymbol{r})$ from the variational principle. This is hard to do in general. FEYNMAN and COHEN [2] have shown that $g(\boldsymbol{r})$ is indeed exactly the classical velocity potential for large $\boldsymbol{r}$. The variational equations are much harder to solve for small $\boldsymbol{r}$.

b) Even if we accept the classical form for g given by eq. (34) we can leave A as a free parameter and determine it from the variational principle. This leads to a value of $A = -3.8 \cdot 10^{-24}$ cm³; which is only 6 % off the classical value.

With this wave function we find

$$\varepsilon(\boldsymbol{p}) \leqslant \boldsymbol{p}^2/2M^* . \tag{38}$$

For a helium III atom $M^* = 1.67 M_3$. This is an improvement of about 50 %. Probably the main defect in the calculation is that the form we have been choosing for $g(\boldsymbol{r})$ is not very good for small r. This could be very important if the impurity were a positive or negative ion.

7. – What is a roton?

The improvement that FEYNMAN and COHEN found for the effective mass of a helium three atom led them to believe that a similar trial function might lead to a large improvement in to the excitation spectrum. The trial function given by eqs. (3) and (6) also does not conserve current locally. It can be put right in a similar way — all we have to do is to symmetrize the trial function we used for the impurity. This leads at once to the function given by eq. (21). The choice of $g(\boldsymbol{r})$ could be left quite free however, the success achieved in the impurity problem encourages us to try exactly the same form in this case. The constant A is determined variationally and for $p \sim 2$ Å^{-1} is equal to $-3.6 \cdot 10^{-24}$ cm³. Unfortunately it has never been shown that $g(\boldsymbol{r})$ exactly satisfies the variational equation for large r. It would be interesting to see if this is so. The velocity field in the fluid as seen from a particular atom is now rather more complicated; we find that it is given by

$$v(\boldsymbol{r}) = -(1/M) p_2(\boldsymbol{r}) \nabla g(\boldsymbol{r}) + \\ + \text{(correlation terms involving three or more atoms)} . \tag{39}$$

The first term is the same as for the impurity, the second contains more complicated correlation effects. It is the first term which leads one to say that the velocity in the fluid is like that of a small smoke ring. For it is just the velocity potential for a classical smoke ring with its plane normal to the momentum $\boldsymbol{p}$. The higher order correlation terms can be shown to be small whenever $|\boldsymbol{p}|r \ll 1$. For $p \sim 2$ Å^{-1} they are small for r of the order of one interatomic distance. Consequently for values of $\boldsymbol{r}$ large enough that $p_2(\boldsymbol{r})$ is a constant, $\boldsymbol{v}(r)$ is exactly the same as the velocity field $\boldsymbol{v}(r)$ a classical smoke ring. Notice:

a) $\boldsymbol{v}(r)$ is *not* the local velocity field in the fluid; it is the velocity field in the fluid as seen from an atom in the fluid. The exact equation for $\boldsymbol{v}(r)$ is

$$(40) \qquad \boldsymbol{v}(\boldsymbol{x}-\boldsymbol{x}') \equiv \boldsymbol{v}(\boldsymbol{r}) = \big(\psi_T(\boldsymbol{p}),\, \varrho(x) \otimes \boldsymbol{v}(x')\,\psi_T(\boldsymbol{p})\big)\;,$$

where $\varrho(\boldsymbol{x})$ is the local density operator and $\varrho(\boldsymbol{x}')$ the local velocity operator and the symbol $\otimes$ stands for a symmetrized product. Since $\psi_T(\boldsymbol{p})$ is an eigenfunction of the momentum, the expectation value of $\varrho(\boldsymbol{x}) \otimes \boldsymbol{v}(\boldsymbol{x}')$ depends only on $\boldsymbol{x}-\boldsymbol{x}'$.

b) There is also a « smoke ring » velocity field representing the backflow round an impurity atom; see eq. (37). There is now no correlation term so the « picture » is exact.

Let us now turn to a discussion of the nature of the excited states, eq. (21). For very small $|\boldsymbol{p}|$ these are phonon states. That is they represent the propagation of longitudinal sound waves in the liquid. The calculated spectrum gives the standard phonon form $\varepsilon(p) = c|\boldsymbol{p}|$. The minimum occurs at $|\boldsymbol{p}| \sim 1.8$ Å^{-1}. Now one would normally suppose that it is not sensible to talk of longitudinal phonons in a liquid for $|\boldsymbol{p}|$ values $\geqslant a^{-1}$ where a is the interatomic spacing (this is the magnitude of the Debye's cut off). If we take $a = 3.7 \cdot 10^{-8}$ cm we see that the k-values near the minimum are about twice the Debye's cut off.

If we reject the notion of « phonons » for these $|\boldsymbol{p}|$-values, we are left with the problem of deciding what motions are present in the fluid when we have an excitation with this momentum and energy. The conjecture I want to put forward is that in the region of the minimum of the energy spectrum we are dealing with the motion of a single helium atom moving fast through the others. (It may be permissible to talk of the motion of one atom for a Bose system if the atom is moving fast enough.) If this statement is correct then the symmetrization of the wave function should not matter for this part of the spectrum. In other words the energy should be the same whether we calculate with a symmetrized or an unsymmetrized wave function. In Fig. 4, we compare the two spectra. The parabolic spectrum is calculated with an impurity type of trial wave function for an atom of mass equal to $1.5 M_4$. We see that in the region of the minimum they are only about 10 % apart. The divergence of the spectrum at higher momentum is due to the approximation

$$\exp[ig] \simeq 1 + ig\;,$$

that we made in calculating with the symmetrized function. The comparison we have made is purely theoretical. We can make an approximate experimental one in the following way. The effective mass of helium three atom moving in helium has been determined experimentally and is about $2.5m_3$.

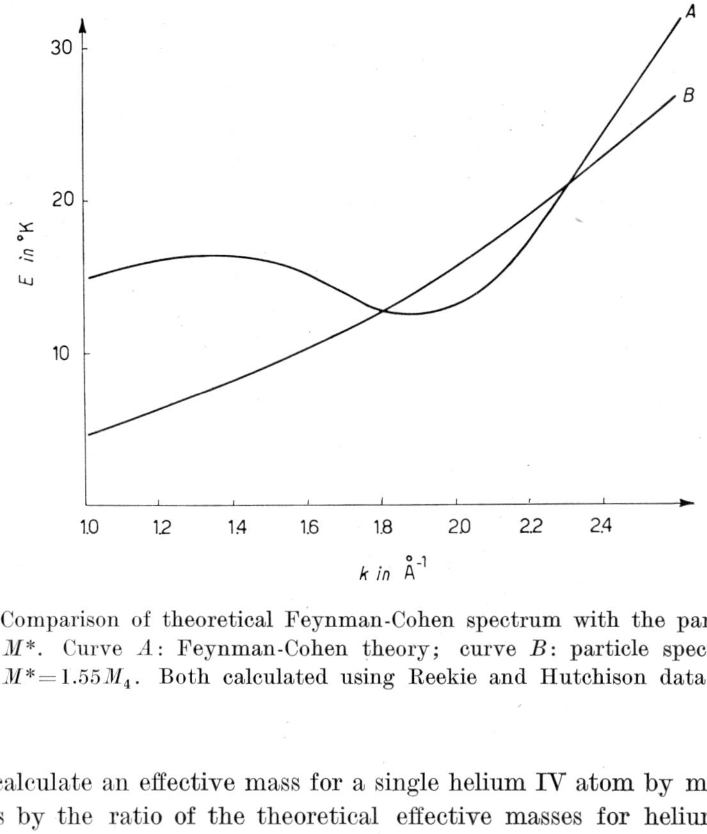

Fig. 4. – Comparison of theoretical Feynman-Cohen spectrum with the particle spectrum $p^2/2M^*$. Curve A: Feynman-Cohen theory; curve B: particle spectrum with $M^*=1.55M_4$. Both calculated using Reekie and Hutchison data.

We can calculate an effective mass for a single helium IV atom by multiplying this mass by the ratio of the theoretical effective masses for helium IV and helium III. This is a crude approximation, — but is should give one something like the correct order of magnitude. This theoretical ratio is about 0.9, so we find that the effective mass for helium four is $2.3m_4$. If we now compare Fig. 5, the measured excitation spectrum, with the spectrum $p^2/2m_4^*$ we see that they agree within about 5 % at the minimum.

There are two further points that are worth mentioning in this connection. First a particle with momentum 2 Å^{-1} has a much higher momentum than the average momentum of the particles. The mean momentum is in fact only about one half of this value. Consequently the mean number of particles with this momentum will be small. This means that the effects due to Bose-Einstein statistics can be expected to be small. Secondly, the energy associated with this particle is probably sufficiently high to allow it to get over the potential barrier due to the neighbouring atoms. If this is so we may reasonably expect that it will be able to propagate fairly easily through the fluid.

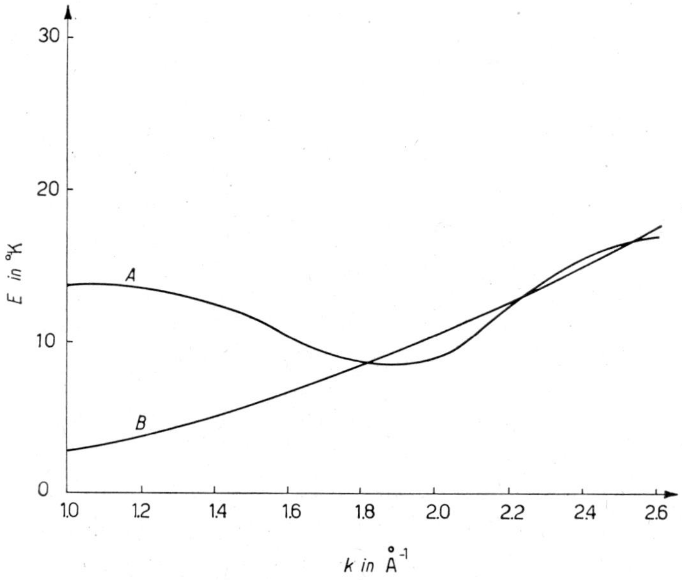

Fig. 5. – Comparison of experimental spectrum, HENSHAW [1], with the « particle » spectrum $\varepsilon(p)=p^2/2M_4^*$; with $M_4^*=2\cdot 3M_4$; curve A: ^{4}He energy spectrum (HENSHAW); curve B: effective mass $M_4^*=2\cdot 3M_4$ spectrum.

The picture we have formed so far is as follows. From $|\boldsymbol{p}|=0$ up to $|\boldsymbol{p}|\simeq 1$ Å^{-1} the motions in the fluid consist of longitudinal density waves. In fact it can be shown that for this range of momentum the density at a point $\boldsymbol{r}'$ in the fluid as seen from an atom at $\boldsymbol{r}$ in the fluid behaves like

$$\exp[i\boldsymbol{p}\cdot(\boldsymbol{r}-\boldsymbol{r}')]\,p_0(\boldsymbol{r}-\boldsymbol{r}')\,, \tag{41}$$

where $p_0(\boldsymbol{r}-\boldsymbol{r}')$ is the correlation function in the ground state. Whenever $|\boldsymbol{r}-\boldsymbol{r}'|$ is larger than the interatomic distance $p_0\simeq 1$ and we are left with a simple sinusoidal density wave in the fluid relative to the atom at the region. On the other hand for $p\simeq 2$ Å^{-1} we find that there is essentially no local density fluctuation and we have a single particle propagating fast through the liquid. The increase in effective mass of this atom comes from a large scale hydrodynamic backflow—with little or no charge in the fluid density. The region between $|\boldsymbol{p}|=1$ Å^{-1} and $|\boldsymbol{p}|=1.8$ Å^{-1} must then be interpreted as a rather complicated transition region in which we pass from a high frequency phonon mode at 1 Å^{-1} to a single particle propagation at 2 Å^{-1}. This is clearly a possible transition in a liquid with an open structure and with low potential barriers.

There is a general way in which we can see that for large $|\boldsymbol{p}|$ the symmetrization of the wave function does not matter when we calculate expectation values. When we calculate the expectation values of any operator two types of integrals will arise.

a) Those which contain factors $\exp[i\boldsymbol{p}\cdot(\boldsymbol{r}_\alpha - \boldsymbol{r}_\beta)]$ and come from terms in the modulating factors $F(\boldsymbol{p})$ for which $\alpha \neq \beta$.

b) Those which do not contain factor like this and come from the $\alpha=\beta$ terms. These terms are exactly the ones which arise with an unsymmetrized wave function.

Now the first type of integrals will be small if $p\ a \gg 1$, where a is of the order of the mean distance between particles. The exponent of the exponential factor $\exp[i\boldsymbol{p}\cdot(\boldsymbol{r}_\alpha - \boldsymbol{r}_\beta)]$ is never less than pa. We can see this, because this factor is always multiplied by a correlation function for the ground state wave function; which vanishes if $|\boldsymbol{r}_\alpha - \boldsymbol{r}_\beta| < a$. Thus whenever the integrand is not zero it contains an exponent which oscillates at least as fast as $\exp[i|\boldsymbol{p}|a]$. Integrals like this will be small if $|\boldsymbol{p}|a \gg 1$. At the « roton minimum » $|\boldsymbol{p}|a \simeq 7$ so our condition is reasonably well satisfied. The integrals without these factors will not show this dependence. For $|\boldsymbol{p}|a \simeq 1$ we would expect both types of integrals to be equally important in any expectation value. This is borne out by our calculations of the energy spectrum for $|\boldsymbol{p}| \simeq 1\ \text{Å}^{-1}$.

A final point is that we would not expect any intrinsic angular momentum to be associated with these excitations. It has been shown by LEE and MOHLING [10] that if the excitations do posses intrinsic angular momentum then the small angle neutron scattering will have special properties. The data are not yet good enough to decide this question experimentally.

It is interesting to notice that for $|\boldsymbol{p}|$ higher than $\sim 2\ \text{Å}^{-1}$ the particle mode can decay into lower energy phonons [11]. It is therefore highly unstable.

So far I have said nothing at all about the motion of other impurities; namely, positive and negative ions in helium. It seems very likely that the effective mass of an ion could be calculated by Feynman's method. There are two difficulties.

a) The function $g(\boldsymbol{r})$ must be considerably modified for small $\boldsymbol{r}$. This modification should come out of the variational equation.

b) To calculate m^* we need to know the pair correlation function for an ion embedded in liquid helium. This is quite unknown. Nevertheless it would be interesting to « guess » a suitable form for this and calculate m^*. It is easy to choose pair correlation functions so that very large effective masses come out of the calculation.

II. - Superfluidity.

1. – Introduction. Basic physical ideas.

In this Section I shall discuss some of the basic physical and theoretical ideas connected with superfluidity. It is commonly accepted that helium below the λ-temperature displays superfluidity. Another way in which this is often stated is the helium exhibits frictionless transport of mass and energy. Two examples will suffice to demonstrate the idea; *a*) in flow through a tube helium appears to flow in a frictionless manner up to a fairly well defined critic velocity; *b*) it is possible to propagate a heat pulse through helium with only very small attenuation. Heat pulses cannot be propagated in an ordinary fluid.

Notice: *a*) that from the two experiments we have mentioned it has not been rigorously shown that helium ever flows in a *completely* frictionless fashion. It is in fact hard to find an experiment in which the existence of completely frictionless flow is convincingly demonstrated. Perhaps the most convincing evidence comes from rotation experiments on helium, where persistent currents have been observed for periods of up to 1 h. These persistent circulating currents seem to be very similar to those observed in superconductors.

2. – The stability of superflow.

Let us concentrate on the transport of mass for the moment. The existence of persistent circulating currents shows that there are flow states in helium that are extraordinarly stable. The forces which tend to slow down such a current are very large. For example the interaction energy of a helium atom with a solid boundary is many times its interaction energy with the other atoms in the fluid. We, therefore, have to explain how these forces are unable to slow liquid helium down. These boundary forces are largely responsible for the absence of persistent current in other fluids. If we imagine a simple flow situation (in a long tube or cylindrical vessel) then at absolute zero we can suppose that the helium is in a definite quantum state and no excitations are present. If the helium is to slow up it will have to make a transition to another state. We can, therefore, picture the slowing up process in terms of transitions of the helium from one state to another. But we are familiar in quantum mechanics with the idea that transitions between states can be forbidden. That is to say the transition probability between certain states of a system can sometimes be (indeed quite often is) identically zero — no matter how strong the force that acts to cause the transition. This analogy suggests that there may be some general principle that ensures that the transition between different states of superflow are in fact « forbidden transitions ». If this idea is correct then we shall have the beginning of a theory of superfluidity. But we must be careful not to prove too much. If we were able to show that all transition probabilities between different flow states were zero, then it would be hard to see how helium could be brought into motion at all.

3. – Landau's criterion for superfluidity.

In his fundamental paper on superfluid helium LANDAU [12] put forward an argument to show why helium exhibits superfluidity. This argument is just of the type we have discussed. That is LANDAU shows that certain transitions from one state of flow to another are impossible in helium. The original argument due to LANDAU can be found in LANDAU and LIFSCHITZ [13]. I should like to begin with a slightly simpler argument.

Consider a large volume of helium at absolute zero, and imagine that we drag a solid body through it. If helium were a normal fluid, viscous forces would act on the body and we should have to do work against them. The work we do would go to heat the fluid; that is, the state of the fluid would change to one of higher energy. If we do the experiment on a normal fluid at absolute zero we should observe the creation of elementary excitations in the fluid. Landau's arguments shows that it is impossible for this to happen in helium because of the « shape » of the energy spectrum. What exactly we mean by the « shape » of the spectrum will become clear in a moment.

The argument is as follows. Initially we suppose that the liquid contains no excitations. If the body moving through it creates a certain number of excitations, say n_p with momentum $\boldsymbol{p}$, then the increment in energy of the fluid is

$$\delta E_f = \sum_{\boldsymbol{p}} n_{\boldsymbol{p}} \varepsilon(\boldsymbol{p}) \,, \tag{42}$$

and the increment in momentum is

$$\delta \boldsymbol{P}_f = \sum_{\boldsymbol{p}} n_{\boldsymbol{p}} \boldsymbol{p} \,. \tag{43}$$

On the other hand the charge in energy of the body is

$$\delta E_b = \boldsymbol{p}_i^2/2M - \boldsymbol{p}_f^2/2M \,, \tag{44}$$

where $\boldsymbol{p}_\mathrm{i}$ and $\boldsymbol{p}_\mathrm{f}$ are the initial and final momenta of the body. If we set

(45) $$\boldsymbol{p}_\mathrm{i} - \boldsymbol{p}_\mathrm{f} = \delta \boldsymbol{P}_\mathrm{b} \,,$$

where $\delta\boldsymbol{P}_\mathrm{b}$ is the change in momenta of the body, we find

(46) $$\delta E_\mathrm{b} = \boldsymbol{v}_\mathrm{i} \cdot \delta \boldsymbol{P}_\mathrm{b} - (\delta \boldsymbol{P}_\mathrm{b})^2/2M \,,$$

where $\boldsymbol{v}_\mathrm{i}$ is the initial velocity of the body. Now from eq. (42) we see that

(47) $$\delta E_\mathrm{f} \geqslant \sum_{\boldsymbol{p}} n_{\boldsymbol{p}} |\boldsymbol{p}| \cdot \left| \frac{\varepsilon(\boldsymbol{p})}{\boldsymbol{p}} \right|_{\min} ,$$

where $|\varepsilon(\boldsymbol{p})/\boldsymbol{p}|_{\min}$ is the minimum value that this ratio can take on *for any value of* $\boldsymbol{p}$. We assumed that all the $n_p \geqslant 0$; we only create excitations. From eq. (43)

(48) $$|\delta \boldsymbol{P}_\mathrm{f}| \leqslant \sum_{\boldsymbol{p}} n_{\boldsymbol{p}} |\boldsymbol{p}| \,,$$

and combining eqs. (41) and (48) we have

(49) $$\delta E_\mathrm{f} \geqslant |\delta \boldsymbol{P}_\mathrm{f}| \left| \frac{\varepsilon(\boldsymbol{p})}{\boldsymbol{p}} \right|_{\min} .$$

We now assume that in the transition, in which the excitations are created in the fluid, energy and momentum and conserved. This implies that

(50a) $$\delta E_\mathrm{f} = \delta E_\mathrm{b} \,,$$

(50b) $$\boldsymbol{P}\delta_\mathrm{f} = \delta \boldsymbol{P}_\mathrm{b} \,.$$

If we substitute eqs. (50) into (44) and use (49) then we have at once

(51) $$\boldsymbol{v}_\mathrm{i} \cdot \delta_\mathrm{f} \boldsymbol{P}_\mathrm{f} - (\delta \boldsymbol{P}_\mathrm{f})^2/2M \geqslant |\delta \boldsymbol{P}_\mathrm{f}| \left| \frac{\varepsilon(\boldsymbol{p})}{\boldsymbol{p}} \right|_{\min} .$$

For a massive body we can neglect the term $(\delta \boldsymbol{P}_\mathrm{f})^2/2M$; eq. (51) then implies that

(52) $$|\boldsymbol{v}_\mathrm{i}| \geqslant \left| \frac{\varepsilon(\boldsymbol{p})}{\boldsymbol{p}} \right|_{\min} .$$

If we do not neglect the term $(\delta \boldsymbol{P}_\mathrm{f}^2)^2/2M$ then the condition becomes

(53) $$|\boldsymbol{v}_i + \boldsymbol{v}_f|/2 \geqslant \left| \frac{\varepsilon(\boldsymbol{p})}{\boldsymbol{p}} \right|_{\min} .$$

This equation tells us that if we create excitations in the fluid then we can only satisfy the conservation of the energy and momentum if the minimum value of $|\varepsilon(\boldsymbol{p})/\boldsymbol{p}|\geqslant v_i$. The general principles that prevent transitions are the conservation of energy and momentum. Now the « shape » of the energy spectrum in helium tells that $\varepsilon(\boldsymbol{p})/\boldsymbol{p}_{\min}=60$ m/s. On this simple argument helium should be a « super » superfluid. If helium had a particle-like spectrum with $\varepsilon(\boldsymbol{p})=\boldsymbol{p}^2/2M$ then $|\varepsilon(\boldsymbol{p})/\boldsymbol{p}|_{\min}$ would be zero. Such a system could not be a superfluid. The ratio $|\varepsilon(\boldsymbol{p})/\boldsymbol{p}|_{\min}$ is usually called the Landau's critical velocity. The following points are important:

a) The argument shows that there is a critical velocity for the creation of excitations no matter what forces act between the helium and the body.

b) If excitations are already present in the fluid these can exchange energy and momentum with the solid body. This is just how the viscosity of the « gas » of excitation arises.

c) We have given the argument in terms of the linear momentum. It can be easily generalized to situations where other constants of the motion exists. If A is the constant we are interested in and the energies of the excitations are labelled with the eigenvalues of this quantity then the criterion is that

$$\frac{\partial E_\alpha}{\partial\alpha}\geqslant\left|\frac{e(\alpha)}{\alpha}\right|_{\min}, \tag{54}$$

where E_α is the energy of the body.

d) We have assumed that the only way the body can loose energy and momentum is by the production of excitations. If we want to use the argument to prove that helium is a superfluid, then we have to assume that the liquid *as a whole* could not take up the energy and momentum. We also have to show that all possible types of excitations obey our criteria, eqs. (53) and (54).

The original argument given by Landau is based on the same physical principles. He considered the flow of helium through a long tube and used the translational invariance of the system and the conservation of energy to produce the same criterion. Again an explicit assumption is that the liquid can only slow up in the tube if elementary excitations are created. This assumption is rather important; it is one of the ways in which a liquid is distinguished from a solid. Because if its rigidity a solid may plausibly be expected to slow up as a whole as well as by the creation of excitations.

We can illustrate the importance of the assumption by the following discussion. If helium flows through a metal tube, why could it not loose energy and momenta by exciting an electron in the wall? If this happened it would have made a transition from one state to another without creating excitations in the fluid — that is it would have slowed up as a whole.

Finally it should be noticed that we do not have to associate the momentum communicated to the fluid, eq. (43), with the excitations created. If, for example, the momentum is transmitted to infinity as an impulse (as LIN [14] has pointed out, this is sometimes the case) then our argument is still correct. For quite generally if an amount of energy δE is communicated to the fluid and an amount of momentum communicated is δP then for the transition to be possible we must have

$$v_i \geqslant \left|\frac{\delta E}{\delta P}\right|_{\min}. \tag{55}$$

Previously we assumed that δE was a function of δP; that is, that the momentum transferred was permanently associated with the excitations. However eq. (55) is true quite generally if we consider δE to be some function of a set of parameters $\gamma_1, \gamma_2 \dots \gamma_n$ $(\equiv\gamma)$ that define the type of excitations present. Likewise we can regard δP to be a function of $\gamma_1, \gamma_2, \dots \gamma_n$. Then the minimization in eq. (55) is to be performed over all possible values of the $\gamma_1, \gamma_2, \dots, \gamma_n$.

4. – The application of Landau's criterion to various systems.

It is instructive to apply the Landau's argument to various systems. We have already seen that it provides a partial explanation for the superfluidity of liquid helium. What about a superconductor? These certainly exhibit persistent current flow. We can use the Landau's argument to show why a supercurrent cannot be effected by the presence of impurities in the superconductor. For the spectrum of elementary excitations is given by

$$\varepsilon(\boldsymbol{p}) = (\varepsilon_0^2 + \Delta^2)^{\frac{1}{2}}, \tag{56}$$

and we find that [15]

$$v_c = |\varepsilon(\boldsymbol{p})/\boldsymbol{p}|_{\min} \simeq \frac{\Delta}{p_F}. \tag{57}$$

Finally let us consider phonons in an insulating crystal. It is tempting to write

$$v_c = \left|\frac{\varepsilon(\boldsymbol{p})}{\boldsymbol{p}}\right|_{\min} = c = \text{velocity of sound}. \tag{58}$$

This ignores the fact that $\varepsilon(p)$ is a periodic function of $\boldsymbol{p}$ and therefore vanishes for finite values of p. Consequently $v_c \equiv 0$. This corresponds to the possibility of having an Umklapp transition. If no Umklapp transitions were possible then the phonons might be able to transmit a temperature wave.

5. – The experimental values of critical velocities.

The experimental critical velocities are never greater than 50 cm/s and usually much less. The values given by Landau's argument when applied to the phonon roton spectrum are about $6 \cdot 10^3$ cm/s. If we want to explain the measured critical velocities by Landau's method we shall have to think of some new type of excitation. We have, however, no direct evidence that they have to be explained in this way [16]. We shall come back to this point in the last section.

6. – Discussion.

The discussion we have just given of the stability of superflow does not seem to provide a complete answer to the questions: *a*) what is superflow? *b*) why does helium exhibit superflow while no other substances do? While no detailed analysis of these questions has been given, it seems worth-while to mention some of the problems that arise from them.

Firstly if an external time-dependent perturbation is applied to liquid helium then in general currents will flow. However, these currents are radically different in liquid helium from those that flow under the same conditions in a normal liquid. For example, if a liquid is contained in a cylindrical container, with a small protuberance on the inside of the wall, then when the container is set into a state of steady rotation the whole liquid will also come — eventually — into steady rotation. However for liquid helium we would expect, at least for very small speeds of rotation, that only a small portion of the liquid near the wall will be disturbed by the rotation and the liquid as a whole will not rotate. Can we state the conditions under which one or another of these two different flows will take place?

Secondly a close analysis of the assumption that a liquid cannot slow down as a whole is necessary. Although it is intuitively appealing, this assumption has never been formulated in a precise mathematical fashion.

Finally we have to understant how we can in some sense « divide » liquid helium at a finite temperature into a « superfluid » and « normal fluid ». In terms of our previous discussion we have to explain how at a finite temperature the response of the system is partially that of a superfluid and partially that of a normal fluid. We shall return to a discussion of this point in later sections.

III. - The Two-Fluid Model Using Feynman's Wave Functions.

1. – Introduction.

Professor De Boer [17] has already described to you how the two fluid models arises naturally from the picture of elementary excitations propagating in liquid helium. In this Section I want to see how far we can interpret this model by using Feynman's methods [5].

As Feynman [5] and others have emphasised there are a number of difficulties in the physical interpretation of even the simplest from of the standard two-fluid equations. These equations have been derived for you by Professor De Boer [17] and they have an extremely simple structure. Moreover they are known to be valid, for small velocities, over a wide temperature range right up to the λ-point. This latter fact means that any physical interpretation of these equations should be valid over the whole range of temperature. Now the main difficulties in interpreting these equations are, *a*) what is the physical meaning of the normal (or superfluid) density ϱ_n, and *b*) what is the precise definition of the superfluid velocity field $\boldsymbol{v}_s$. The normal fluid density arises as a purely mathematical quantity in the excitation picture of the fluid. The superfluid velocity field is usually defined as the velocity of the background fluid in which the excitations are propagating. At very low temperatures when few excitations are present one can reasonably identify $\boldsymbol{v}_s$ with the mean velocity of the fluid before any excitations are present. When a small number of excitations are introduced then we can argue that the mean velocity will be very little altered and so we can talk of the excitations propagating in the moving a fluid. However, it is difficult to maintain this interpretation at temperatures above 1.5 °K, when many excitations are present. Indeed close to the λ-point nearly all the effective mass of the fluid is contained in ϱ_n and it is then very difficult to see what $\boldsymbol{v}_s$ can be. The idea that $\boldsymbol{v}_s$ is the velocity of the moving ground state again makes good sense when only a small number of excitations is present.

Since Feynman's method of discussing the excited states leads us to definite wave functions in configuration space for these states one might hope to extend the basic idea to consider states which correspond to superfluid flow. It is in fact possible to do this [5], but we shall see that we are still unable to give any general answers to either of our questions. The only general answer to *b*) is due to Onsager and Penrose [4]. We shall discuss their work in the next section.

2. – Wave functions for excitations. Interactions between excitations.

Let us begin by writing down wave functions

a) for a state with a single excitation;

b) for a state with several excitations.

We already know how to write down states of type *a*);

$$\psi(\boldsymbol{p}) = F(\boldsymbol{p})\varphi_0 . \tag{59}$$

Where $F(\boldsymbol{p})$ is given by eq. (6). Notice that two wave functions $\psi(\boldsymbol{p}_1)$ and $\psi(\boldsymbol{p}_2)$ are orthogonal if $\boldsymbol{p}_1 \neq \boldsymbol{p}_2$; this follows from the fact they are both eigenfunctions of the momentum operator.

A state with several excitations present can be represented by the wave function

$$\psi(\boldsymbol{p}_1, \boldsymbol{p}_2 \ldots) = F(\boldsymbol{p}_1) \ldots F(\boldsymbol{p}_2)\ \varphi_0 . \tag{60}$$

The energy of this state is

$$E(\boldsymbol{p}_1, \boldsymbol{p}_2 \ldots) = E_0 + \sum_\alpha \varepsilon(\boldsymbol{p}_\alpha) + 0\,(1/U) , \tag{61}$$

the momentum is given exactly by the equation,

$$P = \sum_\alpha \boldsymbol{p}_\alpha . \tag{62}$$

Clearly such a wave function is a good representation of a state with several excitations.

Two states $\psi(\boldsymbol{p}_1, \boldsymbol{p}_2, \ldots)$; $\psi(\boldsymbol{p}_1', \boldsymbol{p}_2', \ldots)$ will be orthogonal only if th e tota momentum associated with each is different:

$$\sum_\alpha \boldsymbol{p}_\alpha \neq \sum \boldsymbol{p}_\alpha' , \tag{63}$$

for example the wave function for two excitations

$$F(\boldsymbol{p}_1)F(\boldsymbol{p}_2)\varphi_0 \tag{64a}$$

is not orthogonal to that for two different excitations

$$F(\boldsymbol{p}_3)F_4(\boldsymbol{p}_4)\varphi_0 \tag{64b}$$

unless

$$\boldsymbol{p}_1 + \boldsymbol{p}_2 \neq \boldsymbol{p}_3 + \boldsymbol{p}_4 \,. \tag{65}$$

This implies that both states are in fact unstable; the first state can decay into the second and vice versa. In such transitions the relevant matrix element is

$$M_{\mathrm{if}} = (\psi_{\mathrm{i}}, H\psi_{\mathrm{f}}) - E(\psi_{\mathrm{i}}, \psi_{\mathrm{f}}) \,, \tag{66}$$

where ψ_{i} is the initial state and ψ_{f} the final state. Since we are interested in real transitions which conserve energy E is either E_{i} or E_{f}. Clearly these matrix elements are very complicated if we use the symmetrical function with the backflow put in. For phonons they are fairly simple. The matrix elements for 3- and 4-phonon processes can be expressed fairly explicitly in terms of the liquid structure factor $S_0(\boldsymbol{p})$. The interaction of excitations and impurities can be treated by the same method.

3. – Wave functions for superflow.

We can attempt to construct wave function which will represent states of the fluid when it is flowing in some way. The simplest case is when the flow is irrotational and incompressible. It is likely that these states are well represented by wave functions of the type

$$\psi = \exp[iS]\varphi_0 \,, \tag{67}$$

where

$$\omega = \sum_{\alpha} \mathscr{S}(\boldsymbol{r}_\alpha) \,, \tag{68}$$

and is taken to be real.

The mean current density in the fluid is

$$\langle j(\boldsymbol{x})\rangle = (1/M)\,\nabla\mathscr{S}(\boldsymbol{x})\cdot\varrho_0 \,, \tag{69}$$

where ϱ_0 is the density in the fluid for the ground state and is a constant. The velocity

$$\boldsymbol{v}(x) = \langle j(\boldsymbol{x})\rangle/\varrho_0 = (1/M)\,\nabla\mathscr{S}(x) \,, \tag{70}$$

clearly $\nabla_{\Lambda}\boldsymbol{v}(x) = 0$; and the flow satisfies the continuity equation if

$$\nabla^2\mathscr{S}(\boldsymbol{x}) = 0 \,. \tag{71}$$

This is the standard equation for a classical velocity potential. Consequently we can immediately make a wave function to represent an irrotational flow situation by substituting (apart from a factor $1/M$) the classical velocity potential for $\mathscr{S}(\boldsymbol{x})$. It has not, however, been shown that all irrotational flows are represented by wave functions of this type. The extra energy associated with the flow is

$$\frac{1}{2} M\varrho_0 \int \boldsymbol{v}^2(x)\, \mathrm{d}^3x \,, \tag{72}$$

as we would expect.

Now since $\boldsymbol{v}(\boldsymbol{x}) \propto \nabla \mathscr{S}(\boldsymbol{x})$ we can make this energy difference very small by choosing a slowly varying $\mathscr{S}(x)$. The simplest $\mathscr{S}(x)$ is $\boldsymbol{p}\cdot x$, then $\boldsymbol{v}(x)=\boldsymbol{p}/M$ and is a constant. This is a state of completely uniform flow. In general wave functions like (61) are *a*) not exactly orthogonal to the ground state (unless of course $\mathscr{S}(x)=\boldsymbol{p}\cdot\boldsymbol{x}$), *b*) do not satisfy Schrödinger's equation exactly and *c*) cannot be rigorously used in the variational principle unless they are eigenfunctions of some operator that commutes with H. (This last remark does not mean that we cannot gain insight into the nature of these states by taking the expectation value of H and calculating the energy.) The only operators that are likely to be exact constants of the motion are the linear momentum and angular momentum. The wave function is an eigenfunction of linear momentum if

$$\mathscr{S}(\boldsymbol{x}) = \boldsymbol{p}\cdot\boldsymbol{x} \,, \tag{73}$$

and of angular momentum about the z-axis if

$$\mathscr{S}(\boldsymbol{x}) = m\varphi \qquad (m = 0, \pm 1, \pm 2) \,, \tag{74}$$

where φ is the azimuthal angle about the z-axis. We shall discuss wave functions of this latter type in Section **6**.

States of compressible flow can be constructed by allowing $\mathscr{S}(\boldsymbol{x})$ to be complex. Such states again lead to irrotational flow. One of the basic questions in the field is whether there are any states of low energy for which the flow field is not irrotational. No strict proof of the existence or non-existence of such states has been given. Plausible arguments have been given by FEYNMAN [18] to show that the low-lying energy states correspond to irrotational flow.

4. – Excitations in the moving fluid-wave packets for excitations.

We can now combine the two basic types of wave functions eqs. (60) and (61), to obtain one which represents an excitation propagating while the fluid as a whole is flowing in an irrotational manner. The wave function

$$F(\boldsymbol{p})\exp[iS(\boldsymbol{x})]\varphi_0\,, \tag{75}$$

represents a state with an excitation of momentum p propagating in the fluid which is flowing with the velocity

$$\boldsymbol{v}(\boldsymbol{x}) = (1/M)\nabla\mathscr{S}(\boldsymbol{x})\,. \tag{76}$$

It is interesting to calculate the energy of such an excited state. We find

$$E = E_0 + \varepsilon(\boldsymbol{p}) + \boldsymbol{p}\cdot\overline{\boldsymbol{v}(x)} + \tfrac{1}{2}\,M\varrho_0\,\overline{\boldsymbol{v}^2(x)}\,. \tag{77}$$

Clearly $\varepsilon(p)$ represents the usual energy of the excitation, the last term is the energy of the moving fluid. The « cross term » $p\cdot\overline{\boldsymbol{v}(x)}$ is the extra energy associated with excitation when it propagates in the moving fluid. Notice also that this equation is not quite of the same form as that used in Kronig's [19] derivation of the two fluid equations. In his method the cross term is $p\cdot v(x)$ where $v(x)$ is the average velocity *in the neighbourhood* of the excitation. This implies that we are dealing with a localized excitation. We, however, calculated the energy for a state that represents a completely non-localized excitation. Let us localize our excitation by constructing a wave packet. Consider the wave function

$$\psi^{(l)}(p) = F^{(l)}(\boldsymbol{p})\varphi_0\,, \tag{78}$$

with

$$F^{(l)}(p) = \sum_\alpha \exp[i\boldsymbol{p}\cdot\boldsymbol{r}_\alpha]\,h(\boldsymbol{r}_\alpha)\,, \tag{79}$$

where $h(\boldsymbol{r}_\alpha)$ is a smoothly varying function of r_α; perhaps like a Gaussian function. The width of $h(\boldsymbol{r}_\alpha)$ should be large compared with p^{-1} but small compared to the size of any macroscopic gradients in the volume of helium. A straightforward calculation shows that the expectation value of current density is now

$$\langle \boldsymbol{j}(\boldsymbol{x})\rangle = \frac{p}{M}|h(\boldsymbol{x})|^2\,. \tag{80}$$

On the other hand the particle density is essentially a constant $\varrho \simeq \varrho_0$. The energy of such a localized excitation in the moving fluid is now

$$\varepsilon(\boldsymbol{p}) + \boldsymbol{p}\cdot v(\boldsymbol{x})\,, \tag{81}$$

where $\boldsymbol{x}$ is the position of the wave packet. This is the standard result used in deriving the two fluid-hydrodynamical equations. The fact that with this simple wave packet the current is not conserved ($\nabla\cdot j\neq 0$) is of course the basic difficulty that lead Feynman to construct the more complicated wave function we discussed earlier.

The local current density for a state like (75) is given by

$$\langle j(\boldsymbol{x})\rangle = \varrho_0 \boldsymbol{v}(\boldsymbol{x}) + \boldsymbol{p}/M\,. \tag{82}$$

Thus the current consists of a part $\varrho_0\boldsymbol{v}(\boldsymbol{x})$ arising from the fluid as a whole and a part p/M coming from the excitations. It is interesting to note that the energy of an impurity propagating in moving liquid helium has exactly the same form as that given by eq. (77).

For a general state with many excitations present we find that the energy

$$E(\boldsymbol{p}_1, \boldsymbol{p}_2 \dots) = E_0 + \sum_{\boldsymbol{p}} n_{\boldsymbol{p}}\varepsilon(\boldsymbol{p}) + \Big(\sum_{\boldsymbol{p}} \boldsymbol{p} n_{\boldsymbol{p}}\Big)\cdot\overline{\boldsymbol{v}} + \tfrac{1}{2}M\varrho_0\overline{\boldsymbol{v}^2}\,, \tag{83}$$

while the momentum $\boldsymbol{P}$ is given by,

$$\boldsymbol{P} = \sum_{\boldsymbol{p}} n_{\boldsymbol{p}}\boldsymbol{p} + \overline{\boldsymbol{v}}\varrho_0\,. \tag{84}$$

5. – Connection with the two fluid model.

The results we have obtained so far in this section suggest a connection with the two fluid model. However, they do not provide a complete interpretation of all the two-fluid variables. In the expressions for the energy and momentum there is an obvious though somewhat arbitrary division into superfluid and normal fluid contributions. The most obvious identification is that the velocity field $(1/M)\nabla\mathscr{S}(\boldsymbol{x})$ should be equated to the superfluid velocity. Unfortunately this quantity does not appear to have any simple interpretation except when there are no excitations present in the fluid. It does not appear to be the expectation value of any simple operator. We can of course proceed from the Feynman picture of elementary excitations and discuss the statistical mechanics and kinetic theory of them. This tells us nothing new, [5] and we arrive at the standard set of two-fluid equations. We can also identify $\sum_{\boldsymbol{p}} \boldsymbol{p} n_{\boldsymbol{p}}$ with the momentum carried by the excitations. If we do this, then we still do not have any means of giving a physical definition of ϱ_n. Finally the whole basis of these calculations rests on the assumption that *all* the superfluid flow states can be accurately represented by wave functions of the type given by eq. (67). Such states lead to irrotational flow for the superfluid. But it is not easy to decide whether we have included a wide enough class of flow states. In the next section we shall discuss the idea of Bose-Einstein condensation in liquid helium. This concept almost certainly provides the missing link between the excitation picture and the particle picture of superfluid helium.

IV. - Bose-Einstein Condensation in Liquid Helium.

The idea of Bose-Einstein condensation (B.E.C.) is familiar from the behaviour of the ideal Bose gas. In this lecture I want to discuss two questions.

a) Whether Bose-Einstein condensation occurs in liquid helium.

b) If it does occur is there a connection between it and superfluidity in helium?

The present theoretical situation is that firstly ONSAGER and PENROSE [4] have proved that B.E.C. occurs in liquid helium at absolute zero and secondly, there are strong arguments but no strict proof that there is a fundamental connection between superfluidity and B.E.C.

The idea that there was a connection between B.E.C. and superfluidity was first put forward by LONDON [20] and TISZA [21]. The importance of this idea has been emphasized by ONSAGER and PENROSE. And as we mentioned in the last section it is likely that the concept of B.E.C. can provide the « missing link » between the particle viewpoint and the excitation model of helium.

1. – The meaning of Bose-Einstein condensation in liquid helium.

In an ideal Bose gas below the condensation temperature the number of the particles in the zero momentum state (we choose periodic boundary conditions) is a finite fraction of the total number of particles. The number in any other momentum state is negligible compared with the total. Mathematically we can express this by the equations,

$$\lim_{N,V\to\infty}\frac{n_0}{N}=\alpha\,,\qquad \lim_{N,V\to\infty}\frac{n_p}{N}=0\,,\qquad \boldsymbol{p}\neq 0\,, \tag{85}$$

where α is a constant which is independent of N and V, except through the ratio N/V. The symbol $\lim\limits_{M,V\to\infty}$ means that limit is to be taken so that while both N and V tend to infinity the ratio N/V is to remain fixed. These equations are true for all temperatures less than the transition temperature T_c. This definition can be easily generalized to systems of particles which interact with one another. First we define the mean number of particles n_p with momentum p by the equation

$$n_{\boldsymbol{p}} = \langle b_{\boldsymbol{p}}^+ b_{\boldsymbol{p}} \rangle = \int \exp[i\boldsymbol{p}\cdot\boldsymbol{v}]\varrho_1(\boldsymbol{r})\,\mathrm{d}^3\boldsymbol{r}\,. \tag{86}$$

We are of course limiting our consideration for the momentum to situations in which the helium is at rest in thermodynamic equilibrium. In eq. (86) the symbol $\langle x \rangle$ means the expectation value of the operator x with the equilibrium density operator for the system, b_p^+ and b_p are the usual second quantization creation and destruction operators. The function $\varrho(\boldsymbol{r})$ is the single particle density matrix for the system in co-ordinate space. Let us confine our attention to absolute zero. Then the N particle density matrix in the co-ordinate representation is simply

$$\varrho_N(\boldsymbol{r}_1\boldsymbol{r}_2 \ldots \boldsymbol{r}_N | \boldsymbol{r}_1'\boldsymbol{r}_2' \ldots \boldsymbol{r}_N') = \varphi_0(\boldsymbol{r}_1\boldsymbol{r}_2 \ldots \boldsymbol{r}_N)\varphi_0(\boldsymbol{r}_1'\boldsymbol{r}_2' \ldots \boldsymbol{r}_N')\,, \tag{87}$$

(φ_0 is real), the density matrix for p particles is defined by equation

$$\begin{aligned}\varrho_p(\boldsymbol{r}_1\boldsymbol{r}_2 \ldots \boldsymbol{r}_p | \boldsymbol{r}_1'\boldsymbol{r}_2' \ldots \boldsymbol{r}_p') &= \\ = \frac{N!}{(N-p)!}\int \varrho_N(\boldsymbol{r}_1\boldsymbol{r}_2 \ldots \boldsymbol{r}_N;\, \boldsymbol{r}_{p+1} \ldots \boldsymbol{r}_N | \boldsymbol{r}_1'\boldsymbol{r}_2' \ldots \boldsymbol{r}_p';\, \boldsymbol{r}_{p+1} \ldots \boldsymbol{r}_N)\,\mathrm{d}^3 r_{p+1} \ldots \mathrm{d}^3 r_N &= \\ = \frac{N!}{(N-p)!}\int \varrho_N(\boldsymbol{r}_1 \ldots \boldsymbol{r}_p;\, \boldsymbol{y} | \boldsymbol{r}_1' \ldots \boldsymbol{r}_p';\, \boldsymbol{y})\,\mathrm{d}^3\boldsymbol{y}\,,\end{aligned} \tag{88}$$

in particular

$$\varrho_1(\boldsymbol{r}_1 | \boldsymbol{r}_1') \equiv \varrho_1(\boldsymbol{r}) = N\int \varrho_N(\boldsymbol{r}_1;\, \boldsymbol{y} | \boldsymbol{r}_1';\, \boldsymbol{y})\,\mathrm{d}^3\boldsymbol{y}\,, \tag{89}$$

where $\boldsymbol{y}$ stands for the entire set of vectors $\boldsymbol{r}_2 \ldots \boldsymbol{r}_N$. Note that ϱ_1 is a function of $\boldsymbol{r}_1 - \boldsymbol{r}_1' = \boldsymbol{r}$ only. That n_p is the mean number of particles with momentum p follows directly from eq. (86), for $b_{\boldsymbol{p}}^+ b_{\boldsymbol{p}}$ is the number operator for the p-th momentum state. We also find that if F is any operator which is a sum of operators $f(\boldsymbol{r}_\alpha)$ referring to single particles then

$$\langle F \rangle = \sum_{\boldsymbol{p}} f(\boldsymbol{p})n(\boldsymbol{p})\,, \tag{90}$$

where $f(\boldsymbol{p})$ is the diagonal element of f in the momentum representation.

With these exact definitions of $n_{\boldsymbol{p}}$ in an interacting system, we can use exactly the same equation as before to define a system in which B.E.C. is present. That is if

$$\lim_{N,V\to\infty}\frac{n_0}{N}=\alpha\,,\qquad \lim_{N,V\to\infty}\frac{n_p}{N}=0\,,\qquad \boldsymbol{p}\neq 0\,, \tag{91}$$

then we shall say that Bose-Einstein condensation has taken place into the zero momentum state. The Pauli exclusion principle of course prevents these equations being true in a Fermi-Dirac system. However, we shall see that analogous equations can be set up in a Fermi system for the two particle density matrix. We have so far talked about Bose-Einstein condensation in terms of momentum states. This is because we are considering complete equilibrium and are using periodic boundary conditions. We shall shortly generalise the idea to include condensation into states which are not momentum eigenfunctions.

2. – The existence of Bose-Einstein condensation in liquid helium.

We know that Bose-Einstein condensation takes place in an ideal Bose gas, and that it exists in weakly interacting [22] or low density [23] Bose systems. The only proof that it exists in general, in liquid Bose systems is due to Penrose and Onsager [4]. I do not propose to discuss this proof in these sections. It is, however, worth mentioning some important points.

(i) The Bose system must be a fluid in the sense that it does not exhibit long range order. A solid Bose system would not show condensation into a single momentum state.

(ii) The proof applies at once to liquid helium at absolute zero — as long as we assume that the ground state wave function for the liquid does not exhibit long range order.

(iii) At finite temperatures no rigorous proof of condensation has been given. If one assumes that the lowest excited states are well represented by phonon states, such as we discussed in Sections **1** and **2**, then one can prove that B.E.C. is present at a finite temperature.

(iv) The fraction of particles in the lowest momentum state may be very much less than unity. A rough estimate by Penrose and Onsager gave only 8 %.

3. – Consequencies of the condensation phenomena.

In order to understand the consequencies of this condensation phenomena we must return to the density matrix formulation that we have just discussed. The single particle density matrix $\varrho_1(\boldsymbol{r}_1|\boldsymbol{r}_1')$ is a Hermitian matrix with real eigenvalues c_λ; we have the eigenvalue equation,

$$\int \varrho_1(\boldsymbol{r}_1 | \boldsymbol{r}_1') \varphi_\lambda(\boldsymbol{r}_1') \, \mathrm{d}^3 r_1' = c_\lambda \varphi_\lambda(\boldsymbol{r}_1) \,. \tag{92}$$

Here φ_λ is the eigenfunction belonging to the eigenvalue c_λ. For a system in complete equilibrium $\varrho_1(\boldsymbol{r}_1|\boldsymbol{r}_1') = \varrho_1(\boldsymbol{r}_1 - \boldsymbol{r}_1')$ and the φ_λ are plane waves;

$$\int \varrho_1(\boldsymbol{r}_1 - \boldsymbol{r}_1') \exp[i\boldsymbol{p}\cdot\boldsymbol{r}_1'] \, \mathrm{d}^3 \boldsymbol{r}_1' = n_{\boldsymbol{p}} \exp[i\boldsymbol{p}\cdot\boldsymbol{r}_1] \,, \tag{93}$$

where

$$n_{\boldsymbol{p}} = \int \exp[i\boldsymbol{p}\cdot\boldsymbol{r}] \varrho_1(\boldsymbol{r}) \, \mathrm{d}^3 \boldsymbol{r} \,, \tag{94}$$

and is consequently the same function as we defined in eq. (86). Thus in equilibrium the eigenvalues of the single particle density matrix are the occupation number of the various momentum states. Therefore, when B.E.C. is present one eigenvalue n_0 is large and all the other $n_{\boldsymbol{p}}$ are small compared to it. The density matrix ϱ_1 can always be written in the form

$$\varrho_1(\boldsymbol{r}_1 - \boldsymbol{r}_1') = \sum_{\boldsymbol{p}} n_{\boldsymbol{p}} \varphi_{\boldsymbol{p}}(\boldsymbol{r}_1) \varphi_{\boldsymbol{p}}^*(\boldsymbol{r}_1') = n_0 + \sum_{\boldsymbol{p}\neq 0} n_{\boldsymbol{p}} \varphi_{\boldsymbol{p}}(\boldsymbol{r}_1) \varphi_{\boldsymbol{p}}^*(\boldsymbol{r}_1') \,, \tag{95}$$

(notice $\varphi_0 = 1$) and, therefore, naturally decomposes into two parts, one coming from the condensed state, the other from all the remaining states.

We can now generalize this idea to nonuniform states which would represent nonequilibrium (flow; $\langle j \rangle \neq 0$) situations. Quite generally the eigenfunctions of ϱ_1 satisfy the equation

$$\int \varrho_1(\boldsymbol{r}_1 | \boldsymbol{r}_1') \varphi_\lambda \, \mathrm{d}^3 r_1' = n_\lambda \varphi_\lambda(\boldsymbol{r}_1) \,, \tag{96}$$

where n_λ is the average number of particles in state φ_λ. We can also write the density matrix in the form,

$$\varrho_1(\boldsymbol{r}_1 | \boldsymbol{r}_1') = \sum_\lambda n_\lambda \varphi_\lambda(\boldsymbol{r}_1) \varphi_\lambda^*(\boldsymbol{r}_1') \,. \tag{97}$$

B.E.C. can now be defined by the equations

$$\lim_{N,V\to\infty} \frac{n_\sigma}{N} = \alpha, \qquad \lim_{N,V\to\infty} \frac{n_\lambda}{N} = 0, \qquad \lambda \neq \sigma, \tag{98}$$

when these two equations are satisfied we shall say that B.E.C. has taken place into the state φ_σ. It is then natural to write

$$\varrho_1(\boldsymbol{r}_1|\boldsymbol{r}_1')=n_\sigma\varphi_\sigma(\boldsymbol{r}_1)\varphi_\sigma^*(\boldsymbol{r}_1')+\sum_\lambda n_\lambda\varphi_\lambda\varphi_\lambda^* . \tag{99}$$

Let us illustrate these ideas with one of Feynman's wave function $\exp[iS]\Phi_0$, where $S=\sum_\alpha\mathscr{S}(\boldsymbol{r}_\alpha)$. It is easy to see that if ϱ_1^0 is the density matrix for the state Φ_0 then that for the state $\exp[i\mathscr{S}]\Phi_0$ is given by the equation

$$\varrho_1=\varrho_1^0\exp\left[i[\mathscr{S}(\boldsymbol{r}_1)-\mathscr{S}(\boldsymbol{r}_1')]\right] . \tag{100}$$

The eigenfunctions are given by,

$$\varphi_\lambda=\exp[i\mathscr{S}]\varphi_p , \tag{101}$$

while

$$n_\lambda=n_p , \tag{102}$$

finally eq. (17) becomes

$$\varrho_1(\boldsymbol{r}_1|\boldsymbol{r}_1')=n_0\exp\left[i(\mathscr{S}(\boldsymbol{r}_1)-\mathscr{S}(\boldsymbol{r}_1'))\right]+\sum_p n_p\varphi_p\varphi_p^*\exp\left[i[\mathscr{S}(\boldsymbol{r}_1)-\mathscr{S}(\boldsymbol{r}_p')]\right] . \tag{103}$$

Equation (102) tells us that if condensation is present in the state Φ_0 then it is also present in the state $\exp[iS]\Phi_0$. There is, therefore, an infinite variety of potential flow states in which B.E.C. is present. However the state into which the particles have condensed is a moving state in the sense that it has a velocity field $(1/M)\nabla\mathscr{S}$ associated with it.

These ideas must also be generalised for density matrices at a finite tems perature. We are now no longer dealing with pure states but with ensemble- and we define ϱ_N^β by the equation

$$\varrho_N^\beta=\sum_l\exp[-\beta E_l]\,\psi_l(\boldsymbol{r}_1\dots\boldsymbol{r}_N)\,\psi_l^*(\boldsymbol{r}_1\dots\boldsymbol{r}_N) , \tag{104}$$

where the ψ_l^s are the eigenstates of the system and the E_l^s the eigenvalues.

This is, of course, the density matrix for a system in thermodynamic equilibrium at temperature $T=(k\beta)^{-1}$. We now define B.E.C. by exactly the same equation as before; if

$$\varrho_p^\beta=\frac{N!}{(N-p)!}\int\varrho_N^\beta(\boldsymbol{r}_1\dots\boldsymbol{r}_p;\boldsymbol{y}\,|\,\boldsymbol{r}_p'\dots\boldsymbol{r}_p';\boldsymbol{y})\,\mathrm{d}^3\boldsymbol{y} , \tag{105}$$

then we say that B.E.C. is present in the state φ_0 at temperautre T if

$$\lim_{N,V\to\infty} \frac{n_0(T)}{N} = \alpha(T), \qquad \lim_{N,V\to\infty} \frac{n_p(T)}{N} = 0 . \tag{106}$$

Finally we must consider non equilibrium flow situations at a finite temperature. This last statement implies that even though we talk of nonequilibrium states of the system we always assume that the departure from *overall* equilibrium is such that the concept of *local* equilibrium is still valid. We can, therefore, talk in terms of a local temperature, which is different from zero. In general the density matrix which corresponds to a typical nonequilibrium situation will depend on the forces that are acting on the system. If we assume that at $t=0$ a mechanical driving force, represented by the potential energy $v(\boldsymbol{r}, t)$, is applied to the system, then the density matrix at a later time is given by

$$\varrho = \varrho_0 + \int_0^t [v(r,t-t'), \varrho(t')]\, dt' . \tag{107}$$

In many cases we can confine our attention to the linear response to the driving force and then

$$\varrho = \varrho_0 + \int_0^t [v(r, t-t'), \varrho_0]\, dt' . \tag{108}$$

We have assumed that the system is in equilibrium at $t=0$. The properties of a density matrix such as this have never been investigated. It is important, however, to realise that one can easily write down the density matrix for such non equilibrium situations and we can, therefore, investigate whether or not B.E.C. is present.

Let us return to the general properties of ϱ_1. Any single particle wave function φ_λ can be written $\exp[iS_\lambda] R_\lambda$ consequently there is a velocity field $(1/M)\nabla\mathscr{S}_\lambda$ associated with it. ONSAGER and PENROSE [4] have put forward the important suggestion that the superfluid velocity field $\boldsymbol{v}_s$ in the two fluid model should be identified with the velocity field of the state into which condensation has taken place. That is to say if B.E.C. has taken place into the state φ_σ then

$$\boldsymbol{v}_s = (1/M)\nabla\mathscr{S}_\sigma . \tag{109}$$

This suggestion is intimately linked with their proof that at absolute zero B.E.C. is present for both uniform and nonuniform states. The above definition of $\boldsymbol{v}_s$ agrees with both Feynman's and Landau's [12] when no excitations are present. For example, we have already seen that B.E.C. is present in the state

$$\exp[iS]\Phi_0 .$$

It is important to show that this definition can be maintained when excitations are present. That is to say we must show *a*) that B.E.C. is still present when we consider a state with excitations and *b*) that Onsager and Penrose's definition of v_s agrees with Landau's [12] (which is the same as Feynman's [5]). No rigorous and general proof has been given of these last two points. However, PENROSE [24] has shown that if one takes a Feynman type of wave function

$$F(\boldsymbol{p}_1)F(\boldsymbol{p}_2)\dots F(\boldsymbol{p}_n)\varphi_0\exp[iS], \tag{110}$$

with

$$F(\boldsymbol{p}) = \sum_\alpha \exp[i\boldsymbol{p}\cdot\boldsymbol{r}_\alpha],$$

and then constructs an ensemble in which the excitations have a nonzero group velocity then

a) if $\boldsymbol{v}_s = (1/M)\nabla\mathscr{S}$ it is identical with Landau's definition, and

b) the mean group velocity of the excitations is identical with Landau's definition of $\boldsymbol{v}_n$.

This shows that with Feynman wave functions one can consistently identify the superfluid velocity with the velocity field of the condensed particles. This gives a precise microscopic interpretation to $\boldsymbol{v}_s$; this is really completely absent in the excitation picture. It is tempting to try and interpret the « density of the condensed phase » as the superfluid density ϱ_s. This is obviously incorrect because at $T=0$, $\varrho_s=\varrho$, whereas $R_\sigma^2 n_\sigma/N < \varrho$. We, therefore, do not have a precise microscopic definition of ϱ_s or ϱ_n.

At the moment we cannot go further than this. Namely we have a very general definition of the superfluid velocity field, but it has not been possible to give a general derivation of the two fluid hydrodynamic equations from the basic property of B.E.C. Perhaps the most interesting question is whether there is any simple connection between the superfluid density ϱ_s and the number of particles n_σ in the condensed state.

4. – Bose-Einstein condensation in Fermi systems.

It is possible to generalise the concept of B.E.C. to Fermi systems. We have already pointed out that one cannot have B.E.C. for a one-particle density matrix for a system where particles obey the Pauli exclusion principle.

However, if we turn to the two-particle density matrix and consider its eigenvalues then we can talk of B.E.C. being present if one of its eigenvalues is of $O(N)$. That is to say if $\varrho_2(\boldsymbol{r}_1\boldsymbol{r}_2|\boldsymbol{r}_1'\boldsymbol{r}_2')$ has eigenvalues $n_\lambda^{(2)}$ and eigenfunction $\varphi_\lambda^{(2)}$, then we shall say that Bose condensation has taken place into the pair wave function $\varphi_\sigma^{(2)}(\boldsymbol{r}_1\boldsymbol{r}_2)$ if

$$\lim_{N,V\to\infty}\frac{n_\sigma}{N}=\alpha, \qquad \lim_{N,V\to\infty}\frac{n_\lambda}{N}=0\,, \qquad \lambda\neq\sigma, \tag{111}$$

where α may depend on N and V only through the ratio N/V.

An example of a model Fermi system which displays B.E.C. as we have just defined it is provided by the Bardeen-Cooper-Schrieffer model of a superconductor. One can show [25] that for this model

$$\varrho_2(\boldsymbol{r}_1\boldsymbol{r}_2|\boldsymbol{r}_1'\boldsymbol{r}_2')\to\chi(r_1r_2)\,\chi(r_1'r_2')\,, \tag{112}$$

as r_1 and r_2 become indefinitely separated from $\boldsymbol{r}_1'$ and $\boldsymbol{r}_2'$. This is enough to show that one eigenvalue of ϱ_2 is of order $O(N)$. In simple flow situations this property is likely to be maintained and it is, therefore, plausible that the velocity field associated with a supercurrent should be defined in terms of the phase of the two particle wave function $\varphi_\sigma=\chi$. This in turn would imply that the supercurrent velocity field was irrotational. We shall return to this point in the last section.

V. - Quantum Vortices.

1. – The condition $\nabla\wedge\boldsymbol{v}_s=0$.

In this last Section I want to discuss the present state of the theory of quantum vortices. The theory of quantum vortices is intimately related to the theory of the superfluid velocity field $\boldsymbol{v}_s$. There have been two approaches to this problem, that by PENROSE and ONSAGER [4] and PENROSE [24] and that by FEYNMAN [18] (and somewhat earlier by F. LONDON [20]).

We begin with the ideas of Penrose and Onsager that we discussed in the last section. We saw that they propose that the correct microscopic definition of $\boldsymbol{v}_s$ is that it is the gradient of the phase of the wave function belonging to the condensed state. We should emphasise that this definition for $\boldsymbol{v}_s$ has never been shown to be correct in *all circumstances*. It has, however, been shown to be correct for a variety of wave functions, with and without excitations being present. Moreover it has considerable intuitive appeal — it is difficult to see how there can be any other simple microscopic interpretation of

$\boldsymbol{v}_s$ other than this. If we accept this definition as correct, then it is trivial that

$$\nabla_\Lambda \boldsymbol{v}_s = 0 , \tag{113}$$

for all states of superflow. It is important to notice that the method by which we have derived this result is rather deep from the quantum mechanical point of view. The equation is true because of certain very special properties of the off diagonal elements of the single particle density matrix.

We must, therefore, regard the condition given by eq. (113) as a quantum mechanical restriction on this flow of a Bose-Einstein fluid. If a phase transition takes place at a finite temperature and B.E.C. is not present above the transition temperature then we at once have that $\boldsymbol{v}_s = 0$, and we revert to the ordinary « one fluid » hydrodynamics. Notice that the condition has the meaning only for those regions of the fluid for which B.E.C. is present. For in a nonuniform situation one can imagine that certain regions of the fluid may have B.E.C. while others may not. One should, therefore, be careful in interpreting the condition $\nabla_\Lambda \boldsymbol{v}_s = 0$ in terms of a universal restriction to be valid everywhere in liquid helium. If our basic ideas are, however, correct then we can only avoid the restriction for regions in which the helium is not in a superfluid state.

The argument put forward by Feynman are based on the argument that all low energy flows will be accurately represented by wave functions of the type

$$\exp[iS]\,\Phi_0 . \tag{114}$$

This of course leads to a velocity field which is irrotational, and this velocity field can be identified with $\boldsymbol{v}_s$. This argument can be extended to states with excitations present and hence to a finite temperature. However we now do not have any *general* definition of $\boldsymbol{v}_s$ like that of Onsager and Penrose; we only know how to define it for states like (114). We must, therefore, regard Feynman's arguments as being somewhat weaker than that we have just discussed. FEYNMAN [18] has given an extensive physical discussion of why we may expect these states to be the lowest energy states for a Bose system.

Finally it should he emphasized that in the standard [19] derivation of the two-fluid equations from the excitation model it is implicitly assumed that $\nabla_\Lambda v_s = 0$. This restriction allows one to assume that one may make a Galilean transformation from the frame of reference where the fluid is at rest to the laboratory frame.

2. – The quantization of circulation.

We now turn to the implications of the condition

$$\nabla_\Lambda \boldsymbol{v}_s = 0 .$$

For a simple connected volume of fluid this implies that the circulation Γ_c round any curve c is zero. For

(115) $$\Gamma_c \equiv \int \boldsymbol{v}_s \cdot \mathrm{d}\boldsymbol{l} = \int_{S_c} (\nabla_\Lambda \boldsymbol{v}_s) \cdot \mathrm{d}\boldsymbol{n} = 0,$$

where S_c is any surface spanning the curve c. This is quite contrary to experiment. Helium appears to rotate at moderate speeds in the same way as an ordinary fluid. Hence it must at least be capable of having a nonzero circulation. In a multiply connected region we have no difficulty for $\Gamma_c \neq 0$ even if $\nabla_\Lambda \boldsymbol{v}_s = 0$ everywhere. For example, in a torus we can have circulation with $\nabla_\Lambda \boldsymbol{v}_s = 0$ everywhere. This suggests that a way of achieving nonzero circulation in a singly connected vessel containing liquid helium would be to suppose that there are singular regions in the fluid that make it multiply connected. The simplest way in which this could happen would be for the liquid to contain vortex lines on sheets. For example, the circulation round a classical vortex line is nonzero; the velocity field is inversely proportional to the radial distance from the line and is irrotational everywhere. There is an important quantum restriction on such vortex lines. If we define $\boldsymbol{v}_s$ in terms of the phase of the condensed state wave function then the circulation Γ_c round any closed curve c is given by

(116) $$\Gamma_c = \int_c \boldsymbol{v}_s \cdot \mathrm{d}\boldsymbol{l} = (1/M) \int \nabla \mathscr{S} \cdot \mathrm{d}\boldsymbol{l} = (\Delta S)_c (1/M)\,,$$

where $(\Delta S)_c$ is the change in the phase $\mathscr{S}$ on going once round the contour. But the contensed state wave function must be single valued, hence

(117) $$(\Delta \mathscr{S})_c = 2n\pi \qquad (n = 0 \pm 1 \,...)\,.$$

This implies that

(118) $$\Gamma_c = \frac{2n\pi}{M}\,.$$

We, therefore, arrive at the important conclusion that the *circulation of the superfluid velocity field is quantized.* This is a quite general result and can be at once applied to the vortex lines we have been discussing. We can imagine vortex lines with different quantized values of the circulation corresponding to $n=1, 2, 3$ and so on. A simple wave function that has some of these properties is

(119) $$\exp\left[im \sum_\alpha \varphi_\alpha\right] \Phi_0\,,$$

where φ_α is the azimuthal angle (in cylindrical co-ordinates) of the α-th particle. The Bose condensed state wave function is

$$\exp[im\varphi]\,, \tag{120}$$

and

$$\boldsymbol{v}_s = 1/M\,\frac{m}{r}\,, \tag{121}$$

and of course $\nabla_\Lambda \boldsymbol{v}_s \equiv 0$ everywhere except at $r=0$, where v_s is not defined.

The circulation

$$\Gamma_c = (1/M)\int_0^{2\pi}\frac{m}{r}\,r\,\mathrm{d}\varphi = \frac{2\pi m}{M}\,. \tag{122}$$

This agrees exactly with our general result, eq. (118). The energy associated with this state leads to a logarithmic divergence at small distances from the centre of the wortex line. If we cut off the integral at a short distance (a) from the line then the energy of the vortes line per unit length is

$$\varepsilon = \varrho\,\frac{\Gamma^2}{4\pi}\,\log(R/a), \tag{123}$$

where R is the radius of the vessel containing the line. It is quite uncertain what value we should take for a; Feynman has suggested $a \sim 10^{-8}$ cm.

3. – Wave functions for vortex lines.

Two important questions are *a*) can we construct a wave function that represents reasonably accurately a state with a quantum vortex present, and *b*) can we show that these states have such a low energy that « rotating helium » will prefer to occupy them rather than some other state of quantized circulation. In connection with the last question Feynman has pointed out that a state with a uniform array of vortex lines will on a macroscopic scale look just like a state of uniform rotation. For the macroscopic vorticity is always equal to the circulation per unit area. Thus if we have n quantum vortices per unit area and we want this state to look like one of uniform rotation with angular velocity ω_0 we must have

$$n(1/M) = 2\omega_0\,, \tag{124}$$

or

$$n = 2M\omega_0\,. \tag{125}$$

For a vessel rotating at a few radians per second the vortex lines are about 10^{-2} cm apart.

The two questions we have posed are intimately related to each other and almost nothing is known about the answers to them. We should emphasise that *once we accept the condition* $\nabla_\Lambda \boldsymbol{v}_s = 0$ then helium can only exhibit a non-zero circulation in a simply connected vessel if the helium itself is multiply connected due to the presence of singular regions in the velocity field. We are equally forced to accept that the circulation round any of these singular regions is quantized.

It is easy to write down wave functions that will lead to finite energies for vortex states. For example the function

$$\exp\left[im\sum_\alpha \varphi_\alpha\right]\pi_\alpha f(\boldsymbol{r}_\alpha)\Phi_0\,, \tag{126}$$

where

$$\begin{aligned} f(\boldsymbol{r}_\alpha) &= 0 \qquad |\boldsymbol{r}_\alpha| < \alpha \\ &= 1 \qquad |\boldsymbol{r}_\alpha| > \alpha \end{aligned} \tag{127}$$

is satisfactory. However, until we can estimate the energy associated with such a state we do not know how good it is; calculating the energy associated with this state leads to quite intractable integrals. Here we should notice that a reliable estimate of the energy of a vortex state can be obtained if we use the same form of the variational principle that we used in the first section. We now demand that that the vortex state wave function be an eigenfunction of the z-component of the angular momentum and obtain an upper bound for the lowest eigenvalue corresponding to a given eigenvalue of l_z. Both the wave functions (119) and (120) we have written down are eigenfunctions of l_z.

It is probably a poor approximation to localise the centre of the line of one definite position. One should take a linear combination of states with the line in different places. The « smearing out » of the line will almost certainly lower its energy. Such a state would be represented by

$$\Psi = \sum_{\boldsymbol{r}} g(\boldsymbol{r})\,\psi(\boldsymbol{r})\,, \tag{128}$$

where $\psi(\boldsymbol{r})$ is a wave function like (126) only with the line at $\boldsymbol{r}_0$ instead of the origin. This is as far as one can go theoretically. No one has as yet managed to get a better estimate of the energy than the simple one given by eq. (123).

There is a considerable body of evidence to support the idea of vortex lines in superfluid helium. Detailed reviews of these experiments have been given recently by HALL [26] and VINEN [27]. However, only one experiment [27] seems to show convincingly that the circulation in a multiply connected volume of helium is quantized with a value of $1/M$.

4. – Discussion.

We conclude this section with some brief remarks on various topics connected with quantized circulation and quantum vortices.

GROSS [28] has recently shown that the Hartree equations for a system of bosons with repulsive interactions have quantized vortex-line solutions. He considers an axially symmetric volume of helium so that l_z is a constant of the motion and shows that it is possible to find solutions of the equations such that *a*) the density is zero on the axis of symmetry and *b*) the density goes to a constant at large distances from the axis of symmetry. These solutions differ from the free-particle solutions whose radial dependence is given by a Bessel function. The density for these wave functions is not a constant but oscillates as we go away from the axis of symmetry. GROSS also considers the elementary excitations (phonons) is such a system and the effect of their zero-point energy on the vortex-line solutions.

Finally we turn to the remarks we made in Section **5** about Bose-Einstein condensation in Fermi systems. We pointed out: *a*) if one considered the two-particle density matrix, then one could easily extend the concept of B.E.C. to Fermi systems; *b*) the particular form of wave function used in the current theory of superconductivity [29] does exhibit B.E.C. We can again define a velocity field in terms of the gradient of the phase of the condensed state wave function. Such a velocity field will have a macroscopic number of particles associated with it and will of course be irrotational. It is likely that this velocity field should be identified with the supercurrent velocity field in a superconductor. If this idea is correct then we would conclude that the circulation of a supercurrent should be quantized and that quantum vortices should exist in superconductors. It is possible that the recent experiments [30] and [31], which have demonstrated the existence of quantized circulation in superconducting cylinders, lend some support to these speculations.

REFERENCES

[1] D. G. HENSHAW: *Phys. Rev.*, **121**, 1266 (1961).
[2] R. P. FEYNMAN and M. COHEN: *Phys. Rev.*, **102**, 1189 (1950).
[3] L. I. SCHIFF: *Quantum Mechanics* (New York, 1949).

[4] L. Onsager and O. Penrose: *Phys. Rev.*, **104**, 576 (1956).
[5] R. P. Feynman: *Phys. Rev.*, **94**, 262 (1954).
[6] R. Kronig and A. Thellung: *Physica*, **18**, 749 (1952).
[7] D. G. Henshaw: *Phys. Rev.*, **119**, 9 (1960).
[8] J. Reekie and T. S. Hutchison: *Phys. Rev.*, **92**, 827 (1953).
[9] G. V. Chester and K. Major: to be published.
[10] T. D. Lee and F. Mohling: *Phys. Rev. Lett.*, **2**, 284 (1959).
[11] J. L. Jamell, C. P. Arnold, P. J. Bendt and E. C. Hern: *Phys. Rev.*, **113** 1379 (1954).
[12] L. D. Landau: *Žurn. Ėksp. Teor. Fiz.*, **5**, 71 (1961).
[13] L. D. Landau and E. M. Lifshitz: *Statistical Mechanics* (London, 1958).
[14] C. C. Lin: this volume, p. 93.
[15] S. Nakajima: *Progr. Theor. Phys.*, **21**, 358 (1959).
[16] W. F. Vinen: this volume, p. 336.
[17] J. De Boer: this volume, p. 1.
[18] R. P. Feynman: *Progress in Low Temperature Physics*, vol. **1** (Amsterdam, 1955).
[19] R. Kronig: *Physica*, **19**, 535 (1953).
[20] F. London: *Superfluids*, vol. **2** (New York, 1954).
[21] L. Tisza: *Nature*, **141**, 913 (1938); *Compt. Rend.*, **207**, 1035, 1186 (1938); *Journ. Phys. et Rad.*, **1**, 164, 350 (1940).
[22] N. N. Bogolyubov: *Žurn. Ėksp. Teor. Fiz.*, **11**, 23 (1947).
[23] T. D. Lee and C. M. Yang: *Phys. Rev.*, **105**, 119 (1957).
[24] O. Penrose: *Proc. Intern. Conf. on Low Temperature Physics* (Madison, Wis. 1957).
[25] J. G. Valatin: *Nuovo Cimento*, **7**, 843 (1958).
[26] H. E. Hall: *Adv. in Phys.*, **9**, 89 (1960).
[27] W. F. Vinen: *Progress in Low Temperature Physics*, vol. **3** (Amsterdam, 1961).
[28] E. P. Gross: to be published.
[29] J. Bardeen, L. Cooper and R. Schrieffer: *Phys. Rev.*, **108**, 1175 (1957).
[30] B. S. Deaver and W. M. Fairbank: *Phys. Rev. Lett.*, **7**, 43 (1961).
[31] R. Doll and M. Näbauer: *Phys. Rev. Lett.*, **7**, 51 (1961).

Hydrodynamics of Helium II.

C. C. Lin

Massachusetts Institute of Technology - Cambridge, Mass.

Introduction.

This set of notes was originally prepared for use of the Enrico Fermi International School of Physics held at Varenna, Italy, July 3-5, 1961. It has been slightly revised for purposes of publication. However, it was felt desirable to preserve the informal nature of the lecture notes, especially since there is necessarily a great deal of speculation on the subject of superfluid helium at the present time.

I am indebted to several of my friends for their help and encouragement in the work that forms the basis of these lectures; especially to C. N. Yang for initiating my interest in this subject and for many helpful discussions during my stay at the Institute for Advanced Study, Princeton, for which I wish to thank Professor J. R. Oppenheimer for his hospitality. The work was partly supported by the Office of Naval Research.

When Professor Careri asked me to give a few lectures at this Course, he assigned two duties to me. The first is to present a brief up-to-date account of our knowledge of hydrodynamics of ordinary fluids, especially those aspects that might relate to the current studies of superfluid helium. The second is to discuss some of my thoughts on the subject itself. This dual assignment is a fortunate one, because I cannot make my points clear in the discussion of the hydrodynamics of superfluids without first reviewing some known results in the ordinary case. I shall attempt to do this in the six hours allotted to me with perhaps two hours devoted to the first subject and four hours devoted to the latter. The topics of my discussions are as follows:

I) Hydrodynamics of ordinary fluids – reversible processes.

II) Hydrodynamics of ordinary fluids – irreversible processes.

III) Hydrodynamics of superfluids – reversible processes.
IV) Hydrodynamics of superfluids – irreversible processes.
V) Theoretical basis for the hydrodynamics of superfluid helium.
VI) Experimental evidence for quantization of superfluid flow.

In the last two chapters, I shall attempt to explain why I feel that the concept of quantized vortex lines with *macroscopic* lengths lacks a satisfactory foundation from the theoretical point of view. Questions will also be raised about the experimental evidence for their existence.

Before I begin to discuss the details, I should perhaps make some general points clear about my line of thinking. I wish to elaborate on a point which is obvious in classical hydrodynamics but wich becomes blurred in the study of superfluid helium. This is the possibility of studying macroscopic and microscopic aspects of a problem almost independently of each other, as is usually done in the hydrodynamics of ordinary fluids. To be sure, it would be very satisfying to develop the macroscopic theory from the molecular theory. However, this has not yet been completed even in the case of ordinary liquids. On the other hand, there is no question that the macroscopic hydrodynamics of water is well developed. Indeed, the same theory applies to air, which has an entirely different molecular structure. Thus, the advantage of a macroscopic theory is to avoid the difficulties of the complicated molecular structure and yet at the same time provide a framework for the description of the macroscopic phenomena.

Such a sharp distinction is questionable in the case of superfluid helium, because we are specifically concerned with *macroscopic quantum-mechanical effects*. However, this does not mean that the development of a continuum theory must be handicapped. Rather, the attempt should be made to decide which experimental observations can be described by a unified macroscopic theory and which *must* be described in terms of some molecular model. Clearly, experiments performed with beams of neutrons and ions are usually directly concerned with the internal molecular structure of the liquid, and belong to the second category. Others, in which only macroscopic effects are observed, can hopefully be entirely described by continuum theories. Indeed, the most prominent and basic phenomenon in superfluid helium is the reversible heat flux, which is directly related to the phenomenon of Bose-Einstein condensation—a decidedly quantum-mechanical effect. Yet its macroscopic effect can be handled entirely on the basis of continuum concepts (see Part III). The mechanism for its origin is of course an entirely different matter. If one would then take an optimistic point of view, one might hope that, even in the case of superfluid helium, there is the possibility of separately developing a macroscopic theory of fluid motion and a microscopic theory of molecular structure.

I. - Hydrodynamics of Ordinary Fluids. Reversible Processes.

1. – Fundamental equations. Variational principle.

The conventional hydrodynamic theory is based on fields of macroscopic quantities. The thermodynamic state of the fluid is characterized by the mass density ϱ and the entropy per unit mass s. A physical description of the system at any instant of time t is given by the specification, at every point $P(\boldsymbol{x})$ of these quantities together with the macroscopic velocity $\boldsymbol{v}$: $\varrho(\boldsymbol{x}, t)$, $s(\boldsymbol{x}, t)$, and $\boldsymbol{v}(x, t)$. For reversible processes, the conservation of mass and energy can be expressed by the following equations:

(I.1.2) $$\frac{\partial \varrho}{\partial t} + \operatorname{div}(\varrho \boldsymbol{v}) = 0\,,$$

(I.1.2) $$\frac{\mathrm{D}s}{\mathrm{D}t} = 0\,,$$

where $\mathrm{D}/\mathrm{D}t$ is the substantial derivative $\partial/\partial t + (\boldsymbol{v}\cdot\nabla)$. The equations of motion can be derived by means of the variational principle (*) (Hamilton's principle)

(I.1.3) $$\delta\iint L(\boldsymbol{v}, \varrho, s)\,\mathrm{d}\tau\,\mathrm{d}t = 0,$$

with (I.1.1) and (I.1.2) as subsidiary conditions. In (I.1.3), the Lagrangian density function is

(I.1.4) $$L = \tfrac{1}{2}\varrho \boldsymbol{v}^2 - \varrho(e + \Omega)\,,$$

where $e = e(\varrho, s)$ is the internal energy, and Ω is the potential of the external force. However, in this formulation, the Euler equation associated with $\delta\boldsymbol{v}$ gives

(I.1.5) $$\boldsymbol{v} = \operatorname{grad}\varphi + \beta \operatorname{grad} s$$

and restricts the flow to be irrotational, when the entropy is uniform. To remove this difficulty, we introduce the condition for the conservation of the identity of particles

(I.1.6) $$\frac{\mathrm{D}\boldsymbol{X}}{\mathrm{D}t} = 0\,.$$

(*) For further details, see the discussion of this writer as quoted by SERRIN [1], p. 148 ff.

Then (I.1.5) becomes (*)

$$\boldsymbol{v} = \operatorname{grad} \varphi + \beta \operatorname{grad} s + \operatorname{grad} \boldsymbol{X}\cdot\boldsymbol{\gamma}\,, \tag{I.1.5a}$$

where φ, β, $\boldsymbol{\gamma}$ are Lagrangian multipliers associated with (I.1.1), (I.1.2) and (I.1.6) respectively. It is possible to show that (I.1.3) is equivalent to the set of equations of motion

$$\varrho \frac{\mathrm{D}\boldsymbol{v}}{\mathrm{D}t} = -\operatorname{grad} p\,, \tag{I.1.7}$$

provided (I.1.1), (I.1.2) and (I.1.6) are assumed. In (I.1.7), the pressure p is defined by the thermodynamic identity

$$T\,\mathrm{d}s = \mathrm{d}e + p\,\mathrm{d}(1/\varrho)\,, \tag{I.1.8}$$

where T is the absolute temperature.

Potential motions. – An important theorem in the hydrodynamics of perfect fluids is the following: a motion will remain irrotational if it is so at a given instant and if the entropy is uniform. Singularities in the form of vortex lines and vortex rings are however permitted in this theory. As we shall note in Chapter II, such singularities cannot exist when irreversible processes are considered.

2. – Rotating co-ordinates.

It is well known in particle dynamics that, in terms of quantities relative to a rotating system, with a prescribed angular velocity $\boldsymbol{\omega}$, the acceleration of a particle is given by

$$\boldsymbol{a}_0 = \frac{\mathrm{d}\boldsymbol{v}}{\mathrm{d}t} + 2\,[\boldsymbol{\omega}\times\boldsymbol{v}] + \boldsymbol{\omega}\times[\boldsymbol{\omega}\times\boldsymbol{r}]\,. \tag{I.2.1}$$

The same principle finds its application in hydrodynamics. It is interesting to note that the centrifugal term can be expressed in terms of a gradient. Specifically, if we take the z-axis to coincide with the axis of $\boldsymbol{\omega}$, then

$$\boldsymbol{\omega}\times[\boldsymbol{\omega}\times\boldsymbol{r}] = -\omega^2\times\boldsymbol{i} - \omega^2 y\boldsymbol{j} = -\nabla\left[\frac{\omega^2 r^2}{2} - \frac{(\boldsymbol{\omega}\cdot\boldsymbol{r})^2}{2}\right], \tag{I.2.2}$$

(*) In eq. (I.1.5a), grad $\boldsymbol{X}\cdot\boldsymbol{\gamma}$ has components $(\partial X_k/\partial x_i)\gamma_k$.

and this must hold in general. The Coriolis term is expected to exhibit other interesting properties.

One of the first attempts to study the effect of rotation was made by LORD KELVIN (*) in 1879. He showed that the shallow water waves of small amplitudes are now governed by equations

$$\frac{\partial u}{\partial t} - 2\omega v = -g\frac{\partial \zeta}{\partial x} - \frac{\partial \Omega}{\partial x}, \tag{I.2.3}$$

$$\frac{\partial v}{\partial t} + 2\omega u = -g\frac{\partial \zeta}{\partial y} - \frac{\partial \Omega}{\partial y}, \tag{I.2.4}$$

where ζ is the elevation of water above the equilibrium height

$$Z_0 = \frac{1}{2}\frac{\omega^2}{g}(x^2 + y^2), \tag{I.2.5}$$

and Ω is the potential of some disturbing force (*e.g.*, the attraction of the moon). The equation of continuity is

$$\frac{\partial \zeta}{\partial t} + \frac{\partial}{\partial x}(uh) + \frac{\partial}{\partial y}(vh) = 0, \tag{I.2.6}$$

where h denotes the depth from the free surface to the bottom, in the undisturbed condition. If the bottom has the same form as (I.2.5), the depth is uniform. In this case, with Ω absent, we have a set of equations with constant coefficients. If the solutions have periodic dependence $\exp[i\sigma t]$, then

$$\left\{\begin{aligned} u &= \frac{g}{\sigma^2 - 4\omega^2}\left(i\sigma\frac{\partial}{\partial x} + 2\omega\frac{\partial}{\partial y}\right)\zeta, \\ v &= \frac{g}{\sigma^2 - 4\omega^2}\left(i\sigma\frac{\partial}{\partial y} - 2\omega\frac{\partial}{\partial x}\right)\zeta, \\ &\frac{\partial^2\zeta}{\partial x^2} + \frac{\partial^2\zeta}{\partial y^2} + \frac{\sigma^2 - 4\omega^2}{gh}\zeta = 0. \end{aligned}\right. \tag{I.2.7}$$

It is clear that there is «resonance» at $\sigma = 2\omega$, and that the character of the solution changes as σ goes through 2ω. This phenomenon will be seen to be relevant to experiments on rotating helium (see Section **3** of Part IV).

(*) Sir W. THOMSON: *On gravitational oscillations of rotating water* [2]. See LAMB [3], p. 317.

3. – Momentum and impulse.

We include here a discussion of the momentum of irrotational fluid motions because the question frequently comes up in connection with the motion of a « quantized vortex ring ».

Let us first observe that for any kind of continuous motion in a long cylindrical tube, the total momentum of the fluid motion along the tube is zero, provided there is no net flux. This is a simple consequence of the equation of continuity. If the cylinder is along the x-axis, then

$$\iint u \,\mathrm{d}y\,\mathrm{d}z = \text{const} = 0\,,$$

and consequently

$$M_x = \iiint \varrho u \,\mathrm{d}x\,\mathrm{d}y\,\mathrm{d}z = 0\,.$$

The same is true if the motion is caused by the translation of a sphere along the axis. Indeed, the symmetry of the pattern of the flow would verify this conclusion (there is no question here of the proper convergence of the integrals, as there would be in the case of infinite space). However, LORD KELVIN has introduced the concept of impulse, which behaves in many ways like the momentum of the fluid. It must be remembered, however, that the momentum is strictly zero. The impulse is that required of the external forces to produce the motion, starting from a state of rest. The momentum created by this external force is actually not retained in the fluid but propagated to infinity. With this subtlety involved in the concept of the impulse, one must exercise care in applying it to complex hydrodynamic situations. In the Appendix to this chapter, we discuss the problem in some detail with the intention that it might help the user in forming his judgment.

4. – Bound vortices and free vortices.

The prevailing theory of quantized vortex lines in the theory of superfluid helium will be discussed in Parts V and VI. It might be well to recall some aspects of the classical theory of the motion of rectilinear vortices here.

KIRCHHOFF has shown that the motion of a system of isolated rectilinear vortices with circulations $\varkappa_1, \varkappa_2, \ldots, \varkappa_N$ and located at $(x_1, y_1), (x_2, y_2), \ldots, (x_N, y_N)$

respectively is governed by the set of equations

$$\varkappa_{(i)} \frac{dx_i}{dt} = + \frac{\partial W}{\partial y_i}, \tag{I.4.1}$$

$$\varkappa_{(i)} \frac{dy_i}{dt} = - \frac{\partial W}{\partial x_i}; \qquad i = 1, 2, \dots, N, \tag{I.4.2}$$

where

$$W = -\frac{1}{2\pi} \sum_{i \neq j} \varkappa_i \varkappa_j \log r_{ij} . \tag{I.4.3}$$

A generalization of this theory to vortices placed in arbitrary cylindrical containers has been developed by LIN [4, 5]. The same formulae (I.4.1) and (I.4.2) are found to hold, but the expression for (I.4.3) is more complicated.

In this theory, the vortices are « *free* » in the sense that they move freely under the influence of *other* vortices with no forces of reaction involved in the process. The vortex line is simply a singularity of the fluid motion and it moves with the velocity of the fluid at the vortex, with the circulatory motion due to the vortex itself substracted. In contrast to this, if we have a fine wire placed in the location of a vortex so as to remove the singularity of the fluid motion, we obtain an irrotational motion with circulation around the wire. The vortex is now « *bound* » to the wire and may be held fixed. In this case, there is a Magnus effect with a cross-wind force given by the Joukowsky formula

$$\boldsymbol{f} = \varrho \boldsymbol{V}' \times \boldsymbol{\varkappa} \tag{I.4.4}$$

per unit length of the cylinder, where $\boldsymbol{V}'$ is the fluid velocity past the wire. Indeed, it is the velocity at which the vortex *would* be moving, had it been *free*.

The significance of this distinction will become clear when we discuss the possible effect of quantized vortex lines in Part VI.

APPENDIX

Momentum and impulse.

One of the subtle concepts in classical hydrodynamic theory is the impulse of the fluid motion introduced by Lord KELVIN. We include a brief discussion of this here, since it is often needed in connection with the motion of vortex rings.

The treatment given by LAMB [3], (Sections 119, 152) should be consulted, but we wish to bring out some points not clearly discussed by him. In particular, we wish to bring into discussion any possible effect of an enclosure such as a straight cylindrical tube.

To begin with, let us consider the classical problem of the motion of a sphere described by the velocity potential

$$\Phi = -\frac{Ua^3}{2r^2}\cos\theta , \tag{I.A.1}$$

where the conventional notation is used. The x-component of the velocity is

$$\Phi_x = -\frac{Ua^3}{2r^3}(1 - 3\cos^2\theta) ,$$

and it is clear that the momentum of the fluid in any concentric spherical shell is zero; for

$$\int_0^\pi (1 - 3\cos^2\theta)\sin\theta \, d\theta = 0 .$$

However, we also know that to accelerate the sphere instantaneously to the velocity U, an impulse of the amount

$$\int F_x \, dt = \frac{1}{2}\left(\frac{4\pi}{3}\varrho a^3\right) U , \tag{I.A.2}$$

is needed. What happened to the momentum produced by this force?

The answer to this question is not difficult to provide, if we examine the total momentum transfer through each spherical surface. Take an arbitrary surface of radius b. The rate of transfer is composed of two parts: (a) the momentum carried by mass flux; (b) the effect of the pressure force. It is easy to see that (a) gives a zero net effect by fore-aft symmetry. The effect of (b) is

$$F_x = \int_0^\pi p\cos\theta \cdot 2\pi b^2 \sin\theta \, d\theta .$$

The total transfer of momentum due to an impulsive action is

$$\int F_x \, dt = \int_0^\pi \{p \, dt\} \cdot 2\pi b^2 \sin\theta\cos\theta \, d\theta .$$

Now, by Bernoulli's equation

$$\frac{p}{\varrho} = -\frac{\partial\Phi}{\partial t} - \frac{1}{2}q^2 + F(t) .$$

Hence, the impulse of pressure is

(I.A.3) $$\tilde{\omega}=\int p\,\mathrm{d}t=-\varrho\Phi\,.$$

By using (I.A.1), this immediately gives (I.A.2). Thus, *the impulse produced by the external force is immediately transmitted through each spherical surface to infinity; no momentum is retained by any spherical region of the fluid.*

In general, if we associate Φ with the impulsive pressure, the momentum of fluid enclosed between an inner surface S_i and an outer surface S_o is given by

(I.A.4) $$\varrho\int\nabla\Phi\,\mathrm{d}\tau=\int_{S_i}\tilde{\omega}\boldsymbol{n}\,\mathrm{d}S-\int_{S_o}\tilde{\omega}\boldsymbol{n}\,\mathrm{d}S\,,$$

stating that the momentum generated by the pressure impulse at the inner surface S_i is partly transmitted through the outer surface S_o and partly retained in the fluid in between.

We shall now turn to the discussion of the momentum and impulse of a fluid motion involving a vortex system of finite dimensions in an incompressible fluid which fills an infinitely long cylindrical tube (Fig. 1). We assume that the vorticity lines close upon themselves and do not terminate on the wall. We shall show that the method of Kelvin (LAMB [3], p. 214) can be easily adapted to this case.

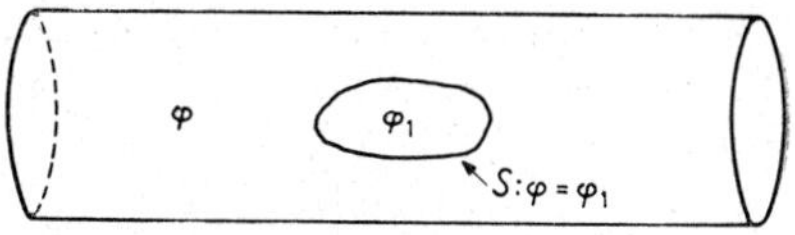

Fig. 1. – Diagram showing two regions of potential flow.

First of all, we notice that *the total x-momentum must be zero.* For

$$\varrho\iiint u\,\mathrm{d}x\,\mathrm{d}y\,\mathrm{d}z=\varrho\int\mathrm{d}x\iint u\,\mathrm{d}y\,\mathrm{d}z\,,$$

and, by continuity

(I.A.5) $$\iint u\,\mathrm{d}y\,\mathrm{d}z=\text{const}\,.$$

If there are no sources, as it is the case in the present problem, then this constant value of flux must be zero, by examining the conditions at infinity. However, the impulse that is required to create the motion from rest is not necessarily zero. Indeed, if we consider the surface S enclosing the vortex system in a simply connected region and denote by φ the single-valued velocity potential which obtains outside of S, then the impulse required is given by

(I.A.6) $$I_x=(\tilde{\omega}_2-\tilde{\omega}_1)A\,,$$

where A is the section area and

$$\tilde{\omega}_2 = \lim_{x\to+\infty} (-\varrho\Phi)\,, \qquad \tilde{\omega}_1 = \lim_{x\to-\infty} (-\varrho\Phi)\,.$$

for the right-hand side of (6) represents the impulse transmitted to infinity.

If we adopt the method described in LAMB [3] (pp. 214-215), we find that (as the following derivation shows)

$$I_x = \frac{1}{2}\varrho\iiint(y\zeta - z\eta)\,d\tau\,, \tag{I.A.7}$$

where (ξ, η, ζ) denotes the components of the vorticity (curl of velocity). It is noteworthy that this formula is *independent* of the presence of the solid boundary, as it is derived by calculations with S alone (*). For a vortex ring symmetrically placed with respect to the tube, we find that

$$I_x = \varkappa\varrho A_R\,, \tag{I.A.8}$$

where $\varkappa$ is the circulation around the filament, and A_R is the area of the vortex ring (**).

The formula (I.A.8) can be checked against (I.A.6). If we consider the case where the vortex ring nearly touches the tube, the flow may be described by placing an image ring outside of the tube. The flow speed at the axis would be effectively zero and the circulation is

$$\varkappa = \oint \boldsymbol{q}\cdot d\boldsymbol{s} = \int_{+\infty}^{-\infty} \boldsymbol{q}\cdot d\boldsymbol{s}\,, \quad \text{along the } \textit{wall}.$$

Thus, (cf. Fig. 1)

$$\varrho\varkappa = \varrho[\Phi(-\infty) - \Phi(+\infty)] = \tilde{\omega}_2 - \tilde{\omega}_1\,.$$

The derivation of (I.A.7) runs as follows: consider the continuation of the potential φ *into* the region S, not analytically but by a *regular* harmonic function *continuous* with φ on S. Denote this function by φ_1.

We now consider a motion produced by the impulsive force J_i per unit mass such that

$$\begin{cases} J_i = u_i - \dfrac{\partial\Phi_1}{\partial x_i}\,, & \text{inside } S, \\ J_i = 0\,, & \text{outside } S, \end{cases} \tag{I.A.9}$$

(*) On the other hand, if we were dealing with the motion of a sphere through a tube, the result would not be the same as (I.A.2). For the « vortex sheet » created at the surface of the sphere would be altered.

(**) Cf. the equivalent magnetic shell of a loop carrying an electric current.

Then, the original flow would be reproduced. For the original flow is compatible with (I.A.9), and the solution should be unique. On the surface S (approaching from inside),

$$J_i = u_i - \frac{\partial \Phi_1}{\partial x_i} = \frac{\partial \Phi}{\partial x_i} - \frac{\partial \Phi_1}{\partial x_i} = J n_i \, ,$$

since $\Phi - \Phi_1 = 0$ or S.

Consider now the integral

$$K_i = \int \varepsilon_{ijk} x_j \omega_k \, \mathrm{d}\tau \, , \tag{I.A.10}$$

over the volume enclosed by S. Then, since

$$\omega_k = \varepsilon_{kpq} \frac{\partial u_q}{\partial x_p} = \varepsilon_{kpq} \frac{\partial J_q}{\partial x_p} \, ,$$

we have

$$K_i = \varepsilon_{kij} \varepsilon_{kpq} \int x_i \frac{\partial J_q}{\partial x_p} \mathrm{d}\tau = \varepsilon_{kij} \varepsilon_{kpq} \int \left[\frac{\partial}{\partial x_p} (x_j J_q) - J_q \delta_{jp} \right] \mathrm{d}\tau =$$

$$= \varepsilon_{kij} \varepsilon_{kpq} \int x_j J_q n_p \, \mathrm{d}S - (\delta_{ip}\delta_{jq} - \delta_{iq}\delta_{jp}) \delta_{jp} \int J_q \, \mathrm{d}\tau \, .$$

Since $J_q = J n_q$, the first term vanishes. Thus,

$$K_i = 2 \int J_i \, \mathrm{d}\tau \, , \tag{I.A.11}$$

which is indeed (I.A.7).

To summarize, *for motions associated with vortices in a long tube, the momentum is always identically zero. But to create such a motion from rest, external forces must be applied to supply a certain amount of momentum to be transmitted to infinity during the transition to the final state from the state of rest.*

II. - Hydrodynamics of Ordinary Fluids. Irreversible Processes.

1. – Navier-Stokes equations.

The principal irreversible processes that occur in fluid motion are *a*) internal friction and *b*) thermal conduction. In the case of homogeneous incompressible fluids, (satisfying an additional restriction to be specified later), the study of fluid motion can be made independently of the equation for thermal energy. For sake of brevity, we shall limit our discussion to such cases. It is important to remember that the internal friction will dissipate mechanical energy into heat, so that the system is not conservative.

Internal friction in many fluids obeys Newton's law of friction

$$\tau = \eta \frac{\mathrm{d}U}{\mathrm{d}y}, \tag{II.1.1}$$

where η is the viscosity coefficient, and the case under discussion is one in which the flow velocity is only in the x-direction and depends on y only. In general, the velocity gradient can be decomposed in terms of a rate of strain and a rotation:

$$\frac{\partial v_i}{\partial x_j} = e_{ij} + \omega_{ij}, \tag{II.1.2}$$

where

$$e_{ij} = \frac{1}{2}\left(\frac{\partial v_i}{\partial x_j} + \frac{\partial v_j}{\partial x_i}\right) \tag{II.1.3}$$

and

$$\omega_{ij} = \frac{1}{2}\left(\frac{\partial v_i}{\partial x_j} - \frac{\partial v_j}{\partial x_i}\right). \tag{II.1.4}$$

The stress tensor associated with internal friction is *assumed*, as a generalization of (II.1.1), as dependent *linearly* on e_{ij}. By a purely mathematical reasoning we can then show that the components of stress tensor are given by

$$\tau_{ij} = 2\eta e_{ij} \tag{II.1.5}$$

for an isotropic incompressible fluid ($e_{kk} = 0$).

The equations of motion for a viscous fluid are then obtained from those for a perfect fluid by the inclusion of the body force (per unit volume)

$$\frac{\partial \tau_{ij}}{\partial x_j} = \eta \Delta v_i. \tag{II.1.6}$$

Here, we have again used the equation of continuity for an incompressible fluid:

$$\frac{\partial v_k}{\partial x_k} = 0. \tag{II.1.7}$$

This equation, together with the Navier-Stokes equation

$$\frac{\mathrm{D}v_i}{\mathrm{D}t} = -\frac{1}{\varrho}\frac{\partial p}{\partial x_i} + \nu \nabla^2 v_i, \qquad \nu = \eta/\varrho, \tag{II.1.8}$$

constitute the basic equations for the four dynamical variables v_i $(i=1,2,3)$ and p. The coefficient of viscosity η has been assumed to be a constant (*). In vector notation, we may write (II.1.7) and (II.1.8) in the form

$$\nabla\cdot\boldsymbol{v}=0\,, \tag{II.1.7$'$}$$

and

$$\frac{\mathrm{D}\boldsymbol{v}}{\mathrm{D}t}=-\frac{1}{\varrho}\nabla p+\nu\nabla^2\boldsymbol{v}\,. \tag{II.1.8$'$}$$

An alternative form of (8′) is

$$\frac{\mathrm{D}\boldsymbol{v}}{\mathrm{D}t}=-\frac{1}{\varrho}\nabla p+\nu\nabla\times[\nabla\times\boldsymbol{v}]\,. \tag{II.1.9}$$

In this form, it can be seen that *the body forces due to viscous friction vanish if the motion is irrotational.* This has often led to the erroneous conclusion that *viscous dissipation is absent in irrotational motion.* We shall explain why this is incorrect in Section **3**.

The vorticity equation. – If we take half of the curl of (II. 1.9), we obtain the vorticity equation

$$\frac{\mathrm{D}\boldsymbol{\omega}}{\mathrm{D}t}=(\boldsymbol{\omega}\cdot\nabla)\boldsymbol{v}+\nu\nabla^2\boldsymbol{\omega}\,. \tag{II.1.10}$$

The first term on the right represents the generation of vorticity by stretching of the vortex lines, while the second term represents the diffusion of vorticity. In two-dimensional motions, only the second term is effective. Thus, a two-dimensional vortex gradually diffuses. *The idealized vortex motion consisting of a potential motion with circulation around a singularity cannot exist in a viscous fluid.*

2. – Boundary conditions.

In the case of perfect fluids, the condition at a solid boundary is that the fluid cannot penetrate the wall. Thus, if V_i is the velocity of the wall, the relative velocity

$$w_i=v_i-V_i \tag{II.2.1}$$

(*) Otherwise, temperature distribution becomes important in the determination of the flow field.

must not have a normal component:

(II.2.2) $$w_i n_i = 0 \, .$$

This condition must continue to hold in the present case. However, additional boundary conditions must now hold for the tangential components of velocity

(II.2.3) $$w_\alpha = w_i t^i_\alpha \qquad (\alpha = 1, 2),$$

where t^i_α is the tangent vector (with two components $\alpha = 1, 2$ in the tangent plane). Mathematically, this is dictated by the fact that the equations of motion are of higher orders in the spatial derivatives. Physically, this is suggested by the fact that a spinning cylinder can bring the air around it into motion. Clearly, the viscous stresses at the boundary must enter the formulation of the boundary conditions as well. This has components

(II.2.4) $$T_\alpha = (\tau_{ij} n_j) t^i_\alpha \, .$$

From the invariance of physical laws with respect to co-ordinate transformations, we should then have

(II.2.5) $$T_\alpha = F(w^2) w_\alpha \, , \qquad w^2 = w_\alpha w^\alpha \, .$$

The simplest assumption is that $F =$ constant. This was the condition suggested by NAVIER, who actually derived it from a certain molecular model (1816, 1823), and used it to account for Girard's experiment for flow of liquid through tubes (1813, 1814, 1815). This slip layer has a thickness $\delta = 0(\eta/F)$. Stokes was inclined to the condition that the fluid should stick to the boundary, *i.e.*,

(II.2.6) $$w_\alpha = 0 \, ,$$

which is a special case of (5) when $F \to \infty$. However, the experiments known to Stokes could not bear (II.2.6) out, and he hesitated (1846) between this hypothesis and Navier's, even though the experimental results of HAGAN (1839) and POISEUILLE (1841) were already available at the time. A few years later (1851), STOKES came to the conclusion that the agreement with observation of results obtained on the assumption of no slip was highly satisfactory; and the hypothesis gradually gained ground. Even then, HELMHOLTZ concluded (1860) that there was a considerable slip for water in contact with a gilt surface, though not with glass.

MAXWELL (1879) concluded from kinetic theory of gases, that the slip at the surface was given by Navier's formula, but the thickness of the slip layer $\delta = \eta/F$ is of the order of a few mean free-paths. Thus, we finally have the modern version of our knowledge, that the slip is negligible in ordinary cases, but must be considered under exceptional circumstances. For example, in the case of rarefied gases, the Navier-Stokes equations still hold in the « slip-flow » range, but the boundary conditions take on the form given by NAVIER rather than that given by STOKES. The period of this investigation, which consists of both theoretical and experimental parts, lasted over quite a long period (1816-1879).

The above historical account is based on the article by S. GOLDSTEIN ([6], II, pp. 676-680). The reader is referred to the original article for further details and the exact references to the literature.

3. – The rate of energy dissipation.

The viscous stresses represent irreversible processes which dissipate mechanical energy into heat. The amount of dissipation may be calculated as follows: Consider any surface enclosing a given mass of fluid. The viscous surface stress $\tau_{ij}n_j\,\mathrm{d}S$ is doing work at the rate of $u_i\tau_{ij}n_j\,\mathrm{d}S$ over the surface element $\mathrm{d}S$. For the whole surface, the rate of work done is

$$\int u_i\tau_{ij}n_j\,\mathrm{d}S = \int \frac{\partial}{\partial x_j}(u_i\tau_{ij}))\,\mathrm{d}V, \tag{II.3.1}$$

showing that the rate of work done per unit volume is

$$\frac{\partial}{\partial x_j}(u_i\tau_{ij}) = u_i\frac{\partial \tau_{ij}}{\partial x_j} + \frac{\partial u_i}{\partial x_j}\tau_{ij} = u_i\frac{\partial \tau_{ij}}{\partial x_j} + \tau_{ij}e_{ij}\,. \tag{II.3.2}$$

The first term (cf. (II.1.6) and (II.1.8)) goes to the increase of the kinetic energy while the second term gives the rate of dissipation per unit volume per second:

$$D = \tau_{ij}e_{ij} = 2\eta\, e_{ij}e_{ij} = 2\eta\left(\omega_{ij}\omega_{ij} + \frac{\partial u_i}{\partial x_j}\frac{\partial u_j}{\partial x_i}\right). \tag{II.3.3}$$

In the case of irrotational motions $\omega_{ij} = 0$, we have

$$D = 2\eta\,\frac{\partial u_i}{\partial x_j}\frac{\partial u_j}{\partial x_i} = 2\eta\left(\frac{\partial u_i}{\partial x_j}\right)^2. \tag{II.3.4}$$

Thus, although there is no net body force from viscous friction in the case of irrotational motions, the rate of energy dissipation is not zero.

As an example, consider the flow between rotating cylinders. If the inner cylinder (radius R_1) is rotating at an angular speed Ω_1 and the outer cylinder (radius R_2) has an angular speed Ω_2, the solution is given by

$$v = Ar + B/r\,, \tag{II.3.5}$$

where

$$\begin{cases} A = (\Omega_2 R_2^2 - \Omega_1 R_1^2)/(R_2^2 - R_1^2)\,, \\ B = -(\Omega_2 - \Omega_1) R_2^2 R_1^2/(R_2^2 - R_1^2)\,. \end{cases} \tag{II.3.6}$$

If $A = 0$, *i.e.*, if the circulation along the two cylinders is the same, then the motion is given by

$$v = B/r\,, \tag{II.3.7}$$

which is an irrotational flow due to a hypothetical vortex. This is also the case if $\Omega_2 = 0$ and $R_2 \to \infty$, *i.e.*, if the inner cylinder is rotating in a viscous fluid of infinite extent. The shear stress is

$$\tau_{r\theta} = \eta\left(\frac{\partial v}{\partial r} - \frac{v}{r}\right) = -2\eta\Omega_1 (R_1/r)^2\,. \tag{II.3.8}$$

The total energy dissipation per unit height of the fluid is

$$\int_{R_1}^{\infty} 4\eta(\Omega_1 R_1^2)^2 \left(\frac{2\pi r\, \mathrm{d}r}{r^4}\right) = 4\pi\eta\Omega_1^2 R_1^2\,. \tag{II.3.9}$$

This is supplied by the work done by the external torque $2\eta\Omega_1 \cdot 2\pi R_1 \cdot R_1$ which has to balance a stress $2\eta\Omega_1$ at the surface of the cylinder (see (II.3.8)). Since the torque is acting on the fluid at the angular velocity Ω_1, it supplies to the fluid exactly the amount of energy required by (II.3.9).

The general case can be discussed by using (II.3.3) or (II.3.4). *When the boundaries are stationary*, one can easily verify, by using (3), that

$$\int D\,\mathrm{d}V = \int 2\eta\omega_{ij}\omega_{ij}\,\mathrm{d}V = 4\eta\int \boldsymbol{\omega}^2\,\mathrm{d}V\,. \tag{II.3.10}$$

There are however no such irrotational motions. Solutions with $e_{ij} = 0$ are also not known, except for the case of uniform rotation. Thus, in general, *the dissipative effect of viscous forces does not vanish.*

4. – Instability of laminar motion; turbulence.

It is well known that laminar flow through a pipe can only be maintained at low Reynolds numbers. At higher Reynolds numbers, the motion is in general turbulent. However, it is also known that if the Reynolds number is reduced to below 10^3, turbulent flow cannot be maintained. Here the Reynolds number is defined by

$$R = \frac{U_m a}{\nu}, \tag{II.4.1}$$

where U_m is the average velocity over a cross-section and a is the radius of the pipe.

For $\nu = 10^{-4}$, and $a = 0.1$ cm, this critical Reynolds number is reached if $U_m = 1$ cm/s.

Flows at higher Reynolds numbers can still be kept in a laminar state, provided extreme care is exercised. Indeed, laminar flows up to Reynolds numbers of the order 10^5 have been reported by various authors. However, these experiments were performed with extreme caution. It is usually necessary to let the water settle for days before the experiment is started (*).

a) Instability with respect to infinitesimal disturbances. Transition to turbulence in a pipe does not seem to be caused by instability of the fully developed Hagen-Poiseuille profile with respect to infinitesimal disturbances. Indeed, such a flow appears to be stable with respect to all infinitesimal disturbances. This is however a rather exceptional case. Most flows are definitely unstable with respect to infinitesimal disturbances at high speeds. The theory for the various types of instability has been fully investigated in recent years (**). The two major types of instability are exemplified by the following two cases, where the basic flows are both solutions of the Navier-Stokes equations:

a) pressure flow through a channel (HEISENBERG, LIN);

b) flow between rotating cylinders (TAYLOR).

In the first case, oscillations develop after the flow becomes unstable; the instability is said to be of the « overstable » type (after EDDINGTON). In the second case, a *secondary flow* develops, and the instability is said to be of the marginal type. Transition to turbulence seems to follow directly the oscillations in the first case; in the second case, a stationary state is first reached,

---

(*) See GOLDSTEIN [6], chap. VII and VIII for further details regarding turbulent flow in pipes, channels and along flat plates. Applications of the classical theories of turbulent transport are to be found there.

(**) See LIN [7]; CHANDRASEKHAR [8].

and instability of the « overstable » type eventually develops as the speed is further increased (*). Transition then ensues.

b) *Oscillations at finite amplitudes.* Transition to turbulent flow does not occur as soon as the amplitude of the oscillations becomes finite. There is a stage of regular finite amplitude oscillations. Then turbulent spots occur. As more and more of these spots develop, the pressure drop along the channel must be correspondingly increased to maintain the same rate of flow. Eventually, the flow becomes fully turbulent.

The study of nonlinear oscillations is one of the main concerns of workers in this field. The reason for this interest is twofold. First, even in fully developed turbulent flow, the *large scale* motions are more in the nature of a continuation of these regular oscillations than random motions satisfying certain universal laws. Secondly, in geophysical problems, the behavior of finite amplitude oscillations in the atmosphere and in the ocean are decidedly of interest.

c) *Turbulent motion* (**). Fully developed turbulent flow has two prominent characteristics. The large scale motions mentioned above are generally responsible for transport of various physical quantities; the small scale motions (with large rate of strain) are chiefly responsible for energy dissipation. As the energy at small scales is destroyed by viscous dissipation, it is being continually replenished by the breakdown of large eddies by a mechanism of stretching of vortex lines (***). Generally speaking, this is similar to the mechanism of ultraviolet catastrophe in the classical theory of electromagnetic radiation. In the present case, there is no quantum effect to cause an equilibrium spectrum to be formed. Rather a quasi-equilibrium is formed according to the theory of Kolmogoroff.

In the Kolmogoroff theory, the total rate of energy dissipation ε (per unit mass per second) is regarded as the fundamental quantity. Thus, the small scale eddies are governed only by ε and the kinetic viscosity ν. From these two quantities, we can form the following scales of length, time, and velocity:

$$\eta_0 = (\nu^3/\varepsilon)^{\frac{1}{4}} , \tag{II.4.2}$$

$$\tau_0 = (\nu/\varepsilon)^{\frac{1}{2}} , \tag{II.4.3}$$

$$v_0 = (\nu\varepsilon)^{\frac{1}{4}} . \tag{II.4.4}$$

(*) The instability of flow through curved channels furnishes a good example for the contrast between instabilities of both types. See LIN [7], p. 48.

(**) For an account of the statistical theory, see article by LIN and REID [9].

(***) Cf. Section II.1.

The spatial spectrum at high wave numbers $\varkappa$ is then of the form

$$E(\varkappa) = v_0^2 \eta_0 f(\varkappa \eta_0) , \tag{II.4.5}$$

where $f(\xi_0)$ is a universale function of ξ_0. In Kolmogoroff's original theory, the small scale eddies are assumed to be isotropic. Lately, there has been some evidence that this need not be true. Some directional characteristics may then have to be introduced as further parameters for describing a specific field.

At the lower end of the universal spectrum, we have

$$E(\varkappa) = C \varepsilon^{\frac{2}{3}} \varkappa^{-\frac{5}{3}} . \tag{II.4.6}$$

For larger scale, there is no universal theory to cover all cases. For the case of decaying isotropic turbulence, HEISENBERG has proposed the following formula to describe the transfer of energy among the various scales:

$$\frac{\partial}{\partial t} \int_0^{\varkappa} E(\varkappa)\, \mathrm{d}\varkappa = -2(\nu_\varkappa + \nu) \int_{\varkappa}^{\infty} \varkappa^2 E(\varkappa)\, \mathrm{d}\varkappa , \tag{II.4.7}$$

where

$$\nu_\varkappa = \int_{\varkappa}^{\infty} \left[\frac{E(\varkappa)}{\varkappa^3} \right]^{\frac{1}{2}} \mathrm{d}\varkappa . \tag{II.4.8}$$

This formula has also been adapted to the study of shear flow.

Another principal idea in the study of turbulence is to ask how closely the theory of Brownian motion can be applied to the phenomena of turbulent motion. It is known that turbulent motion cannot be a completely random process with joint Gaussian distribution for the velocities at various points of the field. Quasi-Gaussian assumptions have been adopted with success. There are still some basic differences between the Kolmogoroff theory and the quasi-Gaussian theory that urgently require reconciliation.

Despite the recent advances in our understanding of the basic mechanisms of turbulent motion, calculations of mean quantities are still made by the use of the older transport theories. This is partly due to the fact that the larger eddies, which determine the transport properties, do not have universal properties, but depend on the boundary conditions to a large extent. Thus, it is difficult to improve upon the crude theories developed earlier. However, it is now possible to give more detailed descriptions in terms of modern concepts. In particular, the problem of turbulent dispersion has now been quite satisfactorily solved by analogy with the theory of Brownian motion. For example, one can now give a theoretical deduction of Richardson's empirical law, which states that the dispersion coefficient associated with the increase of the size of a cloud of particles should itself be increasing as the $\frac{4}{3}$ power of the size. Specifically, if the mean square separation between two particles in the cloud is $\overline{s^2}$, and the dispersion coefficient is defined by

$$\mathscr{D} = \frac{1}{2}\frac{d\overline{s^2}}{dt}, \tag{II.4.9}$$

then one can show theoretically that

$$\mathscr{D} = \left(\frac{3}{2}\right)^{\frac{2}{3}} B^{\frac{1}{3}} \left(\overline{s^2}\right)^{\frac{2}{3}}, \tag{II.4.10}$$

where B may be identified approximately with the rate of energy dissipation ε.

III. - Hydrodynamics of Superfluids. Reversible Processes (*).

1. – General theoretical considerations.

We shall now attempt to develop the continuum theory of the hydrodynamics of liquid helium. The usefulness of such a theory lies in providing a logical structure of a macroscopic hydrothermodynamics which is not specifically dependent on the molecular structure. As we shall see, it can be used to bring out more sharply the basic points that need to be settled by further theoretical and experimental investigations. Specifically, the two-fluid concept introduced in this approach will be somewhat less restrictive than the picture of excitations in a background fluid, with the latter essentially equivalent to a state of the liquid at 0 °K. Of course, the concept must be generally consistent with those developed from the statistical mechanical theory of mole cular structure.

We take this continuum approach, since there does not, as yet, exist a *complete* hydrodynamic theory *derived* from the statistical mechanics of the quasi-equilibrium state of the liquid. The current hydrodynamic theory of liquid helium II is based on the two-fluid concept developed by LONDON,

(*) A more detailed version of this chapter had been distributed in preprint form from the Institute for Advanced Study, Princeton, N. J., April 1960. It will be published elsewhere.

TISZA, and LANDAU (which we shall introduce in a somewhat different manner), and on the Onsager-Feynman theory of quantized vortex lines (which we shall examine critically). In this theory, the supercomponent of the flow must be pointwise irrotational but there can exist line singularities in the fluid around which there is a quantized circulation (cf. Part V). The supercomponent of the fluid can therefore have angular momentum in bulk. Recently, LEE, HUANG, and YANG [10] studied a dilute system of bosons with hardsphere interaction as a model for liquid helium. LEE and YANG [11] have further developed the theory for application at *finite* temperatures. This furnishes a natural connection between the concept of the degenerate phase in the sense of London's work (see LONDON [12]) and the concept of phonon excitation in the sense of Landau's treatment [13]. They have also developed a quasi-equilibrium theory in which Landau's equations of motion are reproduced (with a slight modification). They stated, however, that the question whether the superfluid flow is irrotational was not resolved in their paper (*).

The difficulty is to answer the question: what is superfluid flow? If one adopts the idea of a background fluid plus excitations as the model of He II, and one thinks of a background fluid as the *single* ground state equivalent to the state of the fluid at 0 °K, it is then difficult to see how the fluid can move without excitations being created in it except for a uniform translation. In other words, it is difficult to tell flow (which is hydrodynamical) from internal excitations (which are thermodynamical). It is indeed because of this confusion that in the earlier days the theoretical critical velocity predicted was more than 1000 times the experimental value, because they refer to different physical phenomena (cf. Part V). One should realize that in hydrodynamics we are dealing with extremely low energies. The flow velocity v is of the order of 1 cm/s, whereas the velocity of sound c (second sound) is of the order of 20 m/s. Thus,

$$v^2/c^2 = \left(\frac{1}{2000}\right)^2 = \frac{1}{4}\cdot 10^{-6}\,.$$

There are thus two theories:

a) the theory of fluid structure,

b) the theory of flow.

I shall concentrate my discussions on the second aspect.

In spite of the elegance and appeal of the theory of quantized vortex lines, it is indeed very difficult to establish the irrotationality of the superfluid flow,

(*) A more recent paper by T. T. WU [14] takes up this question again, but also leaves it open in the case of interacting bosons.

and the existence of quantized vortex lines so firmly as to exclude other possibilities. Conclusive theoretical arguments have yet to be given (*). Experimental evidence is often indirect, and does not exclude alternative explanations (such as those we shall present in Part IV). Only VINEN has attempted a direct experimental verification of the existence of single quantization, but the results are as yet inconclusive (**). It is thus desirable to attempt to develop the theory of flow of superfluid helium from the usual continuum approach, and to investigate how far such concepts would carry us. This we shall begin in the next section.

2. – An outline of the perfect fluid theory.

As mentioned above, our aim is to develop a macroscopic theory which would enable us to describe the various observed phenomena by starting out with a minimum of experimental facts, and by using concepts which are compatible with the microscopic theory. In particular, the application of the concept of a classical thermomechanical continuum to liquid helium II should not be accepted uncritically. This point will be taken up in Part V. As far as specific experimental facts are concerned, we shall start out with the single observation that *liquid helium* II *at rest can transmit heat in a reversible manner*. This reversible heat current we associate with a *flux of order* (*disorder*) *in the momentum space* (***).

We shall show that by introducing this extra degree of freedom into the conventional thermohydrodynamics, we can develop a complete theory of perfect fluid flow, which allows the introduction of a two-fluid model as a purely mathematical step. The resultant equations in the two-fluid language will be found to be compatible with the Landau equations (*.*) in the following sense. Our equations permit a class of solutions for which the motion of the super-

---

(*) In the section on « Rotation of the superfluid », FEYNMAN [15] states (p. 40): « We must therefore still consider it conjectural whether the considerations on rotational flow reported here are actually correct ».

(**) Toward the end of his paper, VINEN [16] stated: « But the fact remains that the present measurements do at least *suggest* that the simplest picture of a free vortex line is in some way inadequate, and that the theory behind this picture should be re-examined ».

(***) Historically, as soon as the fountain effect was discovered by ALLEN and JONES [17], the flux of disorder had been suggested as a primary concept by JONES instead of the two-fluid concept, but the latter prevailed. At that time, JONES was probably thinking in terms of order in the configurational space, as suggested by FRÖHLICH. Subsequent X-ray examination of the structure of liquid helium revealed no such configurational order.

(*.*) See LONDON [12], Section **19**.

fluid is irrotational. If then we restrict our interest to this special class of solutions, the equations of motion are identical with those of LANDAU. However, our equations also permit other classes of solutions, and there appears to be no particular basis for restricting the superfluid motion to the irrotational type. This is only natural, because such a restriction must come from quantum-mechanical considerations and should be beyond the scope of purely continuum concepts. We are thus leaving the question of the irrotationality open at the present stage of the theory. In certain experiments, the superfluid component is known to rotate (at least in bulk motion). The present results may then have to be applied. The discussion of such experiments must await the development of the dynamics of a viscous fluid (Chapter IV), where the question of irrotationality of the supercomponent will be taken up more fully.

In the meantime, while we are dealing with phenomena in which irreversible processes are unimportant, and for which the theory of a perfect fluid is therefore applicable, we are in general restricted to irrotational motions; for the fluid generally starts from rest and cannot gain rotation (cf. Section **4**). Thus, the predictions of our theory would in general be identical with those following Landau's equations (*), which are well verified experimentally. Thus, in developing this part of our theory, we need not examine the experimental data. All we need to study would be its compatibility with our concept of the structure of the liquid. This we shall do in Part V.

3. – Fundamental equations. Variational principle.

We shall now proceed to develop a continuum theory of liquid helium II, considering only reversible processes. As mentioned above, we start with the single fact that the fluid is capable of transmitting heat in a reversible manner in the absence of a temperature gradient. To account for this, we introduce the *reversible entropy flux* $\boldsymbol{S}$ *relative to the fluid as a new thermodynamic variable.* The other thermodynamic variables are the mass density ϱ, and the entropy per unit mass s. Alternatively, we shall define a *velocity of internal convection* $\boldsymbol{c}$ by the relation

$$\boldsymbol{S} = \varrho s \boldsymbol{c} \,. \tag{III.3.1}$$

The internal energy of the fluid, being a scalar, can depend on $\boldsymbol{c}$ only through its magnitude. Thus, the thermodynamic state of the system is given by a

(*) Judging from the paper presented at Toronto by BEKAREVICH and KHALATNIKOV [18], it is clear that people are also concerned with the exact form of the equations of motion.

relation of the form

$$e = e(\varrho, s, \zeta)\,, \qquad \zeta = \tfrac{1}{2}\boldsymbol{c}^2 \tag{III.3.2}$$

where e is the internal energy per unit mass.

Next we consider the conservation relations when the liquid is in motion. We adopt the Eulerian description and denote by $\boldsymbol{v}(\boldsymbol{x}, t)$ the velocity field at time t in the Cartesian co-ordinate system $\boldsymbol{x}$. The equation of mass conservation is the well-known one:

$$\frac{\partial \varrho}{\partial t} + \nabla \cdot (\varrho \boldsymbol{v}) = 0\,. \tag{III.3.3}$$

The entropy flux through space is $\varrho s(\boldsymbol{v}+\boldsymbol{c})$, and thus the conservation of thermal energy (for reversible processes) may be written as

$$\frac{\partial(\varrho s)}{\partial t} + \nabla \cdot [\varrho s(\boldsymbol{v}+\boldsymbol{c})] = 0\,. \tag{III.3.4}$$

The equation for the conservation of momentum is more difficult, as we do not know the role of the convection vector $\boldsymbol{c}$. In fact, one needs more than the equation for momentum conservation. There should be one set of equations each governing the behavior of $\boldsymbol{v}$ and $\boldsymbol{c}$ respectively.

To obtain such equations, we shall assume that the Hamiltonian principle is applicable for treating both degrees of frredom. We write down the Lagrangian density function

$$L = \varrho\left[\frac{1}{2}\boldsymbol{v}^2 - e_L(\varrho, s, \zeta) - \Omega\right], \tag{III.3.5}$$

where Ω is the potential energy of an external field, and e_L need not be identical with e of (III.3.2), for reasons explained elsewhere. We then impose the variational relation

$$\delta\int_{t_0}^{t_1}\mathrm{d}t\int_V L\,\mathrm{d}V = 0\,, \tag{III.3.6}$$

subject to suitable subsidiary conditions. Two of these are (III.3.2) and (III.4.3). There is a third one which relates to the fact that the Hamiltonian principle should be applied in the Lagrangian formulation. This requires the introduction of the Lagrangian co-ordinates $\boldsymbol{X}(\boldsymbol{x}, t)$ of the fluid particles. In the Eulerian description, they satisfy the relationship

$$\frac{\mathrm{D}\boldsymbol{X}}{\mathrm{D}t} = \frac{\partial \boldsymbol{X}}{\partial t} + (\boldsymbol{v}\cdot\nabla)\boldsymbol{X} = 0 \,. \tag{III.3.7}$$

It may be recalled that in the application of the Hamiltonian principle in particle mechanics, there are constraints at the end points (t_0, t_1) of the interval of time. It is through these constraints that the Lagrangian co-ordinates $\boldsymbol{X}$ enter the variational relation (III.3.6). A fuller discussion of this point will be made elsewhere.

The variational principle yields the equations for $\boldsymbol{v}$ and $\boldsymbol{c}$. After some calculation, one obtains the following equation of motion

$$\frac{\mathrm{D}\boldsymbol{v}}{\mathrm{D}t} = -\nabla\Omega + \frac{1}{\varrho}\nabla\cdot\tau \,, \tag{III.3.8}$$

where τ is a symmetrical stress tensor given by

$$\tau = -\pi\boldsymbol{I} + z\varrho\boldsymbol{cc} \,, \tag{III.3.9}$$

and π and z are defined by

$$\mathrm{d}e_L = \theta\,\mathrm{d}s + (\pi/\varrho^2)\,\mathrm{d}\varrho + z\,\mathrm{d}\zeta \,, \tag{II.3.10}$$

Thus, *the reversible heat flux gives rise to a momentum transfer*, besides the scalar stress π (which is to be identified with pressure). The form of the additional terms calls to mind the Reynolds stress in the theory of turbulence. The coefficient z is a kind of correlation coefficient, and the turbulence analogy would suggest that it should be negative, as we shall find it to be indeed the case after detailed investigations.

The equation governing the internal convection vector $\boldsymbol{c}$ is most conveniently left in the form (*)

$$\frac{\mathrm{D}\beta}{\mathrm{D}t} = \theta - (2z\zeta)/s \,, \quad \boldsymbol{c} = (s/z)\nabla\beta \,. \tag{III.3.11}$$

Equations (III.3.3), (III.3.4), (III.3.8), (III.3.11) constitute the fundamental equations of the theory of perfect fluids. They govern the behavior of the quantities ϱ, s, $\boldsymbol{c}$ and $\boldsymbol{v}$.

(*) This equation shows that the heat flux S is not subject to Coriolis effect.

4. – Sound waves and thermal waves.

Let us now consider approximate solutions for which both $\boldsymbol{v}$ and $\boldsymbol{c}$ are small, so that the quadratic terms can be neglected in the above equations. One notices immediately that the term $z\varrho\boldsymbol{cc}$ should now be neglected from the stress tensor (III.3.9) and that thermal convection can influence the dynamical motion only in an indirect manner, through its thermal effects. The eqs. (III.3.3), (III.3.4), (III.3.8) and (III.3.11) are now approximately given by

(III.4.1) $$\frac{1}{\varrho_0}\frac{\partial\varrho}{\partial t}+(\nabla\cdot\boldsymbol{v})=0\,,$$

(III.4.2) $$\frac{1}{s_0}\frac{\partial s}{\partial t}+(\nabla\cdot\boldsymbol{c})=0\,,$$

(III.4.3) $$\frac{\partial\boldsymbol{v}}{\partial t}=-\nabla\Omega+\frac{1}{\varrho_0}\nabla\pi\,,$$

(III.4.4) $$(z_0/s_0)\frac{\partial\boldsymbol{c}}{\partial t}=\nabla\theta\,,$$

where π and θ are functions of ϱ and s. Now for very low temperatures,

$$\left(\frac{\partial s}{\partial\pi}\right)_\theta\to 0\,,$$

and hence

$$\left(\frac{\partial\varrho}{\partial\theta}\right)_\pi=\varrho^2\left(\frac{\partial s}{\partial\pi}\right)_\theta=0\,.$$

If this approximation is adopted, namely,

(II.4.5) $$s=s(\theta)\qquad\text{and}\qquad\varrho=\varrho(\pi)\,,$$

the systems (III.4.1)–(III.4.4) can be separated into the two sets (III.4.1), (III.4.3) and (III.4.2), (III.4.4). The first set gives a compression wave with a velocity u_I given by

(III.4.6) $$u_\text{I}^2=\left(\frac{\partial\pi}{\partial\varrho}\right)_s,$$

and the second set gives a thermal wave with velocity u_II given by

(III.4.7) $$u_\text{II}^2=-\frac{s_0^2}{z_0}\left(\frac{\partial\theta}{\partial s}\right)_\varrho.$$

These are the waves discussed by TISZA before they were experimentally discovered. If we compare (III.4.7) with the usual formula for velocity of the second sound (*), we see that

$$z = -\varrho_n/\varrho_s\,, \tag{III.4.8}$$

where ϱ_n and ϱ_s are respectively the densities of the normal and the supercomponents. Their meaning in the present theory will be clarified in Section **5**.

5. – The two-fluid model.

Current macroscopic theories of liquid helium are formulated in terms of a two-fluid concept, which was initiated by LONDON [19] and TISZA [20] on the basis of the phenomenon of Bose-Einstein condensation. LANDAU developed a two-fluid theory from a different approach and obtained the nonlinear thermohydrodynamic equations. These equations were given a variational formulation by ZILSEL.

It has been shown in various ways that the two approaches to the two-fluid theory are equivalent. Recently, LEE and YANG [11, 21] showed that the two approaches to the two-fluid theory are equivalent by giving the detailed microscopic picture of a dilute Bose system of hard spheres. It has also been repeatedly emphasized that the two-fluid concept must be understood from a formalistic point of view. There are two modes of motions, each of which is associated with its own « effective mass » such that the sum of both of these masses is equal to the total true mass of the liquid.

In our present formulation, this point is clearly presented from the macroscopic point of view. We have yet to introduce the two-fluid concept, although we have been able to derive Tisza's theory of thermal waves. We shall now introduce this concept by a purely mathematical process, and show that the equations of Section **3** agree with Landau's equations, *provided* the superfluid component can move only in an irrotational manner. It is not a part of our theory (as it is with LANDAU) that the superfluid component *must* be irrotational. This is one of the central problems in the hydrodynamics of superfluid helium, and it will be taken up in detail in Part V.

We shall now consider a two-component picture of liquid helium: a normal component with density ϱ_n and entropy s_n per unit mass of the normal component and a supercomponent with density ϱ_s and entropy s_s per unit mass of the supercomponent. Each component has its own velocity field $\boldsymbol{v}_n$ and $\boldsymbol{v}_s$ respectively. This model is to be connected with our above formulation by

(*) See LONDON [12], eq. (11), p. 84.

the identification of the densities of mass, entropy, mass flux and entropy flux. Thus, we have

(III.5.1) $$\varrho = \varrho_n + \varrho_s ,$$

(III.5.2) $$\varrho s = \varrho_n s_n + \varrho_s s_s ,$$

(III.5.3) $$\varrho \boldsymbol{v} = \varrho_n \boldsymbol{v}_n + \varrho_s \boldsymbol{v}_s ,$$

(III.5.4) $$\varrho s(\boldsymbol{v} + \boldsymbol{c}) = \varrho_n s_n \boldsymbol{v}_n + \varrho_s s_s \boldsymbol{v}_s .$$

It is customary (though not necessary at this stage) to assume that $s_s = 0$.

With the two-fluid model described above, one can rewrite the basic equations of Section **1** in another form. The equations for the conservation of mass and entropy read

(III.5.5) $$\frac{\partial \varrho}{\partial t} + \nabla \cdot \varrho[x\boldsymbol{v}_n + (1-x)\boldsymbol{v}_s] = 0 ,$$

(III.5.6) $$\frac{\partial(\varrho s)}{\partial t} + \nabla \cdot [\varrho s \boldsymbol{v}_n] = 0 .$$

The derivation of the equations for $\boldsymbol{v}_n$ and $\boldsymbol{v}_s$ is lengthy; but they may be put into the final form:

(III.5.7) $$\frac{D_n \boldsymbol{v}_n}{Dt} = -\frac{1}{\varrho}\nabla\pi - \frac{1-x}{x} s\nabla\theta - \nabla\Phi - \frac{1-x}{2}\nabla q^2 - (1-x)\boldsymbol{q}\times[\nabla\times\boldsymbol{v}_s] ,$$

[III.5.8) $$\frac{D_s \boldsymbol{v}_s}{Dt} = -\frac{1}{\varrho}\nabla\pi + s\nabla\theta - \nabla\Phi + \frac{x}{2}\nabla \boldsymbol{q}^2 + x\boldsymbol{q}\times[\nabla\times\boldsymbol{v}_s] .$$

Equation (III.5.8) can also be written as

(III.5.9) $$\frac{\partial \boldsymbol{v}_s}{\partial t} = -\nabla H + \boldsymbol{v}\times[\nabla\times\boldsymbol{v}_s] ,$$

where

(III.5.10) $$H = g + \Omega + \frac{1}{2}\boldsymbol{v}_s^2 - \frac{x}{2}\boldsymbol{q}^2 , \qquad g = e + \pi/\varrho - Ts .$$

Thus, *the equations permit the class of solutions satisfying the condition*

(III.5.11) $$\nabla\times\boldsymbol{v}_s = 0 .$$

For this class, (III.5.7) and (III.5.8) read

$$\frac{\mathrm{D_n}\boldsymbol{v}_n}{\mathrm{D}t} = -\frac{1}{\varrho}\nabla p - \frac{1-x}{x} s\nabla T - \nabla\Omega - \frac{1-x}{2}\nabla \boldsymbol{q}^2 , \tag{III.5.12}$$

$$\frac{\mathrm{D_s}\boldsymbol{v}_s}{\mathrm{D}t} = -\frac{1}{\varrho}\nabla p + s\nabla T - \nabla\Omega + \frac{x}{2}\nabla \boldsymbol{q}^2 . \tag{III.5.13}$$

Equations (III5.5), (III.5.6), (III.5.11), (III.5.12) and (III.5.13) are the set of equations given by LANDAU.

The difference between the present theory and Landau's theory is now obvious. In Landau's theory, (III.5.11) is an independent *physical* requirement; here it merely characterizes a possible class of solutions. It is interesting to note that in Zilsel's [22, 23] variation formulation, the condition (III.5.11) has been automatically (though unintentionally) incorporated, because of a defect in the variational principle used. It can be shown that if one had introduced the constraint (III.3.7) into Zilsel's derivation, the equations obtained would have been identical with the present ones, without the restriction (III.5.11).

A brief word should be added here regarding the experimental evidence of the irrotationality of the supercomponent, even though we shall take it up in detail in the next two chapters. As is well known, Landau's condition of irrotationality was verified by the experiments of ANDRONIKASHVILI' who found that only the normal component is moving with an oscillating disc system. In the framework of present theory, this is accountable, as we shall see in Part IV by allowing a slip at the boundary in the systems of basic equations for the motion of a viscous field. Thus, we agree to the fact that the superfluid is indeed not rotating. The reason is, however, not that it cannot rotate, but rather that it does not rotate under the conditions specified (cf. ATKINS [24], p. 111).

IV. - Hydrodynamics of Superfluids. Irreversible Processes.

1. – The basic equations and boundary conditions.

The Landau equations discussed above appear to be quite adequate for the discussion of reversible processes. The behavior of superfluid helium is far more complicated when irreversible processes are present. We shall continue to develop our theory in the spirit of continuum mechanics, in a manner analogous to the theory of ordinary fluids (Part II).

For the « viscous stresses », we shall *assume* their *linear* dependence on the rate of strain. Since there are now two components of fluid motion there must be exchange coefficients connecting the stress acting on one component with the rate of strain of the other. Thus, we should have four *exchange coefficients*, defined as follows:

(IV.1.1) $$\tau_{ij}^{(n)} = 2\eta^{(nn)} e_{ij}^{(n)} + 2\eta^{(ns)} e_{ij}^{(s)} ,$$

(IV.1.2) $$\tau_{ij}^{(s)} = 2\eta^{(sn)} e_{ij}^{(n)} + 2\eta^{(ss)} e_{ij}^{(s)} .$$

where $\tau_{ij}^{(n)}$ is the stress tensor acting on the normal component, $e_{ij}^{(n)}$ is the rate of strain of the normal component:

(IV.1.3) $$e_{ij}^{(n)} = \frac{1}{2}\left(\frac{\partial v_i^{(n)}}{\partial x_j}\right) + \left(\frac{\partial v_j^{(n)}}{\partial x_i}\right),$$

and the other symbols have similar meanings. We have used a modified notation for the normal and supercomponents for convenient use of the indicial notation. We have also restricted ourselves to the case of incompressible fluids. The extension to the compressible case can be made in the usual manner.

The equations of motion may be obtained from those for the irreversible case by including the volume forces due to the viscous stresses (IV.1.1) and (IV.1.2). We have the additional force per unit volume

(IV.1.4) $$\frac{\partial \tau_{ij}^{(n)}}{\partial x_j} = \eta^{(nn)} \Delta v_i^{(n)} + \eta^{(ns)} \Delta v_i^{(s)} ;$$

for the normal component, and

(IV.1.5) $$\frac{\partial \tau_{ij}^{(s)}}{\partial x_j} = \eta^{(sn)} \Delta v_i^{(n)} + \eta^{(ss)} \Delta v_i^{(s)} .$$

for the supercomponent. Again, the compressibility of the fluid is neglected.

For boundary conditions, we shall make obvious adaptations of the familiar forms (II.1.2) and (II.1.5). If the solid boundary is impermeable to both components, and there is no supply or withdrawal of heat through the boundary, we have

(IV.1.6) $$w_i^{(n)} n_i = 0 , \qquad w_i^{(s)} n_s = 0 ,$$

where $w_i^{(n)}$ and $w_i^{(s)}$ are the velocities relative to the surface of the normal and the supercomponents respectively. If there is heat supply, then the perpendicular component of mass flux must be zero, while the perpendicular component of the heat flux must be continuous.

For the parallel components, eq. (II.1.5) is to be adopted without modification,

$$T_{\alpha}^{(n)} = F^{(n)}(w^{(n)2})w_{\alpha}^{(n)} , \tag{IV.1.7}$$

where $T_{\alpha}^{(n)} = (\tau_{ij}^{(n)} n_j)t_{\alpha}^{i}$,

$$T_{\alpha}^{(s)} = F^{(s)}(w^{(s)2})w_{\alpha}^{(s)} , \tag{IV.1.8}$$

where $T_{\alpha}^{(s)} = (\tau_{ij}^{(s)} n_j)t_{\alpha}^{i}$.

The functions $F^{(n)}$ and $F^{(s)}$ are of course expected to be different from each other. In fact, we shall assume $F^{(n)}$ to be very large, as in the case of an ordinary fluid, and $F^{(s)}$ to be very small at small velocities, so that there is a complete lack of shear interaction between the superfluid and the solid boundary at low speeds. Thus, to a first appxroximation

$$w_{\alpha}^{(n)} = 0 , \tag{IV.1.9}$$

$$T_{\alpha}^{(s)} = 0 . \tag{IV.1.10}$$

We shall have occasion to use (IV.1.8) in a more general form, *e.g.*,

$$T_{\alpha}^{(s)} = \beta \, | w^{(s)} |^2 w_{\alpha}^{(s)} , \tag{IV.1.11}$$

which is obtained from (IV.1.8) by a power series expansion of $F^{(s)}$ and retaining only the first nonvanishing term. (The velocities of flow are usually much smaller compared with the velocities of sound.)

In the above formulation, we have adopted the simplest possible forms. The actual situation can be more complicated. For example, it might be true that $F^{(s)} \equiv 0$ for $|w^{(s)}|$ smaller than a certain « critical value », and then follows approximately the law (IV.1.9). We shall, however, make no attempt to deviate from the form (IV.1.8) in most applications, since the critical value (if it exists) is very small.

The continuum theory has wide latitudes. Although we have chosen to put it in the simplest form, several other effects could have been included. First of all, the concept of mutual friction (specifically the Gorter-Mellink formula) is consistent with the present theory, and could be added to our basic equations. Secondly, we have adopted a linear relationship between the stresses and the rates of strain. There is no reason why the coefficients cannot be dependent on the rates of strain (with suitable rules of covariance followed). We have also assumed that rotation of the supercomponent does not lead to an essential change in *thermodynamical* properties of the medium. However, Bekarevich and Khalatnikov [18] did suggest a continuum theory in which the internal energy of the fluid depends on the magnitude of the superfluid rotation. Indeed, according to their concept, ϱ_s is a tensor when the fluid is in rotation. We shall not indulge in such generalizations until the experimental evidence indicates that it is absolutely necessary to do so. It is to be hoped that such effects (if there be any) should be « small »; otherwise, the general concept of a continuum would appear to be so far different from the usual ones as to make it rather inconvenient.

2. – Applications to some simple problems.

We shall now apply the theory to a few flow problems. Since the formulation has been made for the incompressible case, we shall exclude for the moment problems dealing with the attenuation of sound waves, which will be discussed in Section **4**, and Part VI.

a) *Flow through a tube.* The first experiment to be examined is the flow through a wide capillary. In this case, the most remarkable feature is the fact that the heat flux is directly proportional to the pressure gradient. In the ordinary theory of mutual friction, this is accounted for by the fact that the pressure gradient is balanced by the viscous forces associated with the normal component alone. In our formulation, the equation of motion for the fluid as a whole contains the viscous force

$$\frac{\partial \tau_{ij}^{(n)}}{\partial x_j} + \frac{\partial \tau_{ij}^{(s)}}{\partial x_j} = [\eta^{(nn)} + \eta^{(sn)}]\Delta v_i^{(n)} + [\eta^{(ns)} + \eta^{(ss)}]\Delta v_i^{(s)} . \tag{IV.2.1}$$

Thus, the same conclusion is obtained if

$$\eta^{(ns)} + \eta^{(ss)} = 0 . \tag{IV.2.2}$$

We shall adopt this relation in our following investigations. If we refer back to (IV.1.1) and (IV.1.2), eq. (IV.2.2) means that the total stress

$$\tau_{ij} = \tau_{ij}^{(n)} + \tau_{ij}^{(s)} = 2\,[\eta^{(nn)} + \eta^{(ns)}]\, e_{ij}^{(n)} = 2\eta^{(n)} e_{ij}^{(n)} , \tag{IV.2.3}$$

i.e., the rate of transfer of mechanical momentum is associated with the normal component alone.

The rate of mean flow (if we adopt (1.13) and put $\eta_{sn} = 0$) is given by a formula similar to that obtained from the Gorter-Mellink formula (ATKINS [24], p. 189), except that the coefficient A is now replaced by a slowly varying function of the diameter [25]. The same dependence of A on channel size was found experimentally by HUNG, HUNT, and WINKEL [26].

Notice that the nonlinear behavior is not caused by a nonlinear term in the equations of motion (as in the Gorter-Mellink theory), but by the nonlinearity in the boundary condition for the superfluid motion.

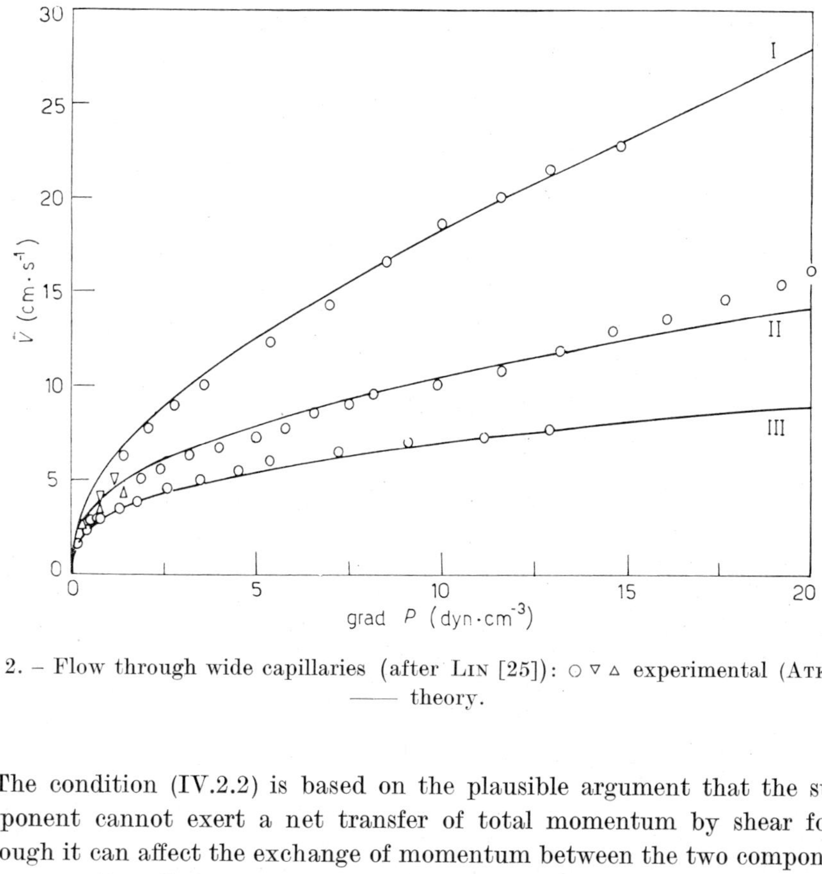

Fig. 2. – Flow through wide capillaries (after LIN [25]): ○ ▽ △ experimental (ATKINS); —— theory.

The condition (IV.2.2) is based on the plausible argument that the supercomponent cannot exert a net transfer of total momentum by shear forces, although it can affect the exchange of momentum between the two components. The assumption that $\eta_{sn} = 0$ again reflects the same idea that the supercomponent is difficult to excite. It appears that the deductions based on both assumptions are well supported by experiments.

b) Flow between rotating cylinders. The next simplest geometrical configuration is a set of two concentric cylinders. Under this category of flow phenomena, we shall also consider the rotating bucket (inner cylinder missing) and the rotating rod (outer cylinder missing).

i) The viscometer. Measurements with the viscometer to determine the relation between the angular speed and the torque have been made by HALLET and his collaborators [27-29] (Fig. 3). At first, it was thought there was a nonlinear dependence. Later experiments show that this is a spurious phenomenon associated with secondary flow, and that careful experiments show complete linearity. With the condition (IV.2.2) imposed, the theory gives linear dependence in all cases, despite the nonlinear boundary condition (IV.1.11). This gives further evidence for the condition (IV.2.2) and the underlying physical ideas.

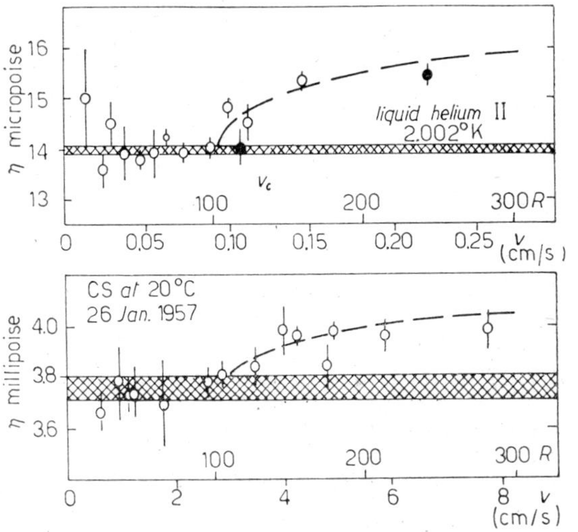

Fig. 3. – « Critical velocity » found with rotating cylinder viscometer in liquid He II and in CS_2 (after WOODS and HALLETT [28]).

ii) Uniform rotation. Whether the inner cylinder is present or absent, the final state of fluid motion is one of uniform rotation, if the boundaries are kept rotating at the same uniform angular speed. This is the prediction of the present theory; it is also found to be true from experimental observations, at reasonably high speeds [30].

In the case of the rotating bucket, there seems to be a critical velocity, below which there is no interaction with the supercomponent [31]. This velocity is extremely small. Once the fluid starts moving, the behavior of the transfer of angular momentum can be described by means of the present theory (for detailed discussions, see Part VI).

iii) Rotating rod. The flow field outside of a rotating rod is given by a pure vortex motion, as in the case of an ordinary fluid. The normal fluid follows the rod, while the superfluid lags behind. Thus, the circumferential velocities are given respectively by

$$v^{(n)} = V\left(\frac{a}{r}\right), \qquad v^{(s)} = V'\left(\frac{a}{r}\right), \tag{IV.2.4}$$

while $\beta(V-V')^3$ gives the stress acting on the supercomponent. The total stress acting on the rod is given by (IV.2.3),

$$\tau = 2\eta^{(n)}(V/a)\,, \tag{IV.2.5}$$

This is doing work at the ratio of $\tau\cdot 2\pi a\cdot V$. The separate components $\tau^{(n)}$

and $\tau^{(s)}_{\square}$ are doing work at the total rate

$$\tau^{(n)}V + \tau^{(s)}V' = \tau V - \tau^{(s)}(V - V') .$$

Thus, there is a dissipation of the amount $\tau^{(s)}(V - V')$ associated with the slip.

iv) General case. In the general case, the flow speed for each component is of the form

$$v = Ar + B/r .$$

The coefficients A and B are determined by certain cubic equations. The interesting point in this problem is a study of its stability. This problem has been studied by CHANDRASEKHAR and DONNELLY [32] on the basis of a set of equations suggested by HALL and VINEN [33]. It would be interesting to repeat the investigation on the basis of the present theory.

3. – Oscillating systems.

One of the basic experiments in the study of superfluid helium is Andronikashvili's oscillating disc system. The continuum theory givenab ove would predict the complete slipping of the superfluid component, not because the superflow must normally remain irrotational, but because of the boundary condition (IV.1.10), which holds for low speeds. At higher speeds, HALLET [34] found that the superfluid component was indeed entrained to move with the

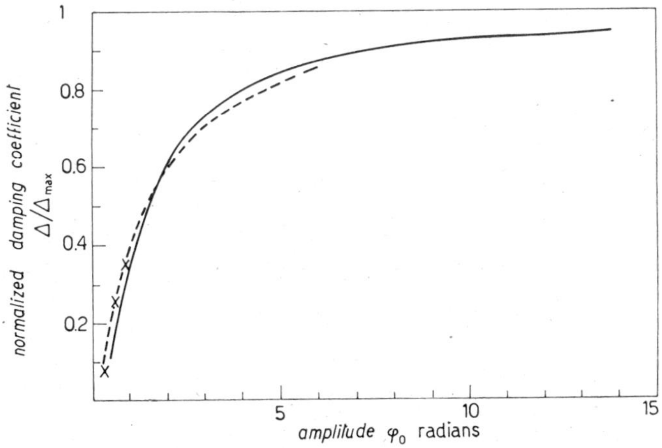

Fig. 4. – Effect of amplitude on damping of an oscillating disk in liquid He II (after GRIBBEN [35]): – × – – – × – experiment (oscillating disc, HALLET [34]); —— theory (GRIBBEN [35]).

normal component. This increasing entrainment with increasing amplitude of oscillation can be accounted for by using the more general boundary condition (IV.1.11).

HALLET also observed the same phenomenon of gradual entrainment with the oscillation of a single disc. The principal features of these observations have been described by using the present theory by R. J. GRIBBEN [35].

In both cases, HALLET found that the mutual friction theory of Gorter and Mellinck does not give a sufficiently large entrainment to account for the observed increase of moment of inertia or damping.

HALL [36] observed a reasonance phenomenon when the above experiments are performed in a rotating system. When the frequency of the oscillation passes through twice the frequency of rotation, there is a change of the nature of the motion. Although HALL was successful in accounting for his observations in terms of « vortex waves », the same phenomenon is expected from the Coriolis effect discussed in the classical case (Section **1**, Part I), and is therefore present also in the continuum theory of superfluid flow. Indeed, the crucial term in Hall's final equations, responsible for this observed change of character, is identical with the Coriolis term in the usual theory. HALL also observed a less pronounced phenomenon of resonance, which we shall discuss in Part VI.

4. – Attenuation of sound waves.

The entrainment of the superfluid into a state of rotation has often been spoken of as the breakdown of superfluidity. This was thought, at one time, to be in analogy with the breakdown of the Meissner effect when the critical magnetic field is exceeded. In that case, the superconducting property is lost. Nothing like that happens in the case of superfluid helium. The propagation of second sound is hardly modified at all when the superfluid is in solid body rotation. The study of the attenuation of the thermal wave under various circumstances, including the state of the superfluid rotation, is therefore a matter of considerable interest.

Although viscosity and heat conduction coefficients furnish sources for the attenuation of sound, they have often been found to be insufficient even in ordinary fluids. For example, the measured absorption coefficient of sound in Benzol (at 17 °C) is about 100 *times* larger than the value predicted on the basis of a simple theory. Although this can be formally described in terms of a second viscosity coefficient, the large numerical factor shows that complicated molecular effects are involved.

As a first attempt, we shall still follow the classical approach of using second viscosity coefficients. Again, there could possibly be four of them. We noticed that the condition (IV.2.2) means that the viscosity coefficient associated with the normal component is alone responsible for the momentum transport. Thus, in the propagation of ordinary sound waves, only the normal coefficient of viscosity exerts a damping influence. If a condition similar to (IV.2.2) is satisfied for the second viscosity coefficients, then the absorption coefficient of the first sound is associated with the behavior of the normal component alone, and should rapidly become small as the temperature decreases, even though the attenuation of the second sound is not.

The effect of rotation on the attenuation of second sound has been observed by HALL and VINEN [33] and by DONNELLY and SNYDER (*). There is no obvious interpretation of such a phenomenon on the basis of a simple continuum theory (**) except for the general suggestion that there could be secondary flow in the fluid, whose exact nature would depend on the rotation. (Cf. Hallet's experience with the viscometer experiments). One must remember that, with $\nu = 10^{-4}$ cm²/s, $\Omega = 9$ rad/s, $L = 1$ cm, the Reynolds number is of the order of 10^5. The secondary motion would be mainly along the axis of rotation, because of Coriolis effect. Thus, the fluid would appear to be anisotropic. We shall return to the discussion of this problem in Part VI.

5. – Turbulence.

There is, as yet, very little established knowledge regarding turbulent flow in superfluid helium. It is however clear that the critical velocities observed in many of the experiments could be caused by transition to turbulence, because the Reynolds number is usually found to be of the order needed for transition in the case of ordinary fluids.

In the present theory, the turbulent flow of the supercomponent may still be described in terms of classical concepts. One should notice the very small scales attained because of the smallness of ν, for both the normal and the supercomponents. With rate of energy dissipation of the order of 1 cm²/s³ per unit mass, the Kolmogoroff length scale is of the order of 10^{-3} cm. Notice that this value is not sensitive to the assumed rate of energy dissipation.

We shall not go into the details of the discussion of possible turbulent flows in the experiments performed so far. Suffice it to say that, in the spirit of the present point of view, we would use classical concepts, including the exchange coefficients, mixture lengths, as well as the more recent statistical concepts, unless the contrary is indicated (***).

(*) Private communication.

(**) One could simply assert, with BEKAREVICH and KHALATNIKOV, that the rotation of the superfluid is a thermodynamic variable.

(***) Cf. Seminar by TACONIS and STAAS given at this Course. The experimental evidence in their paper points toward the soundness of the present point of view.

6. – Critical velocities.

In ordinary hydrodynamics, the resistance to the motion of a solid body in the fluid increases sharpy at certain *critical velocities*. There are at least two such velocities:

a) the velocity marking substantial conversion of the energy of mechanical motion into internal energy of the fluid (actually characterized by the Mach number), and

b) the velocity marking transition from laminar to turbulent flow (actually characterized by the Reynolds number).

In the case of superfluids, the situation is expected to be more complicated. In the usual theory of superfluid structure, the critical velocity calculated (for the production of phonons) is of the same general character as that of type *a*). In the theory of quantized vortex lines, there is a critical velocity associated with the production of these vortex lines, and this is sometimes interpreted as the critical velocity marking the production of turbulence in the superfluid (which is alternatively described as a flow with « tangled vortex lines »).

In terms of the present concepts, we recognize *three* types of critical velocities:

a) the critical velocity for internal excitations;

b) the critical velocity for transition to turbulence;

c) the critical velocity for the incidence of « laminar » friction of the superfluid.

Within each type, there may be several distinct possibilities. For example, within type *a*), there are four possible phenomena and each might be associated with a critical velocity:

a) the excitation of first sound;

b) the excitation of second sound;

c) the production of phonons;

d) the destruction of superfluid behavior.

If one considers the analogy between superfluid flow and superconductivity, the velocity *d*) would be an angular velocity which might be compared with the critical magnetic field that destroys superconductivity. It would be interesting to examine these possible velocities experimentally. All of them are expected to be of the order of the velocity of sound.

The critical velocity for transition to turbulence is very much lower, of the general order of $(0.1 \div 1)$ cm/s for apparatus of the sizes commonly used. There are possibly two such velocities, one for the supercomponent, one for the normal component. As in the case of ordinary fluids, one may have to distinguish between critical Reynolds numbers of instability and transition. Experiments related to these phenomena are reported by TACONIS at this Course.

The critical velocity for the incidence of « laminar » friction could refer to a value of $|\boldsymbol{v}_n - \boldsymbol{v}_s|$ in the mutual friction theory or to a value of $|\boldsymbol{w}_s| = |\boldsymbol{v}_s - \boldsymbol{V}|$ in the theory proposed here (cf. Section **1**). In any case, there is a complicating factor for the incidence of interaction between the superfluid with the boundary (or with the normal fluid). The experiments of REPPY and LANE [31] suggest that the incidence of interaction may have to be « triggered » in some manner. An analogy with the initiation of crystallization in a supercooled liquid seems to be suggested.

It is to be noted that, in the present theory, this critical velocity (perhaps of the order of 1 mm/s) is a genuine velocity and should be independent of the geometry of the apparatus. If one adopts the point of view of the quantized vortex lines, the initiation of friction for the supercomponent depends on the product of a velocity and a linear dimension. Such a critical quantity would be difficult to distinguish from a critical Reynolds number for turbulent transition, as far as scale effects are concerned. Thus, *a careful determination of the lowest critical velocity for tubes of suitably chosen sizes would furnish valuable evidence relevant to the theory of quantized vortex lines.* (Cf. VINEN's paper at this conference).

It need hardly be added that critical velocities for film flow and for extremely narrow channels are outside of the scope of the present discussion, for simple continuum concepts are not adequate in such cases (cf. Section **2**, Part V).

APPENDIX

Is superfluidity the absence of viscosity?

This question was posed by LONDON ([12], Section **22**) and answered by him as follows: « The apparent absence of viscosity ... would not be the outcome of a mysterious absence of any interactions, which would be quite incredible indeed, but rather the outcome of the absence of vortex motion in the superfluid proper. The term of shearing viscosity $\eta_s \nabla \times \nabla \times \boldsymbol{v}_s$ would be absent not because $\eta_s = 0$ but rather because $\nabla \times \boldsymbol{v}_s = 0$. As we have seen in Part II, the absence of the body force does not mean the absence of shear strain. Thus, there would in general be a shear stress acting on the solid boundary when the superfluid is performing an irrotational motion, *e.g.* in the annular region between two coaxial cylinders. Is this an acceptable state of affairs or should one accept the condition $\eta_s = 0$? The latter contradicts London's strong assertion, and is also at variance with the feeling that superfluid flow is associated with order in the momentum space. A shear strain ought to allow momentum to be transferred in such a case.

V. - Hydrodynamics of Liquid Helium II. Microscopic Aspects.

1. – The molecular basis of a hydrodynamic theory.

We have shown how a continuum theory for the hydrodynamics of superfluid helium can be developed on the basis of the concept of slip of the superfluid at the solid boundary, rather than the concept of basic irrotationality of the supercomponent. This purely phenomenological approach has also the advantage of leaving the molecular picture open. As mentioned before, this approach is adopted because there is as yet no available theory for the study of quasi-equilibrium phenomena of a quantum-mechanical system involving long-range order. Still one can try to check our concepts against the known results in the equilibrium theory. In this chapter, we shall begin to examine this aspect of the theory and compare it with the prevailing theory of quantized vortex lines for the superfluid rotation. Specifically, we shall examine the following questions:

1) Is the continuum concept discussed above justifiable in the light of our knowledge of the structure of the liquid?

2) What are the theoretical arguments in favor of, and against, the basic irrotationality of the supercomponent and the existence of quantized vortex lines?

3) What can one conclude from existing experimental evidence, and what further evidence should be collected to improve the status of our knowledge?

2. – The continuum concept.

A continuum theory of fluid motion presupposes the following facts: 1) that there exists a structure length scale Λ such that all fluid elements having a larger size have well-defined macroscopic properties, *e.g.*, density and temperature, and 2) that the minimum size of the experimental apparatus is much larger than this scale Λ. A well-known case where these conditions are violated is the flow of gas through fine tubes at low pressures, first extensively studied by KNUDSEN.

For ordinary media, one might at first think that whenever we take an element of volume with a linear size much larger than the mean interatomic distance (which is of the order of 10^{-8} cm in the case of liquid helium), we can safely regard it as a macroscopic element. This is easily seen to be too loose a criterion, for the mean free path in a gas is usually much larger than the interatomic distance, and it is the mean free path that determines the minimum size of the macroscopic element. In the case of liquid helium, the problem is complicated by the existence of a co-operative phenomenon, which leads to a *correlation length.* This is the distance within which the momentum space ordering is strongly effective. Unfortunately, there is no definite knowledge of this correlation length at the present time. In the case of the ideal Bose gas, this scale is infinite. According to the model studied by LEE, HUANG, and YANG [10], in which a dilute system of Bose particles with hard sphere interaction is considered, this correlation length is given by

(V.2.1) $$r_0 = (8\pi a n)^{-\frac{1}{2}} ,$$

where n is the particle number density, and a is the radius of the hard sphere.

If we take

(V.2)2) $$a = 10^{-8} \text{ cm} , \qquad n = 2\cdot 10^{22} \text{ cm}^{-3} ,$$

we get

(V.2.3) $$r_0 = 1.5\cdot 10^{-8} \text{ cm} .$$

If this figure is trusted, the continuum concept seems to be fully justified in most cases.

However, this estimate appears to be optimistic. Experiments with the propagation of second sound suggest that, at temperatures of the order of 1 °K, there is a strong dispersive effect when the wavelength is of the order 10^{-4} cm or smaller. This is also the scale suggested by the length of the mean free path of the phonons, according to the theory of Landau and Khalatnikov [37]. Blatt, Butler, and Schafroth [38] made the conservative estimate of 10^{-2} cm or smaller, also on the basis of experiments with propagation of second sound.

If an order of 10^{-4} cm is assumed to represent a typical correlation length (around 1 °K), it is clear that flow in films with a thickness of the order of 10^{-6} cm is beyond the scope of any hydrodynamic theory. Also, if the correlation length is of the order of the dimensions of the vessel, as it might be expected at 0 °K or very near to it, we have a « completely » macroscopic quantum-mechanical behavior. It would then be unnatural to expect that ordinary continuum concepts should have any validity. The variables we used above are presumably inadequate and insufficient for the description of such flows. In many of the hydrodynamic experiments, however, the continuum concept seems to be well justified.

3. – Quantized vortex lines.

The prevailing theory is however not based on a true continuum concept. Although macroscopic quantities are used in the final equations of motion, they are not universal equations but have certain *ad hoc* nature. This is due to the belief of irrotational superflow and the existence of quantized vortex lines of *macroscopic* dimensions in most of the hydrodynamic problems.

It is difficult to trace the origin of the idea that superfluid flow must remain pointwise irrotational. Possibly it was the association of irrotational flows with frictionless flows (which is certainly incorrect in the case of ordinary fluids). Possibly it was by analogy with the Meissner effect in superconductivity (*). Possibly it was associated with the irrotational field of the momentum distribution of a single particle moving in a container with a dimension comparable with the de Broglie wavelength. More likely, the origin can be traced to a combination of these factors. At any rate, all these arguments by analogy are naturally not conclusive. In the discussion of the relationship between irrotationality and lack of resistance, there is even a common misunderstanding (see Appendix, Part IV). Nevertheless, Andronikashvili's disc pile experiment gave experimental support to the idea, and it has become quite generally accepted. When actual rotation of the superfluid component has been experimentally discovered, the idea of quantized vortex lines was introduced by ONSAGER [39, 40] and by FEYNMAN [15]. In this picture, vorticity in the fluid is concentrated along vortex lines as in the classical hydrodynamics of perfect fluids. It might be recalled that in the theory of ordinary fluids, such vortex lines cannot persist (Section **1**, Chapter II). Thus, the natural question to be asked is the following: Why can such vortex lines exist in superfluid helium? In other words, *why are these vortex lines stable structures?* (**) A second question is this: What is the mechanism for the vortex line to be formed, which is after all a mammoth structure involving millions of atoms with correlated motions extending over *macroscopic* distances? Why do the vortex lines (if they must be formed for some reason), not form

(*) Cf. Section IV.**6**.

(**) In the recent analysis of GROSS [41], the question of stability was not discussed.

themselves into tiny rings with a diameter (say 10^{-5} cm) much smaller compared with ordinary macroscopic dimensions, but still much larger compared with the diameter of the vortex line (say 10^{-7} cm)?

These questions cannot be easily put aside unless one has a conclusive theoretical evidence that the hydrodynamic rotation of a superfluid component is *absolutely impossible*, or *extremely unlikely* from the point of view of statistical considerations. It is not enough to show that irrotational motion, with or without vortex lines, is a *possible* state of motion of the superfluid component, as has often been attempted.

Before we go into the examination of the detailed arguments made on the basis of Schrödinger's equations, let us first clarify one general point. It is sometimes argued that since quantum mechanical laws are followed by liquid He II, some form of discrete quantization involving Planck's constants must be present. Such an argument is likely to be true at 0 °K (provided the effect in question is large enough to be observed). It does not have to hold when *thermodynamic averages* are involved (*). Since the thermal de Broglie wave length for a helium atom is of the order of 10^{-8} cm at 1 °K, arguments valid for 0 °K need not be applicable at 1 °K.

Let us now examine the oft-quoted derivation of FEYNMAN [15], which runs roughly as follows: If Ψ is the wave function describing the liquid at rest *at* 0 °K, *i.e.*, it is the wave function representing the ground state of the system, then the wave function

$$\psi_{\text{tr}} = \Psi \exp\left[\frac{im}{\hbar} v \cdot \sum_j \boldsymbol{r}_j\right], \tag{V.3.1}$$

represents a uniform translational motion of the liquid as a whole, where $\sum_j$ represents summation over all the atoms, and $\boldsymbol{v}$ is the uniform translational velocity. Consider now the function

$$\psi = \Psi \exp\left[i \sum_j S(\boldsymbol{r}_j)\right], \tag{V.3.2}$$

where S is a slowly varying function of position and $S(\boldsymbol{r}_j)$ is its value at the location of the j-th atom. This function then represents an irrotational flow, with the velocity at any point given by

$$\boldsymbol{v} = \frac{\hbar}{m} \operatorname{grad} S, \tag{V.3.3}$$

(*) The classical example is Planck's law of radiation. The continuous spectrum involves a decidedly quantum effect but no discrete quantization. We wish to emphasize this point because we believe that some effect of discrete quantization should be macroscopically observable in He II at 0 °K.

as a simple quantum-mechanical calculation will show. In a multiply-connected region, such as the annulus between two coaxial cylinders, an irrotational flow can exist with circulation. But since the wave function (V.3.2) must be single-valued, the phase S for each atom must change by an integral multiple of 2π along a path enclosing the inner cylinder. Hence,

$$\oint \operatorname{grad} S \cdot d\mathbf{s} = 2\pi n ,$$

n being an integer, or

$$\varkappa = \oint \boldsymbol{v} \cdot d\mathbf{s} = 2\pi n \frac{\hbar}{m} . \tag{V.3.4}$$

In the axially symmetrical case,

$$v = n \frac{\hbar}{m} \frac{1}{s} . \tag{V.3.5}$$

This is exactly the velocity distribution for a vortex line in classical hydrodynamics.

If the difficulty with the singularity can be reconciled through a detailed investigation, we may omit the inner cylinder and obtain a solution describing a « quantized vortex line ». The kinetic energy associated with such a vortex line of unit length is

$$\frac{1}{2}\int_{a_0}^{b} \varrho_s v_s^2 \cdot 2\pi r \, dr = \frac{\varrho_s \varkappa^2}{4\pi} \log \frac{b}{a_0} , \tag{V.3.6}$$

if the vortex has a core of radius a_0, and it is placed at the center of a cylindrical container of radius b. The core size is usually thought to be of the order of the interatomic distance, *i.e.*, 10^{-8} cm.

Let us now examine the arguments carefully. First, we notice that, if the temperature had been different from 0 °K, we would have to consider a thermodynamical average. In such a case, the above argument would not hold without a rather elaborate modification, and it is not clear that discrete quantization would be obtained, although macroscopic quantum-mechanical effects can still be present in the same qualitative manner; *e.g.*, the order in momentum space and the resultant superflow of heat.

Secondly, if the above derivation is to be valid, both Ψ and ψ must be the solution of Schrödinger's equation. A little calculation will convince the reader that this is in general impossible, even if we assume the phase angle S

to depend on time. The only obvious exception is a pure translation, which is of no interest here (*).

Thirdly, it is clear that the type of argument used above cannot be used to *exclude* other types of motion. The best that can be accomplished is to show that quantized irrotational motion is a *possibility*. We have just seen that even this has not been accomplished.

The above difficulties arise with most of the attempts to establish the theory of quantized vortex lines. To establish the desired theory, it is necessary to prove the *stability* of the proposed state of motion and its statistical likelihood over other rotational motions. This problem still remains to be a challenge to the theoretical physicist.

VI. - Experimental Evidence

1. – Irrotational superflow.

Let us now turn to the discussion of experimental evidence for irrotational superflow and quantized vortex lines. In view of the theoretical difficulties the experiments must be examined critically. This attitude is supported by the unfortunate incidents where some apparent evidence for discrete quantization turned out to be false upon closer examination. After all, mechanical vibrations tend to show discrete behavior, and it is very difficult to avoid small vibrations entirely that might interfere with the necessarily delicate experiments.

We shall divide experiments into the following categories according to the evidence they were designed to provide:

A) irrotational superflow,

B) quantized circulation,

C) individual quantized vortex lines,

D) collective effect of a large number of quantized vortex lines.

Experiments of Class *A* will be discussed in this section, Class *B* will be discussed in Section **2**, and Class *D* in Section **3**. There are no experiments which provided direct evidence for the existence of individual quantized vortex lines. Indirect evidence for their existence will be examined in connection with the other experiments mentioned above.

(*) The mathematical difficulty in the present derivation should be contrasted with the case of the derivation of the quantization of flux in the case of superconductivity [42]. This shows clearly a difference in character between superconductivity and superflow.

Irrotational superflow has no doubt been observed, but the main question is whether the cause is the slip at the boundary or whether the superflow must be pointwise irrotational. A related question is the critical velocity of the wall required to excite rotation.

The critical velocity estimated from the mechanism of phonon excitation is far too large to be related to the critical velocity observed by REPPY and LANE [31] in a rotating bucket. Their experimental result is as follows:

« The helium is contained in a glass bucket (radius 1 cm, length 10 cm) which is hung from a Beams' type magnetic suspension, the whole being enclosed in a high vacuum. This arrangement is very nearly frictionless and the empty bucket, once set in rotation, will coast freely for several hours.

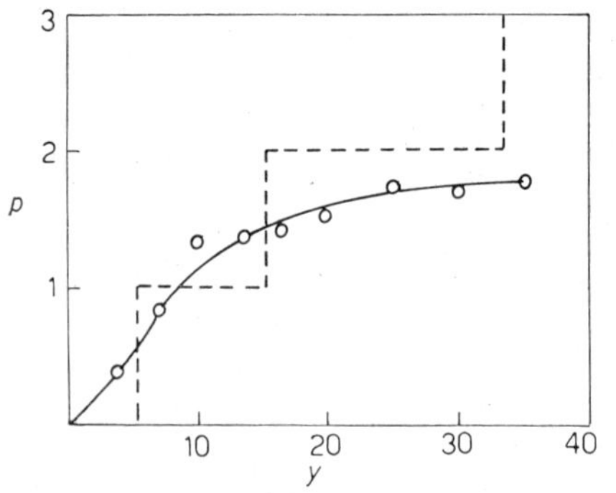

Fig. 5. – Bucket speed as a function of time (after REPPY and LANE [31]).

With the temperature of the helium in the bucket set at 1.2 °K ($\varrho_n/\varrho \approx 3$ per cent) the system is given a short ($\sim\frac{1}{2}$ s) rotational impulse and then permitted to coast freely. The angular velocity of the bucket is then observed as a function of time. In Fig. 5 curve A, the bucket speed is seen to decrease, finally reaching a steady value after some 6 ks. This means that the superfluid has derived angular momentum from the bucket and, further this angular momentum turns out to be just that for solid body type rotation at the final speed. In curve B, taken at a lower initial speed, the same result obtains except that the superfluid is scarcely accelerated at all for the first few thousand seconds. Finally, in curve C, the superfluid remains substantially at rest for the whole course of the experiment which lasts some three hours. The small decline in velocity observed is probably due to the fact that the experiment is not quite isothermal. In the course of 10 ks the helium in the bucket warms up from 1.20 °K ($\varrho_n/\varrho = 3$ per cent) to 1.75 °K ($\varrho_n/\varrho = 27$ per cent) and the normal fluid so created rotates classically thus slowing down the system somewhat ».

REPPY and LANE explained the observed phenomena in terms of vortex lines; but it is just as easy to explain them without a specific vortex line model. All the curious phenomena can be described in terms of a difficulty of interaction between the wall and the superfluid. After the interaction has been initiated, the continuum theory of Part IV is found to be quite adequate to account for the main features of the transition phenomenon, which ends in a state of uniform rotation for the fluid and the bucket alike.

The most conclusive experiment would still be a test of the actual lack of rotation when the bucket speed is extremely small, as originally suggested by LONDON. A specific procedure is as follows. Suppose we have a sealed bucket of superfluid helium suspended in such a manner that it is frictionless with respect to rotation. Let the system be at rest in the laboratory for a *long* time. Since the superfluid helium cannot rotate, the bucket must take on the rotation of the earth by itself. (There is less than one quantum of circulation for a quantized vortex line to be produced for a bucket of ordinary size.) The superfluid helium is then heated up (by radiation, for example). According to the prevailing theory, there should be an observable « spontaneous rotation » of the system, as a calculation of the angular momentum would indicate.

Originally, the apparatus is at rest in the laboratory. It has an angular momentum.

$$J_0 = I_0 \omega_0 \,.$$

After the superfluid becomes ordinary fluid, the whole system has an angular velocity ω given by

$$I_0 \omega_0 = (I_0 + I)\omega \,,$$

where I is the moment of inertia of the fluid. The difference in angular velocity

$$\Delta\omega = \omega - \omega_0 = -\frac{I}{I_0 + I}\,\omega_0 \,,$$

should be observable as a « spontaneous rotation » in the system of co-ordinates at rest in the laboratory.

There are presumably experimental difficulties to be overcome in the procedure suggested above; but such an experiment or its equivalent is really needed to settle the question completely and to match the observation of Meissner effect in superconductivity. (See also suggestion at end of Section **2**).

2. – Experimental evidence. Quantized circulation.

Pending the establishment of irrotational superflow, let us see if quantized circulation can be directly established. VINEN [16] attempted to detect single quantization by performing the following measurements. He filled He II between a pair of concentric cylinders in the form of a wire of length 5 cm and radius $a = 1.3 \cdot 10^{-3}$ cm placed inside of a cylinder with inner radius $b = 0.2$ cm.

The whole instrument system is rotated as various speeds ω characterized by the dimensionless parameter

$$y = 2\pi b^2\omega/(h/m)$$

which is the number of quanta of the circulation $2\pi b^2\omega$ of the wall speed.

The frequencies of the various modes of vibration of the wire are observed by placing the whole instrument assembly in a stationary magnetic field. It is reasoned that if the wire were circular, there would be a degeneracy of the vibration frequencies; in this case, the circulation around the wire would cause, through the Magnus effect, a split of frequency of

$$\Delta\gamma = \varrho_s\varkappa/2\pi w\ ,$$

where ϱ_s is the density of the circulating fluid, and w is the sum of the mass per unit length of the wire and half of the mass of fluid displaced by unit length of wire. It is estimated that

$$\Delta\nu = 0.45\ \text{Hz}$$

per unit quantum of circulation for the wire at hand.

Unfortunately, the experiment is extremely delicate. In the first place, the wire is never symmetrical and has a natural split in frequency amounting to 5 Hz. This has to be corrected by delicate adjustments. Secondly, if the whole instrument assembly were wobbling somewhat during its rotation, such a wobbling frequency would also be registered (*). Thirdly, the damping of the vibration due to the presence of the normal fluid makes the desired measurement barely on the verge of possibility. VINEN treated these and other difficulties very carefully (except for the possible wobbling). The situation is extremely complicated; the explanations given might be satisfactory to an experienced experimentalist, but it is hardly within the confident comprehension of a mere theoretician. For example, it is difficult to understand why the natural splitting of the frequencies should be the observed split when the liquid helium level is just at the top of the wire during the processes of emptying the vessel, a point noted by VINEN himself.

The measured « apparent circulation » is plotted against the speed of rotation (see Fig. 6). It is found that this apparent circulation does not exhibit quantum jumps. By twanging the wire with a current impulse, one could get one quantum of circulation, but two quanta have never been observed, even when the circulation of the outer wall is as high as 35 quanta.

(*) Although a number of observations were made when the apparatus was no longer rotating, the wire was usually twanged. Indeed, a part of vortex line was sometimes observed when the apparatus had not been in rotation prior to the observation.

In attempting to explain these phenomena, VINEN considers free vortex lines and partly attached vortex lines. Now it is clear that if the vortex line has a core of $a_0 = 10^{-8}$ cm whereas the wire has a radius of $a = 10^{-3}$ cm there is a tremendous energy advantage

$$\Delta E = -\frac{\varrho \varkappa^2}{4\pi} \log \frac{a}{a_0},$$

in attachment. Indeed, free vortices would tend to be excluded altogether in favor of multiple quantization. VINEN was led to conclude, after elaborate calculation, that the *core* radius of the vortex line must exceed 10^{-4} cm. The above formula shows that this value might be anticipated qualitatively, since the free vortex line must be energetically comparable with the state of circulation around the wire of radius 10^{-3}, if it is to survive at all.

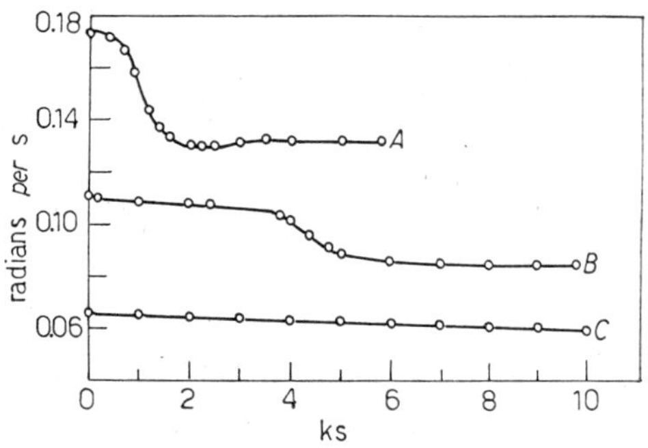

Fig. 6. – Observed apparent circulation around inner cylinder in rotating liquid He II (after VINEN [16]). Solid line connecting experimental points added by the present writer.

If Vinen's estimate of the core radius be accepted, there is either a large hole or a region of distributed vorticity. Either of these would be at variance with the theoretical basis for the existence of vortex lines discussed in Section **3**.

In some of Osborne's experiments, the distance between the vortices is estimated to be of the order of $2 \cdot 10^{-4}$ cm. Thus, according to Vinen's estimate, the vortices are merging into each ither into a continuous distribution of vorticity.

The following quotation from VINEN is noteworthy (*): « But the fact remains that the present measurements do at least *suggest* that the simplest picture of a free vortex line is in some way inadequate, and that the theory behind this picture should be re-examined. It seems possible, for example, that some indeterminacy in the position of a line might appreciably reduce its energy; furthermore, *the value of a_0 might vary with the experimental conditions*, and the value that governs the vortex wave velocity might differ from the one that governs the energy per unit length of a stationary line. It should perhaps be stressed that a large value of a_0 does not imply the existence of a large hole in the liquid; a_0 is to be regarded simply as a parameter that describes the energy per unit length of line, and its value is determined by the

(*) Some italics mine.

detailed conditions near the centre of the line, which may depart considerably from those in the vortex line of classical hydrodynamics ».

The fact that VINEN never observed two quanta of circulation raises the doubt whether the observed apparent single quanta might also be due to extraneous sources. In any case, the extremely large value of a_0 is a serious cause for concern. Experiments of the same type, but performed with a different equipment, should help to remove extraneous effects. A different wire size seems to be extremely desirable.

It is noteworthy that the angular speeds of rotation in some of Vinen's experiments are of the same order of magnitude as those in the experiments of Reppy and Lane. In the latter experiments, the equilibrium state is always that of ordinary solid body rotation, whereas the motion is far from that state in Vinen's theoretically calculated results, even when the wall speed amounts to 35 quanta of circulation. It would be interesting to pursue this point further both experimentally and theoretically. Indeed, a general mathematical theory for the application of the theory of quantized vortex lines has never been published.

3. – Collective effect of quantized vortex lines.

In most of the experiments dealing with rotating helium, a *large* number of quanta of circulation h/m is involved. Thus, one would expect the net effect to be essentially the same as that for continuous vorticity distribution, even when quantized vortex lines do exist. It is therefore all the more remarkable that attenuation of second sound was observed to increase with rotation [33] (*), even though the amount of increase was not large. (In one experiment, it amounted to about 30% of the original attenuation when the angular speed of rotation was 9 rad/s.)

As mentioned before, in the idealized situation, no extra attenuation is to be expected from the continuum theory, but that theory is not always reliable for the prediction of sound attenuation. If one insists on using ordinary continuum concepts, one would perhaps suggest some form of secondary motion. This is supported by the observed fact that there is an initial large transient attenuation comparable to that observed by WHEELER, BLAKEWOOD and LANE [43] in a case where secondary motion is definitely present. The possibility therefore remains that some sort of macroscopic secondary motion was present supported partly by rotation and partly by a steady heat current which is present even in the final state of equilibrium.

(*) Also observed by DONNELLY and SNYDER (private communication).

Another approach to the problem is that given by BEKAREVICH and KHALATNIKOV [18]. As mentioned before, in their picture, the internal energy is supposed to depend on the magnitude of superfluid rotation, making it a new thermodynamic variable. In fact, they were able to obtain the same equations proposed by HALL [36], (also lecture at this Course) without *committing themselves to a definite molecular picture* (cf. alternative suggestions below).

If one accepts the idea of quantized vortex lines, one gets a rather natural interpretation for the fact that the extra attenuation is proportional to the speed of the rotation:

$$\alpha = \alpha_0 + \frac{B\Omega}{2u_{\text{II}}},$$

(where B is a constant (varying with temperature) of the order of unity, Ω is the speed of rotation and u_{II} is the speed of propagation of second sound), since the number of scattering centers is proportional to Ω. However, in the detailed theory, HALL and VINEN considered a force of interaction on a vortex in the form

$$\boldsymbol{f} = C(\boldsymbol{v}_{\text{n}} - \boldsymbol{v}_L)$$

with the normal fluid, but also in the form

$$\boldsymbol{f} = \varrho_{\text{s}}(\boldsymbol{v}_L - \boldsymbol{v}_{\text{s}}) \times \boldsymbol{\varkappa}$$

with the superfluid, where $\boldsymbol{v}_L$ is the velocity of the vortex line. Now if the vortex line is a singularity in the superfluid, it moves with the local superfluid velocity, and is not subjected to the Magnus effect (Section **4**, Part I). Thus, the explanation given by these authors is also somewhat obscure.

If indeed there be a « *granular structure* », the picture of quantized vortex lines does not represent the only possibility. For one thing, the macroscopic convection cells would have such a structure. Also, as mentioned before, small quantized vortex rings might be easier to form than long vortex lines. These rings, if they are formed, would of course be oriented in some manner with a preferential direction in line with the axis of rotation. Qualitatively, the same results would then be obtained as far as attenuation goes. The vortex ring picture would be more readily reconciled with a continuum picture since we may consider a macroscopic element as one containing many such rings. For example, if the ring diameter is 10^{-5} cm, it is still much larger than the probable core diameter 10^{-8} cm, but would still be sufficiently small not to modify the continuum concept, except in the more delicate problems of attenuation and the like. Thus, there is a possibility of reconciling the continuum theory with the experimentally observed phenomena. The necessary appeal to detailed « *sub-macroscopic structures* » for the attenuation processes is similar to that in the case of the attenuation of ordinary sound in ordinary fluids.

It might be interesting to try experiments with the scattering of neutron beams or ion beams (*) in rotating He II to see whether there is a granular structure associated with the vortex lines which are presumed to exist according to the prevailing theory.

Another experiment purporting to support the idea of collective effect of quantized vortex lines was performed by HALL [36]. As mentioned before (Section **3**, Part IV), HALL found, with an oscillating disc in rotating liquid helium, a prominent change of the character of the motion as the oscillating frequency passes through twice the frequency of rotation. This is a classical effect. He also found a slight variation of the frequency of the oscillating system as the level of the liquid is changed, and attributed this to a resonance phenomenon. It should be noted that the oscillating system is damped to half its initial period after about $10 \div 20$ oscillations (**) and that the change of frequency observed is only about 1 %. Thus, it does not constitute a conclusive evidence by itself, especially since very few measurements have been made.

4. – Conclusion.

The fundamental question whether superflow in He II must have a pointwise irrotational character is not resolved. Conclusive theoretical proof and direct experimental evidence are wanting. Although the theory of quantized vortex lines is ingenious and attractive, there is as yet no conclusive evidence for their existence. If something like these vortex lines should exist, there is as yet no clear picture of their size and structure.

The ultimate solution of these basic theoretical problems must await the development of a systematic theory for the non-equilibrium behavior of many-particle systems obeying the laws of quantum mechanics. On the other hand, the continuum theory of hydrodynamics of He II, as developed in this article, can serve the usual purpose of providing an adequate description for most macroscopic problems. Macroscopic behavior which depends more intimately on the molecular character—*e.g.*, the attenuation of sound—cannot be fully accounted for by a simple macroscopic theory, both in ordinary fluids and in superfluids.

(*) Such experiments were reported by Prof. CARERI during this Course.
(**) Private communication from HALL at the conference.

More attention should be paid to the mechanism of interaction between the liquid and the solid boundary, since the boundary conditions are known to have a far reaching significance in fluid motion.

REFERENCES

[1] J. SERRIN: *Handb. d. Phys.*, **8**, 000 (1959).
[2] W. THOMSON: *Proc. Roy. Soc. Edin.*, **10**, 92 (1879).
[3] H. LAMB: *Hydrodynamics* (Cambridge, 1932).
[4] C. C. LIN: *Proc. Nat. Acad. Sci.*, **27**, 570 (1941).
[5] C. C. LIN: *On the Motion of Vortices in Two Dimensions* (Toronto, 1943).
[6] S. GOLDSTEIN: *Modern Development in Fluid Dynamics* (Oxford, 1938).
[7] C. C. LIN: *Theory of Hydrodynamic Stability* (Cambridge, 1955).
[8] S. CHANDRASEKHAR: *Proc. Roy. Soc.*, **241**, 29 (1957); *Hydrodynamic and Hydromagnetic Stability* (Oxford, 1961).
[9] C. C. LIN and W. H. REID: *Handb. d. Phys.*, **8**, 000 (1961).
[10] T. D. LEE, K. HUANG and C. N. YANG: *Phys. Rev.*, **106**, 1135 (1957).
[11] T. D. LEE and C. N. YANG: *Phys. Rev.*, **112**, 1419 (1958).
[12] F. LONDON: *Superfluids*, vol. **2** (New York, 1954).
[13] L. D. LANDAU: *Žurn. Ėksp. Teor. Fiz.*, **5**, 71 (1941).
[14] T. T. WU: *Journ. Math. and Phys.*, **2**, 105 (1961).
[15] R. P. FEYNMAN: *Progress in Low Temp. Phys.*, vol. **1**, ed. by C. J. GORTER (Amsterdam, 1955), chap. II.
[16] W. F. VINEN: *Proc. Roy. Soc.*, **260**, 218 (1960).
[17] J. F. ALLEN and H. JONES: *Nature*, **141**, 75 (1938).
[18] I. L. BEKAREVICH and I. M. KHALATNIKOV: *Proc. VII Intern. Conference on Low Temperature Physics* (1960), p. 459; *Sov. Phys. JETP*, **13**, 643 (1961).
[19] F. LONDON: *Nature*, **141**, 643 (1938); *Phys. Rev.*, **54**, 497 (1938).
[20] L. TISZA: *Nature*, **141**, 913 (1938); *Compt. Rend.*, **207**, 1035 (1938).
[21] T. D. LEE and C. N. YANG: *Phys. Rev.*, **113**, 1406 (1959).
[22] P. R. ZILSEL: *Phys. Rev.*, **79**, 309 (1950).
[23] P. R. ZILSEL: *Phys. Rev.*, **92**, 1106 (1953).
[24] K. R. ATKINS: *Liquid Helium* (Cambridge, 1959).
[25] C. C. LIN: *Phys. Rev. Lett.*, **2**, 245 (1959).
[26] C. S. HUNG, B. HUNT and P. WINKEL: *Physica*, **18**, 629 (1952).
[27] W. J. HEIKKILA and A. C. H. HALLETT: *Can. Journ. Phys.*, **33**, 420 (1955).
[28] A. D. B. WOODS and A. C. H. HALLETT: *Proc. V Intern. Conference on Low Temperature Physics and Chemistry* (Madison, 1957), p. 16.
[29] R. J. DONNELLY and A. C. H. HALLETT: *Ann. Phys.*, **3**, 320 (1958).
[30] D. V. OSBORNE: *Proc. Phys. Soc.*, A **63**, 909 (1950).
[31] J. D. REPPY and C. T. LANE: *Proc. VII Intern. Conference on Low Temperature Physics* (1960), p. 443.
[32] S. CHANDRASEKHAR and R. J. DONNELLY: *Proc. Roy. Soc.*, A **241**, 9 (1957).
[33] H. E. HALL and W. F. VINEN: *Proc. Roy. Soc.*, A **238**, 204, 215 (1956).
[34] A. C. H. HALLETT: *Proc. Roy. Soc.*, A **210**, 404 (1952).
[35] R. J. GRIBBEN: *Journ. Math. and Phys.* **15**, 177, 189 (1961).

[36] H. E. Hall: *Phil. Trans.*, A **250**, 359 (1957).
[37] L. D. Landau and I. M. Khalatnikov: *Žurn. Ėksp. Teor. Fiz.*, **19**, 637, 709 (1949).
[38] J. M. Blatt, S. T. Butler and M. R. Schafroth: *Phys. Rev.*, **100**, 476, 481, 495 (1955).
[39] L. Onsager: *Phys. Rev.*, **64**, 114 (1944).
[40] L. Onsager: *Suppl. Nuovo Cimento*, **6**, 249 (1949).
[41] E. P. Gross: preprint (CERN, 1961).
[42] N. Byers and C. N. Yang: *Phys. Rev. Lett.*, **7**, 46 (1961).
[43] R. G. Wheeler, C. H. Blakewood and C. T. Lane: *Phys. Rev.*, **99**, 1667 (1955).

Elementary Excitations in a System of Interacting Bosons, with Application to Liquid Helium II.

D. Pines

Department of Physics and Electrical Engineering, University of Illinois - Urbana, Ill.

I. – Elementary Excitations in Liquid Helium.

1. – General considerations.

It is convenient to distinguish between the two kinds of elementary excitations in a system of interacting particles. We shall here give a preliminary discussion of this distinction and later, in dealing with field-theoretic methods, present a more mathematical one in terms of the analytic properties of various propagators of the system.

The first kind are related to the single particle excitations. For a non-interacting system, the energy of a single particle with momentum $\boldsymbol{P}$ and mass M is

$$\varepsilon(\boldsymbol{P}) = \frac{\boldsymbol{P}^2}{2M} \cdot$$

As a consequence of the interaction between the particles, the single particle motion is conderably modified. As a particle moves along, it pushes other particles out of its way, drags particles along with it, etc. We may speak of the modified particle as a quasi-particle; it may have a very different energy *vs.* momentum dependence from that for a free particle. At sufficiently low temperatures the system may be regarded as made up of a collection of independent quasi-particles; the specific heat may be simply determined from the quasi-particle spectrum. It is this quasi-particle spectrum which was postulated by Landau to explain the superfluidity and specific heat of liquid helium (cf. the lectures of Professor de Boer).

One does not however, measure the quasi particle spectrum directly. What one measures by means of an external probe (for example by slow neutron scattering in the case of liquid helium) is the energy difference of two states of the system whose momenta differ by, say, $\boldsymbol{k}$, the momentum transfer from the external probe to the system. For a non-interacting system, a typical energy difference, ω, will be

$$\omega(k) = (\boldsymbol{p} + \boldsymbol{k})^2/2m - p^2/2m = \boldsymbol{k}\cdot\boldsymbol{p}/m + k^2/2m$$

corresponding to a single particle making a transition from a state of momentum $\boldsymbol{p}$ to a state $\boldsymbol{p}+\boldsymbol{k}$. (We take $\hbar = 1$ throughout the lectures.) As we shall see, for a longitudinal probe such an excitation spectrum corresponds to that of the density fluctuations of the system.

What is remarkable about liquid helium is that the quasi-particle excitation spectrum, postulated to explain the specific heat, and the density fluctuation spectrum measured by slow neutron scattering, appear to be identical within experimental error. This identity is only possible for interacting boson systems; it certainly does *not* obtain for interacting Fermion systems. For example in the electron gas, the quasi-particles which determine the linear term in the specific heat, possess an excitation spectrum vastly different from the density fluctuation excitations which at long wavelengths are primarily plasmons, the quantized plasma oscillations of the system. We return to the question of the identical quasi-particle, density fluctuation spectrum in a later lecture.

2. – Slow neutrons as an excitation probe.

The experimental determination of the elementary excitations is carried out through inelastic scattering of slow neutrons by liquid helium (PALEVSKY *et al.*; YARNELL *et al.*; HENSHAW and WOODS). One measures the probability per unit time, $W(k, \omega)$, that a neutron transfer energy ω and momentum k, to the liquid; this is given by

$$W(k, \omega) = A\, S(k, \omega)\,, \tag{1.1}$$

where A is a constant which characterizes the neutron-helium atom interaction, and $S(k, \omega)$ describes the elementary excitation spectrum of the density fluctuations of the system (VAN HOVE, COHEN and FEYNMAN). One may define it for the system at absolute zero as

$$S(k, \omega) = \sum_n (\varrho_k^+)_{n0}^2\, \delta(\omega - \omega_{n0})\,. \tag{1.2}$$

Here ϱ_k is the density fluctuation of wave vector k, defined according to

$$\varrho_k = \int d^3r\, \varrho(r) \exp[-i\boldsymbol{k}\cdot\boldsymbol{r}] = \sum_i \exp[-i\boldsymbol{k}\cdot\boldsymbol{r}_i] , \tag{1.3}$$

and $(\varrho_k)_{n0}$ is the exact matrix element of the density fluctuation between the ground state and the n-th excited state. ω_{n0} is the exact excitation frequency appropriate to that state. We are Fourier-analysing in a box of unit volume.

Equation (1.1) may be derived in the following way. The slow neutron-helium interaction is well described by a pseudo-potential, *i.e.* a scattering length. Thus the effective interaction is

$$H_{\text{int}} = \sum_i V_{\text{eff}}(\boldsymbol{R} - \boldsymbol{r}_i) = \sum_i a\, \delta(\boldsymbol{R} - \boldsymbol{r}_i) = \sum_k a\, \varrho_k^+ \exp[-i\,\boldsymbol{k}\cdot\boldsymbol{R}] , \tag{1.4}$$

where $\boldsymbol{R}$ is the position of the neutron, $\boldsymbol{r}_i$ is the co-ordinate of the i-th helium atom and a a is a constant, proportional to the S-wave scattering length. For slow neutrons the scattering is well described by the Born approximation, so that, using the « golden rule » of second-order perturbation theory, one has

$$W = 2\pi \sum_n |(H_{\text{int}})_{n0}|^2\, \delta(E_n - E_0) \tag{1.5}$$

as the probability per unit time that the neutron-helium system undergoes a transition from an initial state 0 to a set of final states n characteristic of the scattering act. Initially one has

neutron:	momentum $\boldsymbol{P}$,	energy $\dfrac{P^2}{2M}$,
helium:	momentum 0,	energy ω_0 .

After the scattering act, one has, for a momentum transfer $\boldsymbol{k}$,

neutron:	momentum $\boldsymbol{P} - \boldsymbol{k}$,	energy $\dfrac{(\boldsymbol{P}-\boldsymbol{k})^2}{2M}$,
helium:	momentum $\boldsymbol{k}$,	energy ω_n ,

(for all states n induced by a scattering act).

Hence, from conservation of energy,

$$E_n - E_0 = \omega_{n0} - \left(\frac{\boldsymbol{k}\cdot\boldsymbol{P}}{M} - \frac{k^2}{2M}\right) = \omega_{n0} - \omega . \tag{1.6}$$

From (1.4), (1.5), and (1.6), one at once obtains (1.1).

3. – Some properties of $S(k, \omega)$; structure factor and F-sum rule.

Two moments of $S(k, \omega)$ are of interest. One has directly from (1.2) that

$$\int_0^\infty \mathrm{d}\omega\, S(k, \omega) = \sum_n (\varrho_k^+)^2_{n0} = \langle 0 | \varrho_k^+ \varrho_k | 0 \rangle = N S(k) , \tag{1.7}$$

where $S(k)$ is the well-known liquid structure factor. $S(k)$ is the Fourier transform of the pair correlation function $p(r)$. Thus,

$$S(k) = \int \mathrm{d}r^3 p(r) \exp[-i\boldsymbol{k}\cdot\boldsymbol{r}] = N^{-1} \langle 0 | \varrho^+(0)\varrho(r) | 0 \rangle . \tag{1.8}$$

We see that $S(k)$ describes an experiment in which one measures only the angular distribution of neutrons emerging from liquid He, and not their energies. Thus from (1.7) we have

$$W(k) = A \int_0^\infty \mathrm{d}\omega\, S(k, \omega) = A S(k) ,$$

where $W(k)$ is the probability time that the neutrons transfer momentum k to the He, and is therefore proportional to the scattering cross-section per unit angle, $\mathrm{d}\sigma/\mathrm{d}\theta$.

The other moment follows from the F-sum rule, which reads

$$\sum_n \frac{2m}{k^2} (\varrho_k^+)^2_{n0} \omega_{n0} = N = \sum_n F_{n0} , \tag{1.9}$$

where F_{n0} is a generalized oscillator strength. Our derivation of (9) follows that of NOZIÈRES and PINES. We consider certain commutators of ϱ_k and H, where H is the liquid helium Hamiltonian,

$$H = \sum_i \frac{p_i^2}{2m} + \sum_{i\neq j} \frac{1}{2} V(\boldsymbol{r}_i - \boldsymbol{r}_j) . \tag{1.10}$$

We assume the potential is velocity-independent. Then

$$[\varrho_k, H] = \sum_i \left(\frac{\boldsymbol{k}\cdot\boldsymbol{p}_i}{m} - \frac{k^2}{2m}\right) \exp[-i\boldsymbol{k}\cdot\boldsymbol{r}_i] ,$$

by direct calculation; and likewise

$$[\varrho_k^+, [\varrho_k, H]] = \frac{Nk^2}{m}.$$

If we apply this to the ground state, we have

$$[\varrho_k^+, [\varrho_k, H]]_{00} = \frac{Nk^2}{m} = \sum_n (\varrho_k^+)_{0n}[\varrho_k, H]_{n0} - [\varrho_k, H]_{0n}(\varrho_k^+)_{n0}.$$

But $[\varrho_k, H]_{n0} = \omega_{n0}(k)(\varrho_k)_{n0}$, etc.; so one finds

$$\sum_n \{(\varrho_k^+)_{0n}(\varrho_k)_{n0}\,\omega_{n0}(k) + (\varrho_k)_{0n}(\varrho_k^+)_{n0}\,\omega_{n0}(k)\} = 2\sum_n \omega_{n0}(k)\,|(\varrho_k^+)_{n0}|^2$$

for this double commutator. In the last equality, we have made use of reflection invariance, which asserts that $E_n(k) = E_n(-k)$. Hence, one obtains eq. (1.9). From (1.9) we see that

$$\int_0^\infty \mathrm{d}\omega\, S(k, \omega)\,\omega = \frac{Nk^2}{2m}, \tag{1.11}$$

on making use of the defining equation (1.2).

Consider $S(k, \omega)$ for an ideal Bose gas. At $T = 0$, all particles have $k = 0$. The excited states of the system therefore have the unique excitation frequency

$$\omega_{\text{ex}} = \frac{k^2}{2m}.$$

From the relation $(\varrho_k^+)^2_{n0} = N$, we see that

$$S(k, \omega) = N\,\delta\left(\omega - \frac{k^2}{2m}\right),$$

and

$$S(k) = 1,$$

which implies no correlation between the atoms.

4. – Feynman variational calculation.

The Feynman excited state wave function $\varrho_k^+|0\rangle$ and excitation energy $E(k) = \hbar^2 k^2/2m\,S(k)$ may be simply obitaned from the preceding considerations. Feynman assumes that the excited state, $\varrho_k^+|0\rangle$, possesses a unique excitation frequency. But this is equivalent to saying that

$$S(k, \omega) = \sum_n (\varrho_k^+)_{n0}^2 \, \delta[\omega - E(k)] = N\, S(k)\, \delta[\omega - E(k)] \,. \tag{1.12}$$

Let us now substitute (1.12) into (1.11). One finds then

$$E(k) = \frac{k^2}{2m\, S(k)} \,. \tag{1.13}$$

Thus the Feynman excitation spectrum is the most general result one can obtain, consistent with (1.7) and (1.11), assuming that the density fluctuations possess a unique excitation frequency. In other words, with the Feynman wave function, a single excited state exhausts the F-sum rule.

5. – Comparison with experiment; Feynman-Cohen theory.

In order to compare (1.13), which holds at $T=0$, with the quantities measured at finite temperature, Feynman estimated $S_T(k)$ at $T=0$ for small k, as explained by Dr. CHESTER. A comparison of the Feynman excitation spectrum with the experimental results obtained from slow neutron scattering (HENSHAW and WOODS) is given in Fig. 1. It may be seen that the Feynman curve has qualitatively the right shape, and indeed exhibits a minimum in the region of the roton excitation, but that the roton minimum is too large by about a factor of two.

Thus it is necessary to improve the wave function describing an excitation in the short wave-length region. An improved wave-function was put forth by FEYNMAN and COHEN, who argued that a correct description of an elementary excitation should yield conservation of current. Thus, as a particle moves through helium, it is surrounded by a return flow of other particles, the backflow associated with the latter bringing about current conservation. The form of the Feynman-Cohen wave-function (which has already been discussed by Dr. CHESTER) is

$$\psi_k = \sum_i \exp[i\, \boldsymbol{k}\cdot\boldsymbol{r}_i] \exp\Big[i \sum_{i\neq j} g(r_{ij})\Big]\, \Phi_0 \,,$$

where $g(r)$ was taken to be the dipolar form

$$g(r) = A\,\frac{\boldsymbol{k}\cdot\boldsymbol{r}}{r^3} \,,$$

which one would expect it to have at large distances. As shown in Fig. 1, this wave-function yields a considerable improvement in the agreement between theory and experiment for the excitation spectrum in the she short wave-length region.

Another way of viewing the lack of agreement between the Feynman excitation spectrum and experiment is to say that the spectral density $S(k, \omega)$ cannot consist of a single discrete line, but is broadened. This broadening corresponds to an interaction between the different excitations. A given excitation can undergo virtual transitions to other configurations, and therefore may be viewed as surrounded by a cloud of other excitations. One would thus expect that the Feynman-Cohen backflow may be regarded as arising from an interaction between the excitations. In a subsequent lecture, we shall demonstrate that this indeed is the case within the Bogoljubov approximation.

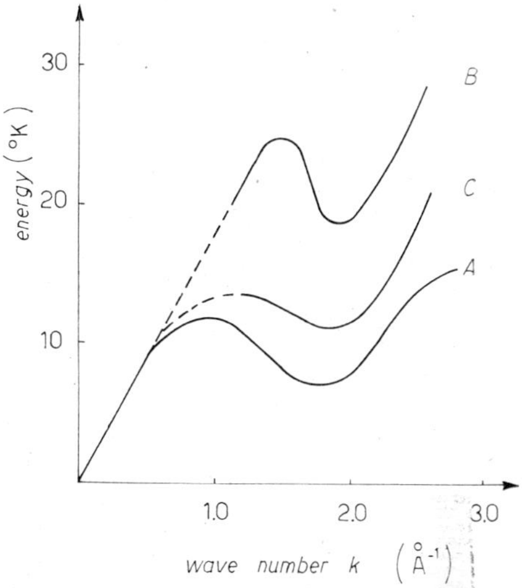

Fig. 1. – Excitation spectrum for ^{4}He: *A*) experimental curve (HENSHAW and WOODS); *B*) FEYNMAN; *C*) FEYNMAN and COHEN.

6. – Nature of the observed elementary excitations in liquid He.

We are led to the following qualitative picture of the excitation spectrum of liquid helium:

1) At long wave-lengths, one has phonons, which possess a linear dispersion relation up to nearly the effective Debye wave-vector k_D (such that there exist N longitudinal modes in the liquid).

2) As already mentioned by Professor DE BOER, the bending over of the excitation curve is what one would expect for a solid—and liquids are surprisingly and notoriously similar to solids in their excitation spectrum. However, for $k \simeq 2k_D$, instead of going to zero, the curve merely displays a minimum at ~ 8.7 °K.

3) The rotons, the excitations in the vicinity of this minimum, behave essentially like single particles, which have a rather large effective mass

($\simeq 2.5\,M_{\text{He}}$). Part of this effective mass may be regarded as arising from the short-range interactions with nearest neighbors; the remainder arises from the virtual cloud of phonons (and to some extent other rotons) carried along by the roton as it moves through the liquid.

II. – Microscopic Theories.

1. – The basic hamiltonian.

In this, and succeeding lectures, we make use of second quantization, which will be discussed more fully by Professor SESSLER. In this representation, the many particle state vector, $|\psi_a\rangle$, is one in which the number of particles, n_p, is specified for every possible momentum state, p. Such a state will be denoted by

$$|n_0, n_p, \dots, n_{p_i}, \dots\rangle .$$

The physical operators can be expressed in terms of the basic creation and annihilation operator, a_p^+ and a_p, defined by the following relation:

$$a_p|\dots, n_p, \dots\rangle = \sqrt{n_p}\,|\dots, n_p - 1, \dots\rangle ,$$

$$a_p^+|\dots, n_p, \dots\rangle = \sqrt{n_p + 1}\,|\dots, n_p + 1, \dots\rangle ,$$

$$[a_{p'}, a_p^+] = \delta_{pp'}, \qquad [a_{p'}, a_p] = 0 , \qquad [a_{p'}^+, a_p^+] = 0 .$$

The second quantization representation is therefore one in which the occupation number operators, $n_p = a_p^+ a_{p'}$ are diagonal.

The Hamiltonian for a system of interacting particles takes the form

$$H = \sum_k \varepsilon_k a_k^+ a_k + \tfrac{1}{2} \sum_{kpq} V_k a_{p+k}^+ a_{q-k}^+ a_q a_p , \tag{2.1}$$

where $\varepsilon_k = k^2/2m$ is the single particle kinetic energy and

$$V_k = \int d^3r\, V(r) \exp[-i\boldsymbol{k}\cdot\boldsymbol{r}] ,$$

is the Fourier transform of the potential energy of interaction, $V(r)$, between the pairs of bosons. The second term in the Hamiltonian has a simple interpretation in terms of an elementary scattering act between two free particles with exchange of a momentum $\boldsymbol{k}$ and overall momentum conservation as shown in the following diagram:

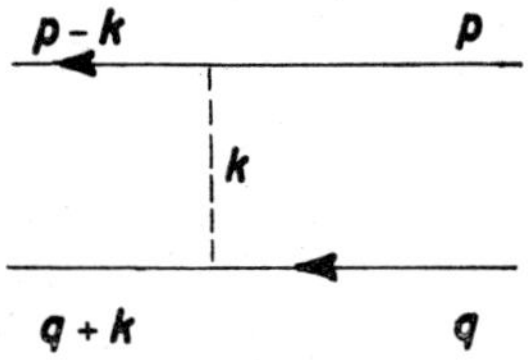

It should be remarked that these is another way to go from states $\boldsymbol{p}$, $\boldsymbol{q}$ to states $\boldsymbol{p}-\boldsymbol{k}$, $\boldsymbol{q}+\boldsymbol{k}$: one in which $\boldsymbol{p}$ scatters to $\boldsymbol{q}+\boldsymbol{k}$ and $\boldsymbol{q}$ scatters to $\boldsymbol{p}-\boldsymbol{k}$, with momentum transfer $\boldsymbol{q}+\boldsymbol{k}+\boldsymbol{p}$. This term is also present in the interaction part of the Hamiltonian (2.1). The two terms correspond to direct and exchange scattering processes respectively.

2. – The role of the zero-momentum state.

1) In the ideal Bose gas, N_0, the number of particles in the zero-momentum condensed state, is a macroscopic number, and is equal to N. Suppose we assume, as was first done by BOGOLJUBOV, that in the presence of interactions N_0 continues to be a finite fraction of N. We shall see that from this assumption there follows directly the prediction of superfluid behavior in Bose systems.

2) Consider a_0 and a_0^+ acting on the ground-state wave function, Ψ_0. One has

$$(2.2)\qquad \begin{cases} a_0|\Psi_0(N_0)\rangle = \sqrt{N_0}\,|\Psi_0(N_0-1)\rangle\,, \\ a_0^+|\Psi_0(N_0)\rangle = \sqrt{N_0+1}\,|\Psi_0(N_0+1)\rangle\,, \end{cases}$$

where $\Psi_0(N_0)$ is the ground-state wave function with N_0 particles in the zero-momentum state, $\Psi_0(N_0-1)$ is that same state with N_0-1 particles in the zero-momentum state, etc. We see that if $N_0 \gg 1$, we may do two things:

i) Replace $\sqrt{N_0+1}$ by $\sqrt{N_0}$;

ii) Assume that the difference between $\Psi_0(N_0-1)$ and $\Psi_0(N_0+1)$ is of order $1/N$; that is, removing or adding a particle to the zero momentum state will not alter the physical properties of the system.

Under these circumstances, a_0 and a_0^+ will commute with each other; since they already commute with all a_p and a_p^+ for $p \neq 0$, they may simply be replaced by the « c » number $\sqrt{N_0}$. This is the famous Bogoljubov prescription; it amounts to neglecting the dynamic behavior of the condensed state.

3) With the Bogoljubov prescription, one no longer has a Hamiltonian which conserves particle number. For example, part of the interaction terms in (2.1) reduce to

$$\sum_p N_0 V_p a_p^+ a_{-p}^+$$

corresponding to the process represented by the diagram

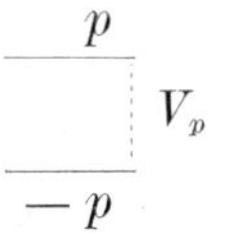

in which a pair of particles is created. Another way of saying this is that the number operator,

$$N_{\text{op}} = \sum_p a_p^+ a_p , \tag{2.3}$$

no longer commutes with the Hamiltonian, once one replaces a_0 and a_0^+ by $\sqrt{N_0}$. A consistent way out of the difficulty is to add a term, $-\mu N_{\text{op}}$, to the Hamiltonian, (2.1). This gives us, to borrow a phrase of Bogoljubov's, a « hunting license » to consider non-particle conserving intermediate states, while offering a mechanism to conserve particle number on the average (HUGENHOLTZ and PINES). μ is, as before, the chemical potential. We find the following Hamiltonian:

$$H' = H_1 + H_2 , \tag{2.4}$$

where

$$H_1 = \sum_p{}' \tilde{\varepsilon}(p) a_p^+ a_p + \sum_p{}' N_0 \frac{V_p}{2} (a_{-p}^+ a_p^+ + a_p a_{-p}) , \tag{2.5}$$

$$H_2 = \sum_{pk}{}' \sqrt{N_0} \frac{V_k}{2} (a_{p+k}^+ a_k a_p + a_{p+k}^+ a_{-k}^+ a_p) + \sum_{kpq}{}' \frac{V_k}{2} a_{p+k}^+ a_{q-k}^+ a_q a_p , \tag{2.6}$$

and

$$\tilde{\varepsilon}_p = \varepsilon(p) + N_0 V_0 + N_0 V_k - \mu \tag{2.7}$$

is the quasi-particle energy in the Hartree-Fock approximation. N_0 and μ

are to be determined by the conditions:

$$N_0 + \langle \Psi_0 | \sum_p{}' a_p^+ a_p | \Psi_0 \rangle = N \,, \tag{2.8}$$

$$\mu = \frac{\partial E_0}{\partial N} = \frac{\partial E_0'}{\partial N_0} \,. \tag{2.9}$$

The prime on the summations indicate that the states of zero momentum are to be omitted. E_0 is the ground-state energy of H and E_0' is that of H_0'. The condition (2.8) guarantees particle conservation on the average.

3. – The weak coupling solution.

1) BOGOLJUBOV assumed that in suitable approximation (which turns out to be the weak coupling limit) the terms of H_1 in (2.4) are large compared to those in H_2. H_1 can be diagonalized by the following canonical transformation:

$$\begin{cases} a_k = u_k \alpha_k - v_k \alpha_{-k}^+ \,, \\ a_k^+ = u_k \alpha_k^+ - v_k \alpha_{-k} \,, \\ u_k^2 - v_k^2 = 1 \,; \qquad u_k^+ = u_k \,; \qquad v_k^+ = v_k \,. \end{cases} \tag{2.10}$$

The relation between u_k and v_k ensures that the transformation is canonical, *i.e.*, that

$$[\alpha_{k'}, \alpha_k^+] = \delta_{kk'} \,,$$

$$[\alpha_{k'}, \alpha_{k'}] = [\alpha_{k'}^+, \alpha_{k'}^+] = 0 \,.$$

One finds, by requiring that the resultant Hamiltonian be diagonal, that

$$H_1 = \sum_k{}' \omega_k \alpha_k^+ \alpha_k + \frac{\omega_k}{2} - \frac{\tilde{\varepsilon}_k}{2} \,, \tag{2.11}$$

with

$$u_k^2 = \frac{1}{2} \left[1 + \frac{\tilde{\varepsilon}_k}{\omega_k} \right] , \tag{2.12}$$

and

$$\omega_k^2 = \tilde{\varepsilon}_k^2 - N_0^2 V_k^2 = \varepsilon_k^2 + 2\varepsilon_k N_0 V_k \,, \tag{2.13}$$

upon making use of the relation

$$\mu = N_0 V_0 \tag{2.14}$$

which leads to

$$\tilde{\varepsilon}_k = \varepsilon_k + N\, V_k\,. \tag{2.15}$$

2) The Hamiltonian (2.11) describes a collection of noninteracting quasi-particles with frequency

$$\omega_k = \left\{\frac{k^2 N_0 V_k}{m} + \frac{k^4}{4m^2}\right\}^{\frac{1}{2}}. \tag{2.16}$$

At long wavelength or small k the quasi-particles are phonons with velocity

$$s = \left(\frac{N_0 V_0}{m}\right)^{\frac{1}{2}}.$$

At short wavelengths, the quasi-particles behave essentially like free particles with an energy $k^2/2m$. The wave-vector at which the transition from phonon to single particle behavior occurs is

$$k_c = 2\sqrt{m N_0 V_k} \simeq 2ms\,.$$

The inverse of this

$$\lambda_c = \frac{1}{2ms}\,,$$

furnishes a natural measure of the distance over which coherence effects are important in the interaction between particles. One may call this a correlation length, provided one understands clearly that it refers to correlations between excitations in the system. These are quite different from the long range correlations associated with the condensation in the $k=0$ mode.

The essential point is that the single particle excitations possess a phonon character, so that already the Bogoljubov approximation satisfies Landau's criterion for superfluidity; that no excitations can be created in a fluid moving with velocity $v<s$.

3) It is instructive to view the canonical transformation from the standpoint of many-body perturbation theory. One may show that it is equivalent to carrying out a summation of the most divergent terms in the perturbation

series expansion for the ground-state energy (BRUECKNER and SAWADA). These have the following character:

2-nd order:

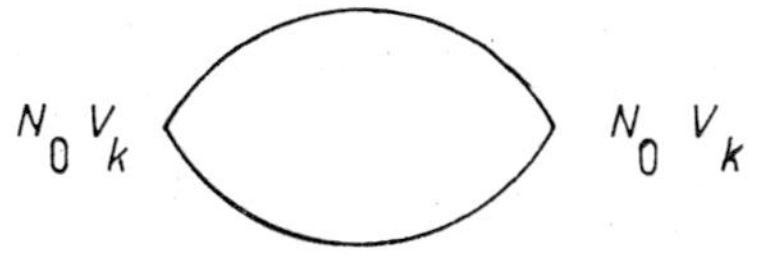

$$\Delta E = -\sum_k \frac{N_0^2 V_k^2}{k^2/m} ,$$

3-rd order:

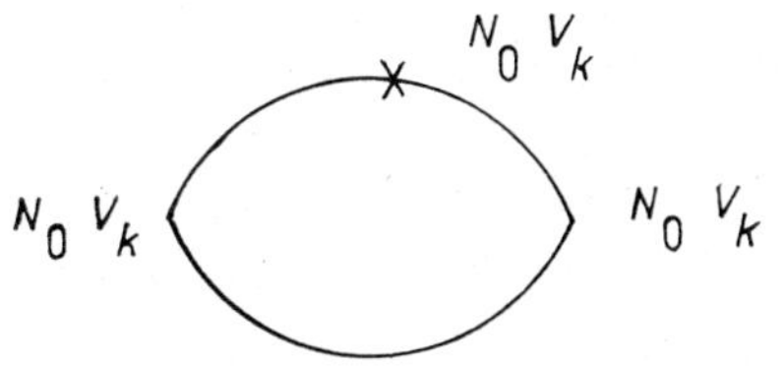

$$\Delta E = -\sum_k \frac{(N_0 V_k)^3}{(k^2/m)^2} .$$

The second order diagram represents a contribution to the ground-state energy from processes in which a pair of particles of momentum k and $-k$ are created, with matrix element $N_0 V_k$, and then subsequently annihilated, with the same matrix element. According to the rules of perturbation theory, one multiplies the matrix elements, divides by the intermediate-state energy, k^2/m, and sums over all possible momentum states, k. The third-order diagram represents pair excitation, followed by forward scattering of one of the particles (denoted by the cross), and then pair annihilation. The forward scattering matrix element is

$$N_0 V_0 + N_0 V_k - \mu = N_0 V_k .$$

The third-order term we have considered is divergent for small momentum transfers (since V_k is then approximately a constant). As one goes to higher orders, the divergence for small k associated with iteration of the terms in the interaction involving the momentum transfer, k, becomes more and more pronounced. If one sums the series, however, one obtains a convergent result

$$E_0 = \sum_k \left[\frac{\omega_k}{2} - \frac{\varepsilon_k + N_0 V_k}{2}\right] + \frac{1}{2} N_0^2 V_0 , \tag{2.17}$$

as may be obtained directly from (2.11), on adding the term $\frac{1}{2}N_0^2 V_0$ to pass from E_0' to E_0.

4) The summation of the series is equivalent to making the random phase approximation introduced for the electron gas by BOHM and PINES.

In this approximation, each momentum transfer in the interaction is treated independently in its effect on the system properties. Thus, if one is calculating $S(k)$, one would keep only the effect of the k-th component of the interaction on the ground-state wave function. Again, in calculating the energy of the k-th single particle excitation mode, one keeps only the k-th momentum transfer in the interaction. It is at once clear from (2.5) and (2.10) that the Bogoljubov approximation is identical to the random phase approximation; we shall denote both by RPA.

5) Let us calculate $S(k\omega)$ in the RPA. We have

$$\varrho_k = \sum_q a_{q-k}^+ a_q = \sqrt{N_0}(a_{-k}^+ + a_k) + \sum_q{}' a_{q-k}^+ a_q \,, \tag{2.18}$$

where the prime denotes the fact that terms with $q=k$ or $q=0$ are to be omitted. In the RPA, we keep only the first term on the right hand side of (2.18), that proportional to $\sqrt{N_0}$. If we now use the transformation (2.10), we have

$$\varrho_k = (u_k - v_k)(\alpha_k + \alpha_{-k}^+)\sqrt{N_0} = (N_0 k^2/2m\omega_k)^{\frac{1}{2}}(\alpha_k + \alpha_{-k}^+) \,. \tag{2.19}$$

We see from (2.19) that the density fluctuation excitation spectrum is identical with the quasi-particle excitation spectrum in the RPA. Thus the excited state n coupled to the ground state 0 possesses the unique excitation frequency, ω_k, and the matrix element

$$(\varrho_k)_{n0} = (N_0 k^2/2m\omega_k)^{\frac{1}{2}} \,. \tag{2.20}$$

We further have

$$S_{\text{RPA}}(k\omega) = \frac{N_0 k^2}{2m\omega_k}\,\delta(\omega - \omega_k) \,. \tag{2.21}$$

On integrating over frequencies, we find

$$S_{\text{RPA}}(k) = N_0 k^2/2m\,\omega_k \,; \tag{2.22}$$

while for the sum rule we obtain

$$\int_0^\infty \mathrm{d}\omega\,\omega\, S_{\text{RPA}}(k\omega) = N_0 k^2/2m \,. \tag{2.23}$$

6) We see from (2.23) that the quasi-particle excitations exhaust the sum rule only if we may take $N_0 = N$. In fact, to the extent that the Bo-

goljubov approximation is valid, one can take $N_0 = N$. The Bogoljubov approximation is a weak coupling theory, and

$$N - N_0 = \langle 0 | \sum_p{}' a_p^+ a_p | 0 \rangle = \sum_p v_p^2 = \tfrac{1}{2} \sum_p \frac{\varepsilon_p + N_0 V_p}{\omega_p} - 1 \simeq 0(V_p) \, . \tag{2.24}$$

7) To summarize: the Bogoljubov approximation, although valid only in a weak coupling limit (as one sees by computing simple diagrams), offers the essential clue to a microscopic theory of superfluidity, in that the low-lying excitations are phonons which obey a linear dispersion relation at long wavelengths.

4. – Dilute boson gas.

1) In order to obtain a satisfactory microscopic model for superfluidity, one which is valid in a given range of densities for an arbitrary force law, it is necessary that one go beyond BOGOLJUBOV, by taking into account certain of the terms in H_2 as well as the RPA terms of H_1. This has been done in a number of ways—by a pseudo-potential method (LEE, HUANG, and YANG), by summing ladder diagrams using the Rayleigh-Schrödinger perturbation theory (BRUECKNER and SAWADA), and by field-theoretic methods (BELIAEV, HUGENHOLTZ and PINES). We shall consider the field-theoretic approach in more detail in a subsequent lecture; let us here consider briefly the series summation approach.

2) By examining the perturbation series we can decide what additional terms must be retained to make the theory valid in a given density region. The point to notice is that corresponding to terms appearing in the energy as calculated in the RPA, there are additional contributions from H_2 which are of the same order in the density. For example we have included the ground state energy diagram

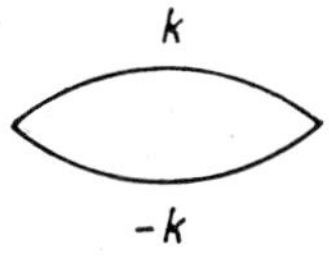

although we have not counted the diagrams

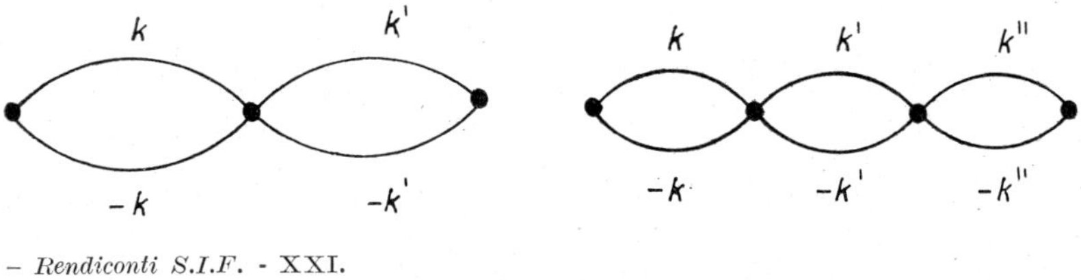

which turn out to be of the same order in N_0. These diagrams correspond to the multiple scattering of a pair of particles in which no other particles in the system play a role. They are just what one might expect to be the most important terms at low densities; that two excited particles would scatter repeatedly against each other, without being affected by the other particles in the system.

3) The summation of the multiple scattering terms may be simply carried out by introducing an effective scattering matrix, or « t » matrix, via a diagrammatic equation which reads:

(2.25)

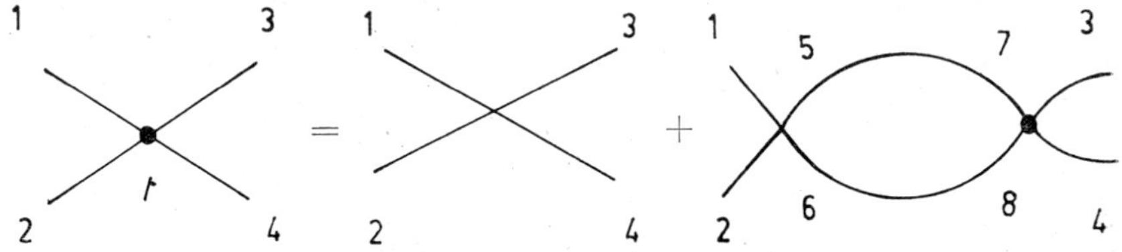

$$t(12;34) = v(12;34) + v(12;56)\,G(57)\,G(68)\,t(78;34)\,, \tag{2.26}$$

where $G(57)$ and $G(68)$ denote the unperturbed propagation of a particle between scattering acts. It is clear, by iteration, that (2.25) represents a sum of all multiple scattering diagrams; to see this, substitute for $t(78;34)$, first $v(78;34)$, then $v(78;34)+v(78;7'8')G(7'3')G(8'4')v(3'4';34)$, etc., corresponding to

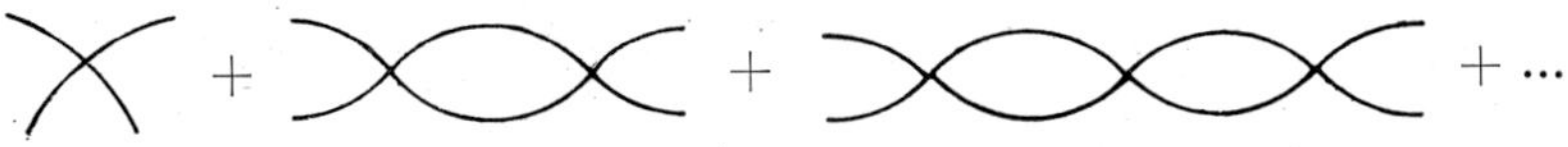

The solution of (2.26) is easily found using the rules for perturbation theory. Let q and ω be the net momentum and energy of the particle pairs; the other free variables are, say, p and p', correspond to a pair of particles initially in state p, and $q-p$, finally in states, p' and $q-p'$. The t-matrix equation then turns out to be

$$t(pp';q\omega) = v_{p-p'} - (2\pi)^{-3}\int dp''\,\frac{v_{p''-p}\,t(pp'';q\omega)}{\omega-\varepsilon(p'')-\varepsilon(p''+q)-i\delta}\,. \tag{2.27}$$

To lowest order in the density, eq. (2.27) is simply the equation describing the s-wave scattering of a pair of isolated particles. It possess the approximate solution

$$t_{\text{free}} \simeq f_0/m\,, \tag{2.28}$$

where f_0 is the « s » wave scattering amplitude.

4) We can proceed to substitute f_0/m for V_k everywhere in our Bogoljubov formulation; in so doing we sum simultaneously RPA and ladder diagrams. The procedure looks very much like the pseudo-potential method; it is not however, because the pseudo-potential is taken to be

$$(2.29) \qquad V_{\text{eff}}(r) = \frac{f_0}{m} \frac{\partial}{\partial r} [r \delta(r)] ,$$

rather than

$$(2.30) \qquad V_{\text{eff}}(r) = (f_0/m) \, \delta(r)$$

the assumption we have used. The difference between (2.29) and (2.30) corresponds to allowing for the possibilityy of diagram duplication in the simultaneous summation of RPA and t-matrix diagrams (NOZIÈRES and PINES). Inspection shows that no excited state diagrams are duplicated thereby, but that the ground state energy diagram: ⬤◯⬤, where ●, now given by f_0/m, is counted twice: once as an RPA diagram, and once as a contribution from the Hartree approximation. The latter term is

$$N_0^2 f_0/m = ● = ● + ◯ + ◯◯ + ◯◯◯ + \cdots$$
$$= ● + ⬤◯⬤ .$$

Hence the term $-\sum_k N_0 f_0^2/mk^2 = -$ ⬤◯⬤ must be subtracted from the ground-state energy expression, (2.17), in which $V_k = f_0/m$.

5) The properties of the dilute boson gas are as follows:

Excitation spectrum: $\qquad \omega_k \quad = \left\{ \dfrac{N_0 k^2 f_0}{m^2} + \dfrac{k^4}{4m^2} \right\}^{\frac{1}{2}}$.

Sound velocity: $\qquad s \quad = \dfrac{1}{m} (N_0 f_0)^{\frac{1}{2}}$.

Correlation length: $\qquad \lambda \quad = \dfrac{1}{2ms} = \dfrac{1}{2 (N_0 f_0)^{\frac{1}{2}}}$.

Structure factor: $\qquad S(k) = \dfrac{k^2}{2m\omega_k} = \dfrac{k^2}{2m} \left\{ \dfrac{1}{N_0 k^2 f_0/m^2 + k^4/4m^2} \right\}^{\frac{1}{2}} =$

$$= \dfrac{1}{(1 + k_c^2/k^2)^{\frac{1}{2}}} \qquad \text{where} \qquad k_c = 2) N_0 f_0)^{\frac{1}{2}} .$$

Depletion of ground state: $\quad N \quad = N_0 \left[1 + \dfrac{(Nf_0^3)^{\frac{1}{2}}}{3\pi^2} + ... \right]$.

Ground state energy: $\qquad E_0 \quad = \dfrac{1}{2} \dfrac{N^2 f_0}{m} \left[1 + \dfrac{16}{15\pi^2} (Nf_0^3)^{\frac{1}{2}} + ... \right]$.

6) The expansion parameter, $(Nf_0^3)^{\frac{1}{2}}$, is the ratio of f_0 to the correlation length, $(Nf_0)^{-\frac{1}{2}}$; it is this ratio which is important because one must *at the outset* include the RPA correlations in order to obtain a sensible theory for a boson system.

7) Calculations to the next highest order in $(Nf_0^3)^{\frac{1}{2}}$ have been made by LEE and YANG, BELIAEV, SAWADA, WU, and HUGENHOLTZ and PINES. The results are

$$E_0 = \frac{N^2 f_0}{2m}\left\{1 + \frac{16}{15\pi^2}(Nf_0^3)^{\frac{1}{2}} + \frac{Nf_0^3}{8\pi^2}\left(\frac{4}{3}-\frac{3}{\pi}\right)\ln(Nf_0^3) + \ldots\right\},$$

for the ground state energy, and

$$\omega_k = \frac{k}{m}(Nf_0)^{\frac{1}{2}}[1 + (1/\pi^2)(Nf_0^3)^{\frac{1}{2}}] - i\,\frac{3}{640\pi}(Nf_0^3)^{\frac{1}{2}}[p/(Nf_0)^{\frac{1}{2}}]^4 + \ldots,$$

for the energy of a low-momentum quasi-particle $[p \ll (Nf_0)^{\frac{1}{2}}]$. The imaginary part in the sound wave frequency results from the fact that in the low density gas one phonon can decay into two, when higher order terms are considered. It should also be noted that the term in $\ln(Nf_0^3)$ in the ground state energy shows that there is not a simple power series expansion in Nf_0^3.

8) The sound velocity obtained from the single particle excitation spectrum agrees with that calculated from the ground-state energy in accordance with the macroscopic relation

$$s = \left(\frac{\partial P}{\partial \varrho}\right)^{\frac{1}{2}}, \qquad \text{where} \qquad P = \varrho^2 \frac{\mathrm{d}}{\mathrm{d}\varrho}\left(\frac{E_0}{\varrho}\right).$$

9) By showing that the dilute hard sphere gas possesses a phonon-like spectrum, we have developed an explicit model for superfluidity. Unfortunately, liquid ^{4}He is not a dilute gas; the parameter (Nf_0^3), (assuming that He behaves like a collection of hard spheres, say) is far from small. It is therefore necessary to go beyond this simple model if we wish to account in a quantitative way for the excitation spectrum and other properties of ^{4}He. We shall return to this question in a subsequent lecture.

III – The Motion of an Impurity Atom (^{3}He) in Liquid He; a Microscopic Picture of Backflow.

1. – Impurity atom motion.

1) I now wish to discuss the way in which the backflow introduced by FEYNMAN and COHEN arises in a microscopic treatment. To do this, I shall first consider the same problem considered by FEYNMAN and COHEN, that of

an impurity atom moving through the boson system. The treatment which I give closely parallels that to be found in a forthcoming paper by MILLER, NOZIÈRES, and PINES. Let us begin by taking the interaction between the impurity atom and the bosons to be the same as that by which the bosons interact with one another. To describe the motion of the impurity atom we therefore add to the basic boson Hamiltonian (2.1) the following terms:

$$(3.1)\qquad \frac{P_{\mathrm{I}}^2}{2M_{\mathrm{I}}}+\sum_i V(\boldsymbol{r}_i-\boldsymbol{R}_{\mathrm{I}})=\frac{P_{\mathrm{I}}^2}{2M_{\mathrm{I}}}+\sum_k V_k\varrho_k\exp[i\boldsymbol{k}\cdot\boldsymbol{R}_{\mathrm{I}}]=$$
$$=\frac{P_{\mathrm{I}}^2}{2M_{\mathrm{I}}}+\sum_k V_k a_{p-k}^+a_p\exp[i\boldsymbol{k}\cdot\boldsymbol{R}_{\mathrm{I}}]=\frac{P_{\mathrm{I}}^2}{2M_{\mathrm{I}}}+H_{\mathrm{int}}\,,$$

$\boldsymbol{P}_{\mathrm{I}}$, $\boldsymbol{R}_{\mathrm{I}}$, and M_{I} are the impurity momentum, co-ordinate, and mass. The impurity atom is coupled to the density fluctuations of the boson system through H_{int}.

As a consequence of the interaction, the impurity is surrounded by a cloud of virtual excitations of the boson system. Hence the impurity will have an energy and a mass which are different from $P_{\mathrm{I}}^2/2M_{\mathrm{I}}$ and M_{I}. Further, the boson cloud will flow about the impurity atom in such a way as to cancel its longitudinal current; this backflow, we shall see, is just that introduced by FEYNMAN and COHEN for the same problem.

2) Let us further assume the interaction, V_k, is weak; we may then calculate the change in energy of the system by using second order perturbation theory. We have

$$(3.2)\qquad \Delta E=-\sum_n\frac{(H_{\mathrm{int}})_{n0}^2}{E_{n0}}\,,$$

the sum is over states n corresponding to momentum transfers q; for a given momentum transfer, one finds

$$(3.3)\qquad E_{n0}(q)=\omega_{n0}(q)-\frac{\boldsymbol{q}\cdot P_{\mathrm{I}}}{M_{\mathrm{I}}}+\frac{q^2}{2M_{\mathrm{I}}}\,,$$

where $\omega_{n0}(q)$ is a boson density fluctuation excitation frequency of momentum q. Remark that $E_{n0}>0$ as long as

$$(3.4)\qquad \omega_{n0}(q)+q^2/2M_{\mathrm{I}}>q\cdot P_{\mathrm{I}}/M_{\mathrm{I}}\,.$$

For small momentum transfers, we have seen that the density fluctuation excitation spectrum will be phonon-like; the condition (3.4) is therefore roughly equivalent to requiring

$$V_{\text{I}} = \boldsymbol{P}_{\text{I}}/M_{\text{I}} < s\,. \tag{3.5}$$

Under these circumstances, the impurity atom does not possess sufficient energy to excite phonons, and all excitations of the boson system are *virtual*. If, on the contrary, $P_{\text{I}}/M_{\text{I}} > s$, real transitions occur, and a probability per unit time of scattering of the impurity atom may be calculated along the lines sketched in Part I. We assume henceforth, that the impurity velocity is so small that no real excitation can occur. On substituting the value of H_{int} from (3.1), we find

$$\Delta E = -\sum_{qn} \frac{V_q^2 (\varrho_q)_{n0}^2}{\omega_{n0}(q) + q^2/2M_{\text{I}} - \boldsymbol{q}\cdot\boldsymbol{P}_{\text{I}}/M_{\text{I}}}\,. \tag{3.6}$$

We make use of the results of the preceding section:

$$(\varrho_q)_{n0}^2 = \frac{N_0 q^2}{2m\omega_q}\,, \tag{3.7a}$$

$$\omega_{n0}(q) = \omega_q = \left\{\frac{N_0 q^2 V_q}{m} + \frac{q^4}{4m^2}\right\}^{\frac{1}{2}}, \tag{3.7b}$$

and so obtain

$$\Delta E = -\sum_q \frac{V_q^2 N_0 q^2}{2m\omega_q} \frac{1}{\omega_q - \boldsymbol{q}\cdot\boldsymbol{P}_{\text{I}}/M_{\text{I}} + q^2/2M_{\text{I}}}\,. \tag{3.8}$$

3) Where the velocity of the impurity atom is sufficiently small, we can expand the denominator in eq. (3.8), and express the energy shift of the impurity atom in a power series expansion in P_{I}^2. If we keep the first two non-vanishing terms, we have

$$\Delta E = E_A + BP_{\text{I}}^2/2M_{\text{I}} + \ldots\,, \tag{3.9}$$

where E_A is the momentum-independent lowering of energy of the boson-impurity system, and B is likewise independent of momentum. On adding to (3.9), the impurity kinetic energy $P_{\text{I}}^2/2M_{\text{I}}$, we find

$$\Delta E = E_A + P_{\text{I}}^2/2M^*\,,$$

where M^*, the effective mass of the impurity, is given by

$$\frac{1}{M^*} = \frac{1}{M_{\text{I}}}\left[1 - \frac{1}{3}\sum_q \left\{\frac{V_q^2 N_0 q^4}{mM_{\text{I}}\omega_q}\frac{1}{(\omega_q + q^2/2M_{\text{I}})^3}\right\}\right]; \tag{3.11}$$

this may be written, upon using (3.7*b*), as

$$\frac{1}{M^*} = \frac{1}{M_\mathrm{I}}\left[1 - \frac{m}{M_\mathrm{I}}\frac{1}{3N}\sum_q \frac{(\omega_q^2 - q^4/4m^2)^2}{\omega_q[\omega_q + q^2/2M_\mathrm{I}]^3}\right], \tag{3.12}$$

when we further set $N_0 = N$, an appropriate approximation for weak interactions.

4) The result (3.12) is likewise valid for an impurity atom in interaction with a dilute boson gas; one need merely substitute for ω_q the appropriate expression

$$\omega_q = \left[\frac{Nq^2 f_0}{m^2} + \frac{q^4}{4m^2}\right]^{\frac{1}{2}}.$$

5) More appropriate for the present considerations is the pattern of flow about the impurity atom. The unperturbed wave function of the system is

$$\Psi_i = \exp[i\boldsymbol{P}_\mathrm{I}\cdot\boldsymbol{R}_\mathrm{I}]\Phi\,, \tag{3.13}$$

where Φ is the boson ground-state wave function. The perturbed wave function may be written as

$$\Psi = \exp[iS]\Psi_i\,. \tag{3.14}$$

To first order in H_{int}, S is determined by

$$i[H_1 + P_\mathrm{I}^2/2M_\mathrm{I},\, S] = -H_{\mathrm{int}}\,, \tag{3.15}$$

where H_1 is that part of the Hamiltonian which is kept in the Bogoljubov approximation. One finds

$$S = i\sum_k V_k\varrho_{-k}\exp[-i\boldsymbol{k}\cdot\boldsymbol{R}_\mathrm{I}]\left\{\omega_k + k^2/2M_\mathrm{I} - \frac{\boldsymbol{k}\cdot\boldsymbol{P}_\mathrm{I}}{M_\mathrm{I}}\right\}^{-1}. \tag{3.16}$$

Hence we have

$$\Psi = \exp\left[i\sum_{j=1}^{N} g(\boldsymbol{r}_j - \boldsymbol{R}_\mathrm{I})\right]\Phi\exp[i\boldsymbol{P}_\mathrm{I}\cdot\boldsymbol{R}_\mathrm{I}]\,, \tag{3.17}$$

where

$$g(r) = i\sum_k (V_k\{\omega_k + k^2/2M_\mathrm{I} - k\cdot P_\mathrm{I}/P_\mathrm{I}\})\exp[i\boldsymbol{k}\cdot\boldsymbol{r}]\,. \tag{3.18}$$

6) We establish the dipolar character of $g(r)$ by expanding the denominator of (3.18) in powers of P_I. The lowest order term is independent of P_I and leads to the constant term, E_A, in the energy of the impurity atom-boson system. The first-order term is

$$g_1(r) = i\sum_k \frac{V_k}{(\omega_k + k^2/2M_I)^2}\,\frac{\boldsymbol{k}\cdot\boldsymbol{P}_I}{M_I}\exp[i\boldsymbol{k}\cdot\boldsymbol{r}]\,. \tag{3.19}$$

In order to determine the form of (19) at large distances, we expand the denominator in powers of k. The leading term is given by

$$g_1(r) = i\sum_k \frac{V_k}{\omega_k^2}\,\frac{\boldsymbol{k}\cdot\boldsymbol{P}_I}{M_I}\exp[i\boldsymbol{k}\cdot\boldsymbol{r}] = \frac{m}{M_I}\,\frac{1}{N}\sum_k \frac{i\boldsymbol{k}\cdot\boldsymbol{P}_I}{k^2}\exp[i\boldsymbol{k}\cdot\boldsymbol{r}] = -\frac{m}{M_I}\,\frac{1}{4\pi N}\,\frac{\boldsymbol{P}_{\cdot}\cdot\boldsymbol{r}}{|r|^3}\,, \tag{3.20}$$

the expected dipolar form. If one combines (3.20) with (3.14) it is straightforward to show that the contribution to the current from the boson response, as described by (3.20), acts to cancel the direct contribution, P_I/M_I, at long distances. Corrections to (3.20) come at distances of the order of $1/M_I s$, the correlation length appropriate to the impurity atom.

7) To summarize: we have seen that in both the weak coupling limit and in the case of the dilute boson gas, the response of the boson system takes the form of dipolar flow about the impurity atom, in such a way as to cancel the longitudinal current associated with its motion at large distances. For the case considered—that of a slowly moving impurity atom—the response consists of a virtual cloud of phonons, which act also to increase the effective mass of the impurity atom according to (3.12).

2. – The relationship to the treatment of Feynman and Cohen; the effective mass of a ^{3}He atom in ^{4}He.

1) We have considered above the same problem as that considered initially by Feynman and Cohen. If we take the impurity atom to have the same mass as those making up the boson system, we find the strength of the dipole to be $(1/4\pi N)$, in agreement with their result and the classical value. FC (Feynman and Cohen) use a trial wave-function of the form

$$\Psi_{FC} = \exp[i\boldsymbol{P}_I\cdot\boldsymbol{R}_I]\exp\Big[i\sum_i g(\boldsymbol{r}-\boldsymbol{R}_I)\Big] \tag{3.21}$$

with

$$g(r) = A\,\frac{\boldsymbol{P}_I\cdot\boldsymbol{r}}{r^3}\,,$$

to calculate the effective mass of an impurity atom moving through liquid helium. They find

$$A \simeq -1/4\pi N ,$$

the effective mass turns out to be:

$$(3.22) \qquad \begin{cases} M^* = 1.5\ m & (M_\mathrm{I} = m_{^4\mathrm{He}}) \\ M^* = 1.67\, m & (M_\mathrm{I} = 3/4\, m_{^4\mathrm{He}}) . \end{cases}$$

2) How can we connect the foregoing microscopic considerations based on perturbation theory to the problem of an impurity atom moving in ^{4}He? In the latter problem the interactions are strong, and the system is dense, so that at first sight a direct connection does not seem possible. However, we can make an approximate and suggestive connection in the following way. Let us suppose that we can include many of the effects of multiple scattering in the helium system by means of an appropriate pseudo-potential. In that case we are led to consider a model Hamiltonian for the boson system, in which the effective interaction, is governed by some $\widetilde{V}_k$. One possible choice for $\widetilde{V}_k$ is such that if we keep only the Bogoljubov terms, H_1, we find the Feynman excitation spectrum for liquid helium,

$$\omega_k = \frac{k^2}{2m\,S(k)} .$$

Thus $\widetilde{V}_k$ is defined according to

$$(3.23) \qquad \omega_k = \left\{\frac{k \cdot N V_k}{m} + \frac{k^4}{4m^2}\right\}^{\frac{1}{2}} = \frac{k^2}{2m\,S(k)} ,$$

where $S(k)$ is the experimentally observed liquid structure factor. If we now take the interaction between the impurity atom and the ^{4}He atoms to be governed by this same effective interaction, $\widetilde{V}_k$, we can make use of the above calculations for the backflow and effective mass. The backflow is of course the same; the Feynman excitations execute a pattern of backflow about the impurity atom which is just such as to cancel its longitudinal current at large distances. The effective mass is given by

$$(3.24) \qquad \frac{M_\mathrm{I}}{M^*} = 1 - \frac{m}{M_\mathrm{I}} \frac{1}{k_D^3} \int_0^\infty \mathrm{d}k\, k^2 [1 - S^2(k)]^2 [1 + (m/M_\mathrm{I}) S(k)]^{-3} ;$$

a numerical integration yields:

$$(3.25)\qquad \begin{cases} M^* = 1.8\, M_{\mathrm{I}} & (M_{\mathrm{I}} = m_{^4\mathrm{He}})\,, \\ M^* = 2.1\, M_{\mathrm{I}} & (M_{\mathrm{I}} = m_{^4\mathrm{He}})\,. \end{cases}$$

These results are in quantitative agreement with those of FC, and with the experimental results for ^{3}He ind ^{4}He ($M^* = 2.5\, M_{^3\mathrm{He}}$).

3) Two words of caution are in order:

i) The change in the effective mass is large—so much so that in fact the use of perturbation theory to compute it is not justified. One can however keep the model, and use instead intermediate coupling theory, to treat the impurity-boson coupling. (The problem turns out to be identical in content to that of the motion of a polaron in an ionic crystal.). One then finds

$$(3.26)\qquad \frac{M^*}{M_{\mathrm{I}}} = 1 + \frac{m}{M_{\mathrm{I}}}\frac{1}{k_D^3}\int_0^\infty \mathrm{d}k\, k^2[1 - S^2(k)]^2[1 + M/M_{\mathrm{I}}S(k)]^{-3}\,,$$

and

$$(3.27)\qquad \begin{cases} M_{\mathrm{I}} = M_{^4\mathrm{He}}\,; & M^* = 1.4\, M_{^4\mathrm{He}} \\ M_{\mathrm{I}} = M_{^3\mathrm{He}}\,; & M^* = 1.5\, M_{^3\mathrm{He}}\,. \end{cases}$$

These are somewhat lower, but rather more accurate values for M^*.

ii) We have assumed in this simple calculation that $N_0 = N$. This turns out *not* to be a consistent assumption, for the depletion one calculates with the model is somewhat greater than 100 %!! However, this defect of the model calculation can be remedied by a somewhat more sophisticated field theoretic approach, which we will discuss in the last lecture. Here we wish only to anticipate that discussion to say that the results of the correct sophisticated calculation are in agreement with those of our naive (and incorrect) simple calculation above.

4) What now about the FC calculation for the excitation spectrum of ^{4}He? What moral can we draw from it?

i) FC determine the excitation spectrum by using a trial wavefunction which is a symmetrized version of their impurity wave function (3.21). As already discussed, the FC spectrum is in consideragly better agreement with experiment, and yields a roton energy (CHESTER: this Course) which is $\simeq 9.5\,^\circ$K. What the FC calculation does is to surround the roton with its appropriate cloud of virtual excitations. The FC picture of this as backflow is successful because in effect the roton is a very slowly moving excitation which is not capable of direct decay into two low-lying excitations; hence the surrounding cloud will consist primarily of virtual excitations.

ii) The obvious parallel in our model problem is to try to calculate directly the interaction between the Feynman excitations of frequency

$$\omega_F = \frac{k^2}{2m\,S(k)}\,;$$

that is, to take into account some of the remaining terms in H_2 in this fashion. Under these circumstances a given excitation will be surrounded by a cloud of other excitations providing the desired backflow. A direct calculation, using, say, the terms in the transformed version of H_2 involving three excitations would be consistent with this program; however, a certain amount of complication ensues because of the necessary presence of the coherence factors in the resultant integrals. An analogous calculation, in which no coherence factors were assumed to enter, was made some time ago by KUPER. He found a roton energy of 11.5 °K, which is in good agreement with the FC result. It can be argued that the coherence effects are perhaps not very important, since one is studying primarily the coupling of a roton to the phonons and the Feynman roton is rather close to a free particle. The use of perturbation theory to treat the coupling between the excitations is open to question, as we have seen. These is, moreover, the rather worrying fact that such a procedure is bound to lead to a rather considerable amount of diagram duplication, since some of the terms in H_2 have already been summed in arriving at $\tilde{V}_k$. Putting this worry aside, one might for instance argue that one can estimate the roton energy from our impurity atom intermediate coupling calculation by regarding the Feynman roton as an impurity of mass such that

$$\frac{k^2}{2M\,S(k)} = \frac{k^2}{2M_{\text{Roton}}}\,,$$

in which case

$$M_{\text{Roton}} = 1.3\,M_{^4\text{He}}\,.$$

One then finds that the coupling between the excitations (primarily to the Feynman phonons) increases the effective mass to

$$M^+_{\text{Foton}} \simeq 1.3 \cdot 1.5\,M_{4\text{He}} \simeq 2\,M_{^4\text{He}}\,.$$

The roton energy is some 12 °K which is not a bad estimate.

iii) The foregoing considerations are certainly qualitative in their nature. However they are suggestive, and when combined with the sum rule considerations of Section **1**, they lead me to the following conclusion. An accurate microscopic calculation of the excitation spectrum of ^{4}He requires that one include the cloud of virtual excitations around a given excitation as it moves through the liquid. If one does this, one may reasonably expect to account for the backflow and may consequently do as well as the FC variational calculation of the excitation spectrum. I will take up in the concluding lecture some microscopic calculations which appear to offer such a possibility.

IV. – Field-Theoretic Formulation.

1. – Introduction.

1) The advantages derived from the application of field-theoretic techniques to the interacting boson problem include the following:

a) They permit a rigorous definition of certain quantities of fundamental physical interest: for example, what we mean by an elementary excitation.

b) They offer a simple compact way of carrying out perturbation theoretic calculations. This is particularly true of the Feynman formulation of time-dependent perturbation theory, in which calculations are carried out by the introduction of diagrams, together with rules for computing the contribution from a given diagram.

2. – Elementary excitations.

1) Elementary excitations are defined by introducing a concept familiar in quantum field theory, that of propagators or Green's functions which determine the propagation in time of a given kind of excitation.

2) The quasi-particle (or single particle-like) excitation spectrum is determined from a study of the single-particle propagator, defined by

$$G(p, \tau) = -i\langle \Psi_0 | T\{a_p(\tau)\, a_p^+(0)\} | \Psi_0 \rangle . \tag{4.1}$$

The operators a_p and a_p^+ are in the Heisenberg representation, which is defined by time dependent operators

$$a_p(\tau) = \exp[+iH\tau]a_p \exp[-iH\tau] \tag{4.2}$$

and a time-independent wave-function, here, Ψ_0, the exact ground-state wave-function. T is the Dyson chronological operator, which orders earlier times to the right; thus, we have

$$\begin{cases} T\{a_p(\tau)a_p^+(0)\} = a_p(\tau)a_p^+(0) & \text{for } \tau > 0, \\ \qquad\qquad\qquad\quad = a_p^+(0)\,a_p(\tau) & \text{for } \tau < 0. \end{cases} \tag{4.3}$$

It is clear then that $G(p, \tau)$ has a discontinuity as we pass from $\tau < 0$ to $\tau > 0$. $G(p, \tau)$ possesses the following physical significance; if at time $t = 0$, one adds a particle of momentum p to the system (the system wave-function is then $\psi = a_p^+|0\rangle$), the probability amplitude that the system will be found in that state at a later time τ is $G(p, \tau)$. We shall be particularly interested in the Fourier transform of $G(p, \tau)$, defined by

$$G(p, \tau) = \frac{1}{2\pi}\int_{-\infty}^{\infty} d\varepsilon\, G(p, \varepsilon)\exp[-i\varepsilon\tau]\,. \tag{4.4}$$

Galitskii and Migdal have shown that the analytic behavior near the real ε-axis of $G(p, \varepsilon)$ determines the energy and life-time of a quasi-particle of momentum p.

3) As an example, consider the single-particle propagator for a non-interacting boson system. The exact ground-state wave-function, $|\psi_0\rangle$ is the vacuum state $|\varphi_0\rangle$. We have from (4.1), (4.2), and (4.3)

$$\begin{cases} G_0(p, \tau) = -i\exp[-i\varepsilon(p)\tau] & \tau > 0, \\ G_0(p, \tau) = 0 & \tau < 0, \end{cases} \tag{4.5}$$

where $\varepsilon(p) = p^2/2m$ is, as before, the free-particle energy. The Fourier transform of $G_0(p, \tau)$ is

$$G_0(p, \varepsilon) = \frac{1}{\varepsilon - \varepsilon(p) + i\delta}, \tag{4.6}$$

where δ is a positive infinitesimal number, specifying the position of the pole in the complex ε plane. To see that (4.6) is correct, consider

$$G_0(p,\tau) = \int_{-\infty}^{\infty} \frac{d\varepsilon}{2\pi} \frac{\exp[-i\varepsilon\tau]}{\varepsilon - \varepsilon(p) + i\delta}\,.$$

We may do the ε integration as a contour integration, as shown in Fig. 2. For $\tau < 0$, we can close the contour above the real axis, since then the exponential factor guarantees convergence of the integral; we obtain zero. For $\tau > 0$ we can close the contour below the real axis. We find then a contribution from the pole at $\varepsilon = \varepsilon(p)$, with residue $-2\pi i \exp[-i\varepsilon(p)\tau]$, and so obtain [5].

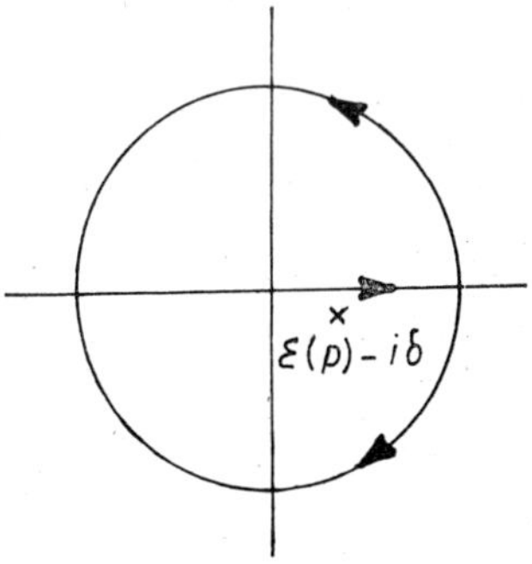

Fig. 2.

$\tau < 0$: Close above.

$\tau > 0$: Close below.

4) The density fluctuation excitation spectrum is determined by the density fluctuation propagator, $F(p, \tau)$, defined as

$$F(p,\tau) = -i\langle \psi_0 | T\{\varrho_p(\tau)\,\varrho_p^+(0)\}|\psi_0\rangle\,. \tag{4.7}$$

The physical significance of $F(p, \tau)$ is similar to that of $G(p, \tau)$. If one introduces a density fluctuation, ϱ_p^+, into the boson system at time $t = 0$, the probability of finding the system (boson ground-state plus density fluctuation) in that same state at time τ is proportional to $F(p, \tau)$.

5) Since in a neutron scattering experiment one excites directly a density fluctuation of the system, we might expect a relationship between $F(p, \tau)$ and the function $S(p, \omega)$, which describes the results of the scattering experiment. To determine the relation, consider first the Fourier transform, $F(p, \omega)$, which is defined through

$$F(p,\tau) = \int_{-\infty}^{\infty} \frac{d\omega}{2\pi} F(p,\omega)\exp[-i\omega\tau]\,. \tag{4.8}$$

If one introduces a complete set of intermediate states in the definition (4.7) of F, one finds

$$F(p,\omega) = \sum_n (\varrho_p)^2_{n0} \left\{\frac{1}{\omega - \omega_{n0} + i\delta} - \frac{1}{\omega + \omega_{n0} - i\delta}\right\}, \tag{4.9}$$

where the $(\varrho_p)_{n0}$ and ω_{n0} are the exact matrix elements and excitation frequencies appropriate to the density fluctuation, ϱ_p. Then using the relation

$$\lim_{\delta\to 0}\frac{1}{x\pm i\delta}+=P\frac{1}{x}\mp\pi i\delta_{(x)}\,,$$

where P denotes the principle value, we find directly

$$\operatorname{Im}F(p,\omega)=-\pi S(p,\omega)\,, \tag{4.10}$$

where $S(p,\omega)$ is the time-dependent structure factor measured in neutron scattering. (We have also used the fact that $\omega_{n0}>0$.) Equation (4.10) furnishes the desired relation between the density fluctuation excitation spectrum and the neutron scattering experiments.

6) With the aid of (4.9) and a similar expression for $G(p,\varepsilon)$ one can establish the analytic properties of these functions and derive dispersion relations which relate their real and imaginary parts (GALITSKII and MIGDAL). One may also derive exact expressions for the distribution function and the ground-state energy in terms of $G(p,\varepsilon)$. The distribution function for the interacting boson system is

$$\langle\psi_0|N_p|\psi_0\rangle=\langle\psi_0|a_p^+a_p|\psi_0\rangle=\frac{G(p,0^-)}{i}=\frac{1}{2\pi i}\int_C\mathrm{d}\varepsilon\,G(p,\varepsilon)\,, \tag{4.11}$$

where 0^- denotes the limit of $\tau\to 0$ from negative values; the contour C runs along the real axis and then is closed in the upper half plane (since the limit is 0^-). The ground state energy is given by

$$E_0=\frac{1}{2}N\mu+\int\frac{\mathrm{d}^3p}{(2\pi)^3}\int\frac{\mathrm{d}\varepsilon}{2\pi}\frac{[\varepsilon+p^2/2m]}{2}G(p,\varepsilon)\,, \tag{4.12}$$

if we work with the extended Hamiltonian, $H-\mu N_{0p}$; to derive (4.12), one simply differentiates in suitable fashion the equation of motion of $G(p,\tau)$ (HUGENHOLTZ and PINES).

3. – Perturbation theoretic formulation.

1) The perturbation theoretic formulation (based on quantum field theory methods) of the boson problem is greatly complicated, at first sight, by the macroscopic occupation of the zero-momentum state. One cannot simply apply the S-matrix development utilized for fermion systems by GOLDSTONE

and others because the linked cluster theorem fails for diagrams involving the zero-momentum particles (BELIAEV, HUGENHOLTZ and PINES). Several methods have been proposed for avoiding this difficulty of which the simplest. I believe, is that due to HUGENHOLTZ and myself. We begin by adding the extra term, $-\mu N_{op}$, to the Hamiltonian, and then follow the Bogoljubov prescription of replacing a_0 and a_0^+ by $N_0^{\frac{1}{2}}$. There are no difficulties with the zero-momentum state because all dynamic effects associated with it are neglected. One can then proceed in the usual way with an S-matrix development, with the important difference that in place of a single scattering diagram, there are now six different kinds:

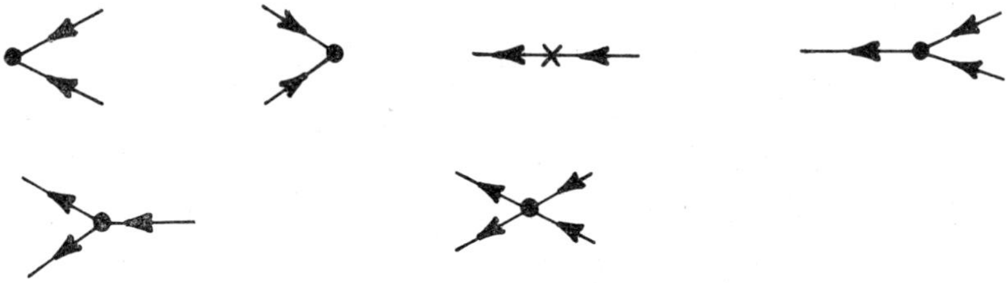

Here the interaction, represented by a dot, is taken to be one-half the sum of the direct and exchange terms, a useful simplification in dealing with short-range interactions. The details of the derivations of the basic formulae and the rules for carrying out perturbation-theoretic calculations may be found in HUGENHOLTZ and PINES. Here we shall merely draw a few pictures to provide a rough indication of the kind of considerations one carries out.

2) Suppose we are interested in calculating $G(p, \varepsilon)$. For a « normal » fermion system, this may be done most simply by means of the Dyson equation, which in terms of diagrams takes the following form:

$$(4.13) \qquad \underset{iG(p,\,\varepsilon)}{=\!=\!=\!=} = \underset{iG_0(p,\,\varepsilon)}{\rule{3cm}{0.4pt}} + \underset{iG_0(p,\,\varepsilon)}{\rule{2cm}{0.4pt}} \left(-i\Sigma\right) \underset{iG(p,\,\varepsilon)}{=\!=\!=\!=}$$

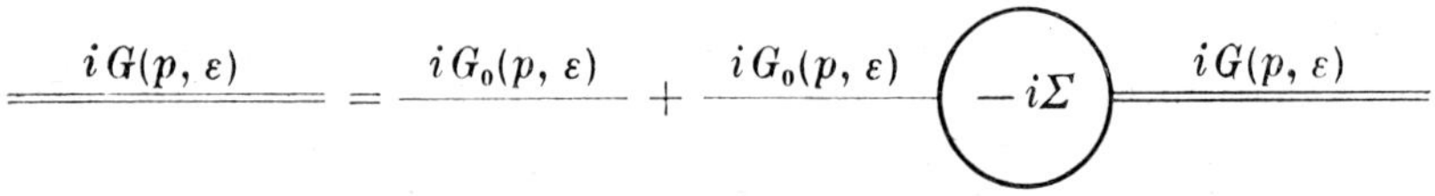

the single line denotes the free particle Green's function, $iG_0(p, \varepsilon)$, the double line denotes the exact Green's function, $iG(p, \varepsilon)$, and we define the circle to represent the sum of all irreducible self-energy diagrams, $-i\Sigma(p, \varepsilon)$. One sees by iteration that in this fashion one sums the following series for $G(p, \varepsilon)$:

$$\underset{}{\rule{2cm}{0.4pt}} = \rule{2cm}{0.4pt} + \rule{1cm}{0.4pt}\!\left(\Sigma\right)\!\rule{1cm}{0.4pt} + \rule{0.8cm}{0.4pt}\!\left(\Sigma\right)\!\rule{0.3cm}{0.4pt}\!\left(\Sigma\right)\!\rule{0.8cm}{0.4pt} + \dots .$$

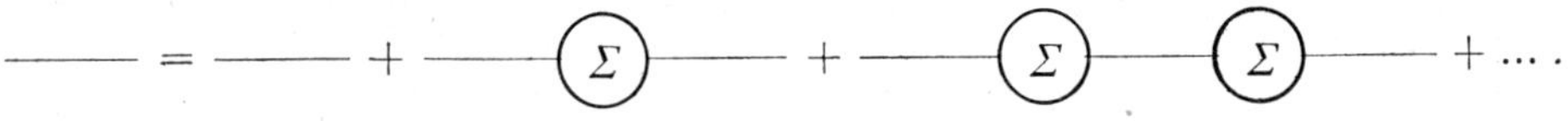

Hence, in order to prevent our counting terms in the perturbation series twice it is necessary to define an irreducible self-energy diagram as one which cannot be split into two self-energy diagrams by cutting a line in two.

The Dyson equation

$$G(p, \varepsilon) = G_0(p, \varepsilon) + G_0(p, \varepsilon)\, \Sigma(p, \varepsilon)\, G(p, \varepsilon)$$

is simply solved, and yields,

$$G(p, \varepsilon) = \frac{1}{\varepsilon - \varepsilon(p) - \Sigma(p, \varepsilon) + i\delta} \,. \tag{4.14}$$

We see that $\Sigma(p, \varepsilon)$ describes in compact fashion the modification in particle propagation brought about by the particle interaction. Because of the iteration procedure of (4.13), by a simple calculation of a primitive self-energy diagram, one effectively sums an infinite set of diagrams in the perturbation-theoretic expansion of $G(p, \varepsilon)$. The irreducible self-energy, $\Sigma(p, \varepsilon)$, will in general have both real and imaginary parts; the quasi-particle energy is obtained by solving the equation

$$\varepsilon = \varepsilon(p) + \Sigma(p, \varepsilon) \,,$$

which may be a very complicated affair indeed.

3) The Dyson equation, (4.13), is not sufficient for a boson system, because in addition to the basic interaction term there are the five additional terms which appear when we carry out the Bogoljubov prescription and neglect the dynamic effects associated with the zero-momentum state. Under these circumstances as was first shown by BELIAEV, the Dyson equation may be replaced by a pair of equations

(4.15)

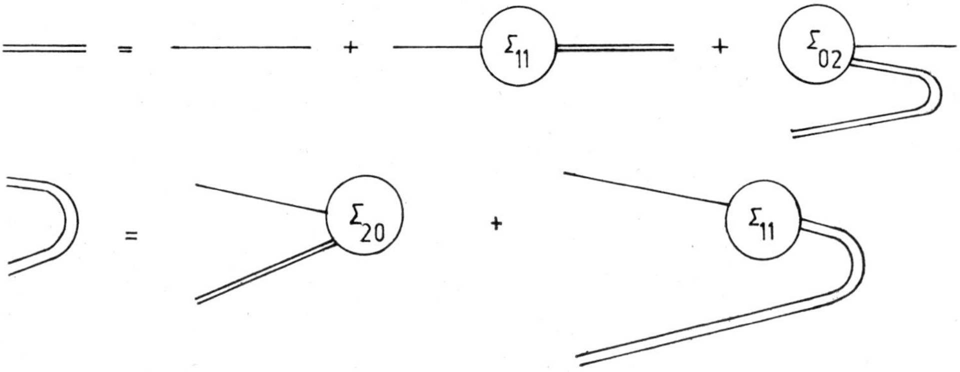

which read

$$
(4.16)\quad \begin{cases} G(p,\varepsilon) = G_0(p,\varepsilon) + G(p\varepsilon)\,\Sigma_{11}^{+}\,G_0(p\varepsilon) + \widetilde{G}(p,\varepsilon)\,\Sigma_{02}(p,\varepsilon)\,G_0(p\varepsilon)\,, \\ \widetilde{G}(p\varepsilon) \;\; = G_0(-p-\varepsilon)\,\Sigma_{20}\,G(p\varepsilon) + G_0(-p-\varepsilon)\,\Sigma_{11}^{-}\,\widetilde{G}(p\varepsilon)\,. \end{cases}
$$

Σ_{11} is the irreducible self-energy part we have introduced earlier; $\Sigma_{02} = \Sigma_{20}$ is the new self-energy part required to take account of the new kind of interactions now present. $\widetilde{G}(p,\varepsilon)$ is the Fourier transform of the propagator,

$$
(4.17)\quad \begin{cases} \widetilde{G}(p,\tau) = \langle\Psi_0|T[a_p^{+}(0)\,a_{-p}^{+}(\tau)]|\Psi_0\rangle \\ \qquad\qquad = \langle\Psi_0|T[a_{-p}(\tau)\,a_p(0)]|\Psi_0\rangle\,. \end{cases}
$$

The pair of coupled equations, (4.16), are algebraic, and may be simply solved to yield

$$
(4.18)\quad G(p,\varepsilon) = \frac{\varepsilon + \varepsilon(p) - \mu + \Sigma_{11}^{-}}{[\varepsilon - (\Sigma_{11}^{+} - \Sigma_{11}^{-})/2]^2 - [\varepsilon(p) - \mu + \frac{1}{2}(\Sigma_{11}^{+} + \Sigma_{11}^{-})]^2 + \Sigma_{02}^{2}}\,,
$$

where $\Sigma_{11}^{+} = \Sigma_{11}(p,\varepsilon)$ and $\Sigma_{11}^{-} = \Sigma_{11}(-p-\varepsilon)$.

4) The structure of $G(p,\varepsilon)$ for the boson system is quite different from that for fermion systems. HUGENHOLTZ and I were able to prove, from a term-by-term examination of the perturbation series, that

$$
(4.19)\qquad \Sigma_{11}(0,0) - \Sigma_{02}(0,0) = \mu\,.
$$

The content of the theorem is that the quasi-particle spectrum cannot exhibit a gap, and that, apart from possible pathological cases, one expects the low momentum quasi-particle excitations will have a linear dispersion relation

$$
\varepsilon(p) = sp\,.
$$

Thus we see that the assumption of macroscopic occupation of the zero momentum state leads directly to a linear phonon dispersion relation, and hence, via Landau's famous argument, to the concept of superfluidity. The proof is dependent on the validity of the perturbation expansion.

5) By means of the summation procedures inherent in the field theoretic formulation, we have removed the divergences in the Rayleigh-Schrödinger perturbation series at low momentum transfer. For example, if in (18) we

insert the weak coupling values:

$\Sigma_{11} =$  $= N_0\,(V_k + V_0)$

$\Sigma_{02} =$  $= N_0\,V_k$

$$\mu = N_0\,V_0\,,$$

we see that we at once obtain the Bogoljubov theory. We find

$$G(p,\varepsilon) = \frac{u_p^2}{\varepsilon - \omega_p + i\delta} - \frac{v_p^2}{\varepsilon + \omega_p - i\delta}\,, \tag{4.20}$$

where u_p, v_p, and ω_p are given by eqs. (4.12) and (4.13) of Section **2**. Thus u_p measures the strength of the free-particle-like pole, and v_p measures the strength of the new pole introduced by the condensed state. The calculation of the low density limit may be performed with equal ease (and with no difficulties about diagram duplication either). For that calculation, and for the calculation of the logarithmic term in the ground-state energy and the corrections to the phonon spectrum, I refer to the original papers (BELIAEV, HUGENHOLTZ and PINES).

6) There is an alternative way of deriving the basic equation, (4.18), which I should like to mention. Suppose instead of the Beliaev equations we consider the following single equation:

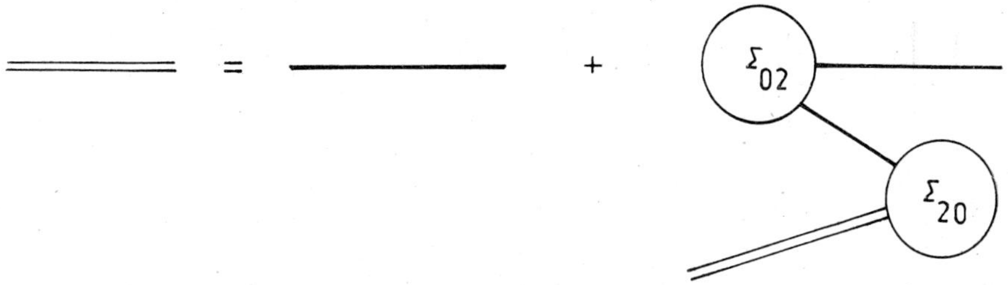

in which the double line denotes the true single particle propagator, and the heavy line denotes a « normal » state propagator,

$$G_n(p,\varepsilon) = \frac{1}{\varepsilon - \varepsilon(p) - \Sigma_{11}(p,\varepsilon) + i\delta}\,, \tag{4.21}$$

that is, one in which all self-energy effects associated with one line in and one line out have been included. The above equation then takes the form

$$G = G_n^+ + G_n^+ G_n^- (\Sigma_{02}^2 G)\,, \qquad G = \frac{(G_n^-)^{-1}}{(G_n^- G_n^+)^{-1} - \Sigma_{02}^2}\,, \tag{4.22}$$

which may easily be seen to reduce to our previous result (4.18). It is, in fact, not difficult to see from the diagrams, by comparing Beliaev's equation with the above, that the two must be identical. In some ways the above equation is more suggestive of the new features in the problem, in that it shows clearly that one need only account in consistent fashion for the scattering of pairs of particles in and out of the condensed state. We see, too, that the lack of particle conservation arises only in intermediate state processes.

7) The Fourier transform $F(k\omega)$ of the density fluctuation propagator may be calculated for the low-density boson system. In lowest approximation it yields just the results we have already described for $S(k\omega)$. A higher-order calculation has not yet been carried out. The calculation is of some interest because it is an open question whether « pure » collective modes exist in the boson system. Thus, one would expect, in general, poles in $F(k\omega)$ associated with the excitation of quasi-particles from the condensed state. These are, in fact, the only poles thus far observed in the neutron scattering experiments (HENSHAW and WOODS), since in the region of the roton minimum and in the phonon region, the energy of the rotons and phonons agrees within experimental error with the quasi-particle energies determined from specific heat data.

8) The dilute boson gas calculation does not furnish a reasonable model for the behavior of actual liquid ^{4}He, because the range of the repulsive interaction and the density are such that $Nf_0^3 \gg 1$. BRUECKNER and SAWADA have attempted an intermediate density calculation along the same lines of approximation as the Brueckner fermion calculations, by taking quasi-particle propagators as the intermediate state propagators for the boson t-matrix, which is then to be determined in self-consistent fashion. Their results are in fair qualitative agreement with the experimental excitation spectrum. However, the results are open to serious question, because BRUECKNER and SAWADA did not allow for depletion of the ground state as a consequence of particle interaction. Recently PARRY and TER HAAR have shown that the depletion of the zero-momentum state with the Brueckner-Sawada theory amounts to some 270 percent. They then do a calculation which allows for the depletion, and find the resulting quasi-particle excitation curves no longer bend over, so that even the qualitative agreement with experiment is lost.

4. – A microscopic model for $G(p, \varepsilon)$ in liquid He.

1) I should now like to return tof the question of the use of a model Hamiltonian to describe the properties of liquid He. In the preceding lecture, we were led to consider the possibility of working with an effective interaction, $\tilde{V}_k$, which yields, in the Bogoljubov approximation (the RPA), the Feynman

excitation spectrum. This procedure, in our present language, is equivalent to taking

$$\left\{\begin{aligned} \Sigma_{11} &= N_0(\widetilde{V}_k + \widetilde{V}_0)\,, \\ \Sigma_{02} &= N_0\widetilde{V}_k\,, \\ \mu &= N_0\widetilde{V}\,. \end{aligned}\right. \tag{4.23}$$

It leads to the following difficulties:

i) If one calculates N_0 from the basic equation:

$$N_0 + \frac{1}{2\pi i}\int_{-\infty}^{\infty} d\varepsilon\, G(p, \varepsilon) = N\,, \tag{4.24}$$

one finds $N_0 > N$.

ii) Even if one puts this difficulty aside by, for example, deciding to concentrate mainly on the long wave length excitations. one at once finds other contradictions arising from the value of $S(k\omega)$ one obtains with (4.23), namely

$$S(k\omega) = N_0\, S(k)\, \delta(\omega - \omega_k)\,. \tag{4.25}$$

Thus one does not obtain the experimental structure factor, and the sum rule

$$\int_0^{\infty} d\omega\, S(k\omega)\,\omega = Nk^2/2m\,, \tag{4.26}$$

is exhausted only to order N_0/N. Where N_0 is very different from N (and there is good reason to think it is; ONSAGER and PENROSE estimate $N_0/N = 0.08$), these departures are so large as to render the model (4.23) meaningless.

2) A partial resolution of the above dilemma may be achieved along the following lines. Suppose we address ourselves to the more modest problem of writing down a sensible value of $G(p, \varepsilon)$, $F(p, \omega)$, and hence $S(p, \omega)$ for the long wavelength excitations, $k \leqslant k_D$, which possess an essentially phonon-like spectrum. For these excitations, the Feynman scheme works rather well so that we know that a single excitation nearly exhausts the f-sum rule. Hence, we know that, to a good degree of approximation,

$$S(k, \omega) = N\, S(k)\, \delta\{\omega - k^2/[2m\, S(k)]\}\,; \tag{4.27}$$

we are led to inquire whether there is a simple choice of Σ_{11} and Σ_{02} which yields (4.27) in suitable approximation. The answer is « yes », provided we are willing to go to a model Green's function $G(p, \varepsilon)$, in which $\Sigma_{11}(p, \varepsilon)$ is taken to be energy-dependent.

Thus suppose we take

$$(4.28)\qquad \begin{cases} \Sigma_{11}^{+} = N_0 \widetilde{V}_k + N_0 \widetilde{V}_0 + a\varepsilon\,, \\ \Sigma_{11}^{-} = N_0 \widetilde{V}_k + N_0 \widetilde{V}_0 - a\varepsilon\,, \\ \Sigma_{02} = N_0 \widetilde{V}_k\,, \end{cases}$$

where a is to ben determined in such a way as to yield (4.27). With this choice the single particle Green's function is

$$(4.29)\qquad G(p, \varepsilon) = \frac{\varepsilon(1-a) + p^2/2m + N_0 \widetilde{V}_p}{\varepsilon^2(1-a)^2 - (\omega_p')^2}\,,$$

where

$$(4.30)\qquad (\omega_p')^2 = \frac{N^0 p^2 \widetilde{V}_p}{m} + \frac{p^4}{4m^2}\,.$$

If, further, we keep only the contribution to $F(p, \varepsilon)$ arising from the excitation of quasi-particles from the condensed state, we find

$$(4.31)\qquad F(p, \varepsilon) = \frac{N_0 p^2}{\varepsilon^2(1-a)^2 - \omega_p'^2 + i\delta}\,.$$

If, therefore, we choose $\widetilde{V}_p$ such that

$$(4.32)\qquad \frac{\omega_p'}{1-a} = \frac{p^2}{2m S(p)}\,,$$

and a such that

$$(4.33)\qquad 1 - a = (N_0/N)^{\frac{1}{2}}\,,$$

we see, on substitution into (4.31), that

$$(4.34)\qquad F(p, \varepsilon) = \frac{N p^2}{\varepsilon^2 - [p^2/2m S(p)]^2 + i\delta}\,,$$

from which, with the aid of (4.10), (4.27) follows.

3) We remark that the response of the boson system to an impurity atom may be expressed directly in terms of $F(p, \varepsilon)$. Moreover, the choice

(4.34), for $F(p, \varepsilon)$ is equivalent to the main calculations we carried out in our previous lecture, so that the above derivation of (4.34) serves to justify the results obtained for the effective mass of a ^{3}He atom by the incorrect arguments of the preceding lecture.

4) A little thought shows that the result, (4.29), cannot be expected to be correct for very large momenta, since it predicts a quasi-particle energy different from that of a free particle. If, however, we confine our attention to low momenta, ($p \ll k_D$), we may expect to use (4.29) to obtain a lower limit on the deplection of the zero momentum state in liquid He as a consequence of particle interaction. We have, from (4.11) and (4.29),

$$N_p = \frac{1}{4S(p)}\left[1 - S(p)\left(\frac{N}{N_0}\right)^{\frac{1}{2}}\right]. \tag{4.35}$$

To calculate the depletion we need merely to solve the equation

$$\frac{N'}{N} = \frac{\sum\limits_{p \neq 0} n_p}{N} = 1 - \frac{N_0}{N} = \frac{3}{4k_D^3}\int\limits_0^\infty \mathrm{d}k\, k^2 \frac{1}{S(p)}\left[1 - S(p)\left(\frac{N_0}{N}\right)^{\frac{1}{2}}\right]^2, \tag{4.36}$$

where $k_D^3 = 6\pi^2 N$. This determines N_0. An approximate calculation can be performed simply if we take

$$S(p) = \frac{p}{2ms}, \tag{4.37}$$

up to $k \simeq k_D$, where s is the sound velocity. Using this value of $S(p)$ in (4.36) will give the contribution of the phonons to the depletion. We find $N_0/N = 0.57$, or a 43 % depletion as our lower limit.

5. – Interaction between ^{3}He atoms.

1) As an another example of the field theoretic methods we can compute the effective interaction between two ^{3}He particles in a ^{4}He bath. If we assume that all particles initially have an effective interaction, $\tilde{V}$, as before, then the bare interaction between the ^{3}He atoms is

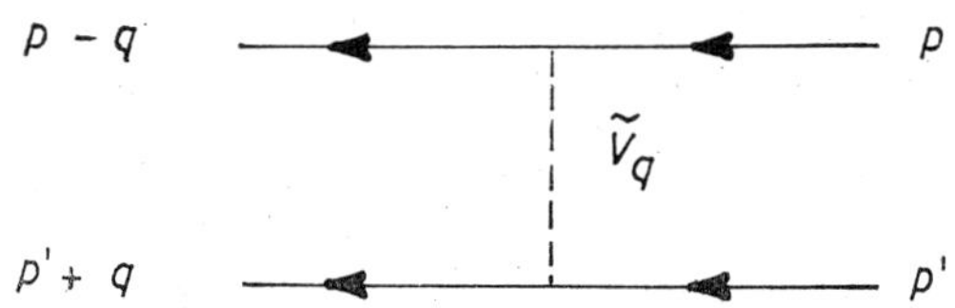

2) One can easily compute the modification in this interaction brought about by screening effects calculated using (29) and (34). The result may be expressed by the following picture:

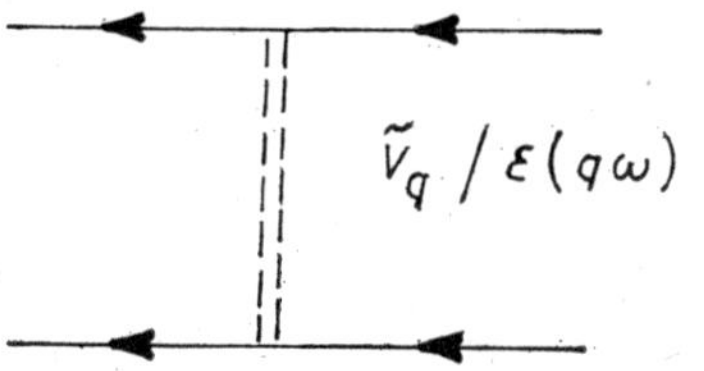

the new interaction is $\tilde{V}_q/\varepsilon(q\omega)$, where $\varepsilon(q\omega)$ is the frequency-dependent screening constant. The frequency appropriate to the scattering act we have sketched is

$$\frac{(p+q)^2}{2m} - \frac{p^2}{2m} = \frac{q^2}{2m} + \frac{\boldsymbol{p}\cdot\boldsymbol{q}}{m} .$$

The effective interaction, $\tilde{V}_q/\varepsilon(q\omega)$ turns out to be

$$\frac{\tilde{V}_q}{\varepsilon(q\omega)} = \tilde{V}_q \left\{ \frac{\omega^2 - (N/N_0)\, q^4/4m^2}{\omega^2 - [q^2/2m\, S(q)]^2} \right\} . \tag{4.38}$$

3) The interaction, (4.38), has certain interesting properties. Suppose that $\tilde{V}_q$ is repulsive. Then for

$$\frac{N}{N_0}\frac{\hbar q^2}{2m} < \omega < \omega_q , \tag{4.39}$$

we find $\varepsilon < 0$, and the effective interaction is attractive. Otherwise it is still repulsive. The physical origin of the attractive interaction is that in the region given by (4.39), the attractive interaction between the backflow of the two ^{3}He atoms outweighs the repulsive interaction between the atoms themselves, and a net attraction results. A possible result of this attractive interaction might be a superfluid ^{3}He phase. Experimentally, one finds instead a phase separation of ^{3}He in ^{4}He which occurs before the ^{3}He is sufficiently degenerate to exhibit superfluidity. One might hope then to obtain an explanation of this phase separation by saying that the attractive interaction leads to an instability in the dilute ^{3}He system.

V. – Conclusions and Speculations.

1. – Conclusions.

I should like to summarize briefly certain of the conclusions that I believe it is sensible to draw concerning the present status of the microscopic theory of ^{4}He.

1) The condensed, zero-momentum state plays a very special role in the formulation of the theory; from the assumption of macroscopic occupation of this state, one is led directly to a whole class of theories which predict superfluid behavior.

2) The dilute hard sphere gas provide us with a specific, well-defined model for an excitation spectrum, which, by the Landau argument, yields an explanation of superfluidity.

3) In a microscopic theory, the backflow introduced by FEYNMAN and COHEN will be accounted for by taking into account the interaction between virtual excitations in the phonon-roton system.

2. – Speculations and open questions.

1) I should like to say a word about the role of the attractive forces in ^{4}He, and about attractive forces in boson systems in general. If one considers V to be negative for small q, one finds that the quasi-particle energies, as well as the density fluctuations excitation frequencies are then pure imaginary, signalling an instability of the system or an incorrect starting point for the calculation. What seems likely, since this is a very long wavelength instability, is that the system will reduce its volume to the point that the « effective » potential, V_q^{eff}, is positive for small momentum transfers. Thus, I am arguing that the condensed state only exists once the system has reached a size such that the long wavelength phonons represent a stable excitation. It seems clear, on the other hand, that the attractive forces do play a considerable role in determining the short wavelength excitation modes in liquid ^{4}He, in that they are undoubtedly responsible for the observed structure factor, and may also be responsible for much of the bending over of the excitation curve toward the roton minimum.

2) What are some useful directions for further investigations concerning a microscopic theory of liquid Helium? One possibility, which is presently

under investigation, is the following. Suppose one considers a t-matrix equation in which the intermediate state propagators are neither those appropriate to non interacting particles (as was the case for the dilute gas) nor single particles with an energy gap (as was the case studied by BRUECKNER and SAWADA), but rather correspond to the actual excitation of the system, *i.e.*, the phonons, for example. This procedure would seem to offer two interesting possibilities:

i) It will allow for the possibility of backflow, in which a single particle may have about it the appropriate phonon cloud.

ii) One may also find collective modes of the boson system, analogous to the collective modes encounteres in the BCS theory of superconductivity.

The approach I am proposing is indeed closely related to one way of including both backflow effects and collective modes in the theory of superconductivity.

3) The main open question is the construction of a theory which will allow properly for the very large depletion of the zero momentum state as a consequence of particle interaction. Whether the sort of improved microscopic theory I have just sketched is adequate to do this is far from clear. What is clear is that in many ways the problem one faces is more difficult than those encountered for electrons in metals or superconductivity, where the analogous effect (the alteration of the free-particle distribution function) does not appear to be of such a substantial magnitude.

* * *

I should like to thank Dr. G. BAYM, Dr. D. MERMIN, and Mr. P. HOHENBERG for assistance in the preparation of these notes. I should also like to acknowledge the support of the Army Research Office (Durham) and the National Science Foundation.

REFERENCES

I. - *Slow Neutron Scattering*

Theory:

[1] M. COHEN and R. P. FEYNMAN: *Phys. Rev.*, **107**, 13 (1957).
[2] L. VAN HOVE: *Phys. Rev.*, **95**, 249 (1954).

Experiment:

[3] D. G. HENSHAW and A. D. B. WOODS: *Phys. Rev.*, **121**, 1266 (1961).
[4] H. PALEVSKY, K. OTNES and K. E. LARSSON: *Phys. Rev.*, **112**, 11 (1959).
[5] J. L. YARNELL, G. P. ARNOLD, P. J. BENDT and E. C. KERR: *Phys. Rev.*, **112**, 1379 (1959).

II. - *Theory of Elementary Excitations in Liquid Helium*

[6] L. D. LANDAU: *Žurn. Ėksp. Teor. Fiz.*, **5**, 71 (1941); **11**, 91 (1947).
[7] R. P. FEYNMAN: *Phys. Rev.*, **94**, 267 (1954).
[8] R. P. FEYNMAN and M. COHEN: *Phys. Rev.*, **102**, 1189 (1956).
[9] C. G. KUPER: *Proc. Roy. Soc.*, **233**, 223 (1955).

III. - *Microscopic Theory of Interacting Boson Systems*

[10] S. T. BELIAEV: *Sov. Phys. JETP*, **7**, 289, 299 (1958).
[11] N. N. BOGOLJUBOV: *Žurn. Ėksp. Teor. Fiz.*, **11**, 23 (1947).
[12] K. A. BRUECKNER and K. SAWADA, *Phys. Rev.*, **106**, 1117 (1959); **106**, 1128 (1959).
[13] V. GALITSKII and A. B. MIGDAL: *Sov. Phys. JETP*, **7**, 96 (1958).
[14] N. M. HUGENHOLTZ and D. PINES: *Phys. Rev.*, **116**, 489 (1959).
[15] T. D. LEE, K. HUANG and C. N. YANG: *Phys. Rev.*, **106**, 1135 (1957).
[16] T. D. LEE and C. N. YANG: *Phys. Rev.*, **112**, 1419 (1958).
[17] A. MILLER, P. NOZIÉRES and D. PINES: *Phys Rev.*, **127**, 1452 (1962).
[18] W. E. PARRY and D. TER HAAR: *Annals of Physics*, **19**, 496 (1962).
[19] K. SAWADA: *Phys. Rev.*, **116**, 1344 (1959).
[20] T. T. WU: *Phys. Rev.*, **115**, 491 (1959).

Theory of Liquid ^{3}He (*).

A. M. SESSLER (**)
Ohio State University - Columbus, Ohio

I. – Introduction.

The theory if liquid ^{3}He has been considerably advanced during the last five years, and a start in the understanding of the properties of the liquid at very low temperatures appears to have been made. This review article will attempt to introduce the reader to the subject. In this spirit it will emphasize the basic ideas and concepts, leaving the complicated applications and difficult computational details to be pursued in the literature by the interested reader.

Attention in this article will be confined to the liquid at low temperatures, where the concept of a quasi-particle is valid. Following the Introduction in which some basic results of Statistical Mechanics are reviewed and the ideal Fermi gas discussed, we turn in Part II to the phenomenological theory as first developed by LANDAU. Part III discusses the microscopic approach. Here, in the first part, the second quantization formalism is developed without assuming a knowledge of field theory. The theory of the ground state is then discussed, and the last part is devoted to statistical mechanics which is developed using the formalism of Green's function and spectral representations. In Part IV the possibility that the ground state of the liquid is a highly correlated state is described, along with some properties of the predicted low temperature phase.

This article has been written so that it may be read without consultation of the literature, but of course it must be appreciated that the serious student

---

(*) Supported in part by the National Science Foundation, U.S.A. This paper was written while the author enjoyed the hospitality of the Institute for Theoretical Physics, University of Copenhagen, Denmark.

(**) Now at the Lawrence Radiation Laboratory, University of California, Berkeley, Cal.

will want to study the original literature. To aid the readers—whose backgrounds will be diverse—in this regard, Section V consists of a guide to the literature.

Finally, the author wishes to acknowledge that he has been assisted by the unpublished notes of R. L. MILLS, R. BROUT, G. BROWN, and R. E. MILLS, as well as by conversations with G. BAYM, G. CHESTER, and D. J. THOULESS. He has, of course, drawn freely from the many papers in the literature, and it is hoped that adequate credit will be clear from the references cited.

1. – Statistical mechanics.

For both the phenomenological theory of liquid helium three (Part II) and the microscopic theory (Part III) one needs an understanding of the basic ideas of Statistical Mechanics. In this section we summarize some of the results we shall need.

1·1. *Canonical distribution.* – Classical statistical mechanics considers a Hamiltonian system characterized by co-ordinates q_i and momenta p_i. One then considers an ensemble of such systems and lets the distribution function $\varrho(p_i, q_i, t)$ be proportional to the probability of finding the system in the configuration specified by (p_i, q_i) at time t. One normalizes the distribution function so that

$$\int \varrho(p_i, q_i, t)\, \mathrm{d}p_i\, \mathrm{d}q_i = 1\,. \tag{1.1}$$

It now follows, from Liouville's theorem, and the requirement that subsystems of a system be statistically independent, that for an ensemble of systems in equilibrium $\log \varrho$ is a function of the additive integrals of the motion. By putting the system in a large box (so that neither linear nor angular momentum is a constant of the motion) one has

$$\varrho = A e^{BE}\,, \tag{1.2}$$

where A and B are constants and $E(p_i, q_i)$ is the energy of the system. Given any function of dynamical variables $f(p_i, q_i)$, identification of time averages with ensemble averages allows the evaluation of the average value of f,

$$\bar{f} = \int f(p_i, q_i)\, \varrho(p_i, q_i)\, \mathrm{d}p_i \mathrm{d}q_i\,. \tag{1.3}$$

The concept of entropy is then introduced, related to the number of different microscopic ways in which essentially the same macroscopic state can be achieved, and it is shown that

$$S = -k\,\overline{\log \varrho}\,. \tag{1.4}$$

Thermodynamics may be employed to identify the temperature from

$$\frac{1}{T} = \left(\frac{\partial S}{\partial \bar{E}}\right)_V, \tag{1.5}$$

which when combined with the definition of the Free Energy $F = \bar{E} - TS$ yields an evaluation of the constants A and B in eq. (1.2). One obtains the Gibbs distribution function:

$$\varrho = \exp\left[\frac{F - E}{kT}\right]. \tag{1.6}$$

Quantum mechanically one introduces a density matrix which is then shown to be diagonal in the energy representation with eigenvalues w_n which are related to the energy eigenvalues E_n by the formula

$$w_n = \exp\left[\frac{F - E_n}{kT}\right]. \tag{1.7}$$

From the normalization condition

$$\sum_n w_n = 1\,, \tag{1.8}$$

one concludes

$$F = -kT \log \sum_n \exp\left[-\frac{E_n}{kT}\right], \tag{1.9}$$

which formula allows the calculation of all thermodynamic properties, given knowledge of the energy spectrum E_n.

1·2. *Grand canonical distribution.* – It is often convenient to work with systems in which the number of particles is not kept fixed. By arguments analogous to those mentioned above, it can be seen that $\log \varrho$ will now depend linearly also on the number of particles in the system. Proceeding directly to the quantum-mechanical case, one has

$$w_{n,N} = \exp\left[\frac{\Omega + \mu N - E_{n,N}}{kT}\right], \tag{1.10}$$

where the constant μ is the chemical potential and Ω is the thermodynamic potential. The normalization condition now yields

$$\Omega = -kT \log \sum_N \left\{ \exp\left[\frac{\mu N}{kT}\right] \sum_n \exp\left[-\frac{E_{n,N}}{kT}\right]\right\}. \tag{1.11}$$

The reader will recall the thermodynamic relations

$$\left\{\begin{aligned} \Omega &= F - \mu N = -PV, \\ N &= -\left(\frac{\partial \Omega}{\partial \mu}\right)_{T,V}, \\ S &= -\left(\frac{\partial \Omega}{\partial T}\right)_{\mu,V}, \end{aligned}\right. \tag{1.12}$$

so that having calculated Ω it can be seen that one can easily deduce the pressure and entropy, while the chemical potential can be eliminated in terms of the average number of particles. All other thermodynamic properties now follow readily. For example, the specific heat is given by

$$C_V = T\left(\frac{\partial S}{\partial T}\right)_V = -T\left(\frac{\partial^2 \Omega}{\partial T^2}\right)_{\mu,V} - T\left(\frac{\partial^2 \Omega}{\partial \mu \partial T}\right)_V^2 \Bigg/ \left(\frac{\partial^2 \Omega}{\partial \mu^2}\right)_{T,V}. \tag{1.13}$$

1·3. *Thermodynamic inequalities.* – From the well known fact that in any process the heat going into the body $\delta Q \leqslant T\,\delta S$, it follows that in any system in which the temperature and volume are held fixed the free energy F will decrease in time until equilibrium is attained—in which state the free energy is at its lowest value. This important principle affords a variational method of calculation which is the generalization to statistical mechanics of the Rayleigh-Ritz variational method of quantum mechanics.

It now follows that if T, V, and μ are held fixed, then Ω attains its minimum at equilibrium. This property proves most convenient for theoretical calculations where these quantities are usually held constant during the course of calculations.

2. – Ideal Fermi gas.

Because of its obvious relevance, we review in this section the theory of an ideal Fermi gas. Actually we shall see in Part II that this is remarkably pertinent to the theory of even the highly nonideal liquid helium three.

2·1. *Entropy of a nonequilibrium ideal Fermi gas.* – For a nonequilibrium gas we can calculate the entropy from its definition in terms of the probability of attaining the macroscopic state. If we have groups of G_j states and N_j particles in the states of group j, then the number of ways of distributing N_j particles in G_j states is (recalling that for Fermi statistics we can put at most one particle in each state):

$$W_j = \frac{G_j!}{N_j!(G_j - N_j)!}\,. \tag{1.14}$$

Now from

$$S = k \log \left(\prod_j W_j\right) \tag{1.15}$$

using Stirling's formula, and introducing $n_j = N_j/G_j$ one obtains

$$S = -k \sum_j G_j[n_j \log n_j + (1 - n_j)\log(1 - n_j)]\,. \tag{1.16}$$

2·2. *Distribution function.* – For a gas of noninteracting particles which obey Fermi statistics we must have a wave function which is an anti-symmetrized product of single particle wave functions. Consequently each momentum state can be either occupied or not with the result that the formula for Ω (eq. (1.11)), when applied to particles with momentum p, becomes

$$\Omega_p = -kT \log\left[1 + \exp\left[\frac{\mu - \varepsilon_p}{kT}\right]\right]. \tag{1.17}$$

The mean number of particles in the p-th state is then

$$n_p = -\frac{\partial \Omega_p}{\partial \mu} = \frac{1}{\exp[(\varepsilon_p - \mu)/kT] + 1}\,. \tag{1.18}$$

For the whole system one has (since the states of an ideal gas are linearly independent)

$$\Omega = \sum_p \Omega_p\,, \qquad \Omega = -kT \sum_p \log\left\{1 + \exp\left[\frac{\mu - \varepsilon_p}{kT}\right]\right\}, \tag{1.19}$$

from which it is a simple matter to calculate the total number of particles (using eq. (1.12)), and so eliminate μ.

2·3. *Low-temperature properties.* – In this section we shall derive just a few of the low-temperature properties of an ideal Fermi gas. We may take

$\varepsilon = p^2/2m$, and by fitting waves into a box deduce that the number of particles in a region of phase space of volume $d^3p\,dV$ is

$$\frac{2\,d^3p\,dV}{(2\pi\hbar)^3}\,, \tag{1.20}$$

where the factor of 2 is associated with the two possible spin orientations in each space-state. Thus from eq. (1.18) we have for the number of atoms with momentum between p and $p+dp$

$$dN_p = \frac{2(4\pi)p^2\,dp\,dV}{(2\pi\hbar)^3\{\exp[((p^2/2m)-\mu)(1/kT)]+1\}}\,. \tag{1.21}$$

Converting to ε as the free variable

$$dN_\varepsilon = \frac{Vm^{\frac{3}{2}}(2\varepsilon)^{\frac{1}{2}}\,d\varepsilon}{\pi^2\hbar^3\{\exp[(\varepsilon-\mu)/kT]+1\}}\,. \tag{1.22}$$

From this we have:

$$N = \frac{Vm^{\frac{3}{2}}\sqrt{2}}{\pi^2\hbar^3}\int\limits_0^\infty \frac{\varepsilon^{\frac{1}{2}}\,d\varepsilon}{\{\exp[(\varepsilon-\mu)/kT]+1\}}\,, \tag{1.23}$$

and

$$\Omega = -\frac{2\sqrt{2}\,m^{\frac{3}{2}}V}{3\pi^2\hbar^3}\int\limits_0^\infty \frac{\varepsilon^{\frac{3}{2}}\,d\varepsilon}{\exp[(\varepsilon-\mu)/kT]+1}\,. \tag{1.24}$$

In the limit as $T\to 0$ we may use the well known approximation that

$$\int\limits_0^\infty \frac{f(\varepsilon)\,d\varepsilon}{\exp[(\varepsilon-\mu)/kT]+1} \approx \int\limits_0^\mu f(\varepsilon)\,d\varepsilon + \frac{\pi^2}{6}(kT)^2 f'(\mu) + \dots\,, \tag{1.25}$$

to obtain

$$\Omega = -\frac{2\sqrt{2}\,m^{\frac{3}{2}}V}{3\pi^2\hbar^3}\left\{\frac{2}{5}\mu^{\frac{5}{2}} + \frac{\pi^2}{6}(kT)^2\left(\frac{3}{2}\right)\mu^{\frac{1}{2}} + \dots\right\}, \tag{1.26}$$

and consequently

$$S = -\left(\frac{\partial\Omega}{\partial T}\right)_{\mu,V} = \frac{\sqrt{2}\,m^{\frac{3}{2}}Vk^2\mu^{\frac{1}{2}}T}{3\hbar^3} + \dots\,,$$

$$N = -\left(\frac{\partial\Omega}{\partial\mu}\right)_{T,V} = \frac{2\sqrt{2}\,m^{\frac{3}{2}}V\mu^{\frac{3}{2}}}{3\pi^2\hbar^3} + \dots\,. \tag{1.27}$$

In the zero temperature limit one sees, from eqs. (1.23) and (1.25), that in momentum space the distribution function corresponds to occupation of a sphere of radius $p_F = (2m\mu)^{\frac{1}{2}}$. This relation defines the Fermi momentum p_F, and it is further evident that a particle at « the top of the Fermi sea » has an energy (« the Fermi energy ») just equal to the chemical potential μ.

Proceeding from eq. (1.27) one can eliminate μ in favor of N/V and thus obtain a formula for the low-temperature specific heat

$$C_V = T\left(\frac{\partial S}{\partial T}\right)_{\mu,V} = \frac{\pi^{\frac{2}{3}} m k^2 N}{3^{\frac{2}{3}}\hbar^2}\left(\frac{V}{N}\right)^{\frac{2}{3}} T + \dots \,. \tag{1.28}$$

The magnetic susceptibility may be readily calculated by ascribing a magnetic moment γ_0 per atom, recalling that in the presence of a field H in the z-direction the magnetic moment $M = \gamma_0 \times$[Net number of atoms lined up] and that the susceptibility $\chi = M/H$. We may employ eq. (1.28) for n_p but now

$$\varepsilon_p = \frac{p^2}{2m} - \boldsymbol{\gamma}_0 \cdot \boldsymbol{H}\,, \tag{1.29}$$

so that the number of atoms with momentum between p and $p+\mathrm{d}p$ and spin up (or down) is

$$\mathrm{d}N_{p(\updownarrow)} = \frac{4\pi V p^2\,\mathrm{d}p}{(2\pi\hbar)^3\{\exp[((p^2/2m) \mp \gamma_0 H - \mu)(1/kT)]\}}\,. \tag{1.30}$$

One has the following two equations, the first to be solved for μ, and the second yielding χ after linearization of the integrals involving H:

$$N = \frac{V m^{\frac{3}{2}}}{\sqrt{2}\pi^2\hbar^3}\left\{\int_0^\infty \frac{\varepsilon^{\frac{1}{2}}\,\mathrm{d}\varepsilon}{\exp[(\varepsilon - \gamma_0 H - \mu)(1/kT)] + 1} + \int_0^\infty \frac{\varepsilon^{\frac{1}{2}}\,\mathrm{d}\varepsilon}{\exp[(\varepsilon + \gamma_0 H - \mu)(1/kT)] + 1}\right\},$$

$$\chi = \frac{V m^{\frac{3}{2}}\gamma_0}{\sqrt{2}\pi^2\hbar^3 H}\left\{\int_0^\infty \frac{\varepsilon^{\frac{1}{2}}\,\mathrm{d}\varepsilon}{\exp[(\varepsilon - \gamma_0 H - \mu)(1/kT)] + 1} - \int_0^\infty \frac{\varepsilon^{\frac{1}{2}}\,\mathrm{d}\varepsilon}{\exp[(\varepsilon + \gamma_0 H - \mu)(1/kT)] + 1}\right\}.$$

Evaluation in the limit that $T \to 0$ yields

$$\frac{\chi}{V} = \left(\frac{3}{\pi^4}\right)^{\frac{1}{3}} \frac{m\gamma_0^2}{\hbar^2}\left(\frac{N}{V}\right)^{\frac{1}{3}}. \tag{1.31}$$

II. – Phenomenological Theory.

In the first section we have discussed the ideal Fermi gas, but of course we have not asserted that this is at all an appropriate description of liquid ^{3}He. In this section we describe a theory of liquid ^{3}He, which may well be an adequate description of the liquid at low temperatures. The pioneer work has been done by LANDAU [LA1, LA2], while further developments of the theory may be found in the references cited in Part V.

In Section **1** we develop the concept of the equilibrium Fermi liquid, Section **2** is developed to the basic idea of zero-sound, while in Section **3** we comment on nonequilibrium properties.

1. – The Fermi liquid.

1·1. *Concepts.* – The basic idea is to realize that the statistical properties of liquid ^{3}He at low temperatures will be determined by the elementary excitation of the system, and furthermore that it is most reasonable to assume that the elementary excitations coincide in number with the atoms in the liquid and obey Fermi statistics. This of course is an assumption, as can be appreciated by considering the double counter example of a gas of diatomic molecules, where each atom obeys Fermi statistics. It can be shown that this assumption concerning the elementary excitation spectrum is equivalent to assuming the validity of perturbation theory starting from a gas of non-interacting atoms. By perturbation theory one does not mean just a low-order calculation, but simply that the theory converges (BL2, DO1, KLE3, LN1, NL1). In Part IV we will explore the possible consequences of nonconvergence, but it should be emphasized that except possibly at extremely low temperatures the convergence of perturbation theory seems well supported by experiment.

The elementary excitations of the system—or the quasi-particles—may be thought of in terms of an atom of the fluid surrounded by its polarization field. Perhaps more usefully, one can think of starting with a system of noninteracting fermions and then slowly turning on the interactions so as to arrive at a state of the true system. Thus one can use the occupation numbers of the noninteracting gas to characterize completely the complicated states of the interacting system. However it is clear that one must ascribe to each noninteracting particle appropriate properties of the true system. For example if we start with the ground state of the noninteracting gas then turning on the interaction between atoms will yield the ground state of the true system (this is the assumed convergence of perturbation theory). Starting with the ground state of the noninteracting gas plus one more particle one will upon turning on the interactions obtain a state of the system which can be designated as the ground state of the true system plus one quasi-particle. To the additional noninteracting particle it is useful to ascribe the differences between the energy of the interacting system with the additional particle and the ground state energy of the interacting system. Thus the quasi-particle energies will be quite different from the energies of free particles.

It must be realized that in a many-particle system the elementary excitations will not be stable, in the sense that a quasi-particle will excite other quasi-particles and thus eventually degrade its energy by the creation of many excitations of a more and more complicated nature—subject only to the restriction of conserved total momentum and energy. In order for a description of the statistical properties of the liquid in terms of quasi-particles of definite momentum and energy to be valid, we must demand that the quasi-particles have a long lifetime. That is the energy of a quasi-particle $\varepsilon(p)$ must be $\gg \hbar/\tau$ where τ is the lifetime. However, the elementary excitations which are near the Fermi surface will have a long lifetime since $1/\tau$ varies as the square of the distance from the Fermi surface (in momentum units). Consequently at low temperatures (where only the low lying excitations need be considered in a calculation of the thermodynamic properties) the quasi-particle theory will become increasingly valid. It is important to realize that the dependence of the lifetime of the elementary excitations upon their momentum has been established from a microscopic theory—assuming only the validity of perturbation theory (GM1, HU1).

At low temperatures, then, we can describe the statistical mechanics in terms of independent quasi-particles moving through a background self-consistent field due to all the atoms. This field in an infinite homogeneous medium can not depend on position, but only upon momentum, and one is thus led to the concept of a single quasi-particle momentum-dependent potential. (This potential is just the energy associated with the quasi-particles as described in the paragraph before the last.) The self-consistent field depends upon the states of all other quasi-particles, and we shall see that this dependence will play an important role in the theory of a Fermi liquid.

In particular, oscillation of the self consistent field corresponds to density waves of a character different from that of ordinary sound—so called « zero-sound », which will be discussed later in this section.

1·2. *Formalism.* – We shall now make the above ideas quantitative, following closely the treatment by LANDAU (LA1, LA2). Let E represent the total energy of the liquid, and $n(\boldsymbol{p})$ the distribution function for quasi-particles. Consider a volume V containing N particles. We may *define* the energy of a quasi-particle $\varepsilon(\boldsymbol{p})$ by the functional derivative of the total energy with respect to the distribution function, *i.e.* it is the change in energy of the system associated with adding one quasi-particle of momentum $\boldsymbol{p}$. Clearly $\varepsilon(\boldsymbol{p})$ is a functional of the distribution function, that is it depends upon the distribution of all the other quasi-particles (*). Thus we have

$$\delta E = \int \varepsilon(\boldsymbol{p})\,\delta n(\boldsymbol{p})\,\frac{2\,\mathrm{d}^3 p}{(2\pi\hbar)^3}\,. \tag{2.1}$$

Because the quasi-particles are independent and further are assumed to obey Fermi statistics, we may use for the entropy (eq. (1.16))

$$S = -k\int [n\log n + (1-n)\log(1-n)]\,\frac{2\,\mathrm{d}^3 p}{(2\pi\hbar)^3}\,. \tag{2.2}$$

It will be recalled that the derivation of this formula required only counting, and since we are characterizing the states of the true system by the quantum

---

(*) We can make this functional dependence of E upon the occupation numbers n_i clear by considering a simple (Hartree) model in which the total energy of the system does not depend only upon the single-particle energies e_i, since there is an assumed interaction energy between particles. We write

$$E = \sum_i e_i n_i + \tfrac{1}{2}\sum_{i,j} n_i n_j f_{ij}\,, \tag{1}$$

so that

$$\frac{\delta E}{\delta n_i} = e_i + \sum_j n_j f_{ij} \equiv \varepsilon_i\,, \tag{2}$$

where it has been assumed that f_{ij} is symmetric. Furthermore

$$\frac{\delta^2 E}{\delta n_j\,\delta n_i} = f_{ij}\,, \tag{3}$$

and it will be seen later that this function f plays a vital role in the theory. Now clearly ε_i depends on n_j so we can not, in particular, integrate eq. (2) to obtain E since

$$\sum_i \left(\frac{\delta E}{\delta n_i}\right) n_i = e_i n_i + \sum_j n_i n_j f_{ij}\,, \tag{4}$$

which is obviously not E. Analogously, in the Fermi liquid theory, we have no simple expression for E but we shall see that we never have need for such an expression, all physical quantities involving only $\delta E/\delta n_i$ and $\delta^2 E/\delta n_i \delta n_j$.

Finally, it is worth noting that in general f_{ij} is a functional of n_k, that is, the interaction between particles i and j depends upon the distribution of other particles. (A simple mechanism which would bring about this dependence is the exclusion principle.) In this case, the situation is more complicated than in the model of this footnote and one has to employ the full theory of a Fermi liquid.

numbers of the noninteracting gas (assumed validity of perturbation theory) the validity of this formula follows.

We may now use a thermodynamic inequality and maximize S subject to the constraint of fixed total energy and fixed total number of particles to deduce the quasi-particle distribution function. Introducing Lagrangian multipliers α', β' we demand:

$$\delta S + \beta' \delta E + \alpha' \delta N = 0 . \tag{2.3}$$

From eqs. (2.1) and (2.2) one obtains

$$\begin{cases} \delta S \simeq -k \left[\dfrac{2}{(2\pi\hbar)^3} \log \left(\dfrac{n}{1-n} \right) \right] \delta n , \\[2ex] \delta E = \dfrac{2}{(2\pi\hbar)^3} \varepsilon(\boldsymbol{p}) \delta n , \\[2ex] \delta N = \dfrac{2}{(2\pi\hbar)^3} \delta n , \end{cases} \tag{2.4}$$

and hence we deduce that

$$n = \frac{1}{\exp[(\alpha' + \beta'\varepsilon)(1/k)] + 1} ,$$

which by a redefinition of the arbitrary constants and identification of the temperature may be written

$$n(\boldsymbol{p}) = \left\{ \exp\left[\frac{\varepsilon - \mu}{kT} \right] + 1 \right\}^{-1} . \tag{2.5}$$

It should be noted that this distribution function for quasi-particles has been rigorously derived. However, it must be remembered that ε is a functional of $n(\boldsymbol{p})$ so that eq. (2.5) is still to be solved for $n(\boldsymbol{p})$.

It is interesting to inquire what happens to $\varepsilon(\boldsymbol{p})$ under a change in the distribution of other quasi-particles. We may write this as

$$\delta\varepsilon(\boldsymbol{p}) = \frac{2}{(2\pi\hbar^3)} \int f(\boldsymbol{p}, \boldsymbol{p}') \delta n(\boldsymbol{p}') \, \mathrm{d}^3 p' , \tag{2.6}$$

where it can be seen that we have defined $f(\boldsymbol{p}, \boldsymbol{p}')$ as the second functional derivative of E. In an ideal gas f is zero, but in the theory of a Fermi liquid f plays an important role. It is considered to be determined phenomenologically, although it can be derived from any microscopic theory of the liquid.

It is useful to define an effective mass of quasi-particles by

$$\nabla_p \varepsilon(\boldsymbol{p})\Big|_{p=p_F} = \frac{p_F}{m^*}\,. \tag{2.7}$$

Now if we increase the momentum of each quasi-particle by an amount $\delta\boldsymbol{p}$ then we have simply set the whole system into motion with velocity $\delta\boldsymbol{p}/m$, and clearly the increase in the energy of a quasi-particle of momentum $\boldsymbol{p}$ must by Galilean invariance, be

$$\delta\varepsilon(\boldsymbol{p}) = \boldsymbol{p}\cdot\frac{\delta\boldsymbol{p}}{m}\,. \tag{2.8}$$

On the other hand, the formalism of the Fermi liquid allows us to calculate the change in $\varepsilon(\boldsymbol{p})$ as arising from two terms, a direct change due to the increase of $\boldsymbol{p}$ by $\delta\boldsymbol{p}$ and an indirect change brought about by the change in the distribution function. Consequently eq. (2.8) becomes, with the aid of eq. (2.6),

$$\frac{\boldsymbol{p}}{m}\cdot\delta\boldsymbol{p} = \nabla_p\varepsilon(\boldsymbol{p})\cdot\delta\boldsymbol{p} + \frac{2}{(2\pi\hbar)^3}\int f(\boldsymbol{p},\boldsymbol{p}')\,\delta n(\boldsymbol{p}')\,\mathrm{d}^3p'\,. \tag{2.9}$$

But displacing the distribution at $\boldsymbol{p}'$ by abount $\delta\boldsymbol{p}'$ *increases* the distribution function $n(p')$ by the amount

$$\delta n(\boldsymbol{p}') = -\nabla_{p'} n(\boldsymbol{p}')\cdot\delta\boldsymbol{p}'\,.$$

Inserting this into eq. (2.9) and employing eq. (2.7) as well as the fact that at $T=0$

$$\nabla_{p'} n(\boldsymbol{p}') = -\frac{\boldsymbol{p}}{p}\,\delta(p-p_F)\,, \tag{2.10}$$

(our result will be valid only for low temperatures, but this range of temperatures is the only one in which the basic assumptions of the theory are valid) we have upon evaluation at $|\boldsymbol{p}| = p_F$,

$$\frac{1}{m} = \frac{1}{m^*} + \frac{2p_F}{(2\pi\hbar)^3}\int f(\theta)\cos\theta\,\mathrm{d}\Omega\,, \tag{2.11}$$

where θ is the angle between the momenta $\boldsymbol{p}$ and $\boldsymbol{p}'$ (both assumed to have magnitude p_F), and $\mathrm{d}\Omega$ is the element of solid angle on the Fermi surface.

1·3. *Specific heat and velocity of sound.* – The specific heat, in the limit as $T\to 0$, may be very easily calculated for a Fermi liquid. The result is that it is given by the formula for an ideal Fermi gas with m replaced by m^*, since the derivation leading to eq. (1.28) holds for quasi-particles, and furthermore involves only the energy spectrum near the Fermi surface, *i.e.* is completely characterized by m^*.

It is a relatively simple matter to calculate the velocity of ordinary sound in a Fermi liquid. However, as $T \to 0$ the mean free path of quasi-particles increases rapidly, which means that the quasi-particles can easily transport momentum over large distances and so (just as in a Knudsen gas) the viscosity of the liquid increases and the attenuation of sound becomes large.

The velocity of sound is given by

$$c^2 = \frac{1}{m}\frac{\partial P}{\partial \varrho}, \tag{2.12}$$

where P is the pressure and ϱ is the number density equal to N/V. This expression can be written as

$$c^2 = \frac{1}{mN}\frac{\partial P}{\partial(1/V)} = -\frac{V^2}{Nm}\frac{\partial P}{\partial V}, \tag{2.13}$$

and using the relation (valid at $T=0$) that

$$\frac{\partial \mu}{\partial N} = -\frac{V^2}{N^2}\frac{\partial P}{\partial V}, \tag{2.14}$$

we obtain

$$c^2 = \frac{N}{m}\frac{\partial \mu}{\partial N}. \tag{2.15}$$

Now $\mu = \varepsilon(p_{\mathrm{F}})$, and $\partial\mu/\partial N$ can be related to $\partial\mu/\partial p_{\mathrm{F}}$, which latter quantity can be calculated as in the argument leading to eq. (2.9). Thus we have two terms:

$$\frac{\partial \mu}{\partial p_{\mathrm{F}}} = \frac{\partial \varepsilon(p_{\mathrm{F}})}{\partial p_{\mathrm{F}}} + \frac{2}{(2\pi\hbar)^3}\int f(p_{\mathrm{F}}, p')\frac{\partial n(p')}{\partial p_{\mathrm{F}}}\,\mathrm{d}^3p',$$

which using eqs. (2.7) and an equation analogous to eq. (2.10) becomes

$$\frac{\partial \mu}{\partial p_{\mathrm{F}}} = \frac{p_{\mathrm{F}}}{m^*} + \frac{2p_{\mathrm{F}}^2}{(2\pi\hbar)^3}\int f(\theta)\,\mathrm{d}\Omega\,. \tag{2.16}$$

Now, since N varies as p_{F}^3 we obtain

$$c^2 = \frac{1}{3}\frac{p_{\mathrm{F}}}{m}\frac{\partial \mu}{\partial p_{\mathrm{F}}},$$

and with eq. (2.11):

$$c^2 = \frac{p_F}{3m}\left[\frac{p_F}{m} + \frac{2p_F^2}{(2\pi\hbar)^3}\int f(\theta)(1-\cos\theta)\,d\Omega\right]. \tag{2.17}$$

1'4. *Magnetic susceptibility.* – We now give an example of the Landau theory for equilibrium processes which is interesting because it explicitly involves the spin of the ^{3}He atoms, a quantity we have so far ignored except for its consequence of allowing two particles in each state. If we wish to consider the effect of an external field then we must generalize the treatment by allowing $f(\boldsymbol{p}, \boldsymbol{p}')$ to depend also on the spins $\boldsymbol{\sigma}$ and $\boldsymbol{\sigma}'$. On the basis of invariance arguments, the most general form is (*)

$$f(\boldsymbol{p}, \boldsymbol{p}', \boldsymbol{\sigma}, \boldsymbol{\sigma}') = F(\boldsymbol{p}, \boldsymbol{p}') + \boldsymbol{\sigma}\cdot\boldsymbol{\sigma}'\, G(\boldsymbol{p}, \boldsymbol{p}'). \tag{2.18}$$

The computation of the susceptibility proceeds analogously to the calculation for the ideal Fermi gas. If we impose a small magnetic field $\boldsymbol{H}$ then the change in the energy of a quasi-particle may be written as

$$\delta\varepsilon(\boldsymbol{p}) = -\gamma_0\boldsymbol{\sigma}\cdot\boldsymbol{H} + \frac{\operatorname{Tr}\sigma'}{(2\pi\hbar)^3}\int f(\boldsymbol{p}, \boldsymbol{p}', \boldsymbol{\sigma}, \boldsymbol{\sigma}')\,\delta n(\boldsymbol{p}', \boldsymbol{\sigma}')\,d^3p', \tag{2.19}$$

where γ_0 is the magnetic moment of a free ^{3}He atom. For small $\boldsymbol{H}$, $\delta\varepsilon(\boldsymbol{p})$ will be proportional to $\boldsymbol{H}$, so we can define a quasi-particle magnetic moment $\gamma(\boldsymbol{p})$ by

$$\delta\varepsilon(\boldsymbol{p}) = -\gamma(\boldsymbol{p})\boldsymbol{\sigma}\cdot\boldsymbol{H}. \tag{2.20}$$

Employing

$$\delta n = \frac{\partial n}{\partial\varepsilon}\delta\varepsilon = -\frac{\partial n}{\partial\varepsilon}\gamma(\boldsymbol{p})\boldsymbol{\sigma}\cdot\boldsymbol{H}, \tag{2.21}$$

we have from eq. (2.19)

$$\gamma(\boldsymbol{p})\boldsymbol{\sigma} = \gamma_0\boldsymbol{\sigma} + \frac{\operatorname{Tr}\sigma'}{(2\pi\hbar)^3}\int f(\boldsymbol{p}, \boldsymbol{p}', \boldsymbol{\sigma}, \boldsymbol{\sigma}')\,\frac{\partial n(p')}{\delta\varepsilon(p')}\gamma(\boldsymbol{p}')\boldsymbol{\sigma}'\,d^3p'.$$

If we multiply this equation by $\boldsymbol{\sigma}$ take $\operatorname{Tr}\sigma$ use eq. (2.18), and recall that

$$\operatorname{Tr}_{\sigma'}\boldsymbol{\sigma}' = 0,$$

$$\operatorname{Tr}_{\boldsymbol{\sigma}}\operatorname{Tr}_{\boldsymbol{\sigma}'}(\boldsymbol{\sigma}\cdot\boldsymbol{\sigma}')^2 = 12,$$

$$\operatorname{Tr}_{\boldsymbol{\sigma}}(\boldsymbol{\sigma}\cdot\boldsymbol{\sigma}) = 6,$$

(*) The well-known «tensor-force» term is excluded here (as contrasted with the two-nucleon problem) by the assumption that liquid ^{3}He is isotropic.

we obtain:

$$6\gamma(\boldsymbol{p}) = 6\gamma_0 + \frac{12}{(2\pi\hbar)^3}\int \frac{\partial n(p')}{\partial \varepsilon(p')}\gamma(\boldsymbol{p}')G(\boldsymbol{p},\boldsymbol{p}')\,\mathrm{d}^3p' \,. \tag{2.22}$$

Now from eqs. (2.10) and (2.17), at $T=0$:

$$\frac{\partial n(p')}{\partial \varepsilon(p')} = -\frac{m^*}{p_\mathrm{F}}\,\delta(p-p_\mathrm{F})\,, \tag{2.23}$$

so eq. (2.22) becomes

$$\gamma(p_\mathrm{F}) = \gamma_0 - \frac{2}{(2\pi\hbar)^3}\frac{m^*}{p_\mathrm{F}}\gamma(p_\mathrm{F})\,p_\mathrm{F}^2\int G(\theta)\,\mathrm{d}\Omega \,. \tag{2.24}$$

The susceptibility χ is defined as the small field limit of $\boldsymbol{M}/\boldsymbol{H}$ where $\boldsymbol{M}$ is the magnetic moment. Assuming $\boldsymbol{H}$ is in the z-direction and that $\chi=0$ for the unperturbed system, we have

$$\chi = \frac{\gamma_0\,\mathrm{Tr}_\sigma\int \delta n_p\sigma_z\,\mathrm{d}^3p}{(2\pi\hbar)^3 H}\,, \tag{2.25}$$

which with eq. (2.21) and evaluation of $\mathrm{Tr}_\sigma\,\sigma_z^2$ as 2 yields

$$\chi = -\frac{2\gamma_0}{(2\pi\hbar)^3}\int \frac{\partial n(p)}{\partial \varepsilon(p)}\gamma(\boldsymbol{p})\,\mathrm{d}^3p\,,$$

which with eq. (2.23) becomes

$$\chi = \frac{8\pi\gamma_0 p_\mathrm{F} m^*}{(2\pi\hbar)^3}\gamma(p_\mathrm{F})\,. \tag{2.26}$$

Now clearly the susceptibility for an ideal gas, χ ideal is just eq. (2.26) with $m^*\to m$ and $\gamma(p_\mathrm{F})\to\gamma_0$, (compare eq. (1.31)), so

$$\frac{\chi}{\chi_\mathrm{ideal}} = \frac{\gamma(p_\mathrm{F})}{\gamma_0}\frac{m^*}{m}\,, \tag{2.27}$$

and solving eq. (2.24) for $\gamma(p_\mathrm{F})$ we have finally

$$\frac{\chi}{\chi_\mathrm{ideal}} = \frac{m^*}{m}\left[1+\frac{2m^*p_\mathrm{F}}{(2\pi\hbar)^3}\int G(\theta)\,\mathrm{d}\Omega\right]^{-1}. \tag{2.28}$$

In this last equation χ_ideal is the zero-temperature susceptibility for an ideal

Fermi gas as given in eq. (1.31). It can be seen that χ may be quite different from χ_{ideal}, and furthermore that this difference involves $G(\theta)$ while the specific heat, for example, involves the *different* function $F(\theta)$ (*). In the theory of a Fermi liquid these are independent, phenomenologically determined functions. A microscopic theory would of course predict both functions, and hence relate them.

1·5. *Comparison with experiment.* – It can be seen that the Landau theory allows a general framework within which to describe liquid ^{3}He. By comparison of our results with experiment, we can determine the phenomenological functions $F(\theta)$ and $G(\theta)$(HO 2).

In particular, the specific heat at low temperatures determines m^* (via eq. (1.28) with $m \to m^*$) and hence (by eq. (2.11)) the first moment of $F(\theta)$; while the velocity of sound (through eq. (2.17)) yields information about the zero-moment of $F(\theta)$. We can make the comparison with experiment more convenient by letting:

$$\left\{\begin{aligned} F(\theta) &= \frac{(2\pi\hbar)^3}{8\pi p_F m^*}\left[F_0 + F_1\cos\theta + F_2\left(\frac{3\cos^2\theta - 1}{2}\right) + \ldots\right],\\ G(\theta) &= \frac{(2\pi\hbar)^3}{8\pi p_F m^*}\left[G_0 + G_1\cos\theta + \ldots\right],\end{aligned}\right. \tag{2.29}$$

in which case eqs. (2.11) and (2.17) become

$$\left\{\begin{aligned} \frac{1}{m} &= \frac{1}{m^*}\left(1 + \frac{1}{3}F_1\right),\\ c^2 &= \frac{p_F^2}{3m^2}\left(\frac{1+F_0}{1+\frac{1}{3}F_1}\right).\end{aligned}\right. \tag{2.30}$$

Taking the density of the liquid as 0.083 g/cm^3 implies $k_F = .785\cdot 10^8$ cm/s and $(p_F/m) = 1.65\cdot 10^4$ cm/s. Using the results of the Illinois Group [1] that $(m^*/m) = 2.82$, and the velocity of sound $c = 1.839\cdot 10^4$ cm/s as measured by LAGUER *et al.* [2], one obtains $F_0 = 9.25$ and $F_1 = 5.46$.

The low temperature susceptibility measurements [3] yield $(\chi/\chi_{\text{ideal}}) \approx 12$, so writing eq. (2.28) as

$$\frac{\chi}{\chi_{\text{ideal}}} = \frac{1+\frac{1}{3}F_1}{1+G_0}, \tag{2.31}$$

(*) In Paragraphs II.1·1, II.1·2, II.1·3 simply replace f with F, in order to be consistent with the notation of this Paragraph.

we obtain $G_0 = -0.70$. It can be seen that liquid ^{3}He is very close to being ferromagnetic, *i.e.* the denominator in eq. (2.31) is reduced from unity almost to zero, by G_0. Thus χ is exceedingly large—even larger than an would expect after considering the effective mass contribution to χ/χ_{ideal}. From a microscopic point of view, the forces between ^{3}He atoms [just the ordinary central spin-independent forces usually approximated by a $(6\div 12)$ potential] yield, when the properties of the liquid are calculated [with appropriate antisymmetry of the wave function], a net tendency for spins to align which is even larger than that associated with the modified single-particle excitation spectrum. It is not possible to isolate this property however, since the spin susceptibility is intimately related to the spacial behavior of the wave function.

Finally, it should be observed that the experimental data are not supplying a quantitative test of the Landau theory. It is consequently very exciting to realize that the Landau theory makes a very definite prediction of a new phenomenon to be expected in liquid ^{3}He, namely zero-sound, the observation of which would afford a striking confirmation of the theory. We turn now to a discussion of zero-sound.

2. – Zero-sound.

As $T \to 0$, ordinary sound in liquid ^{3}He will be strongly attenuated as was explained in Section II.1·3. In the Fermi liquid (as contrasted with a Knudsen gas) the possibility exists of a new type of oscillation which propagates via the self-consistent potential in which quasi-particles move. This phenomenon of zero-sound is a collective motion which is closely related to the collective motion observed in the vibrational and rotational states of atomic nuclei.

In both systems a short-range force (nuclear or van der Waals) gives rise to an effective long-range interaction (called SU3 force in nuclei). The phenomenon of zero-sound is mathematically similar to that of plasma oscillation. The similarity does not transcend the mathematics because of the difference between the long-range Coulomb force and the short-range van der Waals force.

The criterion for propagation of zero-sound is that the mean free path of quasi-particles be large compared to the wavelength of the collective disturbance. In contrast, ordinary sound requires for its propagation local thermal equilibrium. The velocity of ordinary sound (in the absence of interactions) is from eq. (2.17) $c = p_F/m\sqrt{3}$ the $1/\sqrt{3}$ being connected with the thermalization in 3 dimensions. On raising the frequency or lowering the temperature the criterion for zero-sound will be satisfied, and zero-sound will be propagated with the velocity of quasi-particles, *i.e.* p_F/m where the absence of $1/\sqrt{3}$ is related to the lack of local three-dimensional thermalization in zero-sound.

We realize that in the presence of interaction both velocities get modified, and in fact they are predicted to be quite close together in liquid ^{3}He (*). The criterion for zero-sound propagation, implying high frequencies and low temperatures, has so far prevented its experimental observation.

We turn now to a discussion of zero-sound at the absolute zero. Our treatment follows LANDAU (LA2), but we consider in detail only the simplest possible situation, in which the oscillations are spin-independent. Assuming that the oscillations are of small amplitude, we may obtain linear equations. We write the distribution function and quasi-particle energies as

$$\begin{cases} n(\boldsymbol{p}, \boldsymbol{r}, t) = n_0(\boldsymbol{p}) + \delta n(\boldsymbol{p}, \boldsymbol{r}, t)\,, \\ \varepsilon(\boldsymbol{p}, \boldsymbol{r}, t) = \varepsilon_0(\boldsymbol{p}) + \delta\varepsilon(\boldsymbol{p}, \boldsymbol{r}, t)\,, \end{cases} \tag{2.32}$$

where $\varepsilon_0(\boldsymbol{p})$ and $n_0(\boldsymbol{p})$ correspond to the equilibrium quasi-particle distribution. Now we may take

$$\delta\varepsilon(\boldsymbol{p}, \boldsymbol{r}, t) = \frac{2}{(2\pi\hbar)^3}\int f(\boldsymbol{p}, \boldsymbol{p}')\,\delta n(\boldsymbol{p}, \boldsymbol{r}, t)\,\mathrm{d}^3p'\,, \tag{2.33}$$

in analogy with eq. (2.6).

At low temperatures the distribution function $n(\boldsymbol{p}, \boldsymbol{r}, t)$ will obey a collisionless Boltzmann equation since the mean free path of quasi-particles is very long. We may thus write

$$\frac{\partial n}{\partial t} + \boldsymbol{\nabla}_r n \cdot \boldsymbol{\nabla}_p \varepsilon - \boldsymbol{\nabla}_p n \cdot \boldsymbol{\nabla}_r \varepsilon = 0\,, \tag{2.34}$$

where it can be seen that the energy of quasi-particles appears in the role of a Hamiltonian. The linearization of eq. (2.34) yields:

$$\frac{\partial\,\delta n}{\partial t} + \boldsymbol{\nabla}_r \delta n \cdot \boldsymbol{\nabla}_p \varepsilon_0 - \boldsymbol{\nabla}_r \delta\varepsilon \cdot \boldsymbol{\nabla}_p n_0 = 0\,. \tag{2.35}$$

(*) The same argument may be applied to liquid ^{4}He II, where second sound involves thermalized excitations of rotons. That it does is clear from the fact that it may be derived from hydrodynamic equations, just as ordinary sound may be derived in liquid ^{3}He. (Zero-sound, on the other hand, must be derived from a Boltzman equation and not a hydrodynamic approximation to the Boltzmann equation.) Second sound may thus be expected (in the absence of interactions) to have a velocity $1/\sqrt{3}$ times that of « ordinary sound ». Correspondingly, « ordinary sound » in liquid ^{4}He must be quite different from ordinary sound in either liquid ^{3}He or liquid ^{4}He at high temperatures. In particular, a local temperature should not be well defined, and the local stress tensor not diagonal.

We may assume, without further loss of generality, that δn is harmonically analysed in space and time. It follows from eq. (2.33) that $\delta\varepsilon$ has the same behaviour and hence we write:

$$\text{(2.36)} \qquad \begin{cases} \delta n(\boldsymbol{p}, \boldsymbol{r}, t) = \delta n(\boldsymbol{p}) \exp[i(\boldsymbol{k}\cdot\boldsymbol{r} - \omega t)] \,, \\ \delta\varepsilon(\boldsymbol{p}, \boldsymbol{r}, t) = \delta\varepsilon(\boldsymbol{p}) \exp[i(\boldsymbol{k}\cdot\boldsymbol{r} - \omega t)] \,. \end{cases}$$

Letting $\boldsymbol{v} = \nabla_p \varepsilon_0(\boldsymbol{p})$, eq. (2.36) allows us to write eq. (2.35) as

$$\text{(2.37)} \qquad [\boldsymbol{k}\cdot\boldsymbol{v} - \omega]\,\delta n(\boldsymbol{p}) = \boldsymbol{k}\cdot\boldsymbol{v}\, \frac{\partial n_0(p)}{\partial\varepsilon(p)}\, \delta\varepsilon(\boldsymbol{p}) \,.$$

Integration of eq. (2.37) over the magnitude of $\boldsymbol{p}$ for fixed orientation of $\boldsymbol{p}$—most easily carried out by integrating over $\varepsilon(p)$—yields an equation for the net change in the distribution function in the direction of $\boldsymbol{p}$. Thus letting

$$\text{(2.38)} \qquad \nu(\boldsymbol{p}) = \int \delta n(\boldsymbol{p})\, \mathrm{d}\varepsilon \,,$$

and using eqs. (2.23) and (2.33) we have

$$[k v_{\mathrm{F}} \cos\theta - \omega]\,\nu(\theta, \varphi) = -\frac{2m^* p_{\mathrm{F}} \cos\theta}{(2\pi\hbar)^3} \int f(\alpha)\,\nu(\theta', \varphi')\, \mathrm{d}\Omega',$$

where (θ, φ) $[(\theta', \varphi')]$ designate the direction of $\boldsymbol{p}$ $[\boldsymbol{p}']$ measured with respect to the direction of $\boldsymbol{k}$, and α is the angle between $\boldsymbol{p}$ and $\boldsymbol{p}'$. Introducing η the ratio of the velocity of zero-sound ω/k to the velocity of a particle or the Fermi surface $v_{\mathrm{F}} = p_{\mathrm{F}}/m^*$, and using eq. (2.29) with the replacement of (footnote (*), p. 10) $f \to F$ we have

$$\text{(2.39)} \qquad [\eta - \cos\theta]\,\nu(\theta, \varphi) = \frac{\cos\theta}{4\pi} \int \mathrm{d}\Omega'\, \nu(\theta', \varphi')[F_0 + F_1 \cos\alpha + \ldots] \,.$$

We may gain insight into eq. (2.39) by recalling that in the absence of quasi-particle collisions a general motion of the system corresponds to the quasi-particle distribution function changing in time but subject to the restrictions that i) quasi-particles are conserved, and ii) the density in the neighbourhood of any quasi-particle is a constant of the motion (*). At the abso-

(*) Both of these restrictions follow from the fact that the collisionless Boltzmann equation eq. (2.34) is just the Liouville equation in $\boldsymbol{r} - \boldsymbol{p}$ space. Note that this is a much different statement than Lioville's theorem, which is rigorously valid in $6N$-dimensional space.

lute zero the unperturbed distribution function corresponds to a uniformly occupied Fermi sphere. Consequently, zero-sound can only correspond to volume-preserving distortions of the Fermi surface. In fact, $\nu(\theta, \varphi)$ is just the displacement of the Fermi surface in the direction having angles (θ, φ) with respect to the direction of the zero-sound wave, and we see from eq. (2.39) that ν is a relatively complicated function of (θ, φ).

Ordinary sound, by comparison, corresponds to a displacement and swelling (and shrinking) of the Fermi surface but not to a distortion from spherical shape, *i.e.* thermalization always occurs. In zero-sound we don't have thermalization and so the Fermi surface must distort along the direction of wave motion.

The solution of eq. (2.39) is in general relatively complicated. If we simplify the situation by keeping only the F_0 term in the expansion of $F(\alpha)$, then clearly

$$\nu(\theta, \varphi) = \frac{c \cos\theta}{\eta - \cos\theta},$$

where c is a constant (*). Inserting this back into eq. (2.39), the constant c cancels and one is left with

$$1 = \frac{F_0}{4\pi}\int \frac{d\Omega' \cos\theta'}{[\eta - \cos\theta']}, \tag{2.40}$$

which is a dispersion relation to be solved for η. Performing the integration one obtains

$$\frac{\eta}{2}\log\left(\frac{\eta+1}{\eta-1}\right) - 1 = \frac{1}{F_0}. \tag{2.41}$$

The left-hand side of eq. (2.41) varies between $+\infty$ and 0 and η varies between 1 and ∞. For $\eta < 1$ the left-hand side is complex. Thus purely oscillatory solutions only exist for $F_0 > 0$ (*i.e.* repulsive forces between quasi-particles). Clearly given F_0, one can readily find the velocity of zero-sound.

LANDAU (LA2) and ABRIKOSOV and KHALATNIKOV (AK1) have considered solutions to eq. (2.39) in which both the F_0 and F_1 terms are retained. It is not wise to take the numerical predictions very seriously however, since the

(*) Note how this solution corresponds to distortion of the Fermi surface in the general direction of the zero-sound wave (and the opposite direction), but not in the direction perpendicular to the zero-sound wave. One thus sees explicitly the lack of thermalization.

further terms in the expansion of $F(\alpha)$ may play an important role. In particular, we may expect F_2 to be important (see Part IV). It does appear however, that the large positive value of F_0 will guarantee the existence of zero-sound. It is possible to have more complicated solutions to eq. (2.39) [in which $\nu(\theta, \varphi)$, for example, is an explicit function φ], but preliminary estimates (AK1) indicate that none of these modes will propagate in liquid ^{3}He.

LANDAU has also investigated possible spin waves in liquid ^{3}He. The analysis is similar to that given above except that it must be generalized by including the spin-dependence of $f(\theta)$, just as in eq. (2.18). The comparison with experiment is not easy as we only have information about G_0, The fact that G_0 is negative $[-1 < G_0 < 0]$ tends to indicate that spin waves will not occur, but it is realized that this result is not a firm prediction.

Finally, it should be mentioned that there exists the possibility of finding complex solutions to the equation for zero-sound (BRO1, GL1, GHW2). For example, for the simplified case leading to eq. (2.41) there are three cases (*) i) $F_0 > 0$ (repulsive inter-quasi-particle forces), real solution and propagation of zero-sound, ii) $-1 < F < 1$ (weak attractive inter-quasi-particle forces), complex solution corresponding to the damping of the collective oscillation, iii) $F < -1$ (strongly attractive forces), complex solution with collective mode growing in time, corresponding to the fact that the unperturbed state is unstable and hence an inappropriate description of the system. It may be seen that the third case is most interesting, as it would indicate the invalidity of the Fermi liquid theory. This point will be discussed further in Section V, where the instability of the Fermi liquid against a collective excitation (but not of the zero-sound type) will be taken to indicate the onset of a new highly correlated phase.

We can conclude that the observation of zero-sound will give us considerable information concerning the inter-quasi-particle forces. On the other hand a careful theoretical study would seem, at the present time [due to the lack of sufficient information about $F(\theta)$], to demand a microscopic approach. Such a study is now being undertaken by Dr. YIH PWU [4].

3. – Nonequilibrium properties.

Nonequilibrium properties of a Fermi liquid have been studied by LANDAU (LA1), ABRIKOSOV and KHALATNIKOV (AK2, AK1, KA3), and HONE (HO1).

(*) A causality argument must be employed to define the singularity in eq. (2.40). One finds that the denominator should be defined by an infinitesimal negative imaginary constant.

The analysis has been made in a semi-classical approximation, but is nevertheless rather complicated. We shall give a derivation of the conservation laws, and then describe the calculations of kinetic coefficients.

3·1. *Conservation laws.* – In this section we derive the equations which describe conservation of number, momentum, and energy in a Fermi fluid. Limiting attention to the case in which there is no external magnetic field, we have in a semi-classical approximation that the distribution function $n(\boldsymbol{p}, \boldsymbol{\sigma}, t)$ will satisfy the Boltzmann equation

$$\frac{\partial n}{\partial t} + \boldsymbol{\nabla}_r n \cdot \boldsymbol{\nabla}_p \varepsilon - \boldsymbol{\nabla}_p n \cdot \boldsymbol{\nabla}_r \varepsilon = I(n) , \tag{2.42}$$

where $I(n)$ is the collision integral (which will be given in detail subsequently), and ε is a function of $\boldsymbol{p}$, $\boldsymbol{r}$ and t. For our present purposes it suffices to appreciate that conservation of momentum and energy in elementary quasi-particle collisions will be an integral part of $I(n)$.

To find the conservation law associated with quasi-particle number we integrate eq. (2.42) over phase space to obtain

$$\frac{\partial}{\partial t}\int \frac{2\,\mathrm{d}^3 p}{(2\pi\hbar)^3}\, n + \sum_j \int \frac{2\,\mathrm{d}^3 p}{(2\pi\hbar)^3}\left[\frac{\partial n}{\partial x_j}\frac{\partial \varepsilon}{\partial p_j} - \frac{\partial n}{\partial p_j}\frac{\partial \varepsilon}{\partial x_j}\right] = 0 , \tag{2.43}$$

where the vanishing of the term involving the collision integral follows from the conservation of number of quasi-particles in an elementary collision. Rewriting the first part of the second term, and integrating the second part of the second term by parts we have

$$\frac{\partial}{\partial t}\int \frac{2\,\mathrm{d}^3 p}{(2\pi\hbar)^3}\, n + \sum_j \int \frac{2\,\mathrm{d}^3 p}{(2\pi\hbar)^3}\left\{\left[\frac{\partial}{\partial x_j}\left(n\,\frac{\partial \varepsilon}{\partial p_j}\right) - n\,\frac{\partial^2 \varepsilon}{\partial x_j\,\partial p}\right] + n\,\frac{\partial^2 \varepsilon}{\partial p_j\,\partial x_j}\right\} = 0 ,$$

which can be recognized as the law of particle conservation

$$\frac{\partial N}{\partial t} + \sum_j \frac{\partial J_j}{\partial x_j} = 0 , \tag{2.44}$$

where N is the spacial density of quasi-particles and the particle current $\boldsymbol{J}$ has components

$$J_j = \int \frac{2\,\mathrm{d}^3 p}{(2\pi\hbar)^3}\, n\,\frac{\partial \varepsilon}{\partial p_j} . \tag{2.45}$$

To find the conservation law associated with energy we multiply eq. (2.42) by ε and integrate over phase-space to obtain

$$\int \frac{2\,\mathrm{d}^3p}{(2\pi\hbar)^3}\,\varepsilon\,\frac{\partial n}{\partial t}+\sum_j\int\frac{2\,\mathrm{d}^3p}{(2\pi\hbar)^3}\left[\frac{\partial n}{\partial x_j}\frac{\partial \varepsilon}{\partial p_j}-\frac{\partial n}{\partial p_j}\frac{\partial \varepsilon}{\partial x_j}\right]=0\,, \tag{2.46}$$

where the collision integral again vanishes, this time because of energy conservation in elementary collisions. Rewriting the first part of the second term, and integrating the second part of the second term by parts we have

$$\int \frac{2\,\mathrm{d}^3p}{(2\pi\hbar)^3}\,\varepsilon\,\frac{\partial n}{\partial t}+\sum_j\int\frac{2\,\mathrm{d}^3p}{(2\pi\hbar)^3}\left\{\left[\frac{\partial}{\partial x_j}\left(n\varepsilon\,\frac{\partial\varepsilon}{\partial p_j}\right)-n\,\frac{\partial \varepsilon}{\partial x_j}\frac{\partial \varepsilon}{\partial p_j}-n\varepsilon\,\frac{\partial^2\varepsilon}{\partial x_j\,\partial p_j}\right]+n\varepsilon\,\frac{\partial}{\partial p_j}\left(\varepsilon\,\frac{\partial\varepsilon}{\partial x_j}\right)\right\}=0,$$

which we recognize as the law of conservation of energy:

$$\frac{\partial E}{\partial t}+\sum_j\frac{\partial Q_j}{\partial x_j}=0\,, \tag{2.47}$$

where eq. (2.42) has been employed to identify the time derivative of the spacial energy density E, and the energy current $\boldsymbol{Q}$ has components

$$Q_j=\int\frac{2\,\mathrm{d}^3p}{(2\pi\hbar)^3}\,n\varepsilon\,\frac{\partial\varepsilon}{\partial p_j}\,. \tag{2.48}$$

To find the conservation law associated with momentum we proceed in an analogous manner, multiplying eq. (2.42) by a component of momentum p_i and integrating over phase-space. Again the collision term vanishes and performing the same manipulation as before we have

$$\frac{\partial}{\partial t}\int\frac{2\,\mathrm{d}^3p}{(2\pi\hbar)^3}\,np_i+\sum_j\int\frac{2\,\mathrm{d}^3p}{(2\pi\hbar)^3}\left\{\left[\frac{\partial}{\partial x_j}\left(p_i n\,\frac{\partial\varepsilon}{\partial p_j}\right)-p_i n\,\frac{\partial^2\varepsilon}{\partial x_i\,\partial p_j}\right]+\right.$$
$$\left.+\left[n\,\frac{\partial\varepsilon}{\partial x_i}+np_i\,\frac{\partial^2\varepsilon}{\partial p_j\,\partial x_j}\right]\right\}=0\,,$$

since p_k and χ_j are independent variables. This last equation may be written as

$$\frac{\partial}{\partial t}\int\frac{2\,\mathrm{d}^3p}{(2\pi\hbar)^3}\,np_i+\sum_j\int\frac{2\,\mathrm{d}^3p}{(2\pi\hbar)^3}\left\{\frac{\partial}{\partial x_j}\left(p_i n\,\frac{\partial\varepsilon}{\partial p_j}\right)+\left[\frac{\partial}{\partial x_i}(n\varepsilon)-\varepsilon\,\frac{\partial n}{\partial x_i}\right]\right\}=0\,.$$

We may employ eq. (2.1) to identify the last term as $\partial E/\partial x_i$, and thus arrive at the law of conservation of momentum

$$\frac{\partial}{\partial t}M_i+\frac{\partial}{\partial x_j}\Pi_{ij}=0\,, \tag{2.49}$$

where the momentum density $\boldsymbol{M}$ and the stress tensor Π have components given by (*):

$$\left\{\begin{aligned} M_i &= \int \frac{2\,\mathrm{d}^3p}{(2\pi\hbar)^3}\, n p_i\,, \\ \Pi_{ij} &= \int \frac{2\,\mathrm{d}^3p}{(2\pi\hbar)^3}\left[p_i n\frac{\partial\varepsilon}{\partial p_j} + \delta_{ij} n\varepsilon\right] - \delta_{ij}E\,. \end{aligned}\right. \tag{2.50}$$

3·2. *Viscosity and heat conduction coefficients.* – The calculation of these coefficients is rather involved, necessitates numerous approximations, and has no features peculiar to a Fermi liquid. We will outline the method of Abrikosov and Khalatnikov as applied to the viscosity coefficient, but not attempt to give details.

One starts by assuming that the fluid is flowing with a velocity $\boldsymbol{u}$ which is a slowly varying function of position, so that the fluid is almost in local equilibrium. Thus we write

$$n = n_0 + \delta n\,, \tag{2.51}$$

where n_0 is the distribution function corresponding to local equilibrium [obtained from eq. (2.5) by use of Galilean invariance]

$$n_0 = \left\{\exp\left[\frac{1}{kT}[\varepsilon_0 - \boldsymbol{p}\cdot\boldsymbol{u} - \mu]\right] + 1\right\}^{-1}, \tag{2.52}$$

and only terms linear in δn are retained in the calculation. The quantity δn is found from the Boltzman equation eq. (2.42), by assuming that the left-hand side may be evaluated in the approximation of neglecting δn. Employing particle conservation eq. (2.44) one obtains:

$$-\frac{1}{2}\frac{\partial n_0}{\partial\varepsilon}\left(p_i\frac{\partial\varepsilon}{\partial p_k} - \frac{1}{3}p_j\frac{\partial\varepsilon}{\partial p_j}\delta_{ik}\right)\left(\frac{\partial u_i}{\partial x_k} + \frac{\partial u_k}{\partial x_i} - \frac{2}{3}\frac{\partial u_j}{\partial x_j}\delta_{ik}\right) = I(n)\,, \tag{2.53}$$

where repeated indices are to be summed. To make further progress the collision integral must be examined in detail. It can be written as

$$I(n) = -\int w[n_1 n_2(1-n_1')(1-n_2') - (1-n_1)(1-n_2)n_1'n_2']\cdot \\ \cdot\,\delta(\boldsymbol{p}_1 + \boldsymbol{p}_2 - \boldsymbol{p}_1' - \boldsymbol{p}_2')\,\delta(\varepsilon_1 + \varepsilon_2 - \varepsilon_1' - \varepsilon_2')\left[\frac{2}{(2\pi\hbar)^3}\right]^3 \mathrm{d}^3p_2\,\mathrm{d}^3p_1'\,\mathrm{d}^3p_2'\,, \tag{2.54}$$

(*) Note that if we had an ideal gas then the last term in the integral in Π_{ij} would just cancel the $E\delta_{ij}$ term, but for a Fermi liquid there is no such cancellation.

where the conservation of momentum and energy has been explicitly indicated, and $w(\boldsymbol{p}_1'\boldsymbol{p}_2'\boldsymbol{p}_1\boldsymbol{p}_2)$ is the collision probability for two quasi-particles of momentum $\boldsymbol{p}_1$ and $\boldsymbol{p}_2$ to scatter to states of momentum $\boldsymbol{p}_1'$ and $\boldsymbol{p}_2'$. Now the collision integral vanishes if the equilibrium function $n_0(\varepsilon)$ is inserted into it, but does not vanish if $n_0(\varepsilon_0)$ is inserted. These expressions differ because in a Fermi liquid ε is a functional of n. Of course the term δn will also contribute to $I(n)$ and it is easy to show that to lowest approximation these two contributions combine in such a way that:

$$I(n) = \frac{1}{kT}\int w\, n_{01} n_{02}(1-n_{01}')(1-n_{02}')(\psi_1+\psi_2-\psi_1'-\psi_2')\cdot \\ \cdot\delta(\boldsymbol{p}_1+\boldsymbol{p}_2-\boldsymbol{p}_1'-\boldsymbol{p}_2')\,\delta(\varepsilon_1+\varepsilon_2-\varepsilon_1'-\varepsilon_2')\left[\frac{2}{(2\pi\hbar)^3}\right]^3 \mathrm{d}^3p_2\,\mathrm{d}^3p_1'\,\mathrm{d}^3p_2' \,, \tag{2.55}$$

where we have defined a quantity ν by

$$\delta n = \frac{\partial n_0}{\partial \varepsilon_0}\,\nu \,, \tag{2.56}$$

made the usual low temperature approximations, and let

$$\psi_1 = \nu_1 + \frac{2p_{\mathrm{F}}m^*}{(2\pi\hbar)^3}\int f(\chi)\nu_2\,\mathrm{d}\Omega_2 \,. \tag{2.57}$$

Now eq. (2.53) is solved by arguing that ψ must have the functional form of the left-hand side and so may be written as

$$\psi = \frac{1}{2}q_1\left(p_i\frac{\partial\varepsilon}{\partial p_k}-\frac{1}{3}p_j\frac{\partial\varepsilon}{\partial p_j}\delta_{ik}\right)\left(\frac{\partial u_i}{\partial x_k}+\frac{\partial u_k}{\partial x_i}-\frac{2}{3}\frac{\partial u_j}{\partial x_j}\delta_{ik}\right), \tag{2.58}$$

in which case eq. (2.53) becomes an equation for q_1. A rather lengthy calculation by Abrikosov and Khalatnikov yields an approximate expression for q_1:

$$q_1 = \frac{64\pi^2\hbar^6}{3m^{*3}k^2T^2}\left\{\left[\frac{w(\theta,\varphi)}{\cos\theta/2}(1-\cos\theta)^2\sin^2\varphi\right]_{\mathrm{av}}\right\}^{-1}, \tag{2.59}$$

where θ is the angle between $\boldsymbol{p}_1$ and $\boldsymbol{p}_2$, and φ is the angle between the planes determined by $(\boldsymbol{p}_1, \boldsymbol{p}_2)$ and $(\boldsymbol{p}_1', \boldsymbol{p}_2')$. The expression for q_1 (eq. (2.59)), when inserted into eq. (2.58) yields an explicit formula for ψ. From eqs. (2.57), (2.56) one can then find δn, thus having arrived at an approximate distribution function. It will be seen in the next paragraph that solving for δn is not necessary, as the viscosity can be expressed directly in terms of ψ.

The viscosity coefficient is obtained by evaluating the stress tensor Π, which has besides a direct contribution from δn also a term arising from the functional dependence of ε upon δn. It can be shown that

$$\Pi_{ij} = \int p_i \frac{\partial \varepsilon}{\partial p_j} \frac{\partial n_0}{\partial \varepsilon_0} \psi \frac{2\, d^3 p}{(2\pi\hbar)^3}. \tag{2.60}$$

From the fact that Π contains the same function as appears in $I(n)$ (so that f does not explicitly appear either in the Boltzmann equation [eqs. (2.53) and (2.55)], or in the expression for the stress tensor in terms of the solution to that equation), it follows that the viscosity is the same as for a Fermi gas of particles with mass m^* interacting with a scattering law described by w. Identifying the viscosity coefficient η by

$$\Pi_{ij} = -\eta \left[\frac{\partial u_i}{\partial x_j} + \frac{\partial u_j}{\partial x_i} - \frac{2}{3} \frac{\partial u_k}{\partial x_k} \delta_{ij} \right], \tag{2.61}$$

one obtains

$$\eta = \frac{p_F^5 q_1}{15 m^* \hbar^3 \pi^2}. \tag{2.62}$$

Khalatnikov and Abrikosov have given an approximate evaluation of eq. (2.59) for q_1, by approximating $w(\theta, \varphi)$ with $w(\theta, 0)$ [which quantity is approximated by $2\pi f^2(\theta)/\hbar$, and hence obtainable from other experiments], but it is realized that this is only to estimate the magnitude of η. The result, $\eta T^2 \to 1.5 \cdot 10^{-6}$ poise °K^2 is in good agreement with experiment [5] which confirms the predicted T^{-2} temperature-dependence and yields $\eta T^2 \to 2.8 \cdot 10^{-6}$ poise °K^2.

Abrikosov and Khalatnikov have also evaluated the thermal conductivity. They find, by a very similar calculation to that just outlined for the viscosity coefficient, that

$$K = \frac{8\pi^2 \hbar^3 p_F^3}{3 m^{*4} k T} \left\{ \left[\frac{w(\theta, \varphi)(1 - \cos\theta)}{\cos \theta/2} \right]_{\text{av}} \right\}^{-1}. \tag{2.63}$$

Using the same approximations about $w(\theta, \varphi)$ yields $K = (57/T)$ erg/cm s. This theoretical prediction may be compared to the experimental results of the Illinois group [6] which extend down to 0.026 °K, and below 0.04 °K can be fitted by $K = ((48 \pm 3)/T)$ erg/cm s. The temperature-dependence is a fine confirmation of the Landau theory, while the numerical agreement is adequate in view of the approximate treatment of w. An evaluation of w from a microscopic theory would be most interesting, while from a phenomenological approach it is probably prudent to determine $w(\theta, \varphi)$ from the transport coefficients rather than attempting to estimate it from $f(\theta)$.

By methods analogous to those described above, KHALATNIKOV and ABRIKOSOV (KA3) have calculated the attenuation of ordinary sound and zero sound. These computations supply quantitative estimates to the arguments given at the beginning of Section II.2.

3·3. *Self-diffusion coefficient.* – HONE (HO1) has given a derivation of the spin self-diffusion coefficient which is of particular interest in that the result for a Fermi liquid is significantly different than that for a Fermi gas of particles with mass m^* and scattering law described by w.

One starts by assuming a magnetization gradient which is maintained in ^{3}He (in the absence of an external field) by sources of « up » and « down » spins. A steady state diffusive flow is sought by approximately solving the Boltzmann equation (eq. (2.42)) at zero temperatures. The distribution function is written as

$$n = n_0 + \delta n\,, \tag{2.64}$$

with

$$n_0(\varepsilon) = \left\{\exp\left[\frac{\varepsilon - \mu}{kT}\right] + 1\right\}^{-1}; \tag{2.65}$$

μ is a function of position, and δn is treated to lowest order. By considering a point where the magnetization M is zero, the chemical potentials for spin up and spin down are equal while their gradients are opposite. At such a point $\delta n_\uparrow = -\delta n_\downarrow$ and attention can be concentrated on a single spin species. The spacial density of up spins $N_\uparrow$, can be related to the gradient of the up spins chemical potential $\mu_\uparrow$, since (suppressing the spin index):

$$\frac{\partial N}{\partial x} = \frac{\partial}{\partial x}\int \frac{\mathrm{d}^3 p}{(2\pi\hbar)^3}\, n(\varepsilon) = \int \frac{\mathrm{d}^3 p}{(2\pi\hbar)^3}\, \frac{\partial n}{\partial \varepsilon}\left[\frac{\partial \varepsilon}{\partial x} - \frac{\partial \mu}{\partial x}\right], \tag{2.66}$$

which using eqs. (2.18) and (2.23) becomes

$$\frac{\partial N}{\partial x} = \frac{m^* p_{\mathrm{F}}}{\pi^2\hbar^3}\left\{-\frac{1}{4\pi}\,\frac{\partial N}{\partial x}\int G(\theta)\,\mathrm{d}\Omega + \frac{1}{2}\,\frac{\partial \mu}{\partial x}\right\},$$

so that

$$\frac{\partial \mu}{\partial x} = \frac{2\pi^2\hbar^3}{m^* p_{\mathrm{F}}}\,[1 + G_0]\,\frac{\partial N}{\partial x}\,. \tag{2.67}$$

The term in G_0 will result in a diffusion coefficient quite different from that for a Fermi gas of mass m^* and scattering law w, and thus eq. (2.67) contains the important point of this calculation.

The analysis proceeds by solving the Boltzmann equation for δn. As before, one inserts n_0 on the left-hand side but in evaluation the collision integral one must be sure to evaluate n_0 at the true quasi-particle energies ε. Thus writing

$$n(\varepsilon) = n_0(\varepsilon_0) + \delta n = n_0(\varepsilon) - \frac{\partial n_0}{\partial \varepsilon_0} \mathrm{Tr}_{\sigma'} \int f\, \delta n' \frac{\mathrm{d}^3 p'}{(2\pi\hbar)^3} + \delta n\,, \tag{2.68}$$

and expressing δn in terms of ν as in eq. (2.56) yields

$$n(\varepsilon) = n_0(\varepsilon) + \frac{\partial n_0}{\partial \varepsilon_0} \psi\,, \tag{2.69}$$

where

$$\psi = \nu + \frac{2m^* p_{\mathrm{F}}}{(2\pi\hbar)^3} \int G(0) \nu\, \mathrm{d}\Omega\,. \tag{2.70}$$

One now finds that $I(n)$ is given by eq. (2.55) and arguing that ψ must have the functional form of the left-hand side suggests writing

$$\psi = q_2 \frac{\partial \varepsilon}{\partial p_x} \frac{\partial \mu}{\partial x}\,. \tag{2.71}$$

An approximate analysis yields

$$q_2 = \frac{32\pi^2\hbar^6}{m^{*3}k^2T^2} \left\{ \left[\frac{w(\theta, \varphi)(1-\cos\theta)(1-\cos\varphi)}{\cos\theta/2} \right]_{\mathrm{av}} \right\}^{-1}. \tag{2.72}$$

The diffusion coefficient is defined in terms of the (spin up) current J by

$$J = -D \frac{\partial N}{\partial x}\,. \tag{2.73}$$

Evaluating

$$J = \int [n(\varepsilon) - n_0(\varepsilon)] \frac{\partial \varepsilon}{\partial p_x} \frac{\mathrm{d}^3 p}{(2\pi\hbar)^3}\,, \tag{2.74}$$

by use of eq. (2.69) yields

$$J = \int \frac{\partial n_0}{\partial \varepsilon_0} \frac{\partial \varepsilon}{\partial p_x} \psi \frac{\mathrm{d}^3 p}{(2\pi\hbar)^3}\,, \tag{2.75}$$

which by eqs. (2.72) and (2.23) becomes

$$J = -\left(\frac{1}{3}\right)\frac{p_F^3}{2\pi^2\hbar^3 m^*}\, q_2\, \frac{\partial\mu}{\partial x}\,. \tag{2.76}$$

Finally, from eq. (2.73) and eq. (2.76)

$$D = \frac{1}{3}\left(\frac{p_F}{m^*}\right)^2 q_2[1 + G_0]\,. \tag{2.77}$$

Thus one sees explicitly the Fermi liquid correction factor $(1+G_0)$, since the rest of the expression is the same as for a Fermi gas.

Approximate evaluation of q_2, by relating w to f [including a small correction since only collisions between particles of anti-parallel spin contribute to D] yields $DT^2 = 4.2 \cdot 10^{-6}$ cm^2 °K^2/s, in agreement with the experiments [3] which confirm the predicted T^{-2} dependence and yield $DT^2 = 1.54 \cdot 10^{-6}$ cm^2 °K^2/s.

Assuming that $w(\theta, \varphi)$ is independent of φ, one can see from the expression for D and K (eqs. (2.77), (2.72), and (2.63)) that D/K depends only upon m^*/p_F and $(1+G_0)$, and so D/K is determined directly by the specific heat and susceptibility experiments. Comparison with experiment ([14]) indicates a disagreement by a factor of two which probably is due to the fact that the φ-dependence of w is significant.

III. – Microscopic Theory of Normal Systems.

In recent years considerable progress has been made in our theoretical understanding of the quantum statistical mechanics of the many-fermion problem. Much of this work has been described in review articles and several texts (see Section V for references). Our treatment here, although not as comprehensive as some in the literature, will start from an elementary point and combine in one self-contained discussion the basic methods and results which have a bearing on an understanding of liquid ^{3}He at low temperatures.

Section **1** is devoted to introducing the methods of second quantization. In Section **2** the theory of the ground state of the many-fermion problem is developed, and then applied to a study of the zero-temperature properties of liquid ^{3}He. The statistical mechanics of the N-fermion problem is formulated, in Section **3**, employing the methods of Green's functions and spectral representations. A theory is then proposed to describe the low-temperature properties of liquid ^{3}He; and the section ends with a rather comprehensive discussion, from a microscopic point of view, of the concept of quasi-particles.

The reader may continue to Part IV after mastering only Section **1** of this Part, but a complete and deeper understanding of Part IV will necessitate an appreciation of all the contents of this Part.

1. – Second quantization.

The study of a system of many fermions involves complications which can be greatly reduced by the use of the formalism of second quantization. Thus, even when there are no interactions between the particles the wave function of the system will involve a complicated sum over products of single particle wave functions, while in general the wave function will be a linear combination of determinantal wave functions each of which corresponds to the occupation of a set of single particle states. The second quantized formalism is essentially the choice of a new representation—the occupation number representation—which allows the indistinguishability of particles to be built into the theory from the start. The result is a concise formalism in which general theorems can be stated, an orderly perturbation theory formulated, and methods developed which make quantum-statistical mechanics no more difficult than the study of the ground state of the system.

A description in terms of occupation numbers precludes the possibility of using ordinary Schrödinger wave functions since we shall want to describe states with different numbers of particles. We must introduce abstract state vectors to represent the quantum-mechanical states, and linear operators to represent quantum-mechanical dynamical variables.

1·1. *State vectors and operators.* – We briefly review here, the quantum-mechanical formalism for state vectors and operators.

A state vector Ψ may be tought of as column matrix with complex elements, while the hermitean conjugate state vector $\Psi^\dagger$ may be considered a row matrix with elements the complex conjugates of the elements of Ψ. There is a scalar product $\Psi_1^\dagger\Psi_2$ which is the complex number resulting from multiplying together the matrix representation of $\Psi_1^\dagger$ and Ψ_2.

Linear operators A act on a state vector Ψ to form a new state vector $A\Psi$. One can think of A as a matrix so that $A\Psi$ is the column matrix resulting from multiplying the matrix representation of A into the matrix representation of Ψ.

The hermitean conjugate $A^\dagger$ of the operator A is defined by

$$\Psi_1^\dagger A^\dagger \Psi_2 = (A\Psi_1)^\dagger \Psi_2 \tag{3.1}$$

for all Ψ_1, and Ψ_2; and physical variables are represented by hermitean operators—that is, operators for which $A^\dagger = A$. The result of the measurement of the dynamical variable A, in the state Ψ, is

$$\langle A \rangle = \frac{\Psi^\dagger A \Psi}{\Psi^\dagger \Psi}, \tag{3.2}$$

which can easily be shown to be real.

It is assumed that all hermitean operators have a complete set Ψ_n of eigenfunctions which are defined by the equation

$$A\Psi_n = A_n \Psi_n, \tag{3.3}$$

where the real numbers A_n are the eigenvalues of A. The eigenfunctions Ψ_n can be made orthogonal to each other and normalized so that

$$\Psi_n^\dagger \Psi_m = \delta_{n,m} \tag{3.4}$$

or if they form a continuum by

$$\Psi_r^\dagger \Psi_s = \delta(r - s). \tag{3.5}$$

The completeness requirement is equivalent to the statement that

$$\sum_r \Psi_r \Psi_r^\dagger = 1, \tag{3.6}$$

which can easily be seen to be true by acting on the right of eq. (3.6) with the arbitrary state

$$\Psi = \sum_n a_n \Psi_n, \tag{3.7}$$

and then employing eq. (3.4).

1·2. *Creation and annihilation operators.* – Ignoring for the present the dynamics of the problen [that is, considering the system at one time only], we now construct state vectors to describe the different states of the N-body system and linear operators to transform these states into each other.

Consider a system of noninteracting fermions, so the Hamiltonian for the system is the sum of the Hamiltonians for the individual particles. Let the eigenstates of the individual particle Hamiltonian be $\psi_i(\boldsymbol{r}, \sigma)$ and so characterized by the label i. (The states could, for example, be single-particle eigenstates of definite momentum $\boldsymbol{p}_i$ and spin orientation σ_i). The eigenstates of the whole system are determinantal wave functions made up of the functions

$\psi_i(\boldsymbol{r}, \sigma)$ and so characterized by specifying which single-particle states are included, *i.e.* « occupied ».

Let Ψ_0 describe the state with no particles present, *i.e.* the vacuum. Take this state to be normalized so that $\Psi_0^\dagger \Psi_0 = 1$. Let us define an operator $a_i^\dagger$ which acts on Ψ_0 to make the state $a_i^\dagger \Psi_0$ which we want to correspond to one particle is the single-particle state i. Clearly the state $a_i^\dagger a_j^\dagger \Psi_0$ should correspond to the state with two particles present, one in the state i and one in the state j.

We may normalize our states so, for example,

$$\left\{\begin{aligned} &(a_i^\dagger \Psi_0)^\dagger (a_i^\dagger \Psi_0) = 1\,, \\ &\Psi_0^\dagger a_i (a_i^\dagger \Psi_0) = 1\,, \end{aligned}\right. \tag{3.8}$$

which clearly means a_i acts on the state $(a_i^\dagger \Psi_0)$ [of one particle in state i] to bring one back to the vacuum. Hence a_i annihilates particles just as $a_i^\dagger$ creates them. We could have defined the vacuum Ψ_0 as that state for which $a_i \Psi_0 = 0$ where a_i represents any member of the complete set of annihilation operators associated with the complete set of single-particle states labelled by i.

Consider the state corresponding to particles in single-particle states $i, j, k, \ldots$

$$\Psi = a_i^\dagger a_j^\dagger a_k^\dagger \ldots \Psi_0\,. \tag{3.9}$$

Note that the identity of particles has been lost because only state labels and no particle labels appear. But if we interchange two fermions the wave function must change sign, or equivalently we must require that Ψ changes sign under interchange of state labels:

$$a_i^\dagger a_j^\dagger a_k^\dagger \ldots \Psi_0 = - a_j^\dagger a_i^\dagger a_k^\dagger \ldots \Psi_0\,, \tag{3.10}$$

which means that in general we must demand

$$a_i^\dagger a_j^\dagger = - a_i^\dagger a_j^\dagger\,, \qquad i \neq j. \tag{3.11}$$

On the other hand, we clearly cannot put two fermions into the same state so

$$(a_i^\dagger)^2 = 0\,. \tag{3.12}$$

We may summarize eq. (3.11) and (3.12) and their hermitean conjugates by

$$[a_i, a_j]_+ = [a_i^\dagger, a_j^\dagger]_+ = 0\,. \tag{3.13}$$

Now consider the states

$$(3.14)\qquad \begin{cases} \Psi_1 = a_i a_i^\dagger [a_j^\dagger a_k^\dagger \dots \Psi_0] \equiv a_i a_i^\dagger \Psi_B\,, \\ \Psi_2 = a_i^\dagger a_i [a_j^\dagger a_k^\dagger \dots \Psi_0] \equiv a_i^\dagger a_i \Psi_B\,, \end{cases}$$

where the bracketed state Ψ_B either has state i occupied in which case $\Psi_1 = 0$ and $\Psi_2 = \Psi_B$, or it has the state i unoccupied in which case $\Psi_1 = \Psi_B$ and $\Psi_2 = 0$. In either case $\Psi_1 + \Psi_2 = \Psi_B$ and we conclude

$$(3.15)\qquad a_i a_i^\dagger + a_i^\dagger a_i = 1\,.$$

Analogous arguments for $a_i a_j^\dagger$ and $a_j^\dagger a_i$ (employing eqs. (3.13) and (3.15)) lead to the conclusion that

$$(3.16)\qquad a_i a_j^\dagger = a_j^\dagger a_i = 0\,, \qquad i \neq j.$$

We may summarize eq. (3.15) and (3.16) by

$$(3.17)\qquad [a_i, a_j^\dagger]_+ = \delta_{ij}\,.$$

The eq. (3.13) and (3.17) serve to define the operators a_i and $a_i^\dagger$ completely—and are often taken as the starting point in the development of the theory.

1·3. *Construction of operators.* – The eigenstates of our system of noninteracting fermions consist of states of the form

$$(3.18)\qquad \Psi_{\{n\}} = \prod_i (a_i^\dagger)^{n_i} \Psi_0\,,$$

where $n_i = 0$ or 1 and i ranges over the complete set of states of the individual-particle Hamiltonian. To each state $\Psi_{\{n\}}$ corresponds the usual determinantal wave function of ordinary quantum mechanics. We assume these eigenstates of the noninteracting system form a complete set, so that the states of a system of interacting particles can be written as a superposition of these states just as in eq. (3.7) and in exact correspondence to the superposition of determinantal wave functions which would be employed in ordinary quantum mechanics.

We wish now to construct operators in the space of creation and annihilation operators which will correspond to ordinary operators acting on superpositions of determinantal wave functions. Clearly we need not concern ourselves with superpositions of the states $\Psi_{\{n\}}$ and it will suffice to show that we can construct an operator A such that the matrix elements of A between states $\Psi_{\{n\}}$ and $\Psi_{\{n'\}}$ is the same as the matrix element of the ordinary operator corresponding to A between the determinantal functions corresponding to $\Psi_{\{n\}}$ and $\Psi_{\{n'\}}$.

Suppose we have an ordinary operator $Q(\boldsymbol{r}, \boldsymbol{\sigma})$ which is a function of the variables of only one particle. Let its matrix element between single-particle states be

$$Q_{ij} = \langle \psi_i(\boldsymbol{r}, \boldsymbol{\sigma}) | Q(\boldsymbol{r}, \boldsymbol{\sigma}) | \psi_i(\boldsymbol{r}, \boldsymbol{\sigma}) \rangle \, . \tag{3.19}$$

In this case the corresponding second-quantized operator is

$$Q = \sum_{i,j} Q_{ij} a_i^\dagger a_j \, , \tag{3.20}$$

as is clear by comparison of matrix elements according to the prescription of the last paragraph.

If we have a two-particle operator $S(\boldsymbol{r}_1 \boldsymbol{\sigma}_1, \boldsymbol{r}_2 \boldsymbol{\sigma}_2)$ with matrix elements

$$S_{ij,kl} = \langle \psi_i(\boldsymbol{r}_1, \sigma_1) \psi_j(\boldsymbol{r}_2, \sigma_2) | S(\boldsymbol{r}_1 \boldsymbol{\sigma}_1, \boldsymbol{r}_2 \boldsymbol{\sigma}_2) | \psi_k(\boldsymbol{r}_1, \sigma_1) \psi_l(\boldsymbol{r}_2, \sigma_2) \rangle \, , \tag{3.21}$$

[there is *no* anti-symmetrization to be performed in eq. (3.21)], then the corresponding second-quantized operator can be shown to be

$$S = \sum_{ijkl} S_{ij,kl} a_j^\dagger a_i^\dagger a_k a_l \, , \tag{3.22}$$

These two cases exhaust all physically interesting situations.

As a trivial example of eqs. (3.19) and (3.20) we can see that the operator

$$N = \sum_i N_i = \sum_i a_i^\dagger a_i \, , \tag{3.23}$$

is the number operator, as is obvious, since N_i acting on any state vector Ψ will yield either 0 or Ψ depending upon whether Ψ correspond to no particles or one particle in the state i.

As a second example, if the single particle levels of our system of non-interacting fermions have energies ε_i then obviously the Hamiltonian operator H_0 is

$$H_0 = \sum_i \varepsilon_i a_i^\dagger a_i \, , \tag{3.24}$$

It is clear that acting on the states $\Psi_{\{n\}}$ of eq. (3.18) with N and H_0 we have

$$(3.25)\qquad \begin{cases} N\Psi_{\{n\}} = \left[\sum\limits_i n_i\right]\Psi_{\{n\}}\,, \\ H_0\Psi_{\{n\}} = \left[\sum\limits_i n_i\varepsilon_i\right]\Psi_{\{n\}}\,, \end{cases}$$

as expected.

1·4. *The Hamiltonian operator.* – For ^{3}He we may take the ordinary quantum-mechanical Hamiltonian to be

$$(3.26)\qquad \begin{cases} H = H_0 + H_1\,, \\ H_0 = -\dfrac{\hbar^2}{2m}\sum\limits_{i=1}^{N}\nabla^2_{r_i}\,, \\ H_1 = \frac{1}{2}\sum\limits_{i,j=1}^{N} v(|\boldsymbol{r}_i - \boldsymbol{r}_j|)\,, \end{cases}$$

where m is the ^{3}He atomic mass, and we have assumed that there are no three-body forces and that the two-body potential is local and spin-independent. It is usual to approximate $v(r)$ by some convenient potential such as the 6-12-potential. Let us take a complete set of single-particle states $\psi_i(\boldsymbol{r}, \sigma)$ in which case, defining

$$(3.27)\qquad \begin{cases} T_{ij} = -\dfrac{\hbar^2}{2m}\displaystyle\int \psi_i^*(\boldsymbol{r}, \sigma)\,\nabla^2\psi_j(\boldsymbol{r}, \sigma)\,\mathrm{d}^3r\,, \\ V_{ij,kl} = \displaystyle\iint \psi_i^*(\boldsymbol{r}_1, \sigma_1)\,\psi_j^*(\boldsymbol{r}_2, \sigma_2)\,v(|\boldsymbol{r}_1 - \boldsymbol{r}_2|)\,\psi_k(\boldsymbol{r}_1, \sigma_1)\,\psi_l(\boldsymbol{r}_2, \sigma_2)\,\mathrm{d}^3r_1\,\mathrm{d}^3r_2\,, \end{cases}$$

we have from eqs. (3.20) and (3.22) the second-quantized Hamiltonian:

$$(3.28)\qquad \begin{cases} H = H_0 + H_1\,, \\ H_0 = \sum\limits_{ij} T_{ij}\,a_i^\dagger a_j\,, \\ H_1 = \frac{1}{2}\sum\limits_{ijkl} V_{ij,kl}\,a_j^\dagger a_i^\dagger a_k a_l\,. \end{cases}$$

It should be recalled that the index i characterizes the state $\psi_i(\boldsymbol{r}, \sigma)$ and thus includes a spin label.

For a large system of volume v, the single-particle states may be taken as eigenfunctions of momentum and spin orientation:

$$\psi_{\boldsymbol{k}\sigma}(\boldsymbol{r},\sigma') = \frac{1}{v^{\frac{1}{2}}}\exp[i\boldsymbol{k}\cdot\boldsymbol{r}]\,\delta_{\sigma\sigma'}\,, \tag{3.29}$$

in which case

$$\begin{cases} H_0 = \sum\limits_{\boldsymbol{k}\sigma}\dfrac{\hbar^2k^2}{2m}\,a^{\dagger}_{\boldsymbol{k}\sigma}a_{\boldsymbol{k}\sigma}\,, \\[2ex] H_1 = \frac{1}{2}\sum\limits_{\boldsymbol{k}_i\sigma_i} V_{12,34}\,a^{\dagger}_{\boldsymbol{k}_2\sigma_2}a^{\dagger}_{\boldsymbol{k}_1\sigma_1}a_{\boldsymbol{k}_3\sigma_3}a_{\boldsymbol{k}_4\sigma_4}\,, \end{cases} \tag{3.30}$$

with

$$V_{12,34} = \frac{1}{v}\,\delta_{\boldsymbol{k}_1+\boldsymbol{k}_2,\,\boldsymbol{k}_3+\boldsymbol{k}_4}\,\delta_{\sigma_1\sigma_3}\,\delta_{\sigma_2\sigma_4}\,v(|\boldsymbol{k}_3-\boldsymbol{k}_1|)\,, \tag{3.31}$$

and the Fourier component of the potential defined by

$$v(|\boldsymbol{k}|) = \int \mathrm{d}^3r\,\exp[i\boldsymbol{k}\cdot\boldsymbol{r}]\,v(r)\,. \tag{3.32}$$

For actual calculations one would normally work in the momentum and spin representation defined by eq. (3.29). To express general formulas however, it is convenient not to have to explicitly use spin indices. This can be accomplished by using the label l to represent $\boldsymbol{k}$, σ (as usual), but take advantage of the simplicity according from momentum conservation by observing that in the momentum representation H_0 is diagonal. Thus we may write

$$\begin{cases} H = H_0 + H_1\,, \\[1ex] H_0 = \sum\limits_i T_i\,a^{\dagger}_i a_i\,, \\[2ex] H_1 = \frac{1}{2}\sum\limits_{ijkl} V_{ij,kl}\,a^{\dagger}_j a^{\dagger}_i a_k a_l\,, \end{cases} \tag{3.33}$$

where the summation over i includes both a momentum and spin sum as in eq. (3.30), and the matrix elements $V_{ij,kl}$ have the property indicated in eq. (3.31). From here on we will usually use the Hamiltonian of eq. (3.33), *i.e.* work in a momentum representation in which spin indices are suppressed.

1·5. *Equations of motion.* – The formalism so far developed is time-independent and hence suitable for stationary state problems. Thus for equilibrium statistical mechanics we need (eq. (1.11)) the eigenvalues E_n of the Hamiltonian H (eq. (3.33)). One can find approximations to these eigenvalues by approximately solving the Schrödinger equation

$$H\Psi_n = E_n \Psi_n \,, \tag{3.34}$$

and this approach is that of perturbation theory, Brueckner theory, and BCS theory. The newer methods employings Green's functions are formulated in terms of a time-dependent problem, and so we review here various possible formulations of time-dependent quantum mechanics, and in particular the Heisenberg representation (*).

In the usual (or Schrödinger) representation, the time dependence is in the state vector, and no time is explicitly in the operators.

Thus

$$i\hbar \frac{\mathrm{d}\Psi_{\mathrm{S}}}{\mathrm{d}t} = H_{\mathrm{S}} \Psi_{\mathrm{S}} \,, \tag{3.35}$$

where the subscript indicates the representation and H_{S} is the previously derived Hamiltonian (eq. (3.33)). An operator A_{S} in the Schrödinger representation satisfies

$$\frac{\mathrm{d}A_{\mathrm{S}}}{\mathrm{d}t} = 0 \,, \tag{3.36}$$

while it is clear that the matrix element $\Psi_{\mathrm{S}}^{\dagger} A_{\mathrm{S}} \Psi_{\mathrm{S}}$ is time-dependent.

Alternatively, we could put the time-dependence into the operator A, have Ψ time-independent, but keep the time-dependence of $\Psi^{\dagger} A \Psi$ the same as in the Schrödinger representation. This program can be attained by defining an operator $U(t)$ which satisfies

$$i\hbar \frac{\mathrm{d}U(t)}{\mathrm{d}t} = H_{\mathrm{S}} U(t) \,, \qquad U(0) = 1 \,, \tag{3.37}$$

and then taking

$$\begin{cases} \Psi_{\mathrm{H}}(t) = U^{\dagger}(t) \Psi_{\mathrm{S}}(t) \,, \\ A_{\mathrm{H}}(t) = U^{\dagger}(t) A_{\mathrm{S}} U(t) \,. \end{cases} \tag{3.38}$$

(*) Thus an understanding of Section III.2 and most of Part IV does not require mastery of this Part.

The reader can easily verify that in this, the Heisenberg representation,

$$(3.39)\qquad \begin{cases} H_H = H_S = H\,, \\[2mm] \dfrac{d\Psi_H(t)}{dt} = 0\,, \\[2mm] i\hbar\dfrac{dA_H(t)}{dt} = [A(t), H]_-\,, \end{cases}$$

and $\Psi^\dagger_H A_H \Psi_H = \Psi^\dagger_S A_S \Psi_S$.

Applying this formalism to the N-body problem, we can find the equations of motion for the creation and annihilation operators. Thus from eqs. (3.33) and (3.39) by use of eqs. (3.13) and (3.17):

$$i\hbar\frac{da_m}{dt} = [a_m, H]_- = \sum_i T_i a_m,\, a^\dagger_i a_i]_- + \tfrac{1}{2}\sum_{ijkl} V_{ij,kl}[a_m,\, a^\dagger_j a^\dagger_i a_k a_l] =$$
$$= \sum_i T_i \delta_{im} a_i + \tfrac{1}{2}\sum_{ijkl} V_{ij,kl}[\delta_{mj} a^\dagger_i a_k a_l - \delta_{mi} a^\dagger_j a_k a_l]\,,$$

and hence

$$(3.40)\qquad i\hbar\frac{da_m}{dt} = T_m a_m + \tfrac{1}{2}\sum_{jkl}[V_{jm,kl} - V_{mj,kl}]\, a^\dagger_j a_k a_l\,,$$

and a similar equation for $a^\dagger_m$. The equations of motion just derived will play a crucial role in the Green's function formulation of the N-body problem.

In the momentum representation, and for a system of free particles we may integrate eq. (3.40) to obtain

$$(3.41)\qquad a_m(t) = a_m(0)\exp\left[-\frac{i}{\hbar}T_m t\right].$$

2. – Theory of the ground state.

In this section we will consider first the choice of single-particle states, then Rayleigh-Schrödinger perturbation theory, the linked-cluster theorem and diagrammatic methods, Brueckner theory, and finally perturbation results for a dilute nonideal gas. We shall not discuss here many interesting and important results which have been obtained using Green's functions (see GM1, GA2, and LA3), but defer the discussion to Section **3** where we develop the Green's function formalism for statistical ensembles after which the ground state results follow as the special case in which the temperature is zero.

2·1. *Hartree-Fock theory.* – In the development of Section III.1 we considered various single-particle states which were either occupied or not. Of course we realize that in general our state vector will be a linear combination of the state vector $\Psi_{\{n\}}$ corresponding to occupation of the set of single-particle states $\{n\}$. It is interesting to ask how one may « best » choose the set of single particle states. We make this question precise by asking that if we let our state be of the form $\Psi_{\{n\}}$ [and not a linear combination of such states], what set $\{n\}$ will give the best approximation to the ground-state energy.

Given a set of single-particle states we can construct a state $\Psi_{\{0\}}$ which corresponds to the first N-states occupied:

$$\Psi_{\{0\}} = \prod_{i\leqslant N} a_i^\dagger \Psi_0 \,. \tag{3.42}$$

The expectation value of the Hamiltonian (eq. (3.37)) in this state may be evaluated by use of the anti-commutation rules (eqs. (3.31) and (3.17)) and the definition of the vacuum $[a_i\Psi_0 = 0$ for all $i]$. One obtains

$$\begin{cases} E = \Psi_{\{0\}}^\dagger H \Psi_{\{0\}}\,, \\ E = \sum\limits_{i=1}^{N} T_i + \frac{1}{2}\sum\limits_{i,j=1}^{N} [V_{ij,ij} - V_{ji,ij}]\,. \end{cases} \tag{3.43}$$

If we now minimize E, while demanding that the single-particle functions $\psi_i(\boldsymbol{r}, \sigma)$ form an orthonormal set, *i.e.*

$$\int \psi_i^*(\boldsymbol{r}, \sigma)\, \psi_j(\boldsymbol{r}, \sigma)\, \mathrm{d}^3 r = \delta_{ij}\,, \tag{3.44}$$

we obtain by using eqs. (3.43), (3.44), and (3.27) and varying with respect to $\psi_i^*(\boldsymbol{r}, \sigma)$

$$\begin{aligned} -\frac{\hbar^2}{2m}\nabla^2\psi_i(\boldsymbol{r}, \sigma) + \sum_{j=1}^{N} \int \psi_j^*(\boldsymbol{r}_2, \sigma_2)\, v(\boldsymbol{r} - \boldsymbol{r}_2)\, \psi_j(\boldsymbol{r}_2, \sigma_2)\, \psi_i(\boldsymbol{r}, \sigma)\, \mathrm{d}^3 r_2 \,- \\ - \sum_{j=1}^{N} \int \psi_j^*(\boldsymbol{r}_2, \sigma_2)\, v(\boldsymbol{r} - \boldsymbol{r}_2)\, \psi_i(\boldsymbol{r}_2, \sigma_2)\, \psi_j(\boldsymbol{r}, \sigma)\, \mathrm{d}^3 r_2 = \varepsilon_i \psi_i(\boldsymbol{r}, \sigma)\,; \qquad i = 1, \dots N, \end{aligned} \tag{3.45}$$

where the constraint of eq. (3.44) has been inserted through the Lagrangian multiplier ε_i. Because of the spin operators it can be seen that in the first (direct) integral all states j contribute while in the second (exchange) integral only states j with spin direction σ_j the same as σ_i will contribute. This set of simultaneous equations (the Hartree-Fock equations) must be solved simultaneously for the functions $\psi_i(\boldsymbol{r}, \sigma)$. In the momentum representation the solution is trivial however (*), as the eigenfunctions of momentum satisfy eq. (3.45). By multiplying eq. (3.45) by ψ_i^* and integrating one obtains the eigenvalue, or single particle energies as

$$\varepsilon_i = T_i + \sum_{j=1}^{N} [V_{ij,ij} - V_{ji,ij}] , \tag{3.46}$$

while the total energy E is given by eq. (3.43) or

$$E = \sum_{i=1}^{N} \varepsilon_i - \tfrac{1}{2} \sum_{i,j=1}^{N} [V_{ij,ij} - V_{ji,ij}] , \tag{3.47}$$

with the matrix elements evaluated using the plane-wave states of eq. (3.29). Thus in an infinite medium the Hartree-Fock theory gives single-particle states which are still eigenfunctions of momentum, while the single-particle energies are ε_i.

Using the Hartree-Fock theory to define the single-particle energies, we may transform the Hamiltonian of eq. (3.33) by adding $(\varepsilon_i - T_i)$ to H_0 and subtracting it from H_1, to obtain

$$\left\{\begin{aligned} H &= H_0' + H_1' , \\ H_0' &= \sum_i \varepsilon_i a_i^\dagger a_i , \\ H_1' &= \tfrac{1}{2} \sum_{ijkl} V_{ij,kl} a_j^\dagger a_i^\dagger a_k a_l + \sum_i (T_i - \varepsilon_i) a_i^\dagger a_i , \end{aligned}\right. \tag{3.48}$$

where the ε_i are given by eq. (3.46) and $V_{ij,kl}$ by eq. (3.31). The removal from H_i of some diagonal terms means that perturbation theory with eq. (3.48) should converge more quickly than with eq. (3.33)—as in fact is true in atomic-structure problems. For liquid ^{3}He, however, the Hartree-Fock theory is of no value as can be readily seen from eq. (3.31) since for the 6-12-potential $V_{ij,ij}$ (for example) is infinite, and the ε_i are not even defined.

We must develop a more powerful method to be able to treat liquid ^{3}He, and we start by studying perturbation theory.

2·2. *Rayleigh-Schrödinger theory.* – We wish to develop a perturbation theory which will allow us to start with the independent particle states and energies and systematically take into account the interaction terms in the Hamiltonian.

(*) This situation is to be contrasted with atomic problems where an external potential (due to the nucleus) destroys the translational invariance and causes $\psi_i(\boldsymbol{r}, \sigma)$ to be a complicated function of position so that eqs. (3.45) must usually be solved self-consistently.

Let us suppose we write $H = H_0 + H_1$ where we assume that we know the eigenstates Φ_n and corresponding eigenvalues W_n of H_0. We desire to find the eigenvalue E_0 associated with the ground state of H_0. We may do so by simply expanding Ψ_0 and E_0 in power of the strength of the perturbation H_1. Thus we write

$$(3.49)\qquad \begin{cases} H = H_0 + \lambda H_1 , \\ \Psi_0 = \Phi_0 + \lambda \sum_n a_n^{(1)} \Phi_n + \lambda^2 \sum_n a_n^{(2)} \Phi_n + \dots , \\ E_0 = W_0 + \lambda E_1 + \lambda^2 E_2 + \dots , \end{cases}$$

solve the Schrödinger equation (eq. (3.34)) by equating powers of λ, and then set $\lambda = 1$. The result for the energy eigenvalues is, through second order:

$$(3.50)\qquad E_0 = W_0 + \Phi_0^\dagger H_1 \Phi_0 + \sum_{n \neq 0} \frac{(\Phi_0^\dagger H_1 \Phi_n)(\Phi_n^\dagger H_1 \Phi_0)}{W_0 - W_n} + \dots ,$$

where the succeeding terms become more and more complicated to write out.

We may use eq. (3.50) and the Hamiltonian of eq. (3.33) to evaluate explicitly the first few terms in E_0. Thus, if Φ_0 corresponds to the first N levels of H_0 occupied we have

$$(3.51)\qquad \Phi_0 = \prod_{i \leqslant N} (a_i^\dagger) \Phi_{\text{vacuum}} ,$$

where the state of no particles has been indicated explicitly. Using the Hamiltonian (eq. (3.33)) the anticommutation relations (eq. 3.13) and eq. (3.17)). and the definition of the vacuum, one obtains

$$(3.52)\qquad E_0 = \sum_{i=1}^N T_i + \frac{1}{2} \sum_{i,j=1}^N [V_{ij,ij} - V_{ij,ij}] + \frac{1}{2} \sum_{i,j=1}^N \sum_{k,l=N+1}^\infty \frac{|V_{ij,kl}|^2 - V_{ij,kl} V_{lk,ij}}{T_i + T_j - T_k - T_l} + \dots .$$

The perturbation theory of eq. (3.50) may also be applied to the Hamiltonian in the Hartree-Fock representation (eq. (3.48)). Clearly one obtains

$$(3.53)\qquad E_0 = \sum_{i=1}^N \varepsilon_i + \left\{ \frac{1}{2} \sum_{i,j=1}^N [V_{ij,ij} - V_{ji,ij} + \sum_{i=1}^N (T_i - \varepsilon_i) \right\} + \\ + \frac{1}{2} \sum_{i,j=1}^N \sum_{k,l=N+1}^\infty \frac{|V_{ij,kl}|^2 - V_{ij,kl} V_{lk,ij}}{\varepsilon_i + \varepsilon_j - \varepsilon_k - \varepsilon_l} + \dots ,$$

which can be seen to yield a partial cancellation between the zero and first order terms so that eq. (3.53) becomes the same as Eq. (3.52) except that in the second order term the kinetic energies T_i are replaced by single particle energies ε_i. In general (*), (except in first order) the diagonal term in H_1' need not be considered if a special class of terms in eq. (3.52) is omitted. In the remaining terms all T_i's are to be replaced by ε_i's. This corresponds to summing many terms in the expansion of eq. (3.52) and thus combining them into one term. It is exactly this type of manipulation which we want to be able to do more generally and for that purpose we need a systematic and concise characterization of the general term in the perturbation theory. Such is afforded by the « linked-cluster theorem ».

2·3. *Linked-cluster theorem.* – The Rayleigh-Schrödinger theory, if pursued in the manner of the last section, would soon lead to terms in the ground state energy which appear not to depend linearly on the number of particles N. These unphysical terms in fact all cancel, as was first realized by BRUECKNER, and subsequently proved by GOLDSTONE (GD1), HUGENHOLTZ (HU1), and HUBBARD (HUB1). The result of Goldstone's analysis is a convenient graphical representation of the remaining terms which is of particular importance for the insight it affords into the structure of the perturbation theory.

Because of its purely mathematical nature, we will not give a proof of the linked-cluster theorem. The reader should appreciate that it could be established by the pedestrian methods of the last section, but of course the proof of GOLDSTONE is much more concise and elegant. A particularly clear exposition is given in the text by THOULESS (TH1).

We are interested in the result of the linked-cluster theorem, which we state here in a notation which is compatible with the text by THOULESS. First it is convenient to specialize to a system in the momentum representation (H given by eq. (3.33)), in which the unperturbed ground state Φ_0 corresponds to the first N levels occupied. The rules for calculating the n-th order contribution to the ground-state energy eigenvalue E_0 are:

1) Construct a graph by drawing n horizontal dashed lines at different levels, and joining their $2n$ ends with solid lines having arrows to indicate direction in such a way that one directed line enters and one leaves each end of a dashed line. The solid lines may join an end of a dashed line to itself, or one end of a dashed line to its opposite end. Draw only graphs in which there are no unlinked parts, but draw all possible graphs consistent with the above rules.

(¹) This result is shown in Paragraph III.2·3.

2) To calculate the energy contribution associate with each line directed up (a particle line) a state $i > N$ and to each line directed down (a hole line) a state $i \leqslant N$. A line joining a dashed line to itself (either end) is a hole line. Now each dashed line contributes a factor $\frac{1}{2} V_{ij,kl}$, where i and j are the directed lines leaving the vertex on the left and right respectively and k and l are the directed lines entering the vertex on the left and right. Each of the $(n-1)$ intervals between vertices contributes the inverse of the algebraic sum of T_i factors associated with each directed line crossing the interval, particle lines contributing negatively and hole lines positively. There is a factor of (-1) for each closed loop of solid line, and a factor of (-1) for each hole line.

In Fig. 1 the first and second order graphs have been constructed and labelled in accordance with these rules. Application of rule 2) and the relation $V_{ij,kl} = V_{ji,lk}$ shows that the contribution of Fig. 1-*d* and 1*c* are equal as are the contributions of Fig. 1-*e* and 1-*f*. The result is just the first and second order terms of eq. (3.52), as the reader can easily check.

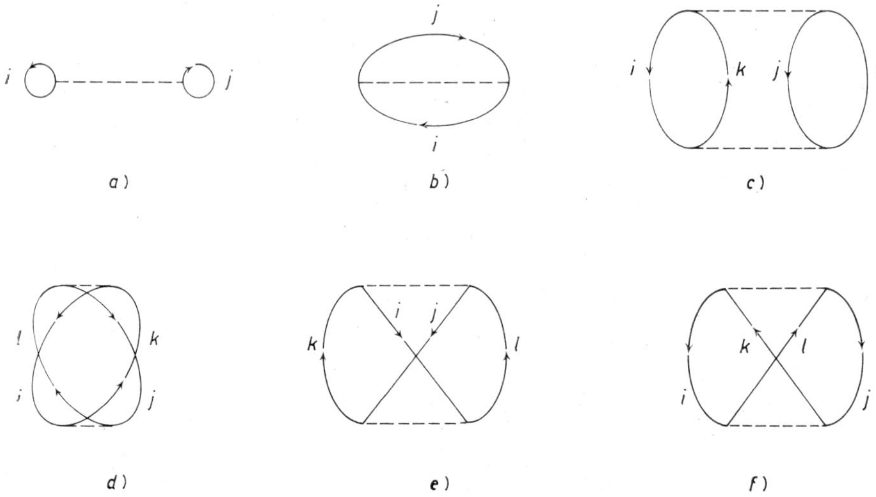

Fig. 1. – Graphs for the ground-state energy in first and second order.

We are now in a position to understand the Hartree-Fock theory from a graphical point of view. A linked-cluster theorem for the Hamiltonian in Hartree-Fock representation (eq. (3.48)) leads to the above rules with T_i replaced by ε_i, but now we have an additional element in the perturbation $(T_i - \varepsilon_i)$ which is represented by a horizontal dashed line terminating in a cross. Thus in first and second order we have the diagrams of Fig. 1 plus the diagram of Fig. 2. This means that if we calculate through second order we just obtain eq. (3.53), the additional contribution in first order corresponding to the graph of Fig. 2. In all orders beyond the first the new element in a solid line just cancels all graphs in which that solid line joins two points at the same level so one can *ignore* this large class of graphs. In Fig. 3 we indicate examples of the class of graphs which the new element in the line l (of the second order graph) of Fig. 1-c will cancel. Thus the net effect is that one has summed many terms in the perturbation expansion of the Hamiltonian of eq. (3.33) and this summation is reflected in the use of single-particle energies ε_i and the restriction in the perturbation analysis to a smaller class of diagrams.

Fig. 2. – Additional graph in first order in the Hartree-Fock representation.

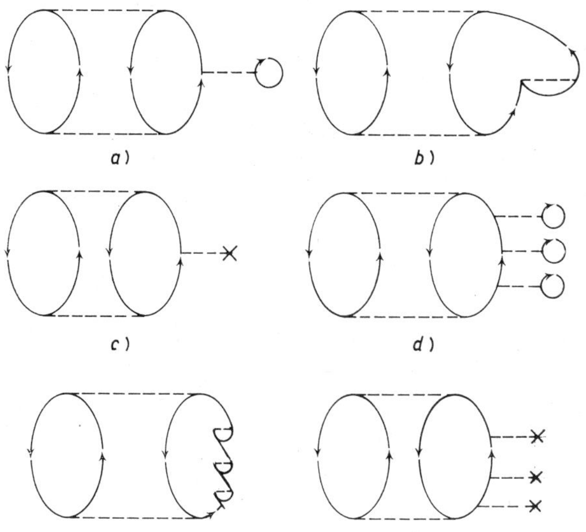

Fig. 3. – In (a) and (b) are indicated some of the modifications (in one line of a second order graph for the ground state energy) which are cancelled by (c), and analogously (d) and (e) are cancelled by (f).

2·4. *Brueckner theory.* – The physical basis of Brueckner's theory (BR1, BG1 and BG2) is the observation that when two ^{3}He atoms interact, they interact strongly; but on the other hand the liquid is sufficiently dilute that while the two atoms interact their interaction with other particles may be considered in an average way.

In this section we shall indicate how this physical picture can be turned into a quantitative theory, and then discuss the results for liquid ^{3}He.

Consider, first of all, two ^{3}He atoms in free space. Because the inter-atomic potential $v(r)$ it very repulsive at short distances the scattering is not given correctly by a perturbation analysis, but one must actually solve the Schrödinger equation. If we expand the solution in powers of the potential we can indicate it by diagrams as in Fig. 4. Any single diagram would be very large, but the sum is well behaved.

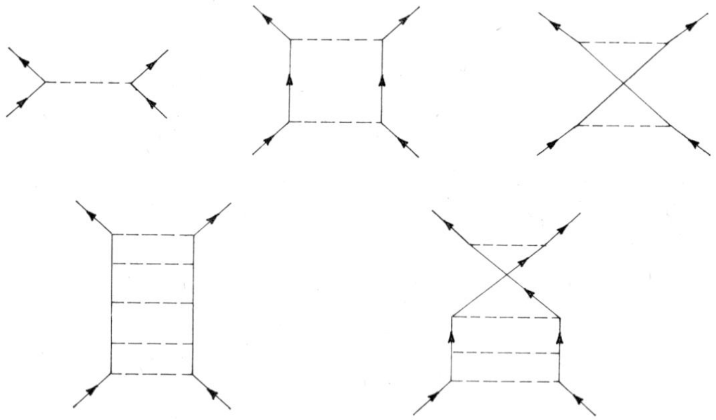

Fig. 4. – Diagrammatic representation of some of the terms which appear in a perturbation-theory expansion of the solution to the Schrödinger equation for the scattering of two ^{3}He atoms.

Now, in the liquid the situation must be comparable, that is, when two ^{3}He atoms interact perturbation theory will not be adequate to describe the interaction. We must sum an infinite class of terms corresponding to the strong interactions between two atoms. The diagrammatic perturbation theory allows us to do so in a systematic way. We start with the Hamiltonian in Hartree-Fock representation (eq. (3.48)), but leave the single-particle energies ε_i still unspecified. The lowest order terms in the ground-state energy correspond to Fig. 1-*a* and 1-*b* and Fig. 2. We wish to augment Fig. 1-*a* and 1-*b* by a selected class of higher-order terms which corresponds to the repeated interaction between the two atoms. Some of these diagrams are indicated in Fig. 1-*c*-*f* and in Fig. 5. The sum of all these terms we may write as

$$E_0 = \sum_{i=1}^{N} T_i + \tfrac{1}{2} \sum_{i,j=1}^{N} K_{ij,ij} , \tag{3.54}$$

where the reaction matrix K is defined by

$$K_{ij,kl} = [V_{ij,kl} - V_{ij,lk}] + \tfrac{1}{2} \sum_{k',l'=N+1}^{\infty} [V_{ij,k'l'} - V_{ij,l'k'}] \frac{1}{\varepsilon_k + \varepsilon_l - \varepsilon_{k'} - \varepsilon_{l'}} K_{k'l'k,l} . \tag{3.55}$$

The reader can check these equations by solving eq. (3.55) by iteration, inserting the result into eq. (3.54) and comparing with the energy E_0 as calculated according to the rules of Paragraph III.3·3.

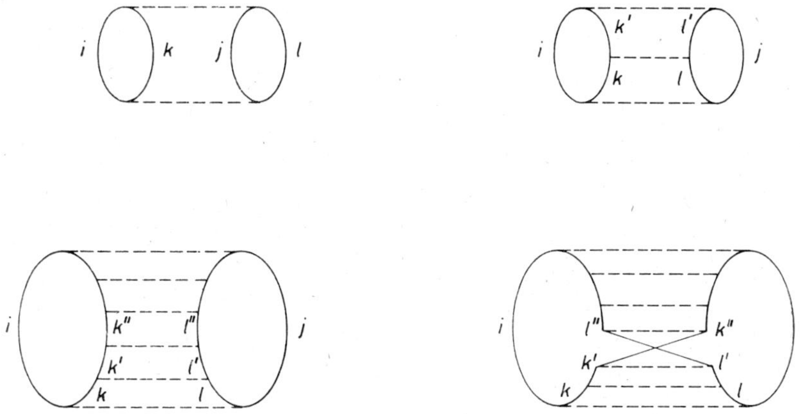

Fig. 5. – Some of the graphs included in Brueckner theory.

There remains the problem of defining the single-particle energies ε_i. Reasoning as in Hartree-Fock theory where (see Fig. 3) the cross-interaction was taken to cancel the first order modification in a particle (or hole) line, the

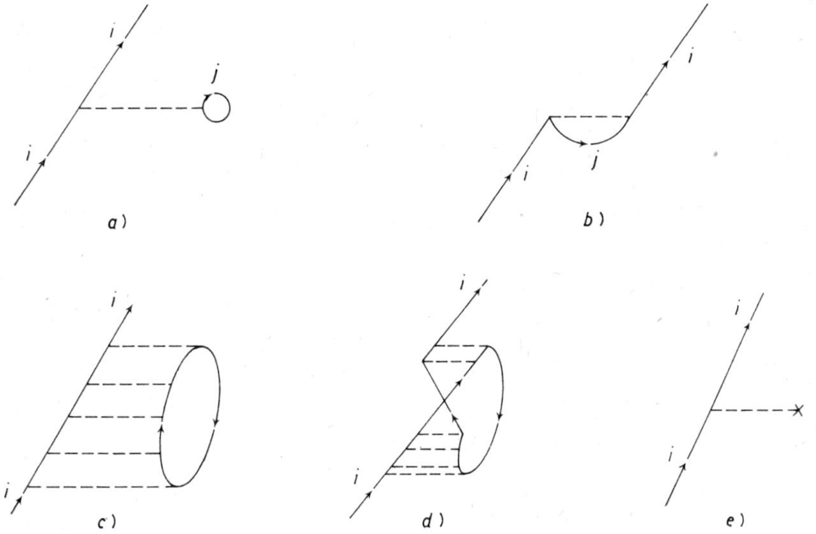

Fig. 6. – The series (a) and (b) are examples of graphs whose contribution will be cancelled by choice of the single-particle energies in accord with eq. (3.56).

single-particle energies in Brueckner theory are taken to be

$$\varepsilon_i = T_i + \sum_{j \leqslant N} K_{ij,ij} \,. \tag{3.56}$$

In Fig. 6 we have indicated the graphical equivalent of eq. (3.56).

It is clear that eq. (3.54) is not a perturbation-theory result as it contains many high-order terms. The remaining terms in a perturbation analysis all correspond to *three* (or more) particles interacting simultaneously. Thus the lowest-order term not already included in eq. (3.56) is indicated in Fig. 7. Physically we can argue that it should be small, and calculation seems to support this view (BG2).

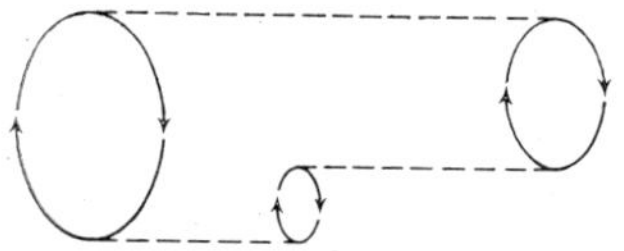

Fig. 7. – Lowest-order term *not* included in the Brueckner theory first-order calculation [(eq. (3.54)].

Returning to the reaction matrix K (eq. (3.55)), we can ask how it differs from an analogous two-particle reaction matrix for scattering of ^{3}He atoms in free space. We can see that there are two important modifications. First the summation over k' and l' is restricted to unoccupied states. This is the exclusion-principle effect (of the other particles in the medium) upon the two strongly interacting particles. Secondly, the single particle energies are ε_i rather than T_i. This is a dispersive effect of the medium which [as can be seen in Fig. 6] corresponds to a particle moving in an effective potential arising from its interaction with the other particles of the medium. Both of these effects are important in that they significantly affect K.

Thus, the mathematics can be seen to reflect the physical hypothesis that one need only consider two particles interacting strongly at one time [ignore theree-body cluster terms as in Fig. 7], while the effect of the remaining particles may be included in an average manner [exclusion principle and dispersive property of the medium].

The theory as defined is finite even if v is singular, and furthermore it is amenable to computation. BRUECKNER and GAMMEL have solved eqs. (3.55) and (3.56) self-consistently, with results which are in remarkably good agreement with experiment for the binding energy per particle (*), density, specific

(*) The binding energy is in poorest agreement with experiment, but most sensitive to higher order terms in the theory. It is easy to think of improvements to the theory such as including hole-hole interactions analogously to the way particle-particle interactions have been treated or by defining the single-particle energies as the functional derivative of E_0 rather than by eq. (3.56) [these are not equivalent since K clearly is a functional of the distribution of particles as can be seen from its definition in eq. (3.55)], which might be expected to improve the agreement with experiment.

heat, susceptibility, compressibility, and coefficient of thermal expansion.

It is clearly possible to formulate the theory so that it can be compared with the Landau theory of a Fermi liquid and so used to determine from a microscopic point of view the phenomenological function $f(\boldsymbol{p}, \boldsymbol{p}', \boldsymbol{\sigma}, \boldsymbol{\sigma}')$. We shall not do this, but limit such a discussion to two other cases, namely the dilute gas at the absolute zero [see the next Paragraph], and a general case where we must employ statistical mechanics [see Paragraph III.3·7].

Finally it should be emphasized that Brueckner theory, although a well defined approximation to the N-body problem, is still intuitive in that a (hopefully!) judicious choice has been made of which (infinite) class of terms in the perturbation theory it is important to retain. Examination of some of the terms not included indicates they are small, but in no sense has it been demonstrated that the remaining (infinite number of) terms make a small contribution.

2·5. *Dilute nonideal gas.* – Liquid ^{3}He is not a dilute system in that $a(N/V)^{\frac{1}{3}} > 1$, where a is the effective interaction radius for two ^{3}He atoms (and may be taken to be of the order of the zero-energy scattering length). On the other hand a dilute gas $a(N/V)^{\frac{1}{3}} \ll 1$ can be studied by perturbation theory in a *systematic* manner. It is thus of interest to briefly consider the results which have been obtained (HY1, AK4) for this system.

The calculations correspond to the limit Brueckner theory approaches as the density is decreased. In this case both the exclusion principle effect and the dispersive effect vanish, and the reaction matrix in the medium is the same as in free space. This matrix may be expressed in terms of the zero energy scattering length a, and the energy computed from eq. (3.54) (*). There results, through second order in the parameter $a(N/V)^{\frac{1}{3}}$, the expression

$$E_0 = \frac{3}{10}(3\pi^2)^{\frac{2}{3}}\frac{\hbar^2}{m}\left(\frac{N}{V}\right)^{\frac{2}{3}} N + \frac{\pi a \hbar^2}{m}\left(\frac{N}{V}\right) N \cdot \left[1 + \frac{6}{35}\left(\frac{3}{\pi}\right)^{\frac{1}{3}} a \left(\frac{N}{V}\right)^{\frac{1}{3}} (11 - 2\log 2)\right]. \tag{3.57}$$

A calculation can also be made of the function $f(\boldsymbol{p}, \boldsymbol{p}', \boldsymbol{\sigma}, \boldsymbol{\sigma}')$ which appears in the Landau theory. For the situation where $|\boldsymbol{p}| = |\boldsymbol{p}'| = p_F$ $f(\theta, \boldsymbol{\sigma}, \boldsymbol{\sigma}')$ is found to be

$$\begin{aligned} f(\theta, \boldsymbol{\sigma}, \boldsymbol{\sigma}') = {} & \frac{2\pi a \hbar^2}{m}\left\{1 + 2\left(\frac{3}{\pi}\right)^{\frac{1}{3}} a \left(\frac{N}{V}\right)^{\frac{1}{3}}\left[2 + \frac{\cos\theta}{2\sin\theta/2}\log\frac{(1+\sin\theta/2)}{(1-\sin\theta/2)}\right]\right\} - \\ & - \frac{2\pi a \hbar^2}{m}(\boldsymbol{\sigma}\cdot\boldsymbol{\sigma}')\left\{1 + 2\left(\frac{3}{\pi}\right)^{\frac{1}{3}} a \left(\frac{N}{V}\right)^{\frac{1}{3}}\left[1 - \frac{1}{2}\sin\theta/2\log\frac{(1+\sin\theta/2)}{(1-\sin\theta/2)}\right]\right\}, \end{aligned} \tag{3.58}$$

(*) For a calculation which proceeds directly from perturbation theory, but yields the same result, see LANDAU and LIFSHITZ (LL1), § 76.

from which it follows that

$$\frac{m^*}{m} = 1 + \frac{8a^2}{15}\left(\frac{3}{\pi}\right)^{\frac{2}{3}}\left(\frac{N}{V}\right)^{\frac{2}{3}}[7\log 2 - 1]\,. \tag{3.59}$$

These formulas are only valid if $a > 0$ (*i.e.* repulsive forces) for otherwise perturbation theory is not convergent and the Landau theory not applicable, all of which will be discussed in Part IV.

3. – Statistical mechanics.

In this Section we will in the first four Paragraphs develop a concise and convenient formulation of quantum-statistical mechanics for the N-fermion problem. In the fifth section a systematic perturbation theory is developed, which is used in the sixth section to extend Brueckner's theory to statistical mechanics. The final section discusses the relation between the microscopic theory and the Landau phenomenological theory. Most of the work to be described is due to GALITSKII and MIGDAL, KLEIN, and MARTIN and SCHWINGER (see Part V for detailed references).

3·1. *Fundamentals.* – We have seen in Part I that we may deduce all the thermodynamic properties of a system if we know the grand potential Ω, where

$$\Omega = -\frac{1}{\beta}\log \mathrm{Tr}\,\{\exp[-\beta(H-\mu N)]\}\,, \tag{3.60}$$

H is the Hamiltonian of the system (eq. (3.33)), and N is the number operator (eq. (3.23)). It proves more convenient in practice to obtain Ω in an alternative manner. Namely, if we calculate the density $\bar{N}$ and the internal energy $\bar{E}$, where these quantities are defined by

$$\left\{\begin{aligned} \bar{N} &= \frac{\mathrm{Tr}\,\{N\exp[-\beta(H-\mu N)]\}}{\mathrm{Tr}\,\{\exp[-\beta(H-\mu N)]\}}\,,\\ \bar{E} &= \frac{\mathrm{Tr}\,\{H\exp[-\beta(H-\mu N)]\}}{\mathrm{Tr}\,\{\exp[-\beta(H-\mu N)]\}}\,, \end{aligned}\right. \tag{3.61}$$

then we can obtain Ω by integration of the relation

$$\bar{E} - \mu\bar{N} = \frac{\partial(\beta\Omega)}{\partial\beta}\,. \tag{3.62}$$

Thus, employing the fact that at zero temperture the entropy is zero we have

$$(3.63)\qquad \Omega = \frac{1}{\beta}\int_{\infty}^{\beta} d\beta'[\bar{E}(\beta') - \mu\bar{N}(\beta') - \bar{E}(\infty) + \mu\bar{N}(\infty)] + \bar{E}(\infty) - \mu\bar{N}(\infty)\,.$$

It is convenient for the subsequent analysis to introduce a modified Hamiltonian $\mathscr{H} = H - \mu N$ and for easy reference we repeat here the expression (obtainable from eqs. (3.23) and (3.33)) for $\mathscr{H}$:

$$(3.64)\qquad \begin{cases} \mathscr{H} = H - \mu N = \mathscr{H}_0 + \mathscr{H}_1\,, \\ \mathscr{H}_0 = \sum\limits_i (T_i - \mu) a_i^\dagger a_i\,, \\ H_1 = \frac{1}{2}\sum\limits_{ijkl} V_{ij,kl}\, a_j^\dagger a_i^\dagger a_k a_l\,, \end{cases}$$

where the creation and annihilation operators satisfy the anti-commutation relation (eq. (3.13) and (3.17))

$$(3.65)\qquad [a_i, a_j]_+ = a_i^\dagger, a_j^\dagger]_+ = 0\,, \qquad [a_i, a_j^\dagger]_+ = \delta_{ij}\,.$$

Working in a modified Heisenberg representation employing $\mathscr{H}$ the equation of motion for an operator $A(t)$ (eq. (3.39)) becomes (*)

$$(3.66)\qquad i\hbar \frac{dA(t)}{dt} = [A(t), \mathscr{H}]_-\,,$$

with the formal solution

$$(3.67)\qquad A(t) = \exp\left[\frac{i\mathscr{H}t}{\hbar}\right] A \exp\left[-i\frac{\mathscr{H}t}{\hbar}\right],$$

as can readily be checked. Define the statistical average of an operator $A(t)$ by

$$(3.68)\qquad \langle A(t)\rangle = \frac{1}{Q}\,\mathrm{Tr}\,\{\exp[-\beta\mathscr{H}]A(t)\}\,,$$

where the partition function Q is

$$(3.69)\qquad Q = \mathrm{Tr}\,\{\exp[-\beta\mathscr{H}]\}\,.$$

(*) We employ time arguments for operators in the modified Heisenberg representation, and no time arguments for operators in the Schrödinger representation (in agreement with the notation in Section III.2).

Thus we can see that

$$\bar{E} - \mu \bar{N} = \langle \mathscr{H} \rangle \,, \tag{3.70}$$

and we may concentrate upon evaluating $\langle \mathscr{H} \rangle$, confident that having obtained it we may evaluate Ω by means of eq. (3.63) and thus determine all of the thermodynamic properties of the systems by means of eq. (1.12). We proceed by introducing correlation functions, and then as an aid in the study of correlation functions we introduce their spectral representations.

3·2. *Correlation functions and spectral functions.* – We define single-particle correlation functions

$$\begin{cases} \mathscr{F}_{ij}^{(+)}(t, t') \equiv \langle a_i(t)\, a_j^\dagger(t') \rangle \,, \\ \mathscr{F}_{ij}^{(-)}(t, t') \equiv \langle a_j^\dagger(t')\, a_i(t) \rangle \,, \end{cases} \tag{3.71}$$

and two-particle correlation functions

$$\begin{cases} \Phi_{ijkl}^{(+)}(t, t') \equiv \langle a_i^\dagger(t)\, a_j(t)\, a_k(t)\, a_l^\dagger(t') \rangle, \\ \Phi_{ijkl}^{(-)}(t, t') \equiv \langle a_l^\dagger(t')\, a_i^\dagger(t)\, a_j(t)\, a_k(t) \rangle \,. \end{cases} \tag{3.72}$$

Note that (in distinction to some treatments) there are only two time arguments. It should be clear that these functions describe the equilibrium properties of the system, and it is even true that some of the nonequilibrium properties can be evaluated in terms of them by means of a generalized Nyquist relation [see ZUBAREV (ZU1) and references cited therein.] We will confine our attention to equilibrium properties.

It is convenient to introduce spectral representations for the correlation functions. We do this by assuming that the (modified) Hamiltonian $\mathscr{H}$ has a complete orthonormal set of eigenstates Ψ_n with associated eigenvalues $\mathscr{E}_n$. Inserting unity in the form of eq. (3.6) between the operators in eq. (3.7), and expressing $a_i(t)$ and $a_j^\dagger(t')$ in terms of a_i and $a_j^\dagger$ by eq. (3.67), we have

$$\mathscr{F}_{ij}^{(+)}(t, t') = \frac{1}{Q} \sum_{n,m} \exp[-\beta \mathscr{E}_n] \exp\left[\frac{i}{\hbar}(\mathscr{E}_n - \mathscr{E}_m)(t - t')\right] \cdot (\Psi_n^\dagger a_i \Psi_m)(\Psi_m^\dagger a_j^\dagger \Psi_n) \,, \tag{3.73}$$

where the complete set Ψ_n has also been employed to evaluate the trace. A similar expression can clearly be obtained for $\mathscr{F}_{ij}^{(-)}(t, t')$. Define a spectral function $\varrho_{ij}(\omega)$ by

$$\varrho_{ij}(\omega) = \frac{1}{Q} \sum_{n,m} (\exp[-\beta \mathscr{E}_n] + \exp[-\beta \mathscr{E}_m]) \cdot \\ \cdot (\Psi_n^\dagger a_i \Psi_m)(\Psi_m^\dagger a_j^\dagger \Psi_n)\, \delta(\mathscr{E}_m - \mathscr{E}_n - \omega) \,, \tag{3.74}$$

which allows us to write eq. (3.73) in the form

$$\mathscr{F}_{ij}^{(\pm)}(t,t') = \int_{-\infty}^{\infty} \mathrm{d}\omega\, \varrho_{ij}(\omega) F_{\pm}(\omega) \exp\left[-\frac{i\omega}{\hbar}(t-t')\right], \tag{3.75}$$

where we have defined

$$F_{\pm}(\omega) = \{\exp[\mp\beta\omega] + 1\}^{-1}. \tag{3.76}$$

The $F_{\pm}(\omega)$ functions satisfy the relation

$$F_{+}(\omega) + F_{-}(\omega) = 1. \tag{3.77}$$

which may be used to establish an important integral relation satisfied by the spectral function. From the definition of $\mathscr{F}^{(\pm)}$ (eq. (3.71)) and the anti-commutation relations (eq. (3.65)) we have:

$$\mathscr{F}_{ij}^{(+)}(t,t') + \mathscr{F}_{ij}^{(-)}(t,t') = \langle \delta_{ij} \rangle = \delta_{ij},$$

which when expressed in terms of spectral functions by eq. (3.75) yields (after use of eq. (3.77)):

$$\int_{-\infty}^{\infty} \varrho_{ij}(\omega)\, \mathrm{d}\omega = \delta_{ij}. \tag{3.78}$$

Proceeding in an analogous manner we can express the two-body correlation function in the form

$$\Phi_{ijkl}^{(\pm)}(t,t') = \int_{-\infty}^{\infty} \mathrm{d}\omega\, \gamma_{ijkl}(\omega)\, F_{\pm}(\omega) \exp\left[-\frac{i\omega}{\hbar}(t-t')\right], \tag{3.79}$$

where the two-body spectral function is defined by

$$\gamma_{ijkl}(\omega) = \frac{1}{Q}\sum_{n,m} (\exp[-\beta\mathscr{E}_n] + \exp[-\beta\mathscr{E}_m])(\Psi_n^{\dagger} a_i^{\dagger} a_j a_k \Psi_m)\cdot \cdot(\Psi_m^{\dagger} a_l^{\dagger} \Psi_n)\,\delta(\mathscr{E}_m - \mathscr{E}_n - \omega). \tag{3.80}$$

We now express $\langle\mathscr{H}\rangle$ in terms of the correlation functions, and then in terms of the spectral functions. From eq. (3.64), (3.71) and (3.72):

$$\langle\mathscr{H}\rangle = \sum_i (T_i - \mu)\mathscr{F}_{ii}^{(-)}(t,t') = \tfrac{1}{2}\sum_{ijkl} V_{il,jk}\, \Phi_{ijkl}^{(-)}(t,t'), \tag{3.81}$$

which by eq. (3.75) and (3.79) becomes

$$(3.82)\qquad \langle \mathscr{H} \rangle = \sum_i (T_i - \mu) \int_{-\infty}^{\infty} \mathrm{d}\omega\, \varrho_{ii}(\omega) F_-(\omega) + \frac{1}{2} \sum_{ijkl} V_{il,jk} \int_{-\infty}^{\infty} \mathrm{d}\omega\, \gamma_{ijkl}(\omega) F_-(\omega)\,.$$

Our program is to replace the impossibly difficult sum over states in Ω, with an alternative process which can be systematically approximated. We have so far succeeded in expressing Ω in terms of $\langle \mathscr{H} \rangle$, and $\langle \mathscr{H} \rangle$ in terms of two spectral functions. We shall, in the next Paragraph (III.3˙3), introduce Green's functions which are similar to correlation functions, but satisfy more convenient equations. We shall in the following Paragraph (III.3˙4) find an identity which allows us to express $\langle \mathscr{H} \rangle$ in terms of only the one-body spectral function, and also express this spectral function in terms of the one-body Green's function. Equations of motion will be obtained for the Green's function, and statistical mechanics reduced to solving these equations in Paragraph III.3˙5, where a formal perturbation solution will also be presented.

3˙3. *Green's functions.* – We define the (retarded) one-body Green's function by

$$(3.83)\qquad G_{ij}(t, t') = \begin{cases} -i\langle [a_i(t), a_j^\dagger(t')]_+ \rangle\,, & t > t'\,, \\ 0\,, & t < t'\,, \end{cases}$$

while the two-body (retarded) Green's function is defined by

$$(3.84)\qquad \Gamma_{ijkl}(t, t') = \begin{cases} -i\langle [a_i^\dagger(t) a_j(t) a_k(t), a_l^\dagger(t')]_+ \rangle\,, & t > t'\,, \\ 0\,, & t < t'\,. \end{cases}$$

From the definition of $\mathscr{F}^{(\pm)}$ (eq. (3.71)), and their spectral representation (eq. (3.75)), we can write the single-particle Green's function G in the form

$$(3.85)\qquad G_{ij}(t, t') = -i \int_{-\infty}^{\infty} \mathrm{d}\omega\, \varrho_{ij}(\omega) \theta(t - t') \exp\left[-\frac{i\omega}{\hbar}(t - t')\right],$$

where the $\theta(t)$ function is defined as unity for t positive and zero for t negative. It is clear from eq. (3.85) that $G_{ij}(t, t')$ is only a function of $(t - t')$.

Defining

$$(3.86)\qquad G_{ij}(E) = \int_{-\infty}^{\infty} \mathrm{d}(t - t') G_{ij}(t - t') \exp\left[\frac{iE}{\hbar}(t - t')\right],$$

we conclude from eq. (3.85) that

$$G_{ij}(E) = \int_{-\infty}^{\infty} \frac{d\omega\, \varrho_{ij}(\omega)}{[E/\hbar - \omega/\hbar + i\delta]}, \tag{3.87}$$

where the infinitesimal positive quantity δ arises because (*)

$$\theta(t) = \frac{i}{2\pi} \int_{-\infty}^{\infty} \frac{\exp[-ixt]}{x + i\delta}\, dx\,. \tag{3.88}$$

Proceeding in an analogous manner one can easily show that

$$\Gamma_{ijkl}(E) = \int_{-\infty}^{\infty} \frac{d\omega\, \gamma_{ijkl}(\omega)}{[E/\hbar - \omega/\hbar + i\delta]}\,. \tag{3.89}$$

3·4. *Equations of motion and spectral relations.* – The equations of motion for the Green's functions may be obtained from their definition (eq. (3.83), (3.84)) by differentiation with respect to time. One must be careful to recall that the derivative of a θ-function is a δ-function, after which it is clear that

$$i\hbar \frac{dG_{ij}(t, t')}{dt} = \hbar\delta(t - t')\langle a_i(t) a_j^\dagger(t') + a_j^\dagger(t') a_i(t)\rangle - \\ - i\langle\{[a_i(t), \mathscr{H}]_-, a_j^\dagger(t')\}_+\rangle\theta(t - t')\,. \tag{3.90}$$

Evaluating the commutator using eq. (3.65) and (3.64) (just as in the derivation of eq. (3.40)) yields

$$i\hbar \frac{dG_{ij}(t, t')}{dt} + (\mu - T_i) G_{ij}(t, t') = \hbar\delta_{ij}\delta(t - t') + \\ + \tfrac{1}{2}\sum_{klm}[V_{ki,lm} - V_{ik,lm}]\Gamma_{klmj}(t, t')\,. \tag{3.91}$$

Thus the equation of motion for the one-body Green's function involves the two-body Green's function, and so on. All approaches to the many-body problem can be described in terms of various approximations which are used to terminate this hierarchy of equations. We shall return to this problem in

---

(*) This can be easily checked by observing that there is a pole in the integrand of eq. (3.88) at $x = -i\delta$. For $t < 0$ the contour must be closed in the upper half-plane to yield zero, while for $t > 0$ closing in the lower half-plane yields unity.

the next section, but continue now to obtain a useful identity and an expression for $\langle\mathscr{H}\rangle$. To this end we Fourier transform eq. (3.91) to obtain

$$(3.92)\qquad \frac{E}{\hbar}\,G_{ij}(E)=\delta_{ij}+\frac{(T_i-\mu)}{\hbar}\,G_{ij}(E)+\frac{1}{2\hbar}\sum_{klm}[V_{ki,lm}-V_{ik,lm}]\Gamma_{klmj}(E)\,.$$

From eq. (3.92) using the spectral representations of eq. (3.87) and (3.89) we obtain the relation:

$$(3.93)\qquad (\omega+\mu-T_i)\,\varrho_{ij}(\omega)=\tfrac{1}{2}\sum_{klm}[V_{ki,lm}-V_{ik,lm}]\gamma_{klmj}(\omega)\,,$$

where we have also used in the derivation, eqs. (3.77) and (3.78). This relation allows us to express $\langle\mathscr{H}\rangle$, as given by eq. (3.82) in terms of only the one-body spectral function. Thus, after employing the fact that $\Phi^{(-)}_{ijkl}=-\Phi^{(-)}_{ljki}$ (obvious from its definition) to break the least term of eq. (3.82) into two terms of the form obtained from eq. (3.93) by setting $j=i$ and summing over i, we have

$$(3.94)\qquad \langle\mathscr{H}\rangle=\frac{1}{2}\sum_i\int_{-\infty}^{\infty}\mathrm{d}\omega(\omega+T_i-\mu)\,\varrho_{ii}(\omega)F_-(\omega)\,.$$

This important result was first obtained (for the case of zero temperature) by Galitski and Migdal (GM1).

It remain to express the spectral function in terms of the Green's function, which is easily accomplished from eq. (3.87). By use of the symbolic identity

$$(3.95)\qquad \frac{1}{E-\omega\pm i\delta}=P\frac{1}{E-\omega}\mp i\pi\,\delta(E-\omega)\,,$$

we obtain

$$(3.96)\qquad \frac{1}{\hbar}\,G_{ij}(E)=P\int_{-\infty}^{\infty}\frac{\mathrm{d}\omega\,\varrho_{ij}(\omega)}{E-\omega}-i\pi\,\varrho_{ij}(E)\,.$$

This equation yields, first of all, an expression for $\varrho_{ij}(E)$ in terms of the single-particle Green's function, namely

$$(3.97)\qquad \varrho_{ij}(E)=-\frac{1}{\pi\hbar}\,\mathrm{Im}\,G_{ij}(E)\,;$$

and, secondly, a dispersion relation for G_{ij}:

$$(3.98)\qquad \mathrm{Re}\,G_{ij}(E)=-\frac{1}{\pi}\,P\int_{-\infty}^{\infty}\frac{\mathrm{d}\omega\,\mathrm{Im}\,G_{ij}(\omega)}{E-\omega}\,.$$

This last relation is a generalized Kramers-Kronig relation.

To summarize, we have reduced statistical mechanics to solving coupled equations for Green's functions, and these equations (eq. (3.91) etc.) don't involve the temperature at all so that methods developed for the ground state will work also in statistical mechanics. We shall in the next section discuss a systematic perturbation theory solution which is the analog of the « linked-cluster expansion ».

3.5. *The single-particle Green's function.* – In this section we shall give a systematic perturbation expansion for the single-particle Green's function. We start from the ideal gas, so it is important to first evaluate the Green's function $G_{ij}^{(0)}(E)$ for a gas of noninteracting fermions. From eq. (3.92), the equation of motion is

$$\left[\frac{E - T_i + \mu}{\hbar}\right] G_{ij}^{(0)}(E) = \delta_{ij} , \tag{3.99}$$

with the obvious result that $G_{ij}^{(0)} = \delta_{ij} G_i^{(0)}$ with

$$G_i^{(0)}(E) = P\left[\frac{\hbar}{E + \mu - T_i}\right] + A\delta(E + \mu - T_i) . \tag{3.100}$$

The arbitrary constant A is determined by the boundary conditions, and they are specified by the dispersion relation of eq. (3.98). Setting $A = A_{\text{Re}} + iA_{\text{Im}}$ we have

$$\operatorname{Im} G_i^{(0)}(E) = A_{\text{Im}}\delta(E + \mu - T_i) ,$$

$$\operatorname{Re} G_i^{(0)}(E) = P\left[\frac{\hbar}{E + \mu - T_i}\right] + A_{\text{Re}}\delta(E + \mu - T_i) ,$$

which when substituted into the dispersion relation implies

$$\begin{cases} A_{\text{Re}} = 0 , \\ A_{\text{Im}} = -\pi\hbar . \end{cases} \tag{3.101}$$

Thus from eq. (3.97) the spectral function for the ideal gas is

$$\varrho_{ij}^{(0)}(\omega) = \delta_{ij}\delta(\omega + \mu - T_i) \tag{3.102}$$

which can easily be seen to satisfy eq. (3.78).

We can now use eq. (3.94) to find the statistical average of $\mathscr{H}$ for the ideal gas, namely

$$(3.103)\quad \begin{cases} \langle \mathscr{H}^0 \rangle = \dfrac{1}{2}\sum_i \int\limits_{-\infty}^{\infty} \mathrm{d}\omega(\omega + T_i - \mu)\,\delta(\omega + \mu - T_i)\,F_-(\omega)\,, \\ \langle \mathscr{H}^0 \rangle = \sum_i \dfrac{T_i - \mu}{\exp[\beta(T_i - \mu)] + 1}\,, \end{cases}$$

from which it follows that the distribution function for the ideal gas $n_i^{(0)}$ is given by

$$(3.104)\qquad n_i^{(0)} = \{\exp[\beta(T_i - \mu)] + 1\}^{-1}\,.$$

The reader should observe that this result has been obtained without evaluation of a partition function, or an equivalent sum over states. We can use eq. (3.104) and the symbolic identity of eq. (3.95) to write the Green's function in the more instructive form

$$(3.105)\qquad G_{ij}^{(0)}(\omega) = \frac{\hbar\,\delta_{ij}}{\omega - T_i + \mu + i\delta}\,,$$

where δ is a small positive constant used to define the singularity. [Clearly we could have arrived at this result directly from eq. (3.83), but the above treatment has been given to illustrate the methods which will be used in the next section.] Note that with the singularity in $G^{(0)}$ defined as in eq. (3.105), $G^{(0)}$ is clearly [by the inverse of eq. (3.86)] retarded, *i.e.* zero for $t<0$.

Turning now to the generation of a perturbation series for G, we must solve the hierarchy of coupled equations for the Green's functions. The first equation in this series has been given in eq. (3.92). It couples G to Γ. We can get an equation for Γ by a process analogous to that used to obtain the equation of motion for G. The result is

$$(3.106)\quad i\hbar\frac{\partial}{\partial t'}\Gamma_{ijkl}(t,t') = -(\Gamma_l - \mu)\Gamma_{ijkl}(t,t') + \hbar\delta(t-t')[\delta_{il}\langle a_i^\dagger a_k\rangle - \delta_{lk}\langle a_i^\dagger a_j\rangle] - \\ - \frac{i}{2}\sum_{i'j'k'}[V_{i'j',k'l} - V_{i'j',kl'}]\langle\{a_i^\dagger(t)\,a_j(t)\,a_k(t),\, a_{j'}^\dagger(t')\,a_{i'}^\dagger(t')\,a_{k'}(t')\}_+\rangle\theta(t-t')\,,$$

which can be seen to couple in a higher order Green's function. We do not attempt in this paper to derive the general term, but only wish to demonstrate the methods one might use to establish a result we shall state without proof. (*)

(*) A derivation of the perturbation-theory rules of Paragraph III.3·5 is given in the report G. BAYM and A. M. SESSLER: *Perturbation-theory rules for computing the self-energy operator in quantum statistical mechanics*, Lawrence Radiation Laboratory Report UCRL-10562, unpublished (1962).

Thus we solve eq. (3.106) to lowest order only. The solution can be surmised by factoring the four operators that appear in the definition of eq. (3.84) into pairs of two operators. Thus one guesses

$$\Gamma_{ijkl}(t, t') \simeq G^{(0)}_{kl}(t, t')\langle a^\dagger_i a_j\rangle - G^{(0)}_{jl}(t, t')\langle a^\dagger_i a_k\rangle , \tag{3.107}$$

which can be easily seen to satisfy eq. (3.106) if the terms of higher order in the interaction are ignored. If we Fourier transform eq. (3.107) and insert it into the equation of motion for G (eq. (3.92)) we have an equation which we can use to obtain a first correction to $G^{(0)}$, namely

$$\left[\frac{E+\mu-T_i}{\hbar}\right] G_{ij}(E) \simeq \delta_{ij} + \frac{1}{2\hbar}\sum_{klm}[V_{ki,lm} - V_{ik,lm}]\cdot \cdot[G^{(0)}_{mj}(E)\langle a^\dagger_k a_l\rangle - G^{(0)}_{lj}(E)\langle a^\dagger_k a_m\rangle] . \tag{3.108}$$

We can see that in general $G_{ij}(E)$ is of the form $\delta_{ij}G_i(E)$. This follows from the same property of $G^{(0)}_{ij}$ and the fact that $V_{ji,kl}$ contains a momentum-conserving factor $\delta_{i+j,k+l}$. Defining a self-energy operator $M_i(E)$ by the equation (*)

$$G_i(E) = G^{(0)}_i(E) + G^{(0)}_i(E)\frac{M_i(E)}{\hbar} G_i(E) , \tag{3.109}$$

which is equivalent to

$$G^{(0)-1}_i G_i = 1 + \frac{M_i}{\hbar} G_i , \tag{3.110}$$

we can upon comparison with eq. (3.108) deduce a first approximation to M_i as

$$M^{(1)}_i(E) = \sum_l [V_{li,li} - V_{il,li}]\langle a^\dagger_l a_l\rangle . \tag{3.111}$$

Equation (3.111) may be written as

$$M^{(1)}_i(E) = \sum_l [V_{li,li} - V_{il,li}] n^{(0)}_l , \tag{3.112}$$

in terms of the zero-order distribution function. We shall return to a discussion of this result in the next section.

Proceeding as indicated; *i.e.* by introducing higher and higher order (retarded) Green's functions which are functions of *only two* time variables, ob-

---

(*) Defining M by eq. (3.109) where G and not $G^{(0)}$ appears in the last term, corresponds to summing many « reducible » diagrams. This is why in the rules to be given for the n-th order terms on M, one omits the class of diagrams corresponding to iterated lower order diagrams.

taining equations of motion by differentiating with respect to one of the times (the functions are only dependent upon the time difference), Fourier transforming to E as an independent variable, and then solving the resulting *coupled* purely *algebraic equations*, one can readily deduce the rule for calculating the n-th order term in the self-energy operator M_k. The result may be convenient described graphically:

Rule for calculating self-energy: Draw all linked graphs in which a particle line K enters and leaves the graph. Indicate the entering and leaving line by a wiggly line. Omit all graphs which contain lower-order graphs as distinct entities along the line K. Calculate the contribution to M_k as in the linked-cluster theorem; to each « particle line » j (upward arrow) assign a factor $[1-n_j^{(0)}]$; to each « hole line » j (downward arrow) assign a factor $n_j^{(0)}$; to the wiggly lines associate no such factor $n_k^{(0)}$ or $[1-n_k^{(0)}]$ but its contribution to the energy denominates is to be calculated as E if no wiggly line is cut, 0 if one is cut, and $-E$ if two wiggly lines are cut. In the sum over states, sum particles and holes over *all* states.

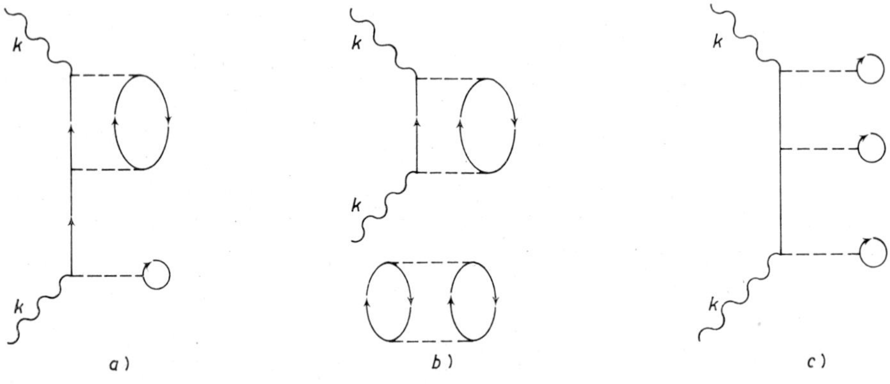

Fig. 8. – Some graphs which are to be omitted in calculating the self-energy $M_k(E)$.

In Fig. 8 we have indicated some graphs which are to be omitted. In Fig. 9 we have indicated the first and second order graphs which contribute to $M_k(E)$, and application of the rule yields, to this order,

$$M_k(E) = \sum_l [V_{kl,kl} - V_{kl,lk}] n_l^{(0)} + \tag{3.113}$$

$$+ \frac{1}{4} \sum_{lmm'} \frac{|V_{kl,mm'} - V_{kl,m'm}|^2 n_l^{(0)} (1 - n_m^{(0)})(1 - n_{m'}^{(0)})}{E + T_l - T_m - T_{m'}} -$$

$$- \frac{1}{4} \sum_{ll'm} \frac{|V_{ll',km} - V_{ll',mk}|^2 n_l^{(0)} n_{l'}^{(0)} (1 - n_m^{(0)})}{T_l + T_{l'} - T_m - E} + \dots,$$

the first term of which checks our previously derived result.

In the calculation of $M(E)$, E must be complex to define the various energy denominators which appear. In accordance with eq. (3.87) we define $G(E)$ in terms of $M(E)$ for E approaching the real axis from above. It is clear [since

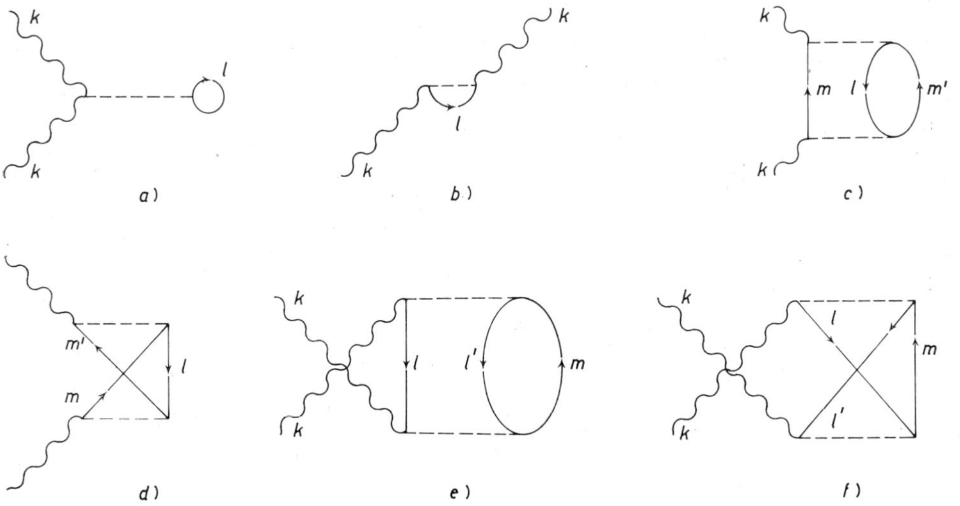

Fig. 9. – The first and second order graphs contributing to the self-energy operator $M_k(E)$.

G so defined is analytic in the complex plane except on the real axis] that the resulting G is zero for $t<t'$ in agreement with its definition. [Defining a G in terms of $M(E)$ for E approaching the real axis from below, would give the advanced Green's function, as all the algebraic analysis, leading to the rules for $M(E)$ is valid for the retarded, advanced, or even the causal Green's function. This definition would afford an equivalently convenient formalism which differs from the one presented here in only a trivial way.]

Thus having a systematic method of obtaining $M_k(E)$ we can find $G_k(E)$ and hence the spectral function $\varrho_k(\omega)$. Then we calculate $\langle\mathscr{H}\rangle$, and hence Ω. In the next section we shall turn to a discussion of various approximation methods.

3·6. *Approximation methods.* – It is interesting to examine the statistical mechanic equivalent of the Hartree-Fock theory. We start with the Hartree-Fock Hamiltonian (eq. (3.48)) rather than the Hamiltonian of eq. (3.64). The resulting modification of the rules of Paragraph III.3·5 is obvious. First of all $T_i \to \varepsilon_i$; secondly $n_k^{(0)} \to n_k^{(\mathrm{HF})}$ which is the same function but with T_k replaced by ε_k, thirdly one omits, except in first order, all graphs in which a solid line joins two points at the same level; and finally in first order there is the additional graph of Fig. 10, with contribution $(T_i - \varepsilon_i)$. Thus we have

$$(3.114)\qquad \begin{cases} \mathscr{H} = \mathscr{H}'_0 + \mathscr{H}'_1 , \\ \mathscr{H}'_0 = \sum_i (\varepsilon_i - \mu) a_i^\dagger a_i , \\ \mathscr{H}'_1 = \frac{1}{2} \sum_{ijkl} V_{ij,kl} a_j^\dagger a_i^\dagger a_k a_l + \sum_i (T_i - \varepsilon_i) a_i^\dagger a_i , \end{cases}$$

and

$$(3.115)\qquad G_{ij}^{(\mathrm{HF})}(\omega) = \frac{\hbar \delta_{ij}}{\omega - \varepsilon_i + \mu + \delta i} ,$$

with

$$(3.116)\qquad n_i^{\mathrm{HF}} = \{\exp[\beta(\varepsilon_i - \mu)] + 1\}^{-1} .$$

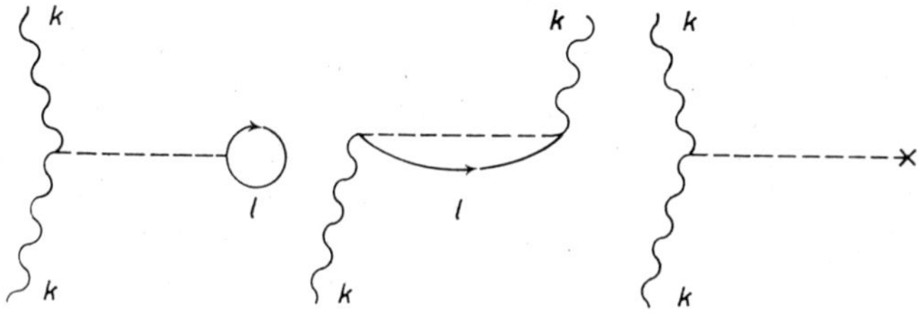

Fig. 10. – The lowest order contribution to $M_k(E)$ in Hartree-Fock theory.

Calculating $M_k(\omega)$ to lowest order, yields according to the rules,

$$(3.117)\qquad M_k^{(1)}(\omega) = \sum_l [V_{kl,kl} - V_{kl,lk}] n_l^{(\mathrm{HF})} + [T_k - \varepsilon_k] .$$

It should be observed that we have not yet specified how to choose the energies ε_i.

Now from eq. (3.109) we have

$$\left[1 - G_k^{(\mathrm{HF})} \frac{M_k}{\hbar}\right] G_k = G_k^{(\mathrm{HF})} ,$$

or

$$(3.118)\qquad [G_k(E)]^{-1} = [G_k^{(\mathrm{HF})}(E)]^{-1} - \frac{M_k(E)}{\hbar} ,$$

which we can use to calculate a first correction to $G_k^{(\mathrm{HF})}$, using $M_k^{(1)}$. Thus

$$G_k^{-1}(E) \simeq \frac{E + \mu - \varepsilon_k - M_k^{(1)}}{\hbar} ,$$

which since $M_k^{(1)}$ is real allows us to easily identify the $\operatorname{Im} G_k(E)$ just as in the argument leading to eq. (3.101) (namely use of the dispersion relation).

We find

$$\operatorname{Im} G_k(E) \simeq -\pi\hbar\, \delta(E+\mu-\varepsilon_k-M_k^{(1)}) , \tag{3.119}$$

and hence

$$\varrho_{kj}(\omega) \simeq \delta(\omega+\mu-\varepsilon_k-M_k^{(1)})\,\delta_{kj} . \tag{3.120}$$

We may write eq. (3.94) as two equations, one for the internal energy and one for the average number of particles:

$$\begin{cases} \bar{E} = \dfrac{1}{2}\sum_i \displaystyle\int_{-\infty}^{\infty} \mathrm{d}\omega(\omega + T_i + \mu)\,\varrho_{ii}(\omega) F_-(\omega) , \\ \bar{N} = \sum_i \displaystyle\int_{-\infty}^{\infty} \mathrm{d}\omega\, \varrho_{ii}(\omega) F_-(\omega) , \end{cases} \tag{3.121}$$

which upon substitution of eq. (3.120) for the spectral function yields (*):

$$\begin{cases} \bar{E} = \sum_i \{T_i + \frac{1}{2}\sum_l [V_{il,il} - V_{il,li}] n_l^{(\mathrm{HF})}\} n_i , \\ \bar{N} = \sum_i n_i , \end{cases} \tag{3.122}$$

where

$$n_i = \{\exp\left[\beta\,(T_i + \sum_l (V_{il,il} - V_{il,li}) n_l^{(\mathrm{HF})} - \mu)\right]_+ + 1\}^{-1} . \tag{3.123}$$

It is now clear that we can choose the single-particle energies ε_i by the requirement that n_i is equal to $n_i^{(\mathrm{HF})}$ *i.e.* by solving *simultaneously*

$$\begin{cases} \varepsilon_i = T_i + \sum_l [V_{il,il} - V_{il,li}] n_l , \\ n_i = \{\exp[\beta(\varepsilon_i - \mu)] + 1\}^{-1} . \end{cases} \tag{3.124}$$

It should be noted that solving for the single particle energies ε_i, and the distribution function n_i is nontrivial—involving the same sort of complication

---

(*) We must *not* replace T_i with ε_i in eq. (3.94) for the derivation of this formula does *not* hold for the Hamiltonian of eq. (3.114) where the perturbation has a term involving only two creation or annihilation operators. Rather we must realize that eq. (3.94) holds in general for two-body forces, and in particular for the approximate spectral function of eq. (3.120).

that we saw in the Landau phenomenological theory where the distribution function was given by an expression like eq. (3.124), but ε_i was a functional of $\{n_k\}$ (see eq. (2.5) and comment following that equation). Having obtained the ε_i and n_i, the internal energy may be calculated from eq. (3.122). The reader should note how our formalism has automatically given the single-particle energies (eq. (3.124)) and the total energy (eq. (3.122)) with the correct factor of unity and ($\frac{1}{2}$) on the interaction terms [compare footnote (*), p. 197].

The Hartree-Fock theory is not adequate to describe liquid ^{3}He because of the singular two-body potential. We must extend to statistical mechanics the theory of Brueckner where the strong interactions between atoms are taken into account carefully. Such an extension is now easy to construct using the formalism we have developed. The discussion parallels that of Paragraph III.2·4 and we will not repeat the various physical arguments in support of the approximation, [which can be expected to have the same general range of validity as Brueckner theory] but turn directly to the formalism.

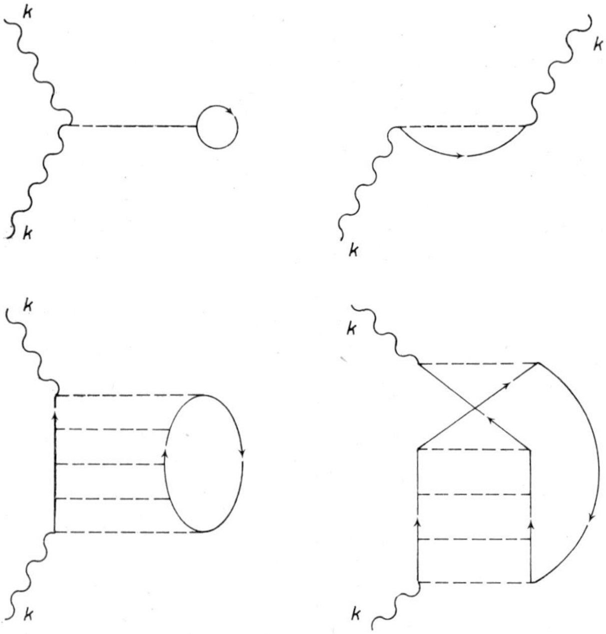

Fig. 11. – Some of the graphs included in the calculation of the self-energy $M_k(E)$ in Brueckner theory.

We start with the Hamiltonian in Hartree-Fock representation (eq. (3.114)) and again leave the single-particle energies ε_i to be specified subsequently. Some of the diagrams we wish to include in the calculation of the self-energy are indicated in Fig. 11. The sum of these terms may be written in terms of a (generalized) reaction matrix $K_{ij,kl}(E)$ defined by

$$K_{ij,kl}(E) = [V_{ij,kl} - V_{ij,lk}] + \frac{1}{2}\sum_{k'l'} \frac{[V_{ij,k'l'} - V_{ij,l'k'}]\, K_{k'l',kl}(E)(1-n_{k'})(1-n_{l'})}{E+\varepsilon_l - \varepsilon_{k'} - \varepsilon_{l'}}, \tag{3.125}$$

where

$$n_k = \{\exp[\beta(\varepsilon_k - \mu)] + 1\}^{-1}. \tag{3.126}$$

We have

$$M_k(E) = \sum_l K_{kl,kl}(E) n_l + (T_k - \varepsilon_k), \tag{3.127}$$

from which one can then calculate $G_k(E)$ by:

$$G_k^{-1}(E) = \frac{E + \mu - \varepsilon_k - M_k(E)}{\hbar}. \tag{3.128}$$

The single particle energies E_k may be determined by the self-consistent criterion that the Re $M_k(E) = 0$. This is easily seen to be a simple generalization of the temperature-dependent Hartree-Fock theory (eq. (3.124)), as well as of the $T = 0$ Brueckner theory (eq. (3.56)). The computational procedure would be the same as in Brueckner's work except that both the single-particle potential and the distribution function are determined self-consistently. There is the additional complication that one must calculate for a range of temperatures and that $K_{ij,kl}(E)$ is a function of E. After self-consistency has been attained, one evaluates $\langle \mathscr{H} \rangle$ from eq. (3.94) using the spectral function determined from the imaginary part of G. The thermodynamic properties of the system then follow readily.

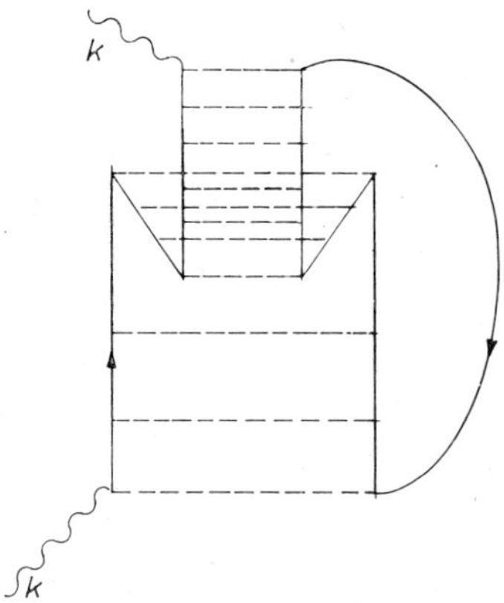

Fig. 12. – A diagram indicating a typical hole-hole interaction which is included in the Galitskii theory.

It is of interest to compare this procedure with Brueckner's work. In particular as $T \to 0$ it can be seen that n_k approaches a step function so that the summation in eq. (3.127) is restricted to $l \leqslant N$, while the summations in eq. (3.125) are restricted to $k', l' > N$. Thus one will only need $K_{ij,kl}(E)$ for values of i, j, k, l in which the denominator in eq. (3.125) is never zero, and consequently K will be real for real E. In that case the analysis given above eqs. (3.119) and (3.120)) for the Hartree-Fock theory will be valid and we can deduce that the spectral function is a δ-function. It now obviously follows that the theory is identical to Brueckner theory (recall eqs. (3.54), (3.55), and (3.56)). For nonzero temperatures the energy E must be kept complex to define $K_{ij,kl}(E)$ from eq. (3.125), and one proceeds via eq. (3.97).

Finally, it can be remarked that the methods employed here suggest a simple generalization of Brueckner theory (GM1), namely the inclusion of « hole-hole interactions » as well as « particle-particle interactions ». In Fig. 12 we have indicated the diagrams we wish to include, and application of the rules shows that if we define a reaction matrix $K_{ij,kl}(E)$ by

$$K_{ij,kl}(E) = [V_{ij,kl} - V_{ij,lk}] + \frac{1}{2}\sum_{k'l'}[V_{ij,k'l'} - V_{ij,l'k'}]K_{k'l',kl}(E)\cdot \left\{\frac{(1-n_{k'})(1-n_{l'})}{E+\varepsilon_l-\varepsilon_{k'}-\varepsilon_{l'}} - \frac{n_{k'}n_{l'}}{E+\varepsilon_l-\varepsilon_{k'}-\varepsilon_{l'}}\right\}, \tag{3.129}$$

and obtain $M_k(E)$ just as in eq. (3.127), then we will have summed all of these new terms. In this case the energy must be kept complex even in the zero-temperature limit of the theory. The inclusion of holes and particles in a symmetric manner is important near the Fermi surface, as will be discussed in detail in Part IV.

The modifications that eq. (3.129) will make in the results of BRUECKNER and GAMMEL for liquid ^{3}He are now under study by R. E. MILLS (*), who is also using the theory outlined above to calculate the low-temperature properties of liquid ^{3}He.

3·7. *Quasi-particles.* – The formalism of Green's functions serves as a convenient framework within which to explore the concept of a quasi-particle from a microscopic point of view. For this purpose we do not need Green's functions defined as ensemble averages, as contrasted with ground state averages, but for purposes of economy of presentation we have only introduced the more general concept and consequently have had to defer the discussion to this juncture. Our treatment will parallel the work of MIGDAL and GALITSKII, DE DOMINICIS, and KLEIN. The reader's attention is called to two recent papers by NOZIÈRES and LUTTINGER [NL1, LN1].

We first give a formal definition of quasi-particles. The rules for perturbation calculation in quantum statistical mechanics, as described in Section III.3·5, allow us to contemplate the following computational procedure:

1) Start with the Hamiltonian in Hartree-Fock representation (eq. (3.114)), but with the single-particle energies ε_k still to be defined. For the distribution function $n_k^{(0)}$ take an ideal Fermi gas function of ε_k (eq. (3.116)).

(*) R. E. MILLS: private communication. Actually, Mr. MILLS is using an improved approximation which is discussed in Paragraph IV.1·1 (in a different context).

2) Use the rules of perturbation theory to calculate the self-energy $M_k(E)$ to any desired order of accuracy—in principle to all orders. From $M_k(E)$ compute the single-particle Green's function $G_k(E)$. From the imaginary part of $G_k(E)$ compute the spectral function $\varrho_k(E)$ using eq. (3.97), and hence obtain the internal energy of the system by the relation of eq. (3.94). The resultant $\bar{E}$ is a functional of the still-to-be-determined single-particle energies ε_k, although if the calculation has been carried to all orders then the *value* of $\bar{E}$ will, of course, be independent of the choice of ε_k.

3) Determine the single-particle energies ε_k by the Landau definition of quasi-particle energy, namely compute the functional derivative of $\bar{E}$ with respect to $n_k^{(0)}$ and demand that it be equal to the initially chosen ε_k.

This procedure serves to define the quasi-particles which are described by the distribution function $n_k^{(0)}$ [which is just a Fermi distribution of ε_k], as in the Landau theory. The quasi-particle energies ε_k are clearly real, as $\bar{E}$ computed by the above procedure is real for *any* distribution function n_k and furthermore are defined exactly as in the phenomenological theory. [Note that in practice this definition of ε_k is quite cumbersome—being much more complicated than the method employed in Brueckner theory to define the single-particle energies. It is superior, however, in that it includes « rearrangement energies »—an improvement which has in fact been incorporated into Brueckner theory for the case of nuclear matter [7].]

It is perhaps unnecessary to comment that the quasi-particle distribution function $n_k^{(0)}$ is in no sense to be confused with the average occupation of free-particle states in the interaction system. This distinction can be seen very clearly at the absolute zero [although this distinction, of course, remains at all temperatures], where the average number of particles in state k is given by

(3.130) $$\bar{n}_k = \Psi_0^\dagger a_k^\dagger a_k \Psi_0 ,$$

and Ψ_0 is the true ground state. The function $\bar{n}_k$ will be nonzero for all k due to the interactions between particles. On the other hand the quasi-particle distribution function $n_k^{(0)}$ is simply a step function at the absolute zero.

In general, the average number of particles $\bar{N}$ can be obtained in terms of the average number of particles in state k

(3.131) $$\bar{N} = \sum_k \bar{n}_k = \sum_k \langle a_k^\dagger a_k \rangle ,$$

[it can be seen that eq. (3.131) reduces to eq. (3.130) as $T \to 0$] or it can be obtained in terms of the quasi-particle distribution function $n_k^{(0)}$:

(3.132) $$\bar{N} = \sum_k n_k^{(0)} = \sum_k \{\exp[\beta(\varepsilon_k - \mu)] + 1\}^{-1} .$$

Analogously, the entropy can be expressed in terms of the average energy $\bar{E}$ using the formula for the grand potential (eq. (3.63)) and then thermodynamic relations. It can also be expressed in terms of the quasi-particle distribution function $n_k^{(0)}$, in which case [by the inverse of the argument used in Part II to derive the Fermi distribution function, eq. (2.5)] it is clearly given by the simple expression:

$$S = -k \sum_k \{n_k^{(0)} \log n_k^{(0)} + (1 - n_k^{(0)}) \log (1 - n_k^{(0)}) \,. \tag{3.133}$$

Although the formal definition of quasi-particles [as given above] gives insight into the concept it can be criticized in that it has not been shown that the single-particle energies are smoothly varying [and physically reasonable] functions. Perhaps a more serious objection is that the above definition is only related to the equilibrium properties of a Fermi liquid and nowhere in the analysis does the « lifetime of a quasi-particle » appear, yet this aspect of a quasi-particle is vital in that it plays a crucial role in the use of the Fermi liquid theory to calculate nonequilibrium properties.

In order to understand the relation between the microscopic theory and quasi-particle lifetime, we must study nonequilibrium situations. The correlation functions allows us to do this in a simple way, and we now turn to this analysis.

The single particle correlation function $\mathscr{F}_{kj}^{(+)}(t, t')$ contains a δ-function in k and j and furthermore is a function only of $(t - t')$. From its definition eq. (3.71) we have

$$\mathscr{F}_k(t) \equiv \mathscr{F}_{kj}^{(+)}(t, 0) = \langle a_k(t)\, a_k^\dagger(0) \rangle \,. \tag{3.134}$$

Now the operator $a_k^\dagger(0)$ creates a particle in state k at time $t = 0$, while $a_k^\dagger(t)$ removes a particle from state k at time t. Thus $\mathscr{F}_k(t)$ is the amplitude for finding the system at time t in equilibrium but with one additional particle in state k, given that at time $t = 0$ it was in equilibrium but with one additional particle in state k (*). Clearly $\mathscr{F}_k(t)$ will decrease as t increases; for equilibrium *plus* one particle in state is *not* an equilibrium state, and the

---

(*) The reader may desire to specialize to the absolute zero, in which case from eq. (3.134)

$$\mathscr{F}_k(t) = \Psi_0^\dagger a_k(t) a_k^\dagger(0)\, \Psi_0 \,, \tag{1}$$

where Ψ is the true ground state. Writing eq. (1) as

$$\mathscr{F}_k(t) = (a_k^\dagger(t)\, \Psi_0)^\dagger (a_k^\dagger(0)\, \Psi_0) \,, \tag{2}$$

it is clear that $\mathscr{F}_k(t)$ is just the overlap amplitude for finding the system in the ground state plus one-particle in state k at time T, if it is put in that state at $t = 0$.

system will tend to approach equilibrium by developing more and more complicated excitations—with a resultant continued decrease in the probability for finding the system in the original equilibrium state *plus* one particle.

Consider now the role of the lifetime of quasi-particles in the Landau theory. The lifetime was important in the evaluation of transport coefficients where it was incorporated in the function $w(\theta, \varphi)$ (Section II.3) which describes the collisions between quasi-particles. To get a microscopic definition of quasi-particle lifetime we must perform the same calculations from a microscopic point of view, but such calculations are yet to be accomplished. It is clear, however, that the response of the equilibrium system to the perturbation of an extra particle is very closely related to the calculation of transport coefficients—and we may take as a rough definition of quasi-particle lifetime the damping of the single-particle correlation function (*).

In order to study $\mathscr{F}_k(t)$, we return to the general perturbation theory for evaluating the self-energy operator $M_k(E)$. The result of a calculation, according to the rules, will generally yield a complex $M_k(E)$ which we may write for E approaching the real axis from above (**) as

$$M_k(E) = M_k^{(1)}(E) + i\,M_k^{(2)}(E)\,, \tag{3.135}$$

where $M_k^{(1,2)}(E)$ are real for real E. We now use the relation for G in terms of M (eq. (3.109)), to deduce the imaginary part of G and hence the spectral function

$$\varrho_k(\omega) = \frac{-(1/\pi)\,M_k^{(2)}(\omega)}{[\omega - T_k + \mu - M_k^{(1)}(\omega)]^2 + [M_k^{(2)}(\omega)]^2}\,. \tag{3.136}$$

From eq. (3.136), we may obtain the internal energy of the system

$$\bar{E} = -\frac{1}{2\pi}\sum_k \int_{-\infty}^{\infty} \frac{(\omega + T_k + \mu)\,F_-(\omega)\,M_k^{(2)}(\omega)\,\mathrm{d}\omega}{[\omega - T_k + \mu - M_k^{(1)}(\omega)]^2 + [M_k^{(2)}(\omega)]^2}\,. \tag{3.137}$$

The above is all exact—and hence completely general.

Let us now suppose that $M_k^{(2)}(\omega)$ is small when evaluated at that value of ω which makes the first term in the denominator of $\varrho_k(\omega)$ vanish. [We shall

(*) It should be emphasized that this is a precise definition of a lifetime — what is « rough » is the relation of this lifetime to the lifetime used in the phenomenological theory.

(**) This properly defines $M_k(E)$ for real E. The arbitrariness is of course removed by the dispersion relation.

discuss the validity of this assumption subsequently.] In this case $\varrho_k(\omega)$ will be very sharply peaked about the root of the equation

$$\omega_k - T_k + \mu - M_k^{(1)}(\omega_k) = 0\,, \tag{3.138}$$

and recalling that (eq. (3.78))

$$\int_{-\infty}^{\infty} \varrho_k(\omega)\, d\omega = 1\,,$$

we may make a rough approximation to the spectral function as

$$\varrho_k(\omega) \simeq \delta(\omega - \omega_k)\,. \tag{3.139}$$

For this spectral function the internal energy and average number of particles are:

$$\begin{cases} \bar{E} \simeq \sum_k [T_k + \frac{1}{2} M_k^{(1)}(\omega_k)]\{\exp[\beta(T_k + M_k^{(1)}(\omega_k) - \mu)] + 1\}^{-1}\,, \\ \bar{N} \simeq \sum_k \{\exp[\beta(T_k + M_k^{(1)}(\omega_k) - \mu)] + 1\}^{-1}\,, \end{cases} \tag{3.140}$$

and we see that the real part of the self-energy, when evaluated at ω_k, can be associated with the energy of a quasi-particle. [To be precise, $T_k + M_k^{(1)}(\omega_k)$ is the energy of the quasi-particle.] To evaluate $\mathscr{F}_k(t)$ we use eq. (3.75)

$$\mathscr{F}_k(t) = \int_{-\infty}^{\infty} d\omega\, \varrho_k(\omega)\, \mathscr{F}_+(\omega) \exp\left[-\frac{i\omega t}{\hbar}\right], \tag{3.141}$$

which allows us to conveniently obtain $\mathscr{F}_k(t)$ from the spectral function $\varrho_k(\omega)$ (*). If we employ the approximate spectral function of eq. (3.139), then we see that the resulting $\mathscr{F}_k(t)$ will be both independent of $M_k^{(2)}(\omega)$ and purely oscillatory in time. We need an improved approximation to the spectral function of Eq. (3.136), and that is achieved by keeping the explicit ω-dependence, but assuming that $M_k^{(1,2)}(\omega)$ are slowly varying functions in the neighborhood

(*) It is important to proceed carefully, and make approximations in $\varrho_k(\omega)$ and then calculate the correlation function from the spectral function, rather than by making approximations directly in the correlation function. This procedure is necessary because approximations in the correlation function, will, in general, lead to a function which has the wrong analytic properties and is noncausal.

of ω_k. In this case $\varrho_k(\omega)$ has two simple poles located at

$$(3.142)\qquad \begin{cases} \omega = \omega_k \pm \dfrac{i\,M_k^{(2)}(\omega_k)}{1 - \mathrm{d}M_k^{(1)}(\omega)/\mathrm{d}\omega|_{\omega=\omega_k}}\,, \\ \omega \equiv \omega_k \pm i\Gamma_k\,; \qquad\qquad \Gamma_k \geqslant 0\,, \end{cases}$$

and by imposing the sum rule of eq. (3.78) we have a second approximation (*) to $\varrho_k(\omega)$ as

$$(3.143)\qquad \varrho_k(\omega) \simeq \frac{\Gamma_k}{\pi[(\omega-\omega_k)^2 + \Gamma_k^2]}\,.$$

With this approximation for $\varrho_k(\omega)$ we can now evaluate $\mathscr{F}_k(t)$ using eq. (3.141). By taking the case $T=0$, the integrals may be readily evaluated to yield the dominant time dependence of $\mathscr{F}_k$ as

$$(3.144)\qquad \mathscr{F}_k(t) \sim \exp\left[-\frac{i}{\hbar}\omega_k t - \frac{\Gamma_k t}{\hbar}\right].$$

Thus we see that a particle put into state k, causes an oscillating amplitude with frequency $\omega_k/\hbar$, and damping (**) constant $\Gamma_k/\hbar$. The oscillatory part corresponds to an energy of excitation $T_k + M_k^{(1)}(\omega_k)$. [We must remove the spurious term in μ associated with the fact that we have used a modified Hamiltonian $\mathscr{H}$ rather than the true Hamiltonian H] while the damping $\Gamma_k/\hbar$ is seen to be proportional to the imaginary part of the self-energy $M_k^{(2)}(\omega_k)$.

The assumption of small imaginary part to the self-energy [when evaluated at ω_k], has led to a microscopic theory in which the particle excitations are described by a quasi-particle distribution function (eq. (3.140)) with quasi-particle energies related to the real part of the self-energy for $\omega=\omega_k$. At the same time the single-particle excitations damp in time at a rate dependent upon the imaginary part of the self-energy for $\omega=\omega_k$. Thus we see *in detail* how the lifetime of quasi-particles may be incorporated into the microscopic theory, playing a vital role in nonequilibrium properties, and yet contributing

---

(*) This assumed functional form for ϱ, plus the normalization requirement, has fixed the residues at the poles in ϱ. It would be a better approximation to determine the residues at the poles by some other means (such as a perturbation theory calculation) and then modify the functional form of ϱ at ω removed from the poles, in order to still satisfy the sum rule. In particular, one can see [as in the derivation of eq. (3.142)] that the variation of $M_k^{(1)}(\omega)$ with ω leads to a residue at ω_k different from that taken in eq. (3.143).

(**) The result that $\mathscr{F}_k(t)$ *damps* is not surprising since the theory is asymmetric with respect to time direction.

only indirectly in the description of equilibrium properties. [Note eq. (3.140), or the *exact* eq. (3.137)].

It is, finally, important to realize that it can be shown, to all orders in perturbation theory, that the imaginary part of the self-energy operator $M_k^{(2)}(E)$ has the following form for E near zero [8] (*):

$$M_k^{(2)}(E) = C_k E^2 , \qquad C_k \leqslant 0 . \tag{3.145}$$

This has the consequence that our assumption about $M_k^{(2)}(\omega_k)$ is increasingly valid as one approaches the Fermi surface. Since, as we saw in Section II, it is only the region near the Fermi surface which is important in describing the low-temperature properties of liquid ^{3}He, the discussion of the latter part of this section is most relevant. We have thus obtained both a microscopic understanding of the equilibrium concept of a quasi-particle [first part of this section], and insight into the lifetime of a quasi-particle near the Fermi surface. A careful microscopic treatment of transport properties remains to be given, but we can feel confident that the damping of the single-particle correlation function will play a vital role in such a theory.

Galitskii has given an explicit calculation of the damping of the single-particle correlation function for a dilute Fermi gas. He finds that near the Fermi surface

$$\Gamma_p = \pm \pi \left(\frac{3}{\pi}\right)^{\frac{2}{3}} \left(\frac{N}{V}\right)^{\frac{2}{3}} \frac{a^2}{m} (p - p_F)^2 , \qquad p \gtrless p_F ; \tag{3.146}$$

where a is the zero energy scattering length. In conjunction with the other properties of the dilute gas (Paragraph III.2·5) we have available a rather complete description of at least one non-trivial model. Zero-sound has also been examined for this system, from a microscopic point of view [see Section V for references], but the results are not sufficiently interesting [in view of the rather elaborate analysis] to merit a discussion in this article.

IV. – **Superfluid Phase.**

A low-temperature superfluid phase of liquid ^{3}He has yet to be observed experimentally. There are, however, compelling reasons for believing that such

(*) Note that eq. (3.145) implies [via eqs. (3.142) and (3.144)] that a quasi-particle at the Fermi surface suffers no damping at $T=0$. A more careful evaluation of eq. (3.141), but still employing the same spectral function, shows that for $T \neq 0$ there is quasi-particle damping even at the Fermi surface, in agreement with physical expectation.

a phase exists. Section **1** is devoted to the arguments in support of the predicted phase, as well as to estimates of the transition temperature. In Section **2** some of the properties of the anisotropic superfluid phase are discussed.

All of the work of this section is based upon the fundamental contributions of BARDEEN, COOPER, and SCHRIEFFER (BCS), and BOGOLIUBOV (BTS), although many other workers have played an active role in applying these general methods to liquid ^{3}He. References may be found in Section V, and the reader's attention is called to the recent review article mentioned there.

1. – The transition temperature.

The theory of Parts II and III assumes the validity of perturbation theory either explicitly as in the microscopic approach, or in the basic assumptions of the phenomenological theory that the quasi-particles are equal in number to the real particles and obey Fermi statistics. As we have seen there is good experimental confirmation of the assumed validity of perturbation theory—in the temperature range so far explored experimentally. On the other hand, there are theoretical reasons for expecting the Landau theory not to be an appropriate description of the liquid at very low temperatures. In this part we shall consider these theoretical arguments; in Section **1** we investigate various criteria for the validity of perturbation theory, Sections **2** and **3** discuss the BCS and BOGOLIUBOV approach through a non-perturbative description of the liquid, while in the fourth section the method of Green's functions is employed to systematize the theoretical analysis. Section **5** is devoted to D-state pairing and numerical estimates of the transition temperature.

The reader who has not mastered all of Part III may concentrate on Sections **2**, **3**, and **5** of this Part with only a loss in the breadth of his understanding—for *all* of the approaches to be described lead to the *same* prediction for the transition temperature.

1·1. *Divergence of perturbation theory.* – In this paragraph we shall « approach the transition temperature from above ». That is, we will *not* start by constructing a theory of the low-temperature phase and then seeing at what temperature the phase vanishes in favor of the normal phase (this is the approach of Sections **2**, **3** and **4**, following), but rather we will examine the theory of the normal liquid and find a temperature below which the theory is not valid. This approach is particularly appealing for it does not depend upon constructing an elaborate theoretical structure for the description of a phase which has never been observed, and then (without at all being able to check the theory with experiment) using it to predict a transition temperature.

It would be most desirable to be able to predict the break-down of the Landau theory by employing the many experimental results obtained in the region where the theory is valid. Such an analysis would be particularly attractive if it contained a minimum of theoretical ideas. No treatment satisfying these desirable criteria has yet been constructed, although it seems likely that experimental results on zero sound and various transport coefficients should be useful in estimating the possible onset of a new phase.

The analysis to be given here is based on work of THOULESS (TH2) and others. Although it is not as « close to experiment » as might be desired, it is no further removed than the methods to be given in Sections **2**, **3** and **4**, and in fact leads to exactly the same theoretical expression for the transition temperature.

We start with the generalization of Brueckner theory to a nonzero temperature. It was shown in Paragraph III.**3**·6 that a large class of diagrams incorporating the strong repeated interactions between particles (and between holes) could be included if the self-energy operator was calculated (as in eqs. (3.12)) from the reaction matrix defined in eq. (3.129). The reaction matrix satisfies an integral equation which was derived by summation. That is, we examined diagrams and derived the equation by noting that if iterated it corresponded term-by-term to the desired diagrams. The equation is only valid, then, if the series it represents converges. We shall now see that there exists a critical temperature below which the series will diverge.

Of course it will be appreciated that the lack of convergence of the particular class of diagrams included in the reaction matrix, does not necessarily imply the invalidity of perturbation theory. However, when coupled with the [intuitive] statement that these particular diagrams should adequately describe the normal fluid [where only two-body correlations are important] the critical temperature takes on additional significance. Finally, it can be shown (TH3) that if a theory of the low-temperature phase is constructed along the lines of Part III, and *then* a similar analysis performed, the analogous series of diagrams forms a *convergent* series. In other words, the divergence of the sum of perturbation-theory diagrams below the critical temperature can be remedied by starting from a new unperturbed system-one not obtainable from noninteracting particles by perturbation theory—and corresponding to a superfluid phase.

The reaction matrix $K_{kl,ij}(E)$ is to be evaluated for $E \to \varepsilon_i$ and $i = k$. The complex energy is only to be used to define any singularities. Thus, we write

$$K_{kl,ij} = [V_{kl,ij} - V_{lk,ij}] + \tfrac{1}{2}\sum_{i'j'} [V_{kl,i'j'} - V_{lk,i'j'}] \cdot K_{i'j',ij} \left\{ \frac{(1-n_{i'})(1-n_{j'}) - n_{i'}n_{j'}}{\varepsilon_i + \varepsilon_j - \varepsilon_{i'} - \varepsilon_{j'} + i\delta} \right\}, \tag{4.1}$$

where

$$n_i = \{\exp[\beta(\varepsilon_i - \mu)] + 1\}^{-1}, \tag{4.2}$$

and the ε_i are the single-particle energies. It is convenient to define an operator R by the equation

$$R_{kl,ij} = V_{kl,ij} + \sum_{i'j'} V_{kl,i'j'} R_{i'j',ij} \left\{ \frac{(1-n_{i'})(1-n_{j'}) - n_{i'}n_{j'}}{\varepsilon_i + \varepsilon_j - \varepsilon_{i'} - \varepsilon_{j'} + i\delta} \right\}, \tag{4.3}$$

from which it can be easily seen that

$$K_{kl,ij} = R_{kl,ij} - R_{lk,ij}. \tag{4.4}$$

Working now with the R-equation, let us transform to center-of-mass frame and take the case in which the particles have opposite spin and the total momentum is zero. It is easy to show (TH2) that this is the « worst case », *i.e.* the series is most likely to be divergent for this situation, but we shall not give the proof since the physical reason for it will become clear in Section **2.** Thus taking i to correspond to $\boldsymbol{i}\uparrow$, and j to corresponds to $-\boldsymbol{i}\downarrow$, and defining

$$V_{\boldsymbol{k},\boldsymbol{i}} = \tfrac{1}{2}[V_{k\,-k,i\,-i} - V_{k\,-k,-ii}], \tag{4.5}$$

and introducing a briefer notation for $R_{kl,ij}$ and remembering that $\varepsilon_{-i} = \varepsilon_i$; eq. (4.3) becomes

$$R_{\boldsymbol{k},\boldsymbol{i}} = V_{\boldsymbol{k},\boldsymbol{i}} + \sum_{\boldsymbol{j}} \frac{V_{\boldsymbol{k},\boldsymbol{j}} R_{\boldsymbol{j},\boldsymbol{i}}}{2(\varepsilon_i - \varepsilon_j) + i\delta} [(1-n_j)^2 - n_j^2]. \tag{4.6}$$

The statistical factor may be re-written using eq. (4.2):

$$(1-n_j)^2 - n_j^2 = \operatorname{tgh}(\beta/2)(\varepsilon_j - \mu). \tag{4.7}$$

Let us define single particle energies ε_i as measured from the Fermi surface, *i.e.* let us introduce $(\varepsilon_i - \mu)$, but continue to call it ε_i for convenience. We now have

$$R_{\boldsymbol{k},\boldsymbol{i}} = V_{\boldsymbol{k},\boldsymbol{i}} + \frac{1}{2} \sum_{\boldsymbol{j}} \frac{V_{\boldsymbol{k},\boldsymbol{j}} R_{\boldsymbol{j},\boldsymbol{i}} \operatorname{tgh}(\beta/2)\varepsilon_j}{\varepsilon_i - \varepsilon_j + i\delta}. \tag{4.8}$$

The nature of eq. (4.8) can be investigated by choosing a separable potential, *i.e.* by assuming

$$V_{\boldsymbol{k},\boldsymbol{i}} = g v_{\boldsymbol{k}}^* v_{\boldsymbol{i}}. \tag{4.9}$$

This is an unphysical assumption which we are making only for mathematical reasons, namely because in this case we can easily solve eq. (4.8) and so learn about its general structure. Letting

$$R_{k,i} = g v_k^* f_i \,, \tag{4.10}$$

we obtain upon substitution

$$f_i = v_i + \frac{g}{2} \sum_j \frac{|v_j|^2 \operatorname{tgh}(\beta/2)\varepsilon_j}{\varepsilon_i - \varepsilon_j + i\delta} f_i \,, \tag{4.11}$$

and hence

$$R_{k,i} = \frac{g v_k^* v_i}{1 - (g/2)\sum_j [|v_j|^2 \operatorname{tgh}(\beta/2)\varepsilon_j]/(\varepsilon_i - \varepsilon_j + i\delta)} \,. \tag{4.12}$$

We will have a reaction matrix singularity, or lack of convergence of the series for $R_{k,i}$ if the denominator is zero. The worst case (*) occurs when $\varepsilon_i = 0$, that is, the particles whose interaction we are considering are located on the Fermi surface [with opposite spin and momenta]. There will exist a value of β, call it β, such that if $\beta < \beta_c$ the denominator in eq. (4.12) is never zero. This critical temperature will only occur if $g < 0$ [corresponding to attractive forces], and in that case the critical temperature is given by

$$\frac{|g|}{2} \sum_j \frac{|v_j|^2 \operatorname{tgh}(\beta_c/2)\varepsilon_j}{\varepsilon_j} = 1 \,. \tag{4.13}$$

To obtain a result which is of more interest for liquid ^{3}He we must be able to handle a general potential and hence return to eq. (4.8). The power series for R will converge if the homogeneous equation obtainable from eq. (4.8) has no nontrivial solution. Again taking the worst case of $\varepsilon_i = 0$, this criterion implies that for $\beta < \beta_c$ the perturbation theory approach will be valid where β_c is the smallest solution of

$$R_k = -\frac{1}{2} \sum_j \frac{V_{k,j} R_j \operatorname{tgh}(\beta_c/2)\varepsilon_j}{\varepsilon_j} \,. \tag{4.14}$$

This equation is identical to one first derived by a slight generalization of the method of BCS. Its solution will be deferred until Paragraph **1**·5.

These results raise the obvious question as to the validity of the work of BRUECKNER and GAMMEL. Actually the singularity only occurs for states

(*) This can be seen by studying eq. (4.10) with the singularity defined in accordance with the discussion of Section III **3**.

with energies *very* close to the Fermi surface and these make a negligible contribution to the general properties of the liquid. Consequently, if these states are ignored, or treated in a manner which is only approximate. [This is the approach of BG who insert a small term in the denominator of the zero-temperature limit of eq. (4.1) in order to approximate some higher terms in the perturbation analysis, and thus incidently remove the singularity], the net error will be small. Thus, although the theory of BG misses the existence of the superfluid phase, it should describe the liquid well, at temperatures above the transition, since the gross features of the liquid change only very slowly with temperature.

It is possible to raise a different question concerning the perturbation-theory description of the liquid, the answer to which is most revealing. Namely one can ask, within the framework of the theory, whether the system is stable against various collective oscillations. A negative answer either will reveal a poor calculation of the frequency of the collective mode, or that the description of the state does not correspond to the phase of lowest free energy. We have already discussed this general approach in Section II.2 with regard to the collective oscillations known as zero-sound. There we saw that the best estimates indicated that some modes of zero sound propagated, and some modes damped, but that no modes had complex frequencies.

When we consider the collective mode corresponding to the two-particle excitations [zero sound is a one-particle excitation in which the interaction of the excited particle and its residual hole must be considered] then we find (TH2) a dispersion relation for the collective mode which has the possibility of having complex frequencies as a solution. The condition that *none* of the modes have complex frequencies may be expressed in terms of the temperature and is exactly equivalent to requiring that $\beta < \beta_c$ where β_c is given by the smallest solution of eq. (4.14). As has been pointed out, this indicates [unless the calculation of the frequencies is inadequate] that for $\beta > \beta_c$ the phase of the system described by perturbation theory will spontaneously convert to a different phase. The same calculation, performed on the superfluid phase [as described by the theory of Part III], yields stable oscillations; once again we have reason to believe that the system will exist in a superfluid phase for $T < T_c$.

A theoretical analysis which brought the above argument closer to experiment, would be most welcome. One can envision employing *experimentally* determined properties of the normal fluid to study the frequency of two-particle collective modes as a function of temperature, but such an analysis has yet to be produced.

Most importantly, it should be realized that the systematic perturbation theory for quantum statistical mechanics (as for example, outlined in Section III.3), allows a straightforward computation of higher order corrections to the criterion for the break-down of perturbation theory. Thus it is possible to exhibit, in a systematic manner, corrections to eq. (4.14). One of the most important improvements would be to formulate the theory for the self-energy operator $M_k(E)$ so that the zero-order correlation functions $\mathscr{F}^{(\pm)(0)}(0)$

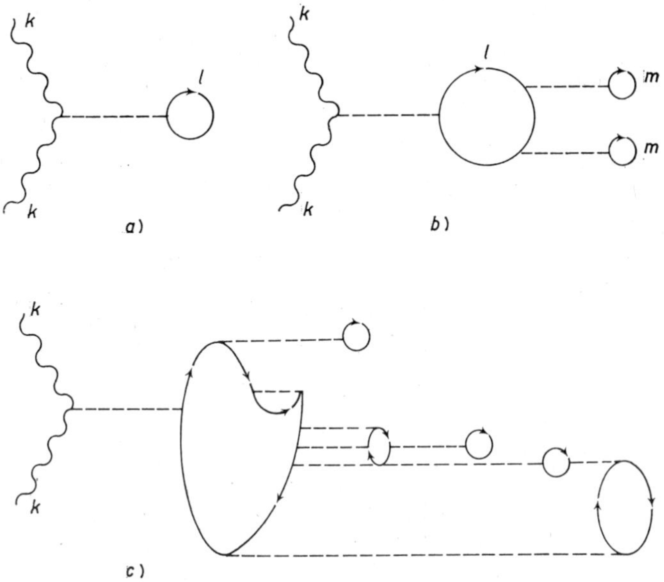

Fig. 13. – Graphs (*b*) and (*c*) are included in the graph (*a*) if $\mathscr{F}_l$ rather than $\mathscr{F}_l^{(0)}$, is employed in evaluating (*a*). According to the rules of Paragraph IV.1·1, diagrams (*b*) and (*c*) are to be omitted in the computation of $M_k(E)$.

[recall eq. (3.111), (3.71)] are replaced by the true correlation functions $\mathscr{F}^{(\pm)}(0)$. This replacement would mean the summation of a large class of diagrams and a consequent reduction in the terms which must be included in calculating the self-energy operators. Thus in Fig. 13 we have indicated some diagrams which are included in the single diagrams of Fig. 13*a*, if $\mathscr{F}_l^{(0)}(0)$ is replaced by $\mathscr{F}_l(0)$. The rule for calculating $M_k(E)$ [Paragraph III.3·5] is now modified (*) by:

1) Leave out all diagrams in the calculation of $M_k(E)$ which are included in replacing $\mathscr{F}^{(0)}$ by $\mathscr{F}$.

(*) The derivation of these rules is easy. From eq. (3.75) we have the $\mathscr{F}_j^{(\pm)}(0)$ functions which are to replace the $n_j^{(0)}$ and $(1 - n_j^{(0)})$ factors, while the replacement of the kinetic energies follows from the relation between G and $G^{(0)}$ [eq. (3.128)]. It can be easily checked that these rules reduce to the rules of Paragraph III.3·3 for the spectral function of the ideal gas [eq. (3.102)].

2) Calculate the contribution of a « particle line » j (upward arrow) by assigning a factor

$$\int_{-\infty}^{\infty} d\omega\, \varrho_j(\omega) F_+(\omega)\,.$$

For a « hole line » j (downward arrow) assign a factor

$$\int_{-\infty}^{\infty} d\omega'\, \varrho_j(\omega') F_-(\omega')\,.$$

The contribution to the energy denominators of the hole line is $(\omega'+\mu)$, and the particle line $-(\omega+\mu)$.

The reader will appreciate that the calculation is now much more complicated in that one must perform additional integrations, and furthermore solve self-consistently for $M_k(E)$ since $\varrho_k(\omega)$ is a function on M (eq. (3.136)). On the other hand, a single term in the perturbation theory now includes an infinite class of the old diagrams.

Let us apply these new rules to obtain an improved equation for the transition temperature. We want to calculate $M_k(R)$ taking into account the repeated interaction of particle k with particle l. Thus we sum the graphs of Fig. 11 and 12, but now replacing $\mathscr{F}^{(0)}$ with $\mathscr{F}$. It is clear that this process does not violate rule 1) (above). Proceeding according to rule 2) we have as an improved description of the normal liquid:

$$M_k(E) = \sum_l \int_{-\infty}^{\infty} d\omega\, \varrho_l(\omega) F_-(\omega) K_{kl,kl}(E+\omega)\,, \tag{4.15}$$

where

$$K_{k'l',kl} = R_{k'l',kl} - R_{l'k',kl}\,, \tag{4.16}$$

and

$$R_{k'l',kl}(E+\omega) = V_{k'l',kl} + \sum_{k''l''} \int_{-\infty}^{\infty} d\omega_{k''} \int_{-\infty}^{\infty} d\omega_{l''} \cdot \tag{4.17}$$

$$\cdot \frac{\varrho_{k''}(\omega_{k''})\, \varrho_{l''}(\omega_{l''})\, V_{k'l',k''l''} R_{k''l'',kl}(E+\omega)}{[E+\omega-\mu-\omega_{k''}-\omega_{l''}]} \cdot \{F_+(\omega_{k''}) F_+(\omega_{l''}) - F_-(\omega_{k''}) F_-(\omega_{l''})\}\,.$$

The expansion for R [which was used to derive eq. (4.17) by summation] will diverge if the homogeneous equation related to eq. (4.16) has a nontrivial solution. Again, the worst case is for particles on the Fermi surface with equal and opposite spin and momenta. Reasoning just as before we get the criterion that $\beta < \beta_c$ where β_c is the smallest solution of (*)

$$(4.18)\qquad R_k = -\sum_j V_{k,j} R_j \int_{-\infty}^{\infty}\int_{-\infty}^{\infty} \frac{d\omega\, d\omega'\, \varrho_j(\omega)\varrho_{-j}(\omega')}{2\mu+\omega+\omega'} \cdot [F_+(\omega)F_+(\omega') - F_-(\omega)F_-(\omega')] .$$

A simple approximation is to choose the spectral function for an ideal gas with single particle energies ε_j:

$$(4.19)\qquad \varrho_j(\omega) = \delta(\omega+\mu-\varepsilon_j) ,$$

in which case eq. (4.18) immediately reduces to eq. (4.14). A significant improvement is afforded by use of the approximate quasi-particle spectral function of eq. (3.143):

$$(4.20)\qquad \varrho_k(\omega) = \frac{\Gamma_k}{\pi[(\omega-\omega_k)^2+\Gamma_k^2]} ,$$

with ω_k the root of

$$(4.21)\qquad \omega_k - \varepsilon_k + \mu - M_k^{(1)}(\omega_k) = 0 .$$

and

$$(4.22)\qquad \Gamma_k = \frac{M_k^{(2)}(\omega_k)}{1-(dM_k^{(1)}(\omega)/d\omega)\,|_{\omega=\omega_k}} .$$

Redefining the single particle energies so that $\omega_j = \omega_{-j} \equiv \varepsilon_j - \mu$ (*i.e.* incorporating the real part of M_j in the single-particle energies), and taking $\Gamma_j = \Gamma_{-j} \equiv \Gamma_j$ we may evaluate the integrals in eq. (4.18) after observing that for small Γ_j the $F_\pm$ functions may to good approximation (**) be evaluated at $\varepsilon_j - \mu$. We have

$$R_k = -\sum_j V_{k,j} R_j [F_+(\varepsilon_j-\mu)F_+(\varepsilon_j-\mu) - F_-(\varepsilon_j-\mu)F_-(\varepsilon_j-\mu)]\, L_j ,$$

where

$$(4.23)\qquad L_j = \left(\frac{\Gamma_j}{\pi}\right)^2 \int_{-\infty}^{\infty} d\omega \int_{-\infty}^{\infty} d\omega' [(\omega-\varepsilon_j+\mu)^2+\Gamma_j^2]^{-1} \cdot [(\omega'-\varepsilon_j+\mu)^2+\Gamma_j^2]^{-1}[2\mu+\omega+\omega']^{-1} .$$

(*) Besides choosing $\boldsymbol{k} = -\boldsymbol{l}$ and defining $V_{k,j}$ as in eq. (4.5) we have evaluated $R(E+\omega)$ for $E=0$, $\omega=-\mu$ which is the appropriate choice to yield the smallest β_c.

(**) A more accurate evaluation of the integrals indicates that this approximation is indeed accurate for liquid ^{3}He.

Changing notation, so $\varepsilon_j - \mu$ is called ε_j and manipulating the $F_\pm$ functions yields:

$$R_k = -\sum_j V_{k,j} R_j \operatorname{tgh} \frac{\beta_c \varepsilon_j}{2} L_j\,,$$

where

$$(4.24)\qquad L_j = \left(\frac{\Gamma_j}{\pi}\right)^2 \int_{-\infty}^{\infty} d\omega \int_{-\infty}^{\infty} d\omega' [(\omega - \varepsilon_j)^2 + \Gamma_j^2]^{-1} [(\omega' - \varepsilon_j)^2 + \Gamma_j^2]^{-1} [\omega + \omega']^{-1}\,.$$

Evaluating the integrals we have

$$(4.25)\qquad R_k = -\sum_j \frac{V_{k,j} R_j \operatorname{tgh}(\beta_c/2)\varepsilon_j}{2\varepsilon_j[1 + \Gamma_j^2/\varepsilon_j^2]}\,,$$

which can be seen to differ from eq. (4.14) by a term involving the damping of quasi-particles. We shall defer discussion of the importance of the damping term until Paragraph IV.1.5, where numerical estimates will be given for the critical temperature, β_c, both with and without quasi-particle damping. It should be noted, however, that Γ_j is not a constant, but rather a function of ε_j (recall eq. (3.147)), and this is very important in determining the effect of quasi-particle damping upon β_c.

It is interesting to ask if one can improve the critical temperature equation even further. The calculation [with true single-particle Green's functions] has included particle-particle and hole-hole interactions, but not particle-hole interactions. Including these by replacing V with a particle-hole reaction matrix corresponds to a systematic improvement. The resulting equations are complicated, and have not yet been studied although they were proposed by SCHRIEFFER (SC1) some time ago.

1.2. *Variational method.* – The BCS description of a superfluid system of fermions [which we will discuss in this section], is based on the observation, by COOPER, that attractive forces between particles near the Fermi surface will lead to a ground state which is significantly different from that of the Fermi-liquid theory. COOPER considered two particles interacting with each other in the presence of many other identical noninteracting particles; and showed that if the interacting particles had equal and opposite spins and momenta as well as energies near the Fermi surface, then no matter how weak the attraction between the particles they would form a bound state—that is, the energy of the particles would be less than the Fermi energy and their relative wave function would decrease as $1/r^2$ for large relative spacing. We recognize that this is exactly the reaction matrix problem which COOPER considered, and his bound state [a « Cooper pair »] is the state associated with the reaction matrix singularity.

To go from the observation of strong correlations between two fermions—in fact such strong correlations that a bound state is formed [and hence perturbation theory will not be applicable, as we saw in detail in Paragraph IV.1·1] to a theory for the N-fermion problem is a nontrivial step. The crucial point is to realize that a particle of momentum and spin $(\boldsymbol{k}, \sigma)$ will « pair » (or strongly correlate) only with *one* other particle, namely the particle $(-\boldsymbol{k}, -\sigma)$. The reason for this is that the pairing of $\boldsymbol{k}$ and $-\boldsymbol{k}+\boldsymbol{q}$ would block the pairing in the total system of particles from *two* other states instead of one, unless *all* particles paired to a net momentum $\boldsymbol{q}$ [which would just correspond to a state with a total nonzero current]. For the ground state, then, we take $\boldsymbol{q}=0$, and consideration of strong correlations only between particles $\boldsymbol{k}$ and $-\boldsymbol{k}$ is a crucial assumption in the BCS theory.

In view of the vital role that this assumption plays in the theory, it is important to realize that the existence of « pairs » has been strongly suggested by the recent experiments demonstrating the quantization of flux in a superconductor.

The mathematical formulation of the above ideas, proceeds by construction of a state vector with a general structure appropriate to pair formation, and then the determination of the parameter in the state vector by variational methods. Thus, BCS chose a trial function of the form

$$\Psi_0 = \prod_{\boldsymbol{k}} [\alpha_{\boldsymbol{k}} + \beta_{\boldsymbol{k}} a^{\dagger}_{\boldsymbol{k}\uparrow} a^{\dagger}_{-\boldsymbol{k}\downarrow}] \Phi_{\text{vacuum}} , \tag{4.26}$$

which can be seen to resemble the Hartree-Fock ground state except that in this state vector if the state $\boldsymbol{k}$ is occupied then $-\boldsymbol{k}$ must also be occupied. The coefficients $\alpha_{\boldsymbol{k}}$ and $\beta_{\boldsymbol{k}}$ determine the occupation of pair states and we can expect $\beta_{\boldsymbol{k}}$ to be unity for $k \ll k_{\text{F}}$ and zero for $k \gg k_{\text{F}}$, and to fall off smoothly for $k \approx k_{\text{F}}$. Normalizing Ψ_0 yields

$$|\alpha_{\boldsymbol{k}}|^2 + |\beta_{\boldsymbol{k}}|^2 = 1 , \tag{4.27}$$

which allows us to write

$$\alpha_{\boldsymbol{k}} = [1 - h_{\boldsymbol{k}}]^{\frac{1}{2}} , \qquad \beta_{\boldsymbol{k}} = h_{\boldsymbol{k}}^{\frac{1}{2}} \exp[i\varphi(\boldsymbol{k})] , \tag{4.28}$$

with $h_{\boldsymbol{k}}$ real positive, and less than unity. The Hamiltonian of eq. (3.30) is now used with Ψ_0 to calculate

$$E = \Psi_0^{\dagger} H \Psi_0 , \tag{4.29}$$

which quantity is then minimized with respect to $h_{\boldsymbol{k}}$ and $\varphi(\boldsymbol{k})$, while keeping the total number of particles fixed. The result of this algebraic computation [see (CMS) or (ES1) for the details] is:

$$\text{(4.30)} \qquad \varphi(\boldsymbol{k}) \text{ is arbitrary}, \qquad h_{\boldsymbol{k}} = \frac{1}{2}\left[1 - \frac{\varepsilon_{\boldsymbol{k}}}{E(\boldsymbol{k})}\right],$$

where

$$\text{(4.31)} \qquad E(\boldsymbol{k}) = [\varepsilon_{\boldsymbol{k}}^2 + |\mathscr{F}(\boldsymbol{k})|^2]^{\frac{1}{2}},$$

and $\mathscr{F}(\boldsymbol{k})$ satisfies the equation

$$\text{(4.32)} \qquad \mathscr{F}(\boldsymbol{k}) = -\frac{1}{2}\sum_{\boldsymbol{k}'} V_{\boldsymbol{k},\boldsymbol{k}'} \frac{\mathscr{F}(\boldsymbol{k}')}{E(\boldsymbol{k}')}.$$

The potential $\mathscr{F}_{\boldsymbol{k},\boldsymbol{k}'}$ is defined as in eq. (4.5), and we have replaced (*) certain diagonal potential matrix elements with the single-particle energies $\varepsilon_{\boldsymbol{k}}$ which are measured from the Fermi surface [as in eq. (4.8)].

Introducing a new function $\chi(\boldsymbol{k})$ by

$$\text{(4.33)} \qquad \chi(\boldsymbol{k}) = [h_{\boldsymbol{k}}(1 - h_{\boldsymbol{k}})]^{\frac{1}{2}} \exp[i\varphi(\boldsymbol{k})],$$

we have from eq. (4.32)

$$\text{(4.34)} \qquad \chi(\boldsymbol{k}) = -\frac{1}{2E(\boldsymbol{k})}\sum_{\boldsymbol{k}'} V_{\boldsymbol{k},\boldsymbol{k}'}\chi(\boldsymbol{k}').$$

One must solve eq. (4.34) for $\chi(\boldsymbol{k})$, while observing that $E(\boldsymbol{k})$ involves $\mathscr{F}(\boldsymbol{k})$ and $\mathscr{F}(\boldsymbol{k})$ is related to $\chi(\boldsymbol{k})$ by

$$\text{(4.35)} \qquad \mathscr{F}(\boldsymbol{k}) = -\sum_{\boldsymbol{k}'} V_{\boldsymbol{k},\boldsymbol{k}'}\chi(\boldsymbol{k}').$$

Assuming that $\mathscr{F}(\boldsymbol{k})$ is slowly varying for $\boldsymbol{k}$ near $\boldsymbol{k}_{\mathrm{F}}$ we may replace $\mathscr{F}(\boldsymbol{k})$ by the constant $\mathscr{F}(k_{\mathrm{F}})$ in eq. (4.31) for $E(\boldsymbol{k})$, in which case we have the *linear* equation for $\chi(\boldsymbol{k})$:

$$\text{(4.36)} \qquad \chi(\boldsymbol{k}) = -\frac{1}{2[\varepsilon_{\boldsymbol{k}}^2 + |\mathscr{F}(k_{\mathrm{F}})|^2]^{\frac{1}{2}}}\sum_{\boldsymbol{k}'} V_{\boldsymbol{k},\boldsymbol{k}'}\chi(\boldsymbol{k}').$$

Here $|\mathscr{F}(k_{\mathrm{F}})|^2$ plays the role of an eigenvalue. The existence of a superfluid ground state is equivalent to obtaining a non-trivial solution to eq. (4.36).

(*) For a singular potential these diagonal terms are not defined, so the identification made here *assumes* that a successful theory has been constructed for the normal liquid. In practice we shall take the $\varepsilon_{\boldsymbol{k}}$ from experiment, and not examine the point in more detail from the BCS point of view. In Paragraph IV.1·1 we had no such difficulty.

The resulting $\chi(\boldsymbol{k})$ describes the extra correlation between paired particles due to the interaction $V_{\boldsymbol{k},\boldsymbol{k}'}$, while the $|\mathscr{F}(k_{\mathrm{F}})|^2$ can be shown to be related to an energy gap in the excitation spectrum of particles.

Having found the ground state, BCS proceed to construct a complete set of excited states which are orthogonal to each other as well as to Ψ_0. These states can be characterized in a straightforward manner, and statistical mechanics undertaken. Introducing a distribution function $f(\boldsymbol{k})$, BCS evaluate the free energy of the system by evaluating the expectation value of the Hamiltonian in the ensemble, and using eq. (1.16) for the entropy. They then minimize the free energy (see Paragraph I.1·3) to determine the distribution function, as well as the best choice of the ground state parameters $\alpha_{\boldsymbol{k}}$, $\beta_{\boldsymbol{k}}$. Note that the ground state is undetermined at *each* temperature. That is a crucial point in the theory since it allows a parameter [namely $|\mathscr{F}(k_{\mathrm{F}})|^2$] to vary with temperature, and so yields a second-order phase transition. This method is not the old fashioned way to do statistical mechanics where one determined states at $T=0$ and then varied only the occupation in numbers as a function of temperature, but is superior in that one does a « self-consistent » calculation at each temperature. It is closely related to the Green's functions techniques discussed in Section III.3.

The details of this calculation are messy [see BCS and ES], but the result is that upon minimizing the free energy one finds:

$$(4.37)\qquad \begin{cases} \varphi(\boldsymbol{k}) \ \text{ is arbitrary}\,, \\ f(\boldsymbol{k}) = \{\exp[\beta E(\boldsymbol{k})]+1\}^{-1}, \end{cases}$$

and

$$(4.38)\qquad \begin{cases} \mathscr{F}(\boldsymbol{k}) = -\dfrac{1}{2}\sum_{\boldsymbol{k}'}\dfrac{\operatorname{tgh}\frac{1}{2}\beta E(\boldsymbol{k}')}{E(\boldsymbol{k}')}V_{\boldsymbol{k},\boldsymbol{k}'}\mathscr{F}(\boldsymbol{k}')\,, \\ E(\boldsymbol{k}) = [\varepsilon_{\boldsymbol{k}}^2+|\mathscr{F}(\boldsymbol{k})|^2]^{\frac{1}{2}}\,. \end{cases}$$

For $\beta>\beta_c$ there will be a nontrivial solution to the equation for $\mathscr{F}$, [$\mathscr{F}$ will be nonzero]; and the distribution function will correspond to a Fermi gas, but with excitation energies $E(\boldsymbol{k})$ which are different from that for an ideal gas. The critical temperature β_c is that temperature at which for $\beta<\beta_c$ there are only trivial solutions for $\mathscr{F}$. At β_c the function $\mathscr{F}$ vanishes, so the transition temperature β_c is given by:

$$(4.39)\qquad \mathscr{F}(\boldsymbol{k}) = -\frac{1}{2}\sum_{\boldsymbol{k}'}\frac{V_{\boldsymbol{k},\boldsymbol{k}'}\operatorname{tgh}(\frac{1}{2}\beta_c)\varepsilon_{\boldsymbol{k}'}}{\varepsilon_{\boldsymbol{k}'}}\mathscr{F}(\boldsymbol{k}')\,,$$

which is identical with eq. (4.14) [if $\mathscr{F}(\boldsymbol{k})$ is identified with $R_{\boldsymbol{k}}$]. It will be recalled that eq. (4.14) was derived [in the simplest approximation], as the criterion for the break-down of perturbation theory, which is now seen to coincide with the point at which the system becomes superfluid.

We shall not pursue the BCS approach any further here, since more convenient methods now exist for studying the superfluid state. The advantages of the BCS approach are that the physical assumptions are clear, and the variational approach is simple in principle [even though complicated in detail]. We turn in the next Paragraph to the formalism of Bogoliubov, after which we will employ the Green's function approach.

1·3. *Canonical transformation.* – Bogoliubov (BTS) has formulated the BCS theory so that its quasi-particle nature is clear. [The reader should appreciate that the quasi-particles in this theory are quite distinct from the quasi-particles of the Landau-Fermi liquid theory.] That, is, the Hamiltonian of the system is approximated by the form

$$H_0 = E_0 + \sum_{\boldsymbol{k}} E(\boldsymbol{k})[\alpha^\dagger_{\boldsymbol{k}}\alpha_{\boldsymbol{k}} + \beta^\dagger_{\boldsymbol{k}}\beta_{\boldsymbol{k}}]\,, \tag{4.40}$$

where the *new* operators $\alpha^\dagger_{\boldsymbol{k}}$ and $\beta^\dagger_{\boldsymbol{k}}$ ($\alpha_{\boldsymbol{k}}$ and $\beta_{\boldsymbol{k}}$) create (destroy) the excitations of the system. These excitations have definite energy $E(\boldsymbol{k})$, definite momentum $\hbar\boldsymbol{k}$, and obey Fermi statistics; but are not obtainable from the free particles of the system by perturbation theory. In fact, reasoning that only particles $(\boldsymbol{k}, \sigma)$ and $(-\boldsymbol{k}, -\sigma)$ interact so strongly that their interaction can not be handled by perturbation theory, Bogoliubov introduced a canonical transformation which allowed him to handle this particular interaction carefully. Thus he chose his quasi-particle operators to be of the form:

$$\begin{cases} \alpha_{\boldsymbol{k}} = u(\boldsymbol{k})a_{\boldsymbol{k}\uparrow} - v(\boldsymbol{k})a^\dagger_{-\boldsymbol{k}\uparrow}\,, \\ \beta_{\boldsymbol{k}} = u(\boldsymbol{k})a_{-\boldsymbol{k}\downarrow} + v(\boldsymbol{k})a^\dagger_{\boldsymbol{k}\uparrow}\,, \end{cases} \tag{4.41}$$

where the coefficients $u(\boldsymbol{k}) = u(-\boldsymbol{k})$ and $v(\boldsymbol{k}) = v(-\boldsymbol{k})$ are still to be determined, but are subject to the restriction

$$|u(\boldsymbol{k})|^2 + |v(\boldsymbol{k})|^2 = 1\,, \tag{4.42}$$

which is imposed to ensure that the operators $\alpha_{\boldsymbol{k}}$ and $\beta_{\boldsymbol{k}}$ correspond to fermions, *i.e.* obey the anti-commutation relations of eq. (3.103) and (3.17).

Writing

$$\begin{cases} u(\boldsymbol{k}) = \cos\gamma(\boldsymbol{k})\exp[i\eta(\boldsymbol{k})]\,, \\ v(\boldsymbol{k}) = \sin\gamma(\boldsymbol{k})\exp[i\zeta(\boldsymbol{k})]\,, \end{cases} \tag{4.43}$$

it can be shown that all physical quantities depend only the difference in phase

$$\varphi(\boldsymbol{k}) = \zeta(\boldsymbol{k}) - \eta(\boldsymbol{k})\,, \tag{4.44}$$

so that the quasi-particle transformation is characterized by the two real functions $\gamma(\boldsymbol{k})$ and $\varphi(\boldsymbol{k})$.

BOGOLIUBOV determined the transformation at $T=0$ by writing the Hamiltonian of eq. (3.30) in terms of the new operators $\alpha_{\boldsymbol{k}}$, $\beta_{\boldsymbol{k}}$, and $\alpha^\dagger_{\boldsymbol{k}}$ and $\beta^\dagger_{\boldsymbol{k}}$ and then rearranging it so that all the creation operators are on the left. The quadrilinear terms are *neglected* and the resulting truncated Hamiltonian is forced to have the form of eq. (4.40), which yields equations for the transformation coefficients $\gamma(\boldsymbol{k})$ and $\varphi(\boldsymbol{k})$. At a nonzero temperature the Hamiltonian is again truncated, a distribution function of Fermi form is taken in terms of the quasi-particle energies $E(\boldsymbol{k})$, and the free energy evaluated and minimized to determine the quasi-particle transformation. One finds that γ and φ are given by

$$(4.45) \qquad \begin{cases} \operatorname{tg} 2\gamma(\boldsymbol{k}) = -\dfrac{|\mathscr{F}(\boldsymbol{k})|}{\varepsilon_{\boldsymbol{k}}}\,, \\ \mathscr{F}(\boldsymbol{k}) = |\mathscr{F}(\boldsymbol{k})| \exp[i\varphi(\boldsymbol{k})]\,, \end{cases}$$

and $\mathscr{F}(\boldsymbol{k})$ satisfies (*) eq. (4.38) with $E(\boldsymbol{k})$ as given in that same equation.

The two theories are consequently equivalent, but the Bogoliubov formulation allows a convenient formalism for computing the properties of the superfluid system, [as we shall see in Section IV.2]. It also exhibits clearly the terms neglected in the Hamiltonian in order to obtain the independent quasi-particle approximation, and thus makes it possible to construct a *systematic* perturbation theory starting with the BCS theory as the zero-order system. We are interested in just such a theory, for we desire as accurate an equation for the transition temperature as can be obtained. Rather than pursue this approach we shall employ the methods of Green's functions starting from the noninteracting system and *not* making a canonical transformation, but taking the important effect of Cooper pairs into account in a different manner.

1·4. *Green's functions.* – In this Paragraph we are going to give a Green's function formulation of the theory of superconductivity, which is important because it allows a systematic presentation of the theory. Our treatment follows that of GOR'KOV [9], but we shall use the time-independent formulation [just as in Section III.3 and Paragraph IV.1·1].

We start with the Hamiltonian of eq. (3.64), namely

$$(4.46) \qquad \begin{cases} \mathscr{H} = \mathscr{H}_0 + \mathscr{H}_1\,, \\ \mathscr{H}_0 = \sum\limits_i (T_i - \mu) a^\dagger_i a_i\,, \\ \mathscr{H}_1 = \frac{1}{2} \sum\limits_{ijkl} V_{ij,kl} a^\dagger_j a^\dagger_i a_k a_l\,, \end{cases}$$

(*) Once again diagonal terms have been identified with the single-particle energies $\varepsilon_{\boldsymbol{k}}$.

and recall that momentum conservation implies that $V_{ij,kl}$ is zero unless $\boldsymbol{i}+\boldsymbol{j}=\boldsymbol{k}+\boldsymbol{l}$. As in Paragraph III.3·3 we introduce the Green's function

$$G_{ij}(t) = -\, i\langle[a_i(t)\; a_j^\dagger(0)]_+\rangle\theta(t)\,, \tag{4.47}$$

where the brackets indicate statistical average. Now, the importance of Cooper pairs means that we must explicitly take into account the possibility of two particles going into a pair state [or coming out of a pair state], and this will be included in the « anomalous » Green's function

$$F_{ij}(t) = -\, i\langle a_i^\dagger(t),\; a_j^\dagger(0)]_+\rangle\,\theta(t)\,. \tag{4.48}$$

We shall see shortly how and why F enters the theory.

As in Paragraph III.3·4 we seek an equation of motion for G. From the commutation relations, and eq. (4.46) we deduce [as in eq. (3.40)]:

$$i\hbar\frac{\mathrm{d}a_i(t)}{\mathrm{d}t} = (T_i-\mu)\,a_i(t) + \frac{1}{2}\sum_{jkl}[V_{ji,kl} - V_{ij,kl}]\,a_j^\dagger(t)\,a_k(t)\,a_l(t)\,, \tag{4.49}$$

which implies

$$i\hbar\frac{\mathrm{d}G_{ij}(t)}{\mathrm{d}t} = \hbar\,\delta_{ij}\,\delta(t) + (T_i-\mu)\,G_{ij}(t) - \frac{i}{2}\sum_{j'k'l'}[V_{j'i,k'l'} - V_{ij',k'l'}]\cdot \tag{4.50}$$
$$\cdot\,\langle[a_{j'}^\dagger(t)\,a_{k'}(t)\,a_{l'}(t),\; a_j^\dagger(0)]_+\rangle\,\theta(t)\,.$$

We now factor the product of four operators, but [unlike in Section III.3] maintain even the anomalous term. Thus

$$\langle[a_{j'}(t)\,a_{k'}(t)\,a_{l'}(t),\; a_j^\dagger(0)]_+\rangle\,\theta(t) \approx \langle a_{j'}^\dagger\,a_{k'}\rangle\frac{G_{l'j}(t)}{(-\,i)} - \langle a_{j'}^\dagger\,a_{l'}\rangle\frac{G_{k'j}(t)}{(-\,i)} + \langle a_{k'}\,a_{l'}\rangle\frac{F_{j'j}(t)}{(-\,i)}\,,$$

where the operators without a time variable are evaluated at equal times, so the bracketed quantities in this last equation are numbers and not dynamical variables.

Transforming to momentum space (via eq. (3.86)), yields

$$\left[\frac{E-T_i+\mu}{\hbar}\right]G_{ij}(E) = \delta_{ij} + \frac{1}{2\hbar}\sum_{j'k'l'}[V_{j'i,k'l'} - V_{ij',k'l'}]\cdot \tag{4.51}$$
$$\cdot\{a_{j'}^\dagger\,a_{k'}\rangle\,G_{l'j}(E) - \langle a_{j'}^\dagger\,a_{l'}\rangle\,G_{k'j}(E) + \langle a_{k'}\,a_{l'}\rangle\,F_{j'j}(E)\}\,.$$

Employing momentum conservation we can see that

$$\left\{\begin{aligned} G_{ij} &= \delta_{ij}G_i\,,\\ F_{ij} &= \delta_{i\,-j}F_i\,, \end{aligned}\right. \tag{4.52}$$

for the only terms coupled to G_i by the potential term are such as to imply this nature of F. Here by i and $-i$ we mean $(\boldsymbol{i}, \sigma)$ and $(-\boldsymbol{i}, -\sigma)$. Consequently, eq. (4.51) becomes

$$\left[\frac{E - T_i + \mu}{\hbar}\right] G_i(E) = 1 + \frac{1}{\hbar}\sum_j [V_{ij,ij} - V_{ij,ji}]\langle a_j^\dagger a_j\rangle G_i(E) + \\ + \frac{1}{2\hbar}\sum_j [V_{i\,-i,j\,-j} - V_{i\,-i,-j\,j}]\langle a_{-j} a_j\rangle F_{-i}(E)\,, \tag{4.53}$$

in which form the importance of Cooper pairs is clear as F_i couples the states i and $-i$, and the potential terms multiplying F just involve the scattering of particles of equal and opposite spins and momenta.

To proceed further we get an equation of motion for F, which may be accomplished by exactly the same process. One finds

$$i\hbar\,\frac{\mathrm{d}F_{ij}(t)}{\mathrm{d}t} = \hbar\langle a_i^\dagger a_j^\dagger + a_j^\dagger a_i^\dagger\rangle - (T_i - \mu)F_{ij}(t) + \frac{i}{2}\sum_{jk'l'}[V_{j'i,k'l'} - V_{ij',k'l'}]\cdot \\ \cdot\langle[a_{l'}^\dagger(t)\,a_{k'}^\dagger(t)\,a_{j'}(t),\, a_j^\dagger(0)]_+\rangle\,\theta(t)\,,$$

and factoring the four operators as

$$\langle[a_{l'}^\dagger(t)\,a_{k'}^\dagger(t)\,a_{j'}(t),\, a_j^\dagger(0)]_+\rangle \approx \langle a_{l'}^\dagger a_{k'}^\dagger\rangle\,\frac{G_{j'j}(t)}{(-i)} + \langle a_{k'}^\dagger a_{j'}\rangle\,\frac{F_{l'j}(t)}{(-i)} - \langle a_{l'}^\dagger a_{j'}\rangle\,\frac{F_{k'j}(t)}{(-i)}\,,$$

and transforming to momentum space yields

$$\left[\frac{E + T_i - \mu}{\hbar}\right] F_i(E) = -\frac{1}{2\hbar}\sum_j [V_{i\,-i,j\,-j} - V_{i\,-i,-j\,j}]\,\langle a_j^\dagger a_{-j'}^\dagger\rangle G_{-i}(E) - \\ - \frac{1}{\hbar}\sum_j [V_{ij,ij} - V_{ij,ji}]\langle a_j^\dagger a_j\rangle F_i(E)\,. \tag{4.54}$$

We can now solve the coupled equation for $G_i(E)$ and $F_{-i}(E)$. First rewrite the equations in the form

$$\begin{cases} \left[\dfrac{E - T_i + \mu - M_i}{\hbar}\right] G_i(E) = 1 + \dfrac{1}{\hbar}\mathscr{V}_i F_{-i}(E)\,, \\[2ex] \left[\dfrac{E + T_{-i} - \mu + M_{-i}}{\hbar}\right] F_{-i}(E) = -\dfrac{1}{\hbar}\mathscr{W}_i G_i(E)\,, \end{cases} \tag{4.55}$$

where we have defined

$$\begin{cases} M_i = \sum_j [V_{ij,ij} - V_{ij,ji}]\langle a_j^\dagger a_j\rangle\,, \\[2ex] \mathscr{V}_i = \frac{1}{2}\sum_j [V_{i-i,j\,-j} - V_{i\,-i,-j\,j}]\langle \alpha_{-j}\, a_j\rangle\,, \\[2ex] \mathscr{W}_i = -\,\frac{1}{2}\sum_j [V_{i-i,j\,-j} - V_{i\,-i,-j\,j}]\langle a_j^\dagger a_{-j}^\dagger\rangle\,. \end{cases} \tag{4.56}$$

But it is clear that

$$\langle a_j^\dagger a_{-j}^\dagger \rangle = \langle a_{-j} a_j \rangle^* \tag{4.57}$$

so that $\mathscr{W}_i = -\mathscr{V}_i^*$. Furthermore $T_{-i} = T_i$ and $M_{-i} = M_i$, and identifying the single-particle energies as measured from the Fermi surface

$$\varepsilon_i = T_i + M_i - \mu \tag{4.58}$$

we may write eq. (4.55) in the form:

$$[E - \varepsilon_i]\, G_i(E) = \hbar - \mathscr{W}_i^* F_{-i}(E)\,,$$

$$[E + \varepsilon_i)\, F_{-i}(E) = -\mathscr{W}_i G_i(E)\,,$$

with the solution (*)

$$\begin{cases} G_i(E) = \dfrac{\hbar(E + \varepsilon_i)}{E^2 - [\varepsilon_i^2 + |\mathscr{W}_i|^2]}\,, \\[2ex] F_{-i}(E) = -\dfrac{\hbar \mathscr{W}_i}{E^2 - [\varepsilon_i^2 + |\mathscr{W}_i|^2]}\,. \end{cases} \tag{4.59}$$

There is still the possibility of adding solutions of the homogeneous equation, and so we turn to the dispersion relation of eq. (3.98) to complete the determination of G and F. It is clear that, just as in the discussion leading to eq. (3.105), we must define the singularities so that the poles are in the lower half-plane.

To complete the analysis, we must obtain $\mathscr{W}_i$, and that can be done by expressing $\langle a_j^\dagger a_{-j}^\dagger \rangle$ in terms of $F(E)$. The relation is evidently just the same as that between the equal-time correlation function $\mathscr{F}_{ij}(t, t)$ (eq. (3.71)) and the retarded Green's function $G_{ij}(t, t')$ (eq. (3.83)). Thus by direct analogy with eq. (3.75) we have

$$\langle a_i^\dagger a_j^\dagger \rangle = \int_{-\infty}^{\infty} \mathrm{d}\omega\, F_+(\omega) \left[-\frac{1}{\pi\hbar} \operatorname{Im} F_{ij} \right],$$

$$\langle a_i^\dagger a_{-i}^\dagger \rangle = -\frac{1}{\pi\hbar} \int_{-\infty}^{\infty} F_+(\omega) \operatorname{Im} F_i \,\mathrm{d}\omega\,, \tag{4.60}$$

(*) Note that we find that F is not necessarily zero. Thus although in perturbation theory (Section III·3) the function F is identically zero in all orders, by assuming a nonzero F we find coupled equations for F and G with the possibility of a nonperturbative, nonzero, solution for F.

so that from eq. (4.59)

$$\langle a_{-i}^{\dagger} a_i^{\dagger}\rangle = -\frac{1}{\pi\hbar}\int_{\infty -}^{\infty} F_{+}(\omega)(\hbar\pi\mathscr{W}_i)\left[\frac{\delta(\omega+\Delta_i)}{\omega-\Delta_i}+\frac{\delta(\omega-\Delta_i)}{\omega+\Delta_i}\right]\mathrm{d}\omega\,,$$

where we have set $\Delta_i = [\varepsilon_j^2 + |\mathscr{W}_i|^2]^{\frac{1}{2}}$. From the definition of the $F_{\pm}(\omega)$ (eq. (3.76)) this last equation becomes

$$\langle a_{-i}^{\dagger} a_i^{\dagger}\rangle = -\frac{1}{2}\frac{\mathscr{W}_i}{\Delta_i}\,\mathrm{tgh}\,\frac{\beta\Delta_i}{2}\,. \tag{4.61}$$

Now, using eqs. (4.56), (4.61) we have

$$\mathscr{W}_i = -\frac{1}{4}\sum_j \frac{[V_{i\,-i,j\,-j} - V_{i\,-i,-j\,j}]\,\mathrm{tgh}\,(\beta\Delta_j/2)\mathscr{W}_j}{\Delta_j}\,, \tag{4.62}$$

and if we identify $\mathscr{F}(\boldsymbol{k})$ as $\mathscr{W}_k$, $E(\boldsymbol{k})$ as Δ_k, and use the definition of eq. (4.5), then eq. (4.62) can be seen to be the same as eq. (4.38).

Having seen the derivation of the BCS theory by means of Green's functions, the reader can probably imagine how to systematically improve the first approximation. Such studies have been made by a number of workers, and in particular by SCHRIEFFER (SC1) who has used this approach to derive, for the first time, an improved equation for the transition temperature. We will not give a detailed discussion of this work here, but will content ourselves with indicating how the above analysis can be extended.

In analogy with the treatment in which we « approached the transition temperature from above », we can improve the theory if we replace the zero-order Green's functions with true Green's functions. The improvement is now to be made in a Green's function, and not an equal-time correlation function, as can be seen by examining eq. (4.59). The insertion of a true Green's function can be made very easily at the transition temperature, where the term in $|\mathscr{W}_i|^2$ can be dropped in the denominator of the expressions for $F_{-i}(E)$ and $G_i(E)$ (eq. (4.59)). Limiting ourselves to this case, we see (from eq. (3.87)) that now $\mathrm{Im}\,F_{-i}(E)$ is (for $T = T_c$):

$$\mathrm{Im}\,F_{-i}(E) = \hbar\pi\mathscr{W}_i\int_{-\infty}^{\infty}\mathrm{d}\omega\int_{-\infty}^{\infty}\mathrm{d}\omega'\,\varrho_i(\omega)\,\varrho_i(\omega')\left\{\frac{\delta(E+\omega')}{E-\omega}+\frac{\delta(E-\omega)}{E+\omega'}\right\},$$

which means that for $T = T_c$:

$$\langle a_{-1}^{\dagger} a_i^{\dagger}\rangle = -\mathscr{W}_i\int_{-\infty}^{\infty}\frac{\mathrm{d}\omega\,\mathrm{d}\omega'\,\varrho_i(\omega)\,\varrho_i(\omega')}{(\omega+\omega')}[F_{+}(\omega)-F_{+}(-\omega')]\,. \tag{4.63}$$

Proceeding as in the derivation of eq. (4.62) we have an improved equation for the transition temperature:

$$\mathscr{W}_i = -\frac{1}{2}\sum_j [V_{i-i,j-j} - V_{i-i,-j\,j}]\mathscr{W}_j \int_{-\infty}^{\infty} \frac{\mathrm{d}\omega\,\mathrm{d}\omega'\,\varrho_j(\omega)\,\varrho_j(\omega')}{(\omega+\omega')}[F_+(\omega) - F_-(\omega')]\,.$$

Upon making the appropriate changes in notation, and manipulating the $F_\pm$ functions, this equation can be seen to reduce to eq. (4.18).

We will leave the further study of this subject to the interested reader, and turn in the next section to direct application of the results of these last four sections to liquid ^{3}He.

1·5. *D-state pairing and numerical estimates,* – The equation for the critical temperature (eq. (4.14) or (4.25)) is a linear equation in the function R_k (or $\mathscr{F}(\boldsymbol{k})$), and has a spherically symmetric kernel. Consequently it can be separated into a set of equations each referring to a definite angular momentum l. In the theory of superconductivity the smallest value of β_c is associated with the solution of the $l=0$, S-wave equation. Correspondingly the correlation function (eq. (4.35)) is spherically symmetric.

For liquid ^{3}He the smallest value of β_c is associated with the $l=2$ equation. This is a consequence of the nature of the two-body potential between ^{3}He atoms which causes two ^{3}He atoms at the Fermi surface to repel each other in S-states. In D-states this interaction is attractive at the Fermi surface (*), and as we saw in the discussion of the separable potential (eq. (4.9) to (4.12)) one can expect a superfluid state in this situation.

The fact that the correlation function varies as $Y_{2m}(\theta, \varphi)$ for T near T_c —where the angles (θ, φ) are measured from an arbitrary axis—implies that for $T < T_c$ the angular dependence of the correlation function will be at least as complicated. (This has important physical consequences which will be described in Section IV.**2**, but in this Paragraph we concentrate on the problem of estimating T_c.

The transition temperature can be found by transforming to configuration space and separating out the $Y_{2m}(\theta, \varphi)$ equation. Letting $R(r)$ be the radial part of the solution. [The equation is degenerate in m, so we can ignore the

(*) We do not have to solve for the interaction between atoms in the medium, in order to see that there is attraction in D-states. We simply realize that a measure of the interaction between ^{3}He atoms at the Fermi surface is afforded by the free space phase shifts at the relative momentum $\hbar k_{\mathrm{F}}$, and they are $-61°$ for $l=0$, $-2°$ for $l=1$, $+19°$ for $l=2$, and $+11°$ for $l=3$. We can expect the smallest β_c for $l=2$ from this indication, and that expectation is indeed confirmed by computation.

A careful study by Pitaevskii (PII) confirms the hypothesis, made here, that many-body effects don't alter this conclusion.

dependence upon m and φ. This is only true at the transition temperature, *i.e.* it is not true for eq. (4.38).] We have:

$$R(r) = -\int_0^\infty G(r, r')\, v(r')\, R(r')\, \mathrm{d}r' \,, \tag{4.64}$$

with

$$G(r, r') = \frac{2}{\pi}\int_0^\infty [kr' j_2(kr')][kr\, j_2(kr)]\, K(k)\, \mathrm{d}k \,, \tag{4.65}$$

and

$$K(k) = \frac{\operatorname{tgh}(\beta_c/2)\,\varepsilon_k}{2\varepsilon_k[1 + \Gamma_k^2/\varepsilon_k^2]} \,. \tag{4.66}$$

The two body potential in configuration space is $v(r)$ and $j_2(kr)$ the spherical Bessel function of order 2.

Solution of eqs. (4.64), (4.65) and (4.66) is a bit complicated, and a number of different approaches have been employed in the literature (BSAM, AM2, ES). Unfortunately there is no simple approximation method known, and the various methods have all employed digital computation. The most careful computation has been made for the no-damping approximation $[\Gamma_k = 0]$, and yields after employing the measured value of effective mass, the results of BRUECKNER and GAMMEL for ε_k [when k is removed from k_F], and the 6-12-potential. There results a value of $T_c \approx 0.08$ °K.

Damping is important, as has been emphasized (*) by BARDASIS and SCHRIEFFER (BS1), ANDERSON and MOREL (AM2), and MOREL and NOZIERES (MN1). In the computation of the effect of damping it is important to realize that $M_k^{(2)}(\omega) \approx c_k\omega^2$ (eq. (3.147)), so that $\Gamma_k \sim M_k^{(2)}(\omega_k) \sim \varepsilon_k$. [Actually this is only valid for $M_k^{(2)}$ small, so our conclusions are not exact for k removed from k_F, but in this region the damping has cut the contribution to G to almost nothing, and a slight error in this contribution is unimportant.]

BARDASIS and SCHRIFFEER have employed an equation which is of the above form, but with

$$K(k) = \frac{\operatorname{tgh}(\beta_c/2)\,\varepsilon_k}{2\varepsilon_k[2(1 + 4\alpha^2\varepsilon_k^2)^{\frac{1}{2}} - 1]} \,, \tag{4.67}$$

the difference between this and eq. (4.66) being a reflection of different ap-

(*) A theoretical calculation of the ffect of damping in superconductivity has been given by O. BETHEBER MATIBET and P. NOZIÈRES: [10] but they do not apply their results to liquid ^{3}He.

proximations employed in inserting the k-dependence of the damping. It can be seen that both equations have the same general effect (*).

From the thermal conductivity data BS estimate that $\alpha \approx 1.77/\mu$. As of this writing, no careful calculation has been made for β_c using either eq. (4.66), or eq. (4.67) with $\Gamma_k \neq 0$. An estimate by BS using a separable equation, indicates that the effect of α in eq. (4.67) is to cut T_c by a factor of 0.32, which when combined with the calculation for $\Gamma_k = 0$, yields an estimate of $T_c \approx \cdot 026$ °K. It is not clear that with the strongly singular potential $v(r)$ this result might not be somewhat modified, and a careful calculation should be performed.

Finally, in regard to numerical estimates, there are a number of comments which can be made:

i) The transition temperature depends *exponentially* on the calculations, so slight errors make a very large change in T_c.

ii) Sources of error include the $v(r)$ employed, and the fact that many-body forces have been ignored. GOR'KOV and PITAEVSKII (GP1) have recently made an estimate of the effect of many-body forces.

iii) Higher order effects—such as replacing v by the particle-hole reaction matrix—have yet to be investigated, and because of the exponential dependence could change T_c considerably even if « small » corrections.

iv) The insertion of the single-particle Green's function from experiment [specific heat to determine the real part, and the thermal conductivity to determine the imaginary part] has not included the « renormalization of the pole effect » [see footnote (*), p. 252] which may well be important. Also, experiment only determines *two* constants of $G_k(\omega)$, and since the functional form is only known for k near k_F and ω near zero, one must depend upon a *calculation* of $G_k(\omega)$ for the normal fluid.

v) It is unclear whether using the experimental one-particle Green's function should not imply a corresponding change in other parts of the calculation. [In effect, the self-energy of G has been calculated to all orders, but the self-energy of F only to lowest order—and this unsymmetrical calculation may be no real improvement at all over proceeding symmetrically.]

(*) This is in contrast with the work of ANDERSON and MOREL, and BETHEBER MATIBET and NOZIÈRES [see footnote (*), p. 272], for these authors take the damping to be k-independent and consequently obtain quite a different result.

2. – The anisotropic superfluid.

In Section IV.**1** we have seen that there are reasons to believe that liquid ^{3}He will exhibit a low-temperature phase which is mathematicly analogous to the superconducting phase of electrons in a metal. This [as yet unobserved] phase is expected to have many curious properties; some analogous to those of superconductors, and some arising from the asymmetry associated with the D-state pairing. In this part, we will discuss some of the predicted properties of the new phase; in Paragraph IV.**2**˙1 we discuss the consequences of the asymmetry, in Paragraph IV.**2**˙2 we comment upon some of the thermodynamic properties of the phase, and in the final Paragraph we discuss the superfluid flow-properties.

2˙1. *Intrisic structure.* – The equation for the transition temperature (eq. (4.39), for example) is a linear equation, and we have seen that it leads to the highest transition temperature when the function $\mathscr{F}(\boldsymbol{k})$ has a $Y_{2m}(\theta, \varphi)$-dependence, where the angles (θ, φ) are measured with respect to an arbitrary axis henceforth called the « quantization axis ». Of course at the transition temperature $\mathscr{F}$ is normalized to zero.

Below T_c, $\mathscr{F}(\boldsymbol{k})$ will be a complicated function since eq. (4.38) is a non-linear equation. We shall return below to some detailed comment concerning $\mathscr{F}(\boldsymbol{k})$, but it is clear that it will in general depend upon the angles (θ, φ); and hence the energy of an excitation will depend upon the *vector* momentum of the excitation, and in particular upon the direction of the excitation with respect to the quantization axis. This anisotropic property of the liquid, in contrast to the original BCS theory which corresponds to spherically symmetric $\mathscr{F}(\boldsymbol{k})$, stems directly from the lack of invariance of the *truncated* Hamiltonian under a rotation. That the truncation is crucial is clear, since the original Hamiltonian is rotation-invariant, and the quasi-particle transformation canonical.

Despite the fact that the Hamiltonian is not rotation-invariant, it corresponds to a physical situation—namely one in which the arbitrary small interactions with the walls of the container [which are *not* included in the original rotationally invariant Hamiltonian] play a crucial role in the orientation of the quantization axis, just as they would in the formation of a crystal. In both cases one can ignore the walls, and work with a Hamiltonian which is not invariant under the rotation group. Thus, at the absolute zero, we expect the quantization axis to orient in some particular direction and the fluid will have a particle density which is uniform and isotropic, while the correlation function (related to $\mathscr{F}$, see eq. (4.35)) will correspond to short-range angle-dependent order. In other words, we have an *anisotropic liquid.*

For nonzero temperature the fluid will break up into zones, each zone being characterized by a quantization-axis direction and a particular mode of solution. We can see that this is so, for as $T \to T_c$ the solutions for $\mathscr{F}$ become more and more approximately linear, with the solutions corresponding to different m-values (in $Y_{2m}(\theta, \varphi)$) becoming degenerate. To determine the extent of the zone structure at any particular temperature, one must do statistical mechanics employing the relative energies of the various modes as well as the energy to be associated with a domain wall.

Associated with the correlation function's angular dependence is not only an anisotropy, but also the possibility of correlation currents (AM1). These currents cancel everywhere in the liquid, but not on the surface of the fluid. Thus at the absolute zero one may expect a surface-sheet current even in the lowest state of the system. The nature of this current, and whether or not it has an associated angular momentum, depends on the detailed nature of $\mathscr{F}(\boldsymbol{k})$. An order-of-magnitude estimate indicates that the effect is exceedingly small and probably beyond experimental observation.

To make these considerations more quantitative we need more information about the function $\mathscr{F}(\boldsymbol{k})$. Actually eq. (4.38) is so complicated, that $\mathscr{F}(\boldsymbol{k})$ will in general be of the form

$$\mathscr{F}(\boldsymbol{k}) = \sum_{l,m} \mathscr{F}_{lm}(y)\, Y_{lm}(\theta, \varphi) \,, \tag{4.68}$$

with all values of l and m coupled together. We know that as $T \to T_c$ the dominant terms in a solution will have $l=2$ with the various m values all leading to the same correlation energy. For $T < T_c$ ANDERSON and MOREL (AM2) have estimated that the $l \neq 2$ terms are very small so that an adequate approximation can be obtained at all temperatures by keeping only the $l=2$ terms in the expansion of $\mathscr{F}$. There is still the m quantum number to be considered, and they find that there are five independent solutions having (at $T=0$) correlation energies that only differ by a few percent. Thus at temperature somewhat below T_c, the liquid will break up into many zones, having different solutions and different directions of the quantization axis. ANDERSON and MOREL conclude that the lowest energy and hence the ground state, probably corresponds to a solution of the form

$$\mathscr{F}(\boldsymbol{k}) = \left[\frac{1}{\sqrt{2}}\, Y_{2,0} + \frac{1}{2}(Y_{2,2} - Y_{2,-2})\right] \mathscr{F}(k) \,, \tag{4.69}$$

but this conclusion is not rigorous.

Having approximate solutions, one can proceed to evaluate the various thermodynamic properties of the system, as has in fact been done by AM. However, those results depending upon the detailed nature of $\mathscr{F}(\boldsymbol{k})$ can not be considered firm, while those results which depend only upon the general properties of $\mathscr{F}(\boldsymbol{k})$ can be obtained by less detailed arguments.

In particular, we can see that in *general* $\mathscr{F}(\boldsymbol{k}_F)$ will vanish for some directions on the Fermi surface. This implies that the single-particle excitation spectrum $E(\boldsymbol{k})$ will have some directions for which there is no energy gap. These directions are dominant in determining the properties of the anisotropic superfluid both in regard to its thermodynamic behavior and its flow properties, and we can estimate most of the expected properties of the liquid using only this one general feature of $\mathscr{F}(\boldsymbol{k})$. The next two Paragraphs are devoted to this analysis.

2·2. *Thermodynamic behavior.* – The thermodynamic properties of the superfluid can be calculated from the distribution function for quasi-particles

$$f(\boldsymbol{k}) = \{\exp[\beta E(\boldsymbol{k})] + 1\}^{-1}, \tag{4.70}$$

and the excitation energy

$$E(\boldsymbol{k}) = [\varepsilon_{\boldsymbol{k}}^2 + |\mathscr{F}(\boldsymbol{k})|^2]^{\frac{1}{2}}. \tag{4.71}$$

For the $\varepsilon_{\boldsymbol{k}}$ we can use the observed single-particle energies for the normal liquid, since the transition to the superfluid phase should not modify these energies significantly.

The specific heat c may be obtained using eq. (1.13), with the result that

$$c = 2k\beta^2 \sum_{\boldsymbol{k}'} f(\boldsymbol{k}')[1 - f(\boldsymbol{k}')] \left| E^2(\boldsymbol{k}') + \beta E(\boldsymbol{k}') \frac{\mathrm{d}E(\boldsymbol{k}')}{\mathrm{d}\beta} \right|. \tag{4.72}$$

Evaluation of eq. (4.72) in the case that $\mathscr{F} = 0$ yields the specific heat of an ideal gas. Evaluating the discontinuity at the transition temperature yields

$$\frac{c_s - c_n}{c_n} = \frac{3}{\pi^2} \beta_c^3 \left. \overline{\mathscr{F}(\boldsymbol{k}) \frac{\mathrm{d}\mathscr{F}(\boldsymbol{k})}{\mathrm{d}\beta}} \right|_{|k| = k_F,\ \beta = \beta_c}. \tag{4.73}$$

where the bar indicates angular average. The general order of magnitude of the discontinuity is $c_s - c_n \approx c_n$ with detailed numbers depending upon the solution $\mathscr{F}(\boldsymbol{k})$.

For $T \to 0$, the specific heat varies linearly for a Fermi liquid, while it vanishes exponentially for a superconductor where the energy gap $|\mathscr{F}(\boldsymbol{k}_F)|^2$ is nonzero on the whole Fermi surface. We can consequently expect that the zeros in $\mathscr{F}(\boldsymbol{k}_F)$ will cause the specific heat to vanish *algebraically* as $T \to 0$.

The paramagnetic susceptibility for a Fermi liquid approaches a constant as $T \to 0$, while for a superconductor with nonzero gap it vanishes exponentially. Once again, we can expect that for liquid ^{3}He the susceptibility should vanish *algebraically* as $T \to 0$. It should not exhibit a discontinuity at the transition temperature.

Proceeding in this manner, one can calculate—or estimate—any desired property of the low-temperature phase. We do not pursue this subject further, but turn now to a discussion of the flow-properties that may be expected.

2·3. *Flow properties.* – The flow properties of the low-temperature phase have been calculated (GS1) employing the general methods of Landau, and Blatt, Butler, and Schafroth.

Consider the uniform flow of the fluid down an infinite channel. Let $\boldsymbol{v}$ be the velocity of the excitations, *i.e.* the velocity of the frame of reference in which the quasi-particle distribution function is that for a fluid at rest. If $\langle \boldsymbol{P} \rangle$ is the mean total momentum per unit volume, we can define the effective mass for flow as

$$M_f(0) = \left. \frac{\partial \langle \boldsymbol{P}(\boldsymbol{v}) \rangle}{\partial \boldsymbol{v}} \right|_{v=0}, \tag{4.74}$$

and take as a convenient definition of a superfluid, a system for which $M_f(0) < nm$ where n is the density and m the particle mass.

Now, by the general principle of statistical mechanics,

$$\langle \boldsymbol{P} \rangle = \frac{\mathrm{Tr}\,\{\boldsymbol{P} \exp[-\beta(H - \mu N - \boldsymbol{P}\cdot\boldsymbol{v})]\}}{\mathrm{Tr}\,\{\exp[-\beta(H - \mu N - \boldsymbol{P}\cdot\boldsymbol{v})]\}}. \tag{4.75}$$

Carrying out the differentiation, in eq. (4.74) and employing the fact that $\boldsymbol{P}$ commutes with $H - \mu N$ and that $\langle \boldsymbol{P} \rangle = 0$ for $\boldsymbol{v} = 0$, yields

$$M_f(0) = \frac{\beta}{v^2} \langle (\boldsymbol{P}\cdot\boldsymbol{v})^2 \rangle ; \tag{4.76}$$

where the brackets indicate an ensemble average in the rest frame $[\boldsymbol{v} = 0]$. Evaluating this expression is relatively easy since

$$\boldsymbol{P} = \sum_{\boldsymbol{k}} \boldsymbol{k}(\alpha^\dagger_{\boldsymbol{k}} \alpha_{\boldsymbol{k}} - \beta^\dagger_{\boldsymbol{k}} \beta_{\boldsymbol{k}}), \tag{4.77}$$

and the statistical average quickly reduces to the occupation numbers for quasi-particles. One finds

$$M_f(0) = 2\beta \sum_{\boldsymbol{k}} \left(\frac{\boldsymbol{k}\cdot\boldsymbol{v}}{v} \right)^2 f(\boldsymbol{k})[1 - f(\boldsymbol{k})], \tag{4.78}$$

in which form the dependence upon the nature of the excitation spectrum is clear.

Evaluating eq. (4.78) for an ideal gas yields $M_f(0) = nm$; while for a superconductor with an energy gap which is nonvanishing on the whole Fermi surface, $M_f(0)$ vanishes exponentially as $T \to 0$. For liquid ^{3}He we may expect $M_f(0)$ to vanish *algebraically* as $T \to 0$ and, in general, to depend upon the angle between the quantization axis and the direction of flow. Different $M_f(0)$, and hence different free energies, are in general to be associated with different modes of solution and different orientations of the axis. Thus allowing the liquid to flow should serve to reduce the zone structure in the fluid (*).

A similar analysis can be given for the moment of inertia, with the result that for the case in which the quantization axis and the rotation axis coincide,

$$\frac{I(0)}{I_0} = \beta \sum_{\boldsymbol{k}} (\boldsymbol{k} \times \hat{\boldsymbol{n}}) f^2(\boldsymbol{k})[1 - f(\boldsymbol{k})] \, , \tag{4.79}$$

where $\hat{\boldsymbol{n}}$ is a unit vector in the rotation direction and I_0 is the rigid body moment of inertia. For an ideal gas one finds $I(0) = I_0$, while for a normal superconductor $I(0)$ vanishes exponentially as $T \to 0$. For the anisotropic superfluid we expect an *algebraic* vanishing of the moment of inertia as $T \to 0$ (**).

V. – A Guide to the Literature and Bibliography.

The relevant literature is so extensive that a mere listing of references would be of limited value. An attempt has therefore been made to organize the papers, in such a way as to aid the worker who is first approaching the subject.

The list of references is not exhaustive, but should serve to quickly bring to the reader's attention any relevant references to the theory of liquid ^{3}He at low temperatures.

(*) An exceptional case is the mode of eq. (4.69) which has an effective mass for flow which is independent of the angle between the flow direction and the quantization direction.

(**) The computations of Paragraph IV.2·3 have been performed in the independent quasi-particle approximation, *i.e.* collective excitations have not been included in the energy spectrum in the evaluation of $M_f(0)$ or $I(0)$. Collective excitations have only been briefly studied for superfluid liquid ^{3}He, but it is expected that the modification they will cause in these results is small; *i.e.* the algebraic vanishing of the inertial parameters should be a correct description as $T \to 0$. Similarly, the general conclusions about the specific heat and susceptibility should not be significantly altered by the collective excitations.

1. – Introduction.

A general review article about liquid ^{3}He can serve as the first introduction to the subject:

(DA1) J. G. Daunt: *Science*, **131**, 579 (1960).
(BB1) N. Bernardes and D. F. Brewer: *Rev. Mod. Phys.*, **34**, 199 (1962).

One needs a firm understanding of the basic ideas of Statistical Mechanics, and this can be obtained readily from the first five chapters of:

(LL1) L. D. Landau and E. M. Lifshitz: *Statistical Physics* (London, 1958).

which book is extremely useful since the notation is the same as that in many of the following references and also because a brief summary of the theory of a Fermi liquid is given in Chapter VI.

2. – Phenomenological theory.

The phenomenological theory of a Fermi liquid can be approached either through one of the following review papers:

(AK1) A. A. Abrikosov and I. M. Khalatnikov: *Usp. Fiz. Nauk*, **56**, 177 (1958); trans. *Sov. Phys. Usp.*, **66** (1), 68 (1958), or *Rep. Progr. Phys.*, **32**, 329 (1959).
(BRO1) R. Brout: *Interacting Fermi Systems* (LASLN, 1960), unpublished.

(The first of which summarizes most of the results of the phenomenological work, and can serve as a convenient guide to the literature, while the second also serves as an introduction to the microscpic theory), or by studying the fundamental paper of Landau:

(LA1) L. D. Landau: *Žurn. Ėksp. Teor. Fiz.*, **30**, 1058 (1956) [trans. *Sov. Phys. JETP*, **3**, 920 (1956)].

Properties of a Fermi liquid are derived in the following references:

(KA2) I. M. Khalatnikov and A. A. Abrikosov: *Žurn. Ėksp. Teor. Fiz.*, **32**, 915 (1957) [transl. *Sov. Phys. JETP*, **5**, 745 (1957)].
(AK2) A. A. Abrikosov and I. M. Khalatnikov: *Žurn. Ėksp. Teor. Fiz.*, **32**, 1083 (1957) [trans. *Sov. Phys. JETP*, **5**, 887 (1957)].
(HO1) D. Hone: *Phys. Rev.*, **121**, 669 (1961) [see Errata *Phys. Rev.*, **121**, 1864 (1961)].
(BK1) I. L. Bekarevich and I. M. Khalatnikov: *Žurn. Ėksp. Teor. Fiz.*, **39**, 1699 (1960) [transl. *Sov. Phys. JETP*, **12**, 1187 (1961)].

The phenomenon of zero sound was first described by LANDAU in:

(LA2) L. D. LANDAU: *Žurn. Ėksp. Teor. Fiz.*, **32**, 59 (1957) [transl. *Sov. Phys.*, **5**, 101 (1957)].

while more detailed studies of the dispersion and absorption of zero sound and ordinary sound may be found in:

(KA3) I. M. KHALATNIKOV and A. A. ABRIKOSOV: *Žurn. Ėksp. Teor. Fiz.*, **33**, 110 (1957) [transl. *Sov. Phys. JETP*, **6**, 84 (1958)].

The possibility of observing zero sound through Rayleigh scattering is explored in:

(AK3) A. A. ABRIKOSOV and I. M. KHALATNIKOV: *Žurn. Ėksp. Teor. Fiz.*, **34**, 198 (1958) [transl. *Sov. Phys. JETP*, **7**, 135 (1958)].

A summary of the properties of ^{3}He is given in:

(HO2) D. HONE: *Phys. Rev.*, **125**, 1494 (1962).

3. – Microscopic theory.

3·1. *Introduction.* – The microscopic theory employs the potential effective between atoms of ^{3}He and this is derived and discussed in the following references:

(YS1) J. L. YNTEMA and W. G. SCHNEIDER: *Journ. Chem. Phys.*, **18**, 646 (1950).
(BO1) J. DE BOER: *Suppl. Physica*, **24**, 90 (1958).

The modern methods in the study of the many-body problem are rather elaborate and may seem somewhat formidable to one first approaching the field. THOULESS has recently published a text which develops the subject without assuming *a priori* knowledge of field theory. It is strongly recommended as an introduction to the microscopic theory:

(TH1) D. J. THOULESS: *The Quantum Mechanics of Many-Body Systems* (New York and London, 1961).

Also of considerable value are the two review articles:

(FW1) N. FUKUDA and Y. WADA: *Progr. Theor. Phys. Suppl.*, **5**, 61 (1960).
(KKN) T. KATO, T. KOBAYASHI and M. NAMIKI: *Progr. Theor. Phys. Suppl.*, **15**, 3 (1960), the first of which introduces the reader to many-body theory, while the second treats Green's functions.

The two sets of unpublished notes:

(MI1) R. L. MILLS: *An Approach to Statistical Thermodynamics Studies by the Methods of Quantum Field Theory* (1961), unpublished.
(BR1) G. E. BROWN: *Lecture Notes on the Many-Body Problem* (Copenhagen, 1961), unpublished),

as well as the notes by BROUT (BRO1) afford possible alternative introductions to the theory.

Articles in the following volume will be most helpful

(DE1) C. DEWITT, ed.: *The Many-Body Problem* (New York, 1959),

Although it is possible to understand many-body theory without a complete understanding of field theory, most readers will want to do at least some reading in a text like

(SC1) S. S. SCHWEBER: *An Introduction to Relativistic Quantum Field Theory* (Evanston, Ill., 1961).

3·2. *Perturbation theory.* – The basic idea of a «linked-cluster expansion» is given in:

(GO1) J. GOLDSTONE: *Proc. Roy. Soc.*, A **239**, 267 (1957).
(HU1) N. M. HUGENHOLTZ: *Physica*, **23**, 481 (1957).
(HU2) J. HUBBARD: *Proc. Roy. Soc. London*, A **240**, 539 (1957).

A number of independent formulations of perturbation theory in the many-body problem, and of quantum statistical mechanics have been given. Two independent formulations are:

(MW1) E. W. MONTROLL and J. C. WARD: *Phys. of Fluids*, **1**, 55 (1958).
(GHW1) A. E. GLASSGOLD, W. HECKROTT and K. M. WATSON: *Phys. Rev.*, **115**, 1374 (1959).

A formulation has been presented by the Saclay Group:

(BL1) C. BLOCH: *Nucl. Phys.*, **7**, 451 (1958).
(BD1) C. BLOCH and C. DE DOMINICIS: *Nucl. Phys.*, **7**, 459 (1958).
(BD2) C. BLOCH and C. DE DOMINICIS: *Nucl. Phys.*, **10**, 181 (1959).
(BD3) C. BLOCH and C. DE DOMINICIS: *Nucl. Phys.*, **10**, 509 (1959).
(BAD1) R. BALIAN and C. DE DOMINICIS: *Nucl. Phys.*, **16**, 502 (1960).
(BL2) C. BLOCH: *Suppl. Physica*, **26**, 562 (1960).
(DO1) C. DE DOMINICIS: *Suppl. Physica*, **26**, 94 (1960).

and also in the papers:

(KL1) W. KOHN and J. M. LUTTINGER: *Phys. Rev.*, **118**, 41 (1960).
(LW1) J. M. LUTTINGER and J. C. WARD: *Phys. Rev.*, **118**, 1417 (1960).

Independent work is described in:

(LY1) T. D. LEE and C. N. YANG: *Phys. Rev.*, **113**, 1165 (1959).
(LY2) T. D. LEE and C. N. YANG: *Phys. Rev.*, **117**, 22 (1960).

An important theorem is proved in the following reference:

(HH1) N. M. HUGENHOLTZ and L. VAN HOVE: *Physica*, **24**, 363 (1958).

3·3. *Applications.* – Calculations of the properties of a low density hard-sphere gas are given in:

(HY1) K. HUANG and C. N. YANG: *Phys. Rev.*, **105**, 767 (1957).
(LY3) T. D. LEE and C. N. YANG: *Phys. Rev.*, **105**, 1119 (1957).

A perturbative treatment of a low density gas with short range forces may be found in:

(AK4) A. A. ABRIKOSOV and I. M. KHALATNIKOV: *Žurn. Ėksp. Teor. Fiz.*, **33**, 1154 (1957) [transl. *Sov. Phys. JETP*, **6**, 888 (1958)].

Brueckner theory is described in the following brief review article:

(BR1) K. A. BRUECKNER: *Helium III*, edited by J. G. DAUNT (Columbus, Ohio, 1960), p. 70.

and also in the extensive rewiev papers by BRUECKNER in reference (DE1). The original papers are:

(BG1) K. A. BRUECKNER and J. L. GAMMEL: *Phys. Rev.*, **109**, 1023 (1958).
(BG2) K. A. BRUECKNER and J. L. GAMMEL: *Phys. Rev.*, **109**, 1040 (1958) [see Errata, *Phys. Rev.*, **121**, 1863 (1963)].

A treatment for liquid ^{3}He at non-zero temperatures, similar in spirit to Brueckner's approach, but based on the Lüttinger-Ward formalism (LW1) is presented in:

(MM1) R. E. MILLS and R. L. MILLS: *Helium Three*, edited by J. G. DAUNT (Columbus, Ohio, 1960), p. 78.

3·4. *Green's functions and spectral functions.* – Although the reader will want to study at least one of the formulations of Quantum Statistical Mechanics described in Section III.3, it is possible to start directly with the formulation in terms of Green's functions. The basic papers are by SCHWINGER:

(SC1) J. SCHWINGER: *Proc. Nat. Acad. Sci.*, **37**, 452 (1951).
(SC2) J. SCHWINGER: *Proc. Nat. Acad. Sci.*, **37**, 455 (1951).

while a review of the applications of the methods to statistical mechanics is given in:

(ZU1) D. N. ZUBAREV: *Usp. Fiz. Nauk*, **74**, 71 (1960) [transl. *Sov. Phys. Usp.*, **3**, 320 (1960)].

A comprehensive treatment is given in the recent text:

(KB1) L. P. KADANOFF and G. BAYM: *Quantum Statistical Mechanics* (New York, 1962)

One would be well advised to start with this paper, the notes by BROWN (BR1), the text by THOULESS (TH1), or the review paper (KKN). A general discussion is also given in

(BK1) V. L. BORCH-BRUEVICH and SH. M. KOGAN: *Ann. of Phys.*, **9**, 125 (1960).

The Russian school has developed the use of Green's functions for the ground state in the series of papers:

(MI1) A. B. MIGDAL: *Žurn. Ėksp. Teor. Fiz.*, **32**, 399 (1957) [transl. *Sov. Phys. JETP*, **5**, 333 (1957)].
(GM1) V. M. GALITSKII and A. B. MIGDAL: *Žurn. Ėksp. Teor. Fiz.*, **34**, 139 (1958) [transl. *Sov. Phys. JETP*, **7**, 96 (1958)].
(GA2) V. M. GALITSKII: *Žurn. Ėksp. Teor. Fiz.*, **34**, 151 (1958) [transl. *Sov. Phys. JETP*, **7**, 104 (1958)].
(LA3) L. D. LANDAU: *Žurn. Ėksp. Teor. Fiz.*, **35**, 97 (1958) [trans. *Sov. Phys. JETP*, **8**, 70 (1959)].

where the basic formulation will be found in the first two papers, and applications in the last two.

LANDAU introduces Green's functions for an ensemble and establishes an important dispersion relation in:

(LA4) L. D. LANDAU: *Žurn. Ėksp. Teor. Fiz.*, **34**, 262 (1958) [trans. *Sov. Phys. JETP*, **7**, 182 (1958)].

An independent (and complete) formulation appears in:

(MS1) P. C. MARTIN and J. SCHWINGER: *Phys. Rev.*, **115**, 1342 (1959).

while an alternative (but less convenient) formalism is presented in:

(MA1) T. MATSUBARA: *Progr. Theor. Phys. Kyoto*, **14**, 351 (1955).

KLEIN has given a self-contained discussion of Green's functions for the ground state, and for ensembles in the series of papers:

(KP1) A. KLEIN and R. PRANGE: *Phys. Rev.*, **112**, 994 (1958).
(KLE1) A. KLEIN: *Theory of Normal Fermion Systems*, Notes International Spring School of Physics (Naples, 1960), (New York, 1963).

(KLE2) A. KLEIN: *Phys. Rev.*, **121**, 950 (1961).
(KLE3) A. KLEIN: *Phys. Rev.*, **121**, 957 (1961).

which can be conveniently followed by:

(KA1) V. I. KARPMAN: *Žurn. Ėksp. Teor. Fiz.*, **39**, 185 (1960) [transl. *Sov. Phys., JETP*, **12**, 133 (1961)].

One may find instructive the following paper on correlation functions:

(FH1) S. FUJITA and R. HIROTA: *Phys. Rev.*, **118**, 6 (1960).

as well as many applications of Green's function to problems other than liquid ^{3}He, as is briefly described (with complete references) in the review paper of ZUBAREV (ZU1).

A discussion of quasi-particles from a microscopic point of view is given in:

(NL1) P. NOZIÈRES and J. M. LUTTINGER: *Phys. Rev.*, **127**, 1423 (1962).
(LN1) J. M. LUTTINGER and P. NOZIÈRES: *Phys. Rev.*, **127**, 1431 (1962).

3·5. *Collective excitations.* – Collective oscillations in an electron gas were first considered in the following papers:

(KS1) Y. KLIMONTOVITCH and V. P. SILIN: *Žurn. Ėksp. Teor. Fiz.*, **23**, 151 (1952).
(SI1) V. P. SILIN: *Žurn. Ėksp. Teor. Fiz.*, **23**, 641 (19 2).
(SI2) V. P. SILIN: *Žurn. Ėksp. Teor. Fiz.*, **27**, 269 (1954).
(SI3) V. P. SILIN: *Žurn. Ėksp. Teor. Fiz.*, **28**, 749 (1955) [transl. *Sov. Phys. JETP*, **1**, 607 (1955)].
(LI1) J. LINDHARD: *Kgl. Dan. Videnskab. Selskab. Mat.-Phys. Medd.*, **28**, no. 8 (1954).
(BP1) D. BOHM and D. PINES: *Phys. Rev.*, **92**, 608 (1953).
(FG-MP) M. FERENTZ, M. GELL-MANN and D. PINES: *Phys. Rev.*, **92**, 836 (1953).

where the methods employed are the «random phase approximation or the self-consistent field» approach. A field-theoretic formulation by SAWADA and others:

(SA1) K. SAWADA: *Phys. Rev.*, **106**, 372 (1957).
(SBFB) K. SAWADA, K. A. BRUECKNER, N. FUKUDA and R. BROUT: *Phys. Rev.*, **108**, 507 (1957).
(BR2) R. BROUT: *Phys. Rev.*, **108**, 515 (1957).
(WE1) G. WENTZEL: *Phys. Rev.*, **108**, 1593 (1957).
(HU3) J. HUBBARD: *Proc. Roy. Soc. London*, A **243**, 336 (1958).

stimulated the application of these methods to neutral systems:

(GHW2) A. E. GLASSGOLD, W. HECKROTTE and K. M. WATSON: *Ann. of Phys.*, **6**, 1 (1959).
(GL1) A. E. GLASSGOLD: UCLR 8889 (1959), unpublished.

The equivalence of all these methods is discussed in:

(EC1) H. EHRENREICH and M. COHEN: *Phys. Rev.*, **115**, 786 (1959).
(GG1) J. GOLDSTONE and K. GOTTFRIED: *Nuovo Cimento*, **13**, 849 (1959).

while Green's function formulations and refinements of the methods are discussed in:

(PI1) D. PINES: *Suppl. Physica*, **26**, 103 (1960).
(SC1) J. R. SCHRIEFFER: *Suppl. Physica*, **26**, 124 (1960).
(GA1) V. M. GALITSKII: *Suppl. Physica*, **26**, 174 (1960).

The notes of BROWN (BR1) and the text by THOULESS both give particularly comprehensive treatments of collective excitations.

4. – Superfluid phase.

The reader might conveniently start with the recent review-paper:

(SE1) A. M. SESSLER: *Helium III*, edited by J. G. DAUNT (Columbus, Ohio, 1960), p. 81.

The theory is based on the BCS theory of superconductivity which may be approached either through the notes:

(CS1) J. M. CORNWALL and A. M. SESSLER: UCRL 9318 (1960), unpublished.

or the following:

(BTS) N. N. BOGOLIUBOV, V. V. TOLMACHEV and D. V. SHIRKOV: *A New Method in the Theory of Superconductivity* (New York, 1959).
(BCS) BARDEEN, L. N. COOPER and J. R. SCHRIEFFER: *Phys. Rev.*, **108**, 1175 (1957).
(BE1) S. T. BELYAEV: *Mat. Fys. Medd. Dan. Vid. Selsk.*, **31**, no. 11 (1959).
(TH2) D. J. THOULESS: *Ann. of Phys.*, **10**, 553 (1960).

Applications of BCS theory to liquid ^{3}He is made in:

(CMS) L. N. COOPER, R. L. MILLS and A. M. SESSLER: *Phys. Rev.*, **114**, 1377 (1959).
(ES1) V. J. EMERY and A. M. SESSLER: *Phys. Rev.*, **119**, 43 (1960).
(BS1) A. BARDASIS and J. R. SCHRIEFFER: *Phys. Rev. Lett.*, **7**, 79 (1961). [see Errata *Phys. Rev. Lett.*, **7**, 472 (1961)].
(BSAM) K. A. BRUECKNER, T. SODA, P. W. ANDERSON and P. MOREL: *Phys. Rev.*, **118**, 1442 (1960).
(PI1) L. P. PITAEVSKII: *Žurn. Ėksp. Teor. Fiz.*, **37**, 1794 (1959) [transl. *Sov. Phys. JETP*, **37** (10), 1267 (1960)].
(GP1) L. P. GOR'KOV and L. P. PITAEVSKII: *Žurn. Ėksp. Teor. Fiz.*, **42**, 600 (1962) [transl. *Sov. Phys. JETP*, **15**, 417 (1962)].
(MN1) P. MOREL and P. NOZIÈRES: *Phys. Rev.*, **126**, 1909 (1962).

Properties of the highly correlated phase are discussed in:

(AM1) P. W. ANDERSON and P. MOREL: *Phys. Rev. Lett.*, **5**, 136 (1960).
(AN1) P. W. ANDERSON: *Suppl. Physica*, **26**, 137 (1960).
(GS1) A. E. GLASSGOLD and A. M. SESSLER: *Nuovo Cimento*, **19**, 723 (1961).
(NV1) L. H. NOSANOV and R. VASUDEVAN: *Phys. Rev. Lett.*, **6**, 1 (1961).
(SV1) T. SODA and R. VASUDEVAN: *Phys. Rev.* (to be published).
(AM2) P. V. ANDERSON and P. MOREL: *Phys. Rev.* (to be published).

REFERENCES

[1] A. C. ANDERSON, G. L. SALINGER, W. A. STEYERT and J. C. WHEATLEY: *Phys. Rev. Lett.*, **7**, 295 (1961).
[2] H. L. LAGUER, S. G. SYDORIAK and T. R. ROBERTS: *Phys. Rev.*, **113**, 417 (1959).
[3] W. M. FAIRBANK and G. K. WALTERS: *Proc. of the Symp. on Solid and Liquid* ^{3}He (Columbus, O., 1957); A. C. ANDERSON, H. R. HART jr. and J. C. WHEATLEY: *Phys. Rev. Lett.*, **5**, 133 (1960); A. C. ANDERSON, W. REESE, R. J. SARWINSKI and J. C. WHEATLEY: *Phys. Rev. Lett.*, **7**, 220 (1961).
[4] Y. PWU: private communication.
[5] K. N. ZINOV'EVA: *Žurn. Ėksp. Teor. Fiz.*, **34**, 609 (1958); W. R. ABEL, A. C. ANDERSON and J. C. WHEATLEY: *Phys. Rev. Lett.*, **7**, 299 (1961).
[6] A. C. ANDERSON, G. L. SALINGER and J. C. WHEATLEY: *Phys. Rev. Lett.*, **6**, 443 (1961).
[7] K. R. BRUECKNER, J. L. GAMMEL and J. T. KUBIS: *Phys. Rev.*, **118**, 1438 (1960).
[8] J. M. LUTTINGER: *Phys. Rev.*, **121**, 942 (1961).
[9] L. P. GOR'KOV: *Žurn. Ėksp. Teor. Fiz.*, **24**, 735 (1959) [transl. *Sov. Phys. JETP*, **7**, 505 (1958)].
[10] O. BETHEBER MATIBET and P. NOZIÈRES: *Compt. Rend.*, **252**, 3943 (1961).

SEMINARI

The Nature of the λ-Transition in Liquid Helium (*).

W. M. FAIRBANK

Stanford University - Stanford, Cal.

One of the subjects which will certainly be discussed at great length at this conference is the extent and nature of the long-range order in momentum space in liquid helium. Included will be a discussion of the validity of curl $v = 0$ for the superfluid ground state of He II and the companion requirement of quantized rotation. I would like to discuss in this seminar two experiments which I think relate indirectly to this problem, even though they do not involve measurements of the flow or rotational properties of helium. First is an experiment on the specific heat of liquid helium to within one-millionth of a degree of the λ-point made by KELLERS, BUCKINGHAM, and FAIRBANK at Duke University [1]. The second experiment is one completed shortly before this conference which demonstrated that magnetic flux is quantized in a superconducting cylinder [2, 3].

Ever since LONDON suggested that liquid helium represents in its superfluid phase a long-range order in momentum space, physicists have been puzzled by exactly what this means. LONDON, LANDAU, ONSAGER, and FEYNMANN have suggested that helium rotating slowly enough would rotate in one single quantized state of rotation regardless of the size of the bucket.

BLATT, BUTLER, and SCHAFROTH have taken another approach to the problem [4-7]. They question whether a real liquid, as contrasted with an ideal Bose gas can exhibit arbitrarily long-range order or correlation in momentum space. They suggest that there may be a finite correlation length of approximately 10^{-5} cm over which helium atoms are correlated in momentum space, but beyond which there is no correlation at all. In particular, they developed a theory of a Bose liquid in which helium atoms are correlated over a distance of about 10^{-5} cm, the correlated atoms representing the super-

(*) Work supported in part by grants from the National Science Foundation, the Office of Ordnance Research (U.S. Army), and the Linde Company.

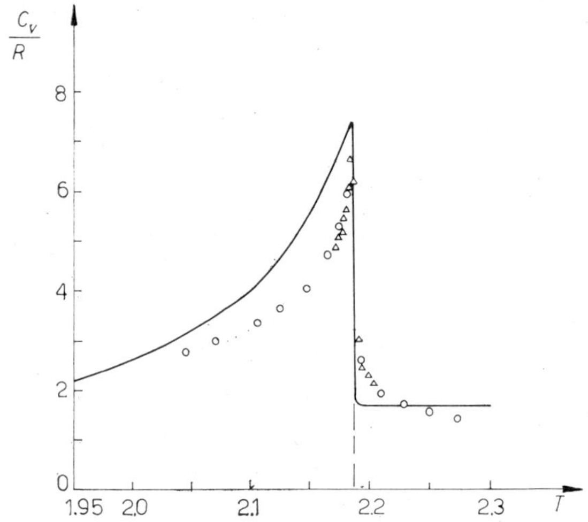

Fig. 1. – The specific heat of liquid helium *vs.* temperature in the neighborhood of the λ-point. The curve is theoretical, the experimental points are taken from KEESOM: *Helium* (Amsterdam, 1942), Table 4.20. The theoretical curve is a continuous and differentiable function of temperature everywhere, represented by the same theoretical expression throughout the temperature region covered in the figure [6].

fluid. By calculating the amount of this correlated phase ϱ, they calculate the ratio of uncorrelated to correlated phase (normal to superfluid, ϱ_n/ϱ_s) as a function of temperature, the correlation length being an adjustable parameter. With a correlation length of 10^{-5} cm, they were able to accomplish something which to my knowledge had not been accomplished before. They were able to derive ϱ_n/ϱ_s as a function of temperature in agreement with experiment up through the lambda point and with the same theory, to derive the specific heat at the lambda point in quite good agreement with experiment both as to shape, magnitude, and temperature of the transition. Their theoretical curves are shown along with the then existing experimental data in Figs. 1 and 2. One characteristic new feature of this specific heat transition is that it is a quasi-transition. Instead of being a second-order transi-

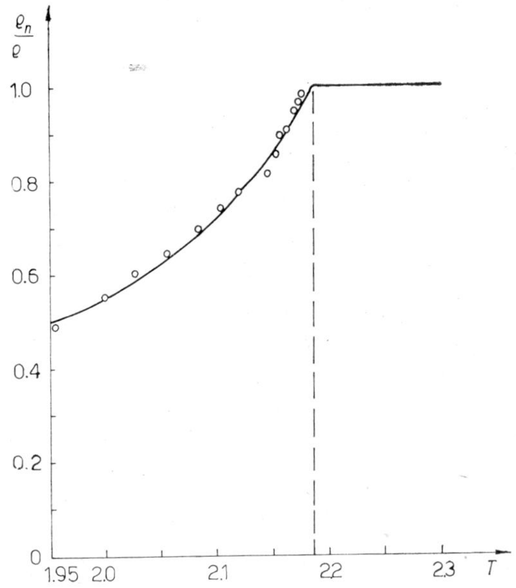

Fig. 2. – The normal fluid concentration ϱ_n/ϱ in liquid helium *vs.* temperature, in the neighbourhood of the λ-point. The curve is theoretical, the experimental points are based on the second sound data of Pellam. The theoretical curve is a continuous and differentiable function of temperature everywhere, represented by the same theoretical expression throughout the temperature region covered in the figure [7].

tion with a discontinuity in the specific heat at the lambda point, the specific heat is rounded at the lambda point for a temperature interval of approximately 10^{-3} degree. This rounding could not have been observed in the then existing data. It also follows from their theory that curl $v_s = 0$ is not an equilibrium property of superfluid helium II.

FEYNMANN [8] took issue with the premise that the correlation length in liquid helium could be cut off at some finite distance, and suggested that the lambda point in liquid helium is not only sharp at one-thousandth of a degree from the λ-point but it is sharp at a hundred-thousandth of a degree. He believed that even though the correlation in liquid helium II gets weaker and weaker as the distance gets bigger, it still remains finite. Thus, no matter how large the bucket, liquid helium would still rotate in a single quantum state if the rotation were sufficiently slow. This then served as the incentive for the experiment which we performed at Duke University.

These conflicting theoretical predictions served as an incentive for us to perform a specific heat measurement to at least 10^{-5} degree of the lambda point. This experiment was performed by BUCKINGHAM, KELLERS, and FAIRBANK, and has been reported in ref. [1]. Ultimately C_s was measured to within 10^{-6} degree of the lambda point, T_λ, with equal precision both above and below T_λ. I would like to recall here just a few highlights of the experiment and comment on the possible significance of the data.

To obtain such high temperature resolution, it is essential that the attainment of equilibrium be unaffected by the drastic change of thermal conductivity of liquid helium at the lambda point, or by the onset of the creeping film. Both of these requirements were met by permanently sealing the

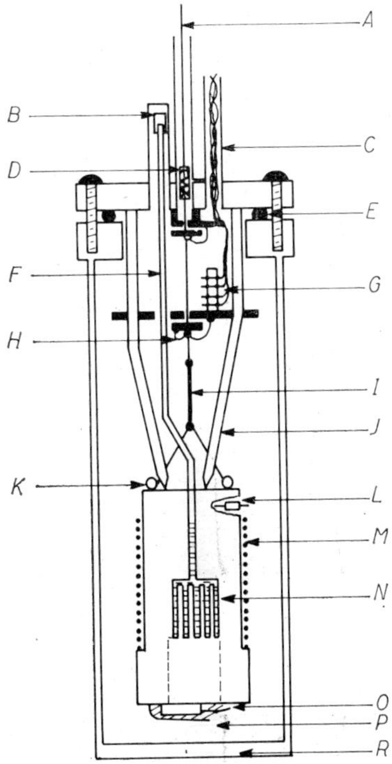

Fig. 3. – Schematic diagram of adiabatic chamber for specific heat measurements [1c]: *A*) stainless steel wire for closing heat switch; *B*) brass cap on filling capillary; *C*) wires connecting to heater and resistors on sample; *D*) cotton plug dyed with carbon black as radiation trap; *E*) indium « O » ring; *F*) filling capillary; *G*) Kovar seal used as thermal short for wires to sample; *H*) radiation shield and thermal short for heat switch; *I*) nylon cord; *J*) three prongs of heat switch (copper); *K*) indium coating and suspension for sample; *L*) temperature sensitive resistor; *M*) heater; *N*) sample cavity; *O*) temperature-sensitive resistor; *P*) copper shield over resistor; *R*) calorimeter wall.

helium (0.0587 g) in a copper container (200 g), the inside of which was in the form of fins so placed that the helium was everywhere within 0.003 in. of a copper surface (Fig. 3). With a heat input of 10 erg/s temperature e-quilibrium of greater than 10^{-6} degree could be obtained even in the helium I region. In order to eliminate the need for removing exchange gas, the sealed container was suspended in a vacuum and a mechanical heat switch provided for contact wiht the bath when required.

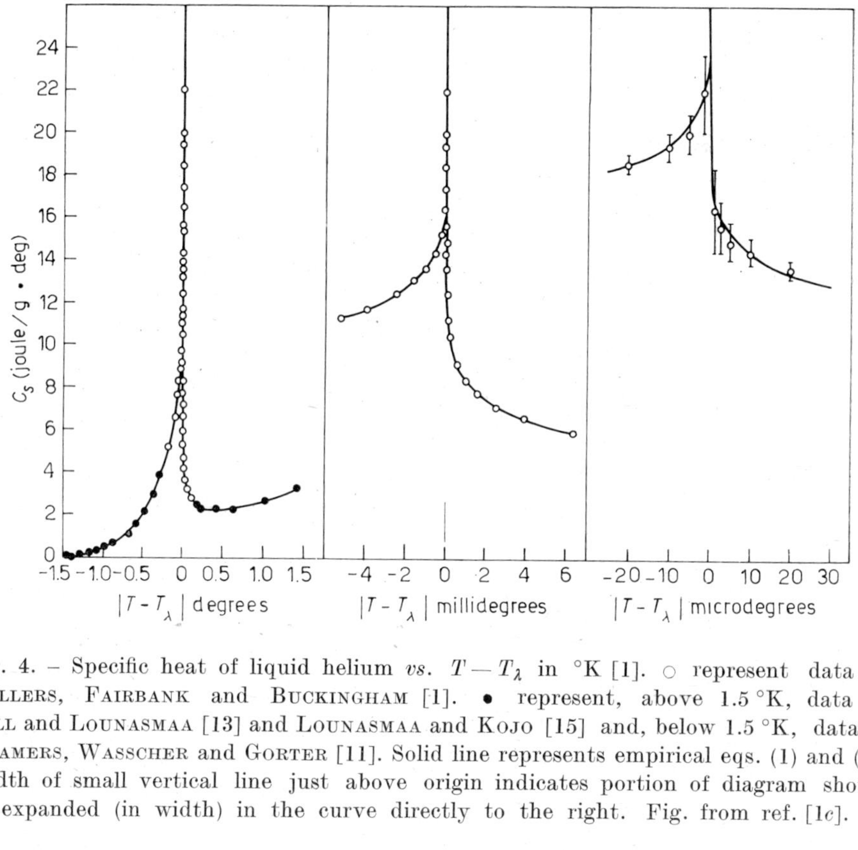

Fig. 4. – Specific heat of liquid helium *vs.* $T - T_\lambda$ in °K [1]. ○ represent data of KELLERS, FAIRBANK and BUCKINGHAM [1]. • represent, above 1.5 °K, data of HILL and LOUNASMAA [13] and LOUNASMAA and KOJO [15] and, below 1.5 °K, data of KRAMERS, WASSCHER and GORTER [11]. Solid line represents empirical eqs. (1) and (2). Width of small vertical line just above origin indicates portion of diagram shown expanded (in width) in the curve directly to the right. Fig. from ref. [1c].

Measurements were made by means of a carbon resistor with a minimal detectable change representing $2 \cdot 10^{-7}$ degree. It was possible to make measurements both while increasing and decreasing the temperature.

Figure 4 is a plot of the experimental results. The specific heat, C_s, is plotted as ordinant versus $T - T_\lambda$. In order to show the nature of the transition very near the lambda point, the data are shown on successively expanded temperature scales. To aid in a visualization of the very large amount of expansion of each successive curve, a small vertical line has been drawn just above the origin. The width of the line indicating the fraction of the curve which is shown expanded in the curve directly to the right.

It can be seen that as the specific heat is displayed on a more and more expanded scale, it maintains the same geometric shape. There is certainly no indication of a rounded quasi-transition as predicted by the theory of Blatt, Butler, and Schafroth.

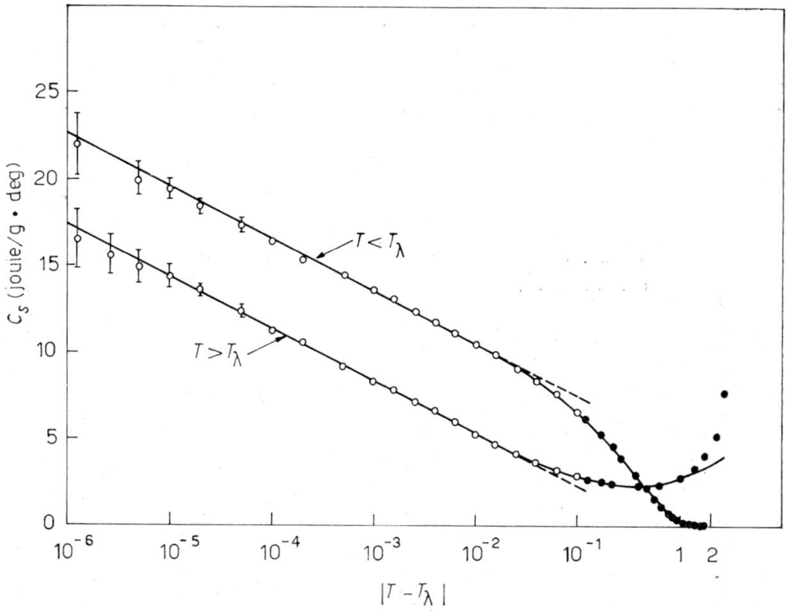

Fig. 5. – The specific heat of liquid helium *vs.* log $|T-T_\lambda|$ [1]. ○ represent data of KELLERS, FAIRBANK and BUCKINGHAM [1]. • represent, above 1.5 °K, data of HILL and LOUNASMAA [13] and LOUNASMAA and KOJO [15] and, below 1.5 °K, data of KRAMERS, WASSCHER and GORTER [11]. Solid line represents empirical eqs. (1) and (2). Fig. from ref. [1*c*].

In Fig. 5, C_s has been plotted as ordinant against $T-T_\lambda$ in degrees Kelvin on a logarithmic scale. It is seen in this kind of a plot that near the lambda point on each side there is a factor of 10^4 in $T-T_\lambda$ over which the data fall in two parallel straight lines which are branches of the expression

$$C_s = 4.55 - 3.00 \log_{10} |T-T_\lambda| - 5.20\,\Delta\,, \tag{1}$$

where $\Delta = 0$ for $T < T_\lambda$, and $\Delta = 1$ for $T > T_\lambda$.

By combining our data near the lambda point with that published by other workers outside this range [9-15] it is possible to obtain an empirical function that fits the specific heat data to within five percent from the lowest temperature at which measurements have been made up to 3.6 °K. The expression is as follows:

$$(2)\quad \begin{cases} C_s = C_D(T) + (130 - 90.9 \log_{10} |T - T_\lambda|) \exp[-7.40/T], & T < T_\lambda \\ C_s = C_D(T) + (23.5 - 16.4 \log_{10} |T - T_\lambda|) \exp[-3.70/T], & T > T_\lambda \end{cases}$$

where C_D represents the Debye function, evaluated at each temperature with the appropriate value of density and velocity of sound.

It was found empirically that the best simple fit to the date both above and below the lambda point could be obtained by inserting, as a factor in eq. (1), a simple exponential term after first subtracting out a term representing the theoretical Debye specific heat. The constants in the exponential term that gave the best fit to the data were found to be 7.40 °K below T_λ and 3.70 °K above the lambda point. When these had been determined, the other constants shown in the equation above all followed directly from eq. (1) setting $T = T_\lambda$ in the exponential term. Thus all of the constants except the « energy gap » in the exponential term were obtained from the data taken within 10^{-2} degree of the lambda point.

It is intereasting that the presence of the logarithmic term leads to a relatively simple empirical expression over the whole temperature range removing the difficulty found in the past of finding such an expression. We do not necessarily imply any particular theoretical significance to the actual expression.

In summary, the specific heat data near the lambda point can be represented by a logarithmic singularity. With this term, which can be determined from the data with 10^{-2} degree of the lambda point, it is possible by multiplying by a simple exponential term with a temperature-independent gap, different on the two sides of the lambda line, to fit the data over the entire range of temperature from the lowest temperature up to 3.6 degrees, due allowance being made for the Debye phonon specific heat. Although this does not necessarily represent the most useful theoretical equation, it points up the need for the logarithmic term in explaining the data. It is interesting to note that the constant difference, Δ, between the two straight lines representing the specific heat above and below the lambda point (Fig. 3) is equal to $\frac{5}{2}R$.

We see from this experimental data that the lambda point, instead of being a rounded quasi-transition as suggested by the theory of Blatt, Butler, and Schaforth assuming finite correlation length, is in fact sharp to at least two orders of magnitude closer in temperature to the lambda-point than predicted in their theory. Furthermore, a plot of the specific heat *vs.* $T - T_\lambda$ indicates a logarithmic dependence which extrapolates to a logarithmic singularity at the lambda point. It is interesting to note that this logarithmic infinity in the specific heat is also characteristic of Onsager's exact solution of the two-dimensional Ising model of cooperative transition [16]. The statistical theory of interacting particles which form a system capable of undergoing a cooperative transition is so difficult that no approximation method has yet been devised which is valid in the neighborhood of the transition and thus capable of yielding the direct nature of the singularity for a

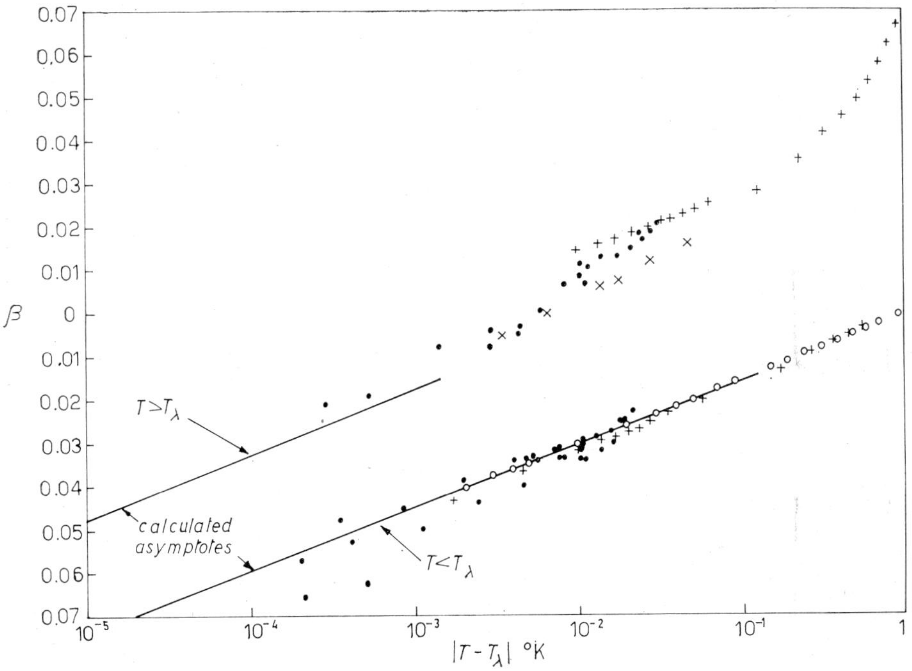

Fig. 6. – Coefficient of expansion *vs.* $|T - T_\lambda|$. The two asymptotic straight lines are determined as discussed in the text using the specific heat data of ref. [1]. Figure from ref. [1*c*]. ○ Atkins and Edwards, + Edwards, • Chase and Maxwell, × Kerr and Taylor.

three-dimensional system. It is a striking fact that the nature of the singularity in Onsager's exact two-dimensional solution is of just the same form as that observed in the liquid helium transition.

The suggestion that the specific heat of helium might become infinite at the transition was first made by Tisza [17]. Also, it was suggested by Atkins and Edwards [18] that a logarithmic term could be used to describe the results of their measures of thermal expansion coefficient below the lambda point.

Buckingham [1*c*] has derived rigorously the thermodynamic consequences of lambda transitions characterized by the absence of a latent heat, but at which the specific heat at constant pressure becomes infinite. Pippard [19] had previously considered such a transition and worked out thermodynamic relationships based on the assumption that the entropy surface is cylindrical near the lambda line. The thermodynamic relationships worked out by BUCKINGHAM and PIPPARD can be used to compare the behavior of various thermodynamic properties in the neighborhood of the lambda line [1c]. Figure 6 is a curve showing the experimental data near the lambda line on the coefficient of expansion [18, 20 -22]. The straight lines are the asymptotic theoretical curves [1c] assuming the results of the specific heat experiment [1], recent experimental values for the slope of the lambda line [23] of 130 atm/degree, and $(\partial S/\partial T)_\lambda$ from data of LOUNASMAA and KOJO [15] and LOUNASMAA and KAUNISTO [24].

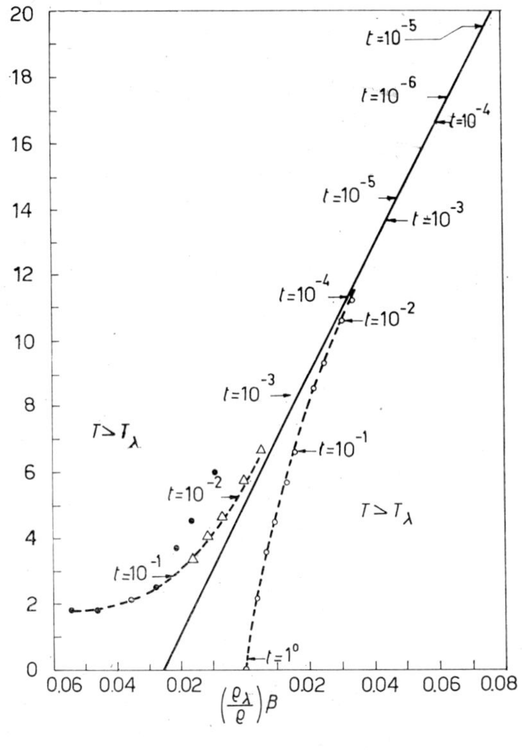

Fig. 7. – Parametric plot of $(T_\lambda/T)\,C_p$ *vs.* $(\varrho_\lambda/\varrho)\,\beta$. Solid straight line represents asymptotic value of C_p *vs.* β calculated from eq. (3) as discussed in the text. Figure from ref. [1c]. ○ C_p *vs.* β, ATKINS and EDWARDS, • C_p *vs.* β, EDWARDS, △ C_p *vs.* β, KERR and TAYLOR.

The relationship between the specific heat and the expansion coefficient can be written in the following convenient form [1]:

$$(3)\qquad \left(\frac{T_\lambda}{T}\right) C_p = T_\lambda \left(\frac{\partial S}{\partial T}\right)_t + \frac{P'_\lambda T_\lambda}{\varrho_\lambda}\left(\frac{\varrho_\lambda}{\varrho}\right)\beta\,,$$

where ϱ represents the density, $P'_\lambda = (\partial \varrho/\partial T)_\lambda$, $\beta = 1/V(\partial V/\partial T)_P$ and $t = T - T_\lambda$. It is convenient to compare the specific heat and coefficient of expansion by means of a parametric plot. Figure 7 is such a parametric plot. The asymptotic straight line is given by the above equation with the slope and intercept determined by the experimental data mentioned in the paragraph above. It is seen that the experimental results on the expansion coefficient and specific heat are both consistent with an infinite logarithmic singularity at the lambda point.

We have seen that measurements very near the lambda point suggest that the specific heat and expension coefficient become infinite at the transition. Of course, actual measurements could never prove an infinite value, particularly in view of the logarithmic behavior found. To obtain a specific heat value, fifty percent higher than the highest obtained would require a temperature resolution better than 10^{-8} degree. Still the lambda point is sufficiently sharp experimentally that BLATT, BUTLER, and SCHAFROTH have been unable to reconcile their theory based on a finite correlation length of about 10^{-5} cm with the experimental results [25].

It is interesting to consider the transition in superconductivity. This is the only transition which shows no lambda curve, but perhaps has a direct

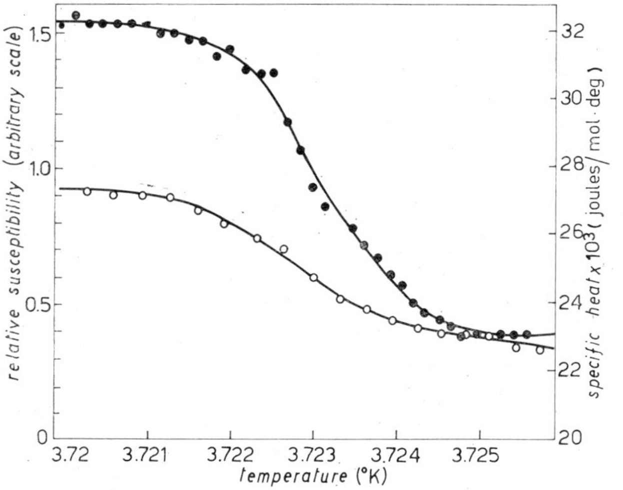

Fig. 8. – The specific heat and relative susceptibility of tin as a function of temperature very close to the transition [26, 1c]. ○ Susceptibility, • specific heat.

second order transition. In ref. [1c], it is suggested that there may be a lambda transition in superconductors if one could get a pure enough sample to look close enough to transition to avoid the broadening effects of impurities, strains, and frozen-in flux. Since the density of electrons that effect the transition is roughly the density of electrons near the Fermi surface, we might expect this lambda transition to be less than a thousandth of a degree wide.

In an attempt to test this suggestion, measurements on a sample of tin were made [26] with the same apparatus which was employed for the high resolution experiments on liquid helium. The temperature resolution was actually more than adequate, the transition being spread over a millidegree. Figure 8 shows the observed specific heat and change of susceptibility for a few millidegrees each side of the transition. There is no evidence for an anomaly in the behavior of either quantity, but we would in any case only expect it to show up in a rather more sharply defined transition. Attempts to find a specimen satisfying this exacting condition are being made.

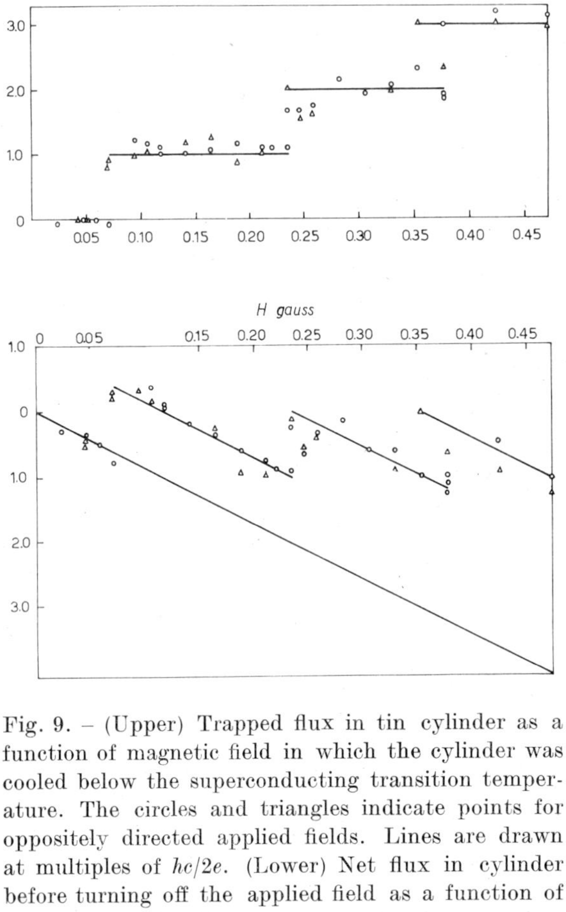

Fig. 9. – (Upper) Trapped flux in tin cylinder as a function of magnetic field in which the cylinder was cooled below the superconducting transition temperature. The circles and triangles indicate points for oppositely directed applied fields. Lines are drawn at multiples of $hc/2e$. (Lower) Net flux in cylinder before turning off the applied field as a function of the applied field. The circles and triangles are points for oppositely directed applied fields. The lower line is the diamagnetic calibration to which all runs have been normalized. The other lines are translated vertically by successive steps of $hc/2e$. Figure from ref. [2].

Finally in this talk I would like to discuss an experiment we have jus- completed at Stanford University [2] on superconductors which is closely related to predicted properties of liquid helium. A similar experiment has also just been reported by DOLL and NÄBAUER [3]. We have observed that flux trapped in hollow superconducting cylinders is quantized. That such an effect might occur was originally suggested by LONDON and ONSAGER, the predicted unit being hc/e. The quantized unit we find experimentally is not hc/e, but $hc/2e$, within experimental error. Similar experimental results have been found by DOLL and NÄBAUER.

The experimental results on a small tin cylinder in the Deaver-Fairbank experiment are shown in Fig. 9. In the lower half of the diagram is shown the net(*) flux within the cylinder when cooled down in the presence of a magnetic field H. In the upper half of the figure is shown the flux trapped after the magnetic field is turned off. It is seen that both before and after the magnetic field is turned off, the flux inside the superconducting cylinder is quantized in units of $hc/2e$.

(*) By net flux we mean the change in the flux passing through the total crosssectional area of the cylinder (walls plus hole) on cooling through the superconducting transition.

The factor of two was predicted to us in a private communication by ONSAGER and seems to indicate that electrons in a superconductor are paired as would be expected from either the Bose-Einstein condensation theory or the Bardeen-Cooper-Schrieffer theory. This experiment has been discussed more completely in [2]. In the same issue will appear a discussion of the experiment of DOLL and NÄBAUER [3] and the theory of ONSAGER [27] based on a Bose model, and BYERS and YANG [28] based on the BCS theory.

In conclusion, it appears to me that the lambda point experiments lend circumstantial evidence to the suggestion that there is a very-long-range order in liquid helium below the lambda point. The experiments on quantized flux suggest by analogy the reasonableness of such an assumption concerning rotating liquid helium. Of course, an experimental proof will have to come from actual experiments on quantized rotation in liquid helium. A long step in this direction has certainly been made by the experiments of HALL and VINEN [29], particularly the vibrating wire experiment of Vinen.

LONDON [30] has suggested that an experiment could be performed which would show that irrotational motion and quantized rotation are equilibrium states for superfluid helium. This experiment would involve cooling a bucket of liquid helium while rotating through the lambda point. If the superfluid were required to be in an irrotatioual state of rotation, then the angular momentum of the liquid helium would have to change and the bucket, if suspended freely without any angular momentum contacts with the outside world either have to gain or lose the angular-momentum lost or gained by the liquid helium. In particular, if the bucket were rotating slowly enough, the perfectly frictionless liquid helium would stop rotating and the bucket would speed up. We are trying to perform such an experiment in our laboratory.

REFERENCES

[1] *a*) W. M. FAIRBANK, M. J. BUCKINGHAM and C. F. KELLERS: *Proc. 5-th Int. Conf. Low Temp. Phys.* (Madison, Wis., 1957), p. 50; *b*) C. F. KELLERS: *Thesis* (Duke, University, 1960); *c*) M. J. BUCKINGHAM and W. M. FAIRBANK: *Progress in Low Temperature Physics*, vol. **3**, edited by C. J. GORTER (Amsterdam, 1961), chap. III.
[2] B. S. DEAVER and W. M. FAIRBANK: *Phys. Rev. Lett.*, **7**, 43 (1961).
[3] R. DOLL and M. NÄBAUER: *Phys. Rev. Lett.*, **7**, 51 (1961).
[4] J. M. BLATT, S. T. BUTLER and M. R. SCHAFROTH: *Phys. Rev.*, **100**, 481 (1955).
[5] S. T. BUTLER and J. M. BLATT: *Phys. Rev.*, **100**, 495 (1955).
[6] S. T. BUTLER, J. M. BLATT and M. R. SCHAFROTH: *Nuovo Cimento*, **4**, 674 (1956).
[7] J. M. BLATT, S. T. BUTLER and M. R. SCHAFROTH: *Nuovo Cimento*, **4**, 676 (1956).
[8] R. P. FEYNMAN: private communication.

[9] W. H. Keesom and K. Clusius: *Proc. Kon. Acad. Amsterdam*, **35**, 307 (1932).
[10] W. H. Keesom and A. P. Keesom: *Physica*, **1**, 128 (1933-34).
[11] H. C. Kramers, J. D. Wasscher and C. J. Gorter: *Physica*, **18**, 329 (1952).
[12] G. Hercus and J. Wilks: *Phil. Mag.*, **45**, 1163 (1954).
[13] R. W. Hill and O. V. Lounasmaa: *Phil. Mag.*, **2**, 145 (1957).
[14] A. H. Markham, D. C. Pearce, R. G. Netzel and J. R. Dillinger: *Proc. Int. Conf. Low Temp. Phys.* (Madison, Wis., 1957), p. 45.
[15] O. V. Lounasmaa and E. Kojo: *Physica*, **36**, 3 (1959).
[16] L. Onsager: *Phys. Rev.*, **65**, 117 (1944); G. G. Newell and E. W. Montroll: *Rev. Mod. Phys.*, **25**, 353 (1953).
[17] L. Tisza: *Phase Transformations in Solids*, edited by Smolychowski, Mayer and Weyl (New York, 1951).
[18] K. R. Atkins and M. H. Edwards: *Phys. Rev.*, **97**, 1429 (1955).
[19] A. B. Pippard: *The Elements of Classical Thermodynamics* (Cambridge, 1957), chap. IX.
[20] M. H. Edwards: *Canad. Journ. Phys.*, **36**, 884 (1958).
[21] E. Maxwell, C. E. Chase and W. E. Millet: *Proc. 5-th Int. Conf. Low Temp. Phys.* (Madison, Wis., 1957), p. 53.
[22] E. C. Kerr and R. Dean Taylor: private communication.
[23] W. E. Keller and E. F. Hammel jr.: *Ann. of Phys.*, **10**, 202 (1960); E. Maxwell and C. E. Chase: *Physica*, **24**, 5139 (1958) and private communication.
[24] O. V. Lounasmaa and L. Kaunisto: *Ann. Acad. Sci. Fennicae*, A **6**, no. 59 (1960).
[25] J. M. Blatt: private communication.
[26] C. C. F. Kellers and W. M. Fairbank: unpublished.
[27] L. Onsager: *Phys. Rev. Lett.*, **7**, 50 (1961).
[28] N. Byers and C. N. Yang: *Phys. Rev. Lett.*, **7**, 46 (1961).
[29] W. F. Vinen: *Progress in Low Temperature Physics*, vol. **3**, edited by C. J. Gorter (Amsterdam, 1961), chap. I.
[30] H. London: *Phys. Soc. 1946 Camb. Conference Report* (London, 1947) p. 48.

Vorticity in the Helium Film.

J. F. Allen

St. Andrews University - St. Andrews

I should like to discuss six problems which are mainly concerned with the helium II film:

1) critical velocities;
2) the motion of the film;
3) changes in the film transfer rate;
4) the effect of mechanical stirring of the liquid;
5) damping of isothermal film oscillations;
6) critical angle of surface tension in helium II.

1. – Critical velocities.

The problem of critical velocities and in particular the critical velocity of the mobile film has remained unsolved since the first detailed study of film transfer was made by Daunt and Mendelssohn [1]. They found that the rate of transfer was nearly independent of the length of film path and that it varied with temperature from zero at the λ-point to a roughly constant value of $7.5\cdot10^{-5}$ cm^3 s^{-1} per cm width of film path below 1.2 °K.

Consideration of the two-fluid theory would lead one to suppose that, in the film, the normal component remains stationary or nearly so on the substrate owing to viscosity and the thinness of the film, while the superfluid component of the film moves with a critical velocity which is more or less independent of temperature.

It was assumed that the liquid in the film could be accelerated with no friction, or at any rate with extremely small friction, up to its critical velocity above which friction rose steeply to some very high value so that the critical velocity could not be exceeded. This and the appearance of what have been called critical velocities in the bulk liquid have been extensively treated by Atkins [2].

The mass of experimental observation is confused and it may be more realistic to speak of characteristic velocities which are not necessarily much more precise than order of magnitude, which are observed in particular geometries under certain conditions, and to restrict the term « critical velocity » to experiments in which either a persistent velocity or a slip velocity is observed. Persistent or slip velocities can be inferred from the experiments of REPPY and LANE [3], VINEN [4] and FAIRBANK [5]. In these cases the persistent or slip velocity was between 0.1 mm s^{-1} and 1 mm s^{-1}. Slip velocities as great as this have been mentioned by LIN in his lectures.

Characteristic velocities which involve minimal friction may possibly be associated with the generation of vortex lines as FEYNMAN [6] has suggested [$v_c \sim \hbar/md \ln d/a$] or with the motion of such lines. I shall endeavour to show that the helium film can be considered to be an example of this type of motion.

The very high critical velocities associated with the generation of rotons or phonons need not concern us here.

2. – The motion of the helium film.

The first few layers of helium atoms near the substrate will be bound fairly strongly to the substrate and so will probably not take part in the motion of the film. The first moving layer might be expected to move sluggishly since the forces holding it to the wall are still considerable. We can, however, imagine a slip velocity for this layer of say, 0.1 mm or 1 mm s^{-1}, since this would accord with observed persistent velocities, and it would be in keeping with ordinary notions of kinetic theory. We are thus faced with three possibilities; either the film, with its average velocity of about 20 cm s^{-1}, possesses a river-type flow with a steep velocity gradient at the bottom and zero gradient at the surface, or it must have a slip velocity equal to its average velocity, or there must be some sort of ordered vorticity present. River-type motion can be ruled out as exhibiting rotational flow. High slip velocity is unlikely since it is not observed elsewhere in helium II, and would require a high degree of structural rigidity in the film, which is implausible. There remains the vorticity case.

It is possible to conceive of a vortex sheet lying between the substrate and the film surface, but GINSBURG [7] has shown that a vortex sheet is energetically costly. As FEYNMAN has suggested, it is possible to substitute for the vortex sheet an array of vortex lines, and ALLEN [8] and KUPER [9] have proposed that such an array of vortex lines may provide the mechanism of movement of the helium film. If one imagines that this plane array of quantized vortex lines, as in Fig. 1a, moves with something like the average film velocity, then it is possible to satisfy the condition that the local velocity at the substrate shall be on the average nearly zero. KUPER [10] has pointed out, however, that such an array is unstable and that the simple picture is therefore not adequate. He has suggested instead that one might start with

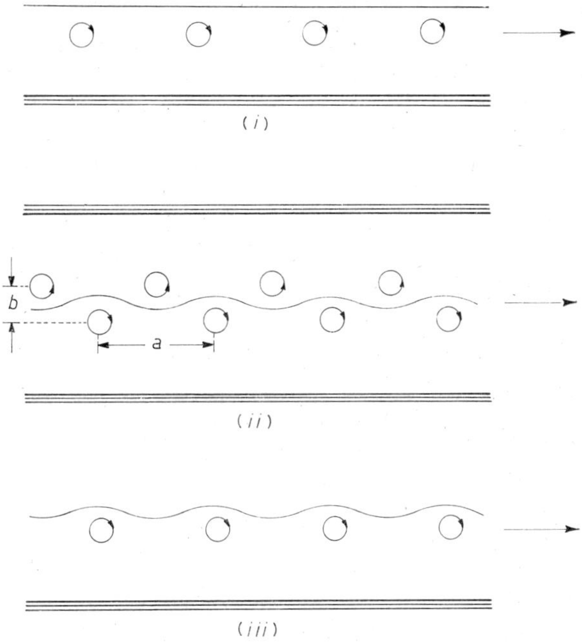

Fig. 1. – Possible structure and profile of the film. *i*) Early Allen-Kuper model. *ii*) Karman street in a parallel channel showing streamline surface. *iii*) Later Kuper model.

a parallel-sided channel in which there exists a Karman street of vortices consisting of two rows of vortices of opposite sign as in Fig. 1*b*. ROSENHEAD [11] has shown that such an arrangement is stable for a range of ratios of the parameters b/a. If one now imagines the flow channel split along the central streamline surface and one half of the channel then removed, the remaining half, containing one row of vortices and a corrugated free surface might constitute a stable situation. Thus the picture of the moving helium film might be as in Fig. 1*c*.

3. – Changes in the film transfer rate.

DAUNT and MENDELSSOHN established that the normal critical rate of transfer was observed in wide beakers with very clean smooth surfaces when the level of the bulk liquid was at least 1 cm below the rim of the beaker and was not close to any change in the width of the perimeter of the beaker. If the surface was contaminated or rough or if the bulk liquid was within a few mm of the rim or within a few mm of a change in perimeter, then a higher than normal transfer rate was always observed.

ATKINS [12] has further observed a considerable variation in the rate of transfer with length of path of the film. The observations were made with fairly narrow beakers of 1 mm to 3 mm diameter. ESELSON and LASEREW [13] reported the dependence of the rate of transfer out of a narrow beaker on the « condition » of the liquid in the beaker. If the beaker had been filled by plunging it empty below the surface of the bath, then the subsequent rate of transfer out of the vessel was faster than if it has been filled initially by film transfer only. I have observed the same thing. An example of it can be seen in Fig. 2. Here the first emptying run of the sequence, marked P,

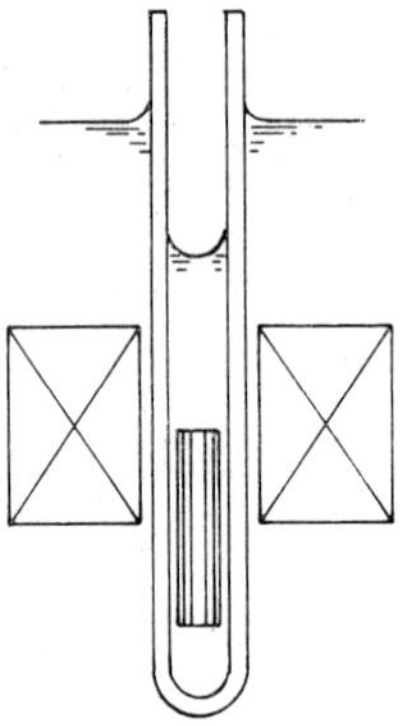

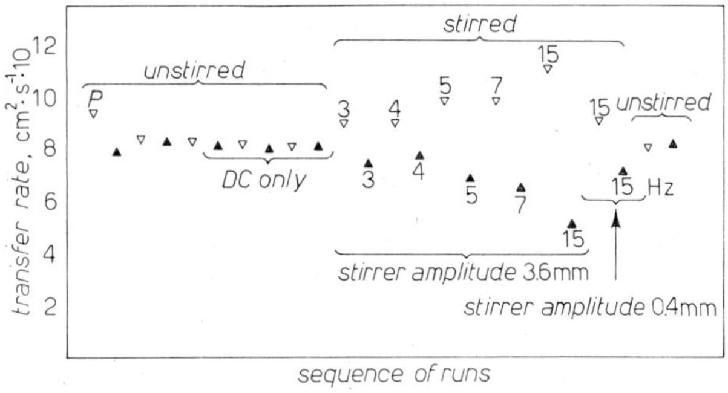

Fig. 2. – Effect of stirring on transfer rate. The inverted open triangles refer to emptying; the solid erect triangles refer to filling. The points marked « DC only » were taken with d.c. only, and no a.c. flowing through the solenoid. « P » indicates plunge-fill. The numerals adjacent to the points refer to the number of up and down cycles per second of the stirrer. Inset shows stirring beaker, stirrer and actuating solenoid.

shows an enhanced transfer rate out of the beaker after a plunge-fill. All subsequent fillings were by film transfer only.

In my experiments with narrow beakers, the rate of transfer out of the beaker is enhanced (*a*) after plunge-filling, and (*b*) after filling by film transfer when the rim of the vessel is close to the bath level. Condition (*b*) is further enhanced if the rim of the beaker is ground with fine emery powder.

We have imagined that normal film transfer occurs because of the presence in the film of an array of vortex lines, and that there is always present in the bulk liquid a sufficient amount of vorticity to supply the needs of any film transfer. Vinen has shown that the creation of a new length of vortex line in the bulk liquid is expensive, and it is most unlikely that the film can make its own array of vortex lines as it goes along. Where the surface is rough, however, as by grinding, then the asperities might well act as spinnerets to spin out vortex lines indefinitely. If there is an excess of vorticity produced in this way or produced by bulk liquid tumbling into an empty beaker then the film transfer rate can be enhanced. Furthermore, where the rim is not far above the bath the film is much thicker so that the pattern of vortex lines may well be more complex and crowded, thus permitting the higher rate of transfer which is observed under such conditions.

4. – The effect of mechanical stirring of the liquid.

Since it seemed likely that, as stated above, excess turbulence in the bulk liquid in the beaker enhanced the transfer rate out of the beaker, the next step was clearly to attempt to create turbulence by stirring the liquid in the beaker. Stirring was successfully achieved by the apparatus shown in Fig. 2. In the beaker there was a roughly cylindrical stirrer of Mullard ferroxcube which was actuated by a solenoid outside the beaker. The actuating current was a steady d.c. on which was superimposed an a.c. whose frequency and amplitude could be altered. This gave a moderately smooth action by the stirrer.

The results are also shown in Fig. 2. It can be seen that stirring the liquid in the beaker not only enhances the transfer out of the vessel but also inhibits transfer into the vessel. The magnitude of the effect on the transfer rate seems to depend on the rate of stirring of the liquid. This appears to imply (*a*) that mechanical stirring produces excess vorticity, (*b*) that the rate of transfer depends on the amount of excess vorticity present, and (*c*) that the rate of transfer depends on the specific quantity of vorticity at *both* ends of the film. (*c*) implies that the vorticity in the film possesses a structure since otherwise the film would not « know » that it must slow down if there is an excess of vorticity at the downstream end of the film. (*c*) also seems to imply that there is an appreciable Gibbs function and therefore entropy due to turbulence. In consequence one might expect to find a static or equilibrium effect, that is a level difference between two vessels connected by film when one vessel has a higher specific turbulence than the other, the vessel with the higher turbulence showing the lower level.

This static effect was looked for and has now been observed. When the liquid in the beaker was stirred continuously, the equilibrium level in it fell by about 0.5 mm below that of the bath outside. There is thus some evidence of a third term, involving specific turbulence, in the expression for the pressure in liquid helium:

$$p = g\varrho h + \varrho s \Delta T - f\,(\text{vorticity})\,.$$

Recently a modification to the apparatus was made so that there could be no possibility of eddy current or other solenoid heating of the liquid in the beaker. A new stirrer was hung in the beaker by a fine nylon thread which passed over a pulley to the ferroxcube rod in its solenoid some distance away. Thus the only possible source of heat inside the beaker was Joule paddle wheel heating, or heating by frictional work due to the slight rubbing of the stirrer on the wall of the beaker.

With the new stirrer it was found to be difficult to « stir » the liquid at all. The amplitude of the stirrer was necessarily much less than in the earlier version. But with a stirring rate of not more than 10 Hz there was no effect on the transfer rate, no equilibrium level difference and no visible agitation of the beaker liquid surface. At 12 Hz, however, « stirring » was successful; the transfer rates began to be affected as in Fig. 2, a noticeable equilibrium level drop in the beaker was produced, and the beaker liquid surface became visibly agitated.

Since any Joule frictional heating might be expected to be proportional to the rate of movement of the stirrer, it seems clear that no unwanted heating effects were being produced. The experiment is moreover a demonstration of how difficult it is to « stir » an ideal liquid like helium II. Calculation showed that helium had to be moved back and forth past the stirrer at about 50 cm s^{-1} in order to create turbulence. When stirring ceased the equilibrium level difference disappeared in about two minutes.

5. – Damping of isothermal film oscillations.

Measurements have been made on the damping of isothermal film oscillations at various temperatures between 1.2 and 1.9 °K. The oscillations were induced in a U-tube as shown in Fig. 3, where the amount of liquid at the two ends of the tube and hence the length of the film path between them could be varied. A small black piece of wire in one of the arms could be illuminated so as to absorb heat and thus produce a level difference and so initiate film oscillations in which the mass of liquid in the film formed the inertial element.

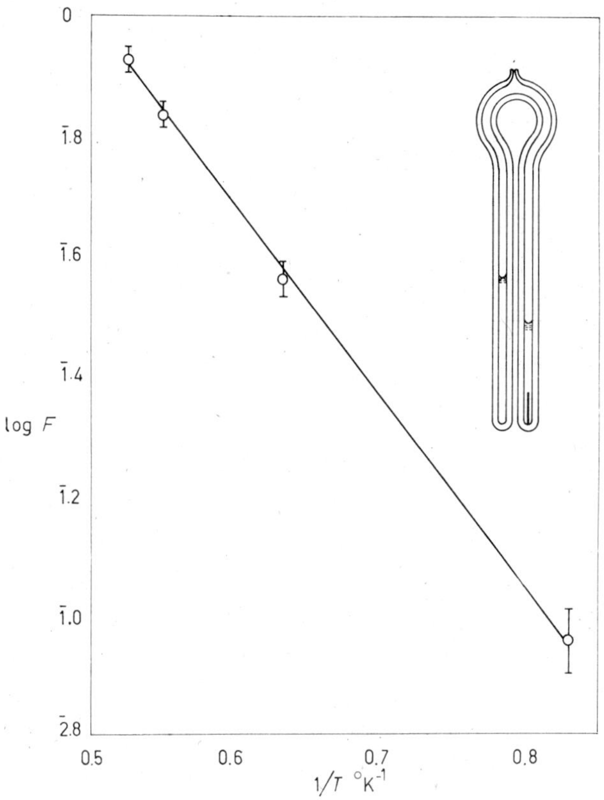

Fig. 3. – Damping of isothermal oscillations in the film. Inset shows U-tube containing film and liquid reservoirs at either end.

If one assumes that motion of the film takes place by means of an array of vortex lines and that all of the damping of the motion of the film takes place in the film itself and not at the ends, then the damping force may be plausibly assumed to be due to scattering of rotons by the vortex lines (*). If this is so, then the decrement in the oscillations will be proportional to

(*) In the discussion, KUPER pointed out that this scattering cannot be a process whereby each vortex line operates independently, since if this were so the damping would be complete in a fraction of a cycle. It seems likely, therefore, that the vortex lattice is a fairly strong and elastic one, so that when a roton is scattered the whole vortex lattice is involved.

the number density of rotons. Thus the damping force $F \propto \exp[-\Delta/kT]$ where Δ is the Landau energy of roton formation. As shown in Fig. 3, a plot of the logarithm of the decrement of the oscillation versus $1/T$ gives a good straight line whose slope (*) is $\Delta = (7.5 \pm 0.4)$ °K.

If the vorticity to operate the film is regarded as being supplied from the bulk liquid, then at the end of each swing of the oscillations the film must be swept free of vorticity since in the next part of the cycle the vorticity is of the opposite sign. Vorticity can leave the film in a time of the order of 1 s since the film path is of the order of 10 cm in length. The period of the oscillations is of the order of 100 s so that there is plenty of time for right-handed vortices to leave the film before left-handed vortices enter. The force to drive them back into the film will be provided by the Gibbs function.

One would not expect this type of damping if the film flow were based on a vortex sheet. This is therefore further evidence in favour of vortex line motion and against that by vortex sheet.

6. – Critical angle of surface tension in liquid helium.

During the course of measurements on film transfer near the rim of the beaker, as described in Section 3, it was found that the rate of transfer into the vessel was greatly enhanced when the bath level approached the rim of the beaker.

In an endeavour to get the bath level as close as possible to the rim of beaker it was found to be possible, with care, to submerge the rim below the bath while still preserving flow in by film transfer. The results are shown in Fig. 4. Two beakers were used, one with a smooth lightly flame-polished rim and the other with a rim ground flat with 320 grade emery. The flame polished rim had slight radii of about 0.2 mm.

With both beakers it was possible to depress the rim about 0.6 mm below the bath while still preserving flow in by film transfer. The situation when the rim is submerged is very stable, since small wavelets which are created in the bath by slight shaking of the cryostat will spill over into the beaker but will not break the surface tension meniscus enough to cause flooding of the beaker.

The effect on the transfer rate of the roughly ground rim is seen not only when the beaker is submerged but also when the rim is some distance above

(*) In the discussion, KUPER stated that the generally accepted value of $\Delta = 8.5$ °K was for free rotons in the bulk liquid. For a roton held by its image force 10 Å from the wall the corresponding value of $\Delta \simeq 7$ °K. In the film, rotons can be considered as something between fixed and free so that an intermediate value of Δ can be considered as very satisfactory.

the bath, although why the curves should possess the observed humps is not clear. In any case, on the assumption that film flow is facilitated by excess turbulence, this is further evidence that the grinding asperities act as vortex spinnerets.

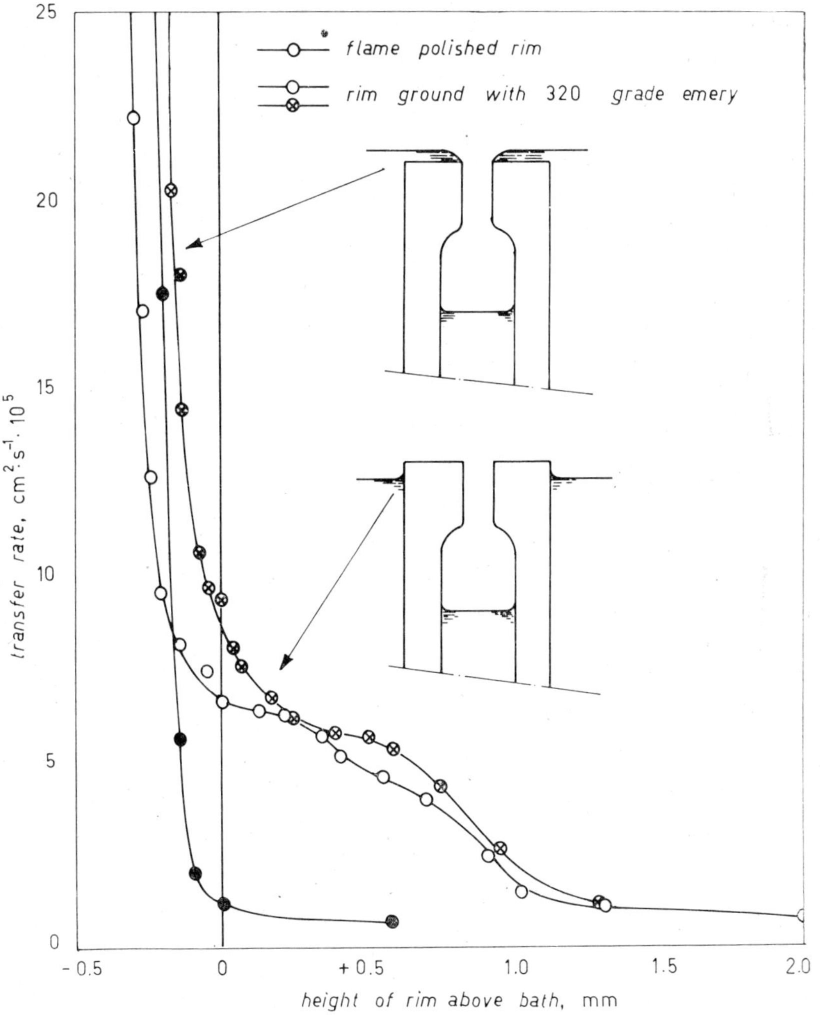

Fig. 4. – Rate of transfer into beaker as function of the height of rim of beaker above the bath. Zero on abscissa indicates rim and bath level are coincident.

The whole experiment gives a strong indication that there must be a finite contact angle of something like 45° in helium II. It may be, therefore, that the meniscus of the bulk liquid and the film contour do not merge into one another as has been assumed heretofore, and that the bulk liquid does not perfectly wet the substrate even though the film is present. It is hoped to make direct measurements of the angle of contact in the near future.

7. – General.

With the exception of the isothermal film oscillations, all of the experiments were made at the bottom bath temperature of 1.15 °K. The beakers were all of internal diameter about 1.5 mm. The outer diameter was usually about 6 mm, but 3 mm in the case of the mechanically stirred beaker.

Precautions against heat entering the helium measuring system appeared effective. The light source was a dim 12 V incandescent bulb. The light traversed a 15 cm long water tank and a Chance ON 22 glass plate before entering the cryostat. The beaker itself was contained in a copper cell which had windows of ON 22 glass 3 mm thick.

REFERENCES

[1] J. DAUNT and K. MENDELSSOHN: *Proc. Roy. Soc.*, A **170**, 423 (1939).
[2] K. R. ATKINS: *Liquid Helium* (Cambridge, 1959).
[3] J. D. REPPY and C. T. LANE: *Proc. of the VII International Conference on Low Temperature Physics, Toronto 1960*, edited by GRAHAM and HALLETT (Toronto and Amsterdam, 1960) p. 443.
[4] W. F. VINEN: *Proc. Roy. Soc.*, A **260**, 218 (1961).
[5] W. FAIRBANK: private communication (*).
[6] R. P. FEYNMAN: *Progress in Low Temperature Physics*, **1** edited by C. J. GORTER (Amsterdam, 1955), chap. II.
[7] V. L. GINSBURG: *Žurn. Ėksp. Teor. Fiz.*, **29**, 244 (1956); *Sov. Fis. JETP*, **1**, 170 (1956).
[8] J. F. ALLEN: *Nature*, **185**, 831 (1960).
[9] C. G. KUPER: *Nature*, **185**, 832 (1960).
[10] C. G. KUPER: *Proc. of the VII International Conference on Low Temperature Physics, Toronto 1960*, edited by GRAHAM and HALLET (Toronto and Amsterdam, 1960) p. 516.
[11] L. ROSENHEAD: *Phil. Trans. Roy. Soc.*, A **208**, 275 (1929).
[12] K. R. ATKINS: *Proc. Roy. Soc.*, A **203**, 240 (1950).
[13] B. N. ESELSON and B. G. LASEREW: *Žurn. Ėxp. Teor. Fiz.*, **23**, 552 (1952).

(*) *Note added in proof.* - J. N. KIDDER and W. M. FAIRBANK: *Phys. Rev.*, **127**, 987 (1962).

On the Theory of Helium Film.

S. FRANCHETTI

Istituto di Fisica dell'Università - Firenze

It is an easy matter to justify the existence of a static helium film, without entering quantitative details. These however are hard to work out. And when it comes to the non-static film, there are problems that as yet have not been even touched from the theoretical viewpoint. Here, I will confine myself for brevity's sake to the static film.

That one should expect to run into difficulties in any quantitative approach, is suggested by the remark that at least for the saturated film, *gravity* plays an essential role. Since gravity is such a weak force, this means that any small disturbance will be important as well. Among such disturbances are on one hand temperature gradients, even very small, the kinetic energy of the flow even if slow, and on the other hand there are — very characteristically (and present in the static film, too) — the small deviations from the normal properties of the liquid due to the fact that we are dealing with *sheets* of liquid one of whose dimensions is no longer very large in comparison with molecular dimensions (say for instance the atomic interdistance).

Let us begin with the simplest case, namely the static film under *saturated* conditions, that is in the presence of an amount of bulk liquid.

In this case, what we wish to know is essentially the law relating the thickness t of the film to the height h above the bulk liquid surface. To get this relationship in the simplest way one can note that the pressure $p(h)$ in the vapour phase at any point at a height h, is related to the pressure $p(0)$ at $h=0$ by

$$p(h) = p(0) \exp\left[-\frac{mgh}{kT}\right]. \tag{1}$$

Of course in the case of the saturated film $p(0)$ is just P_0, the *ordinary* vapour pressure in equilibrium with the bulk liquid at its surface.

Under equilibrium conditions $p(h)$ should similarly be the vapour pressure in equilibrium with the film at a height h. Such a vapour pressure is related to that of the bulk liquid by the equation

$$p(h) = P_0 \exp\left[-\frac{\Delta\lambda}{kT}\right], \tag{2}$$

where $\Delta\lambda$ is the increase in the latent heat per atom (with respect to its value for the bulk liquid), a function of t (and of the nature of the wall) tending to zero when $t \to \infty$.

On equating the right-hand members of (1) and (2) one has

$$mgh = \Delta\lambda(t)\,. \tag{3}$$

This result shows in a quantitative way what was mentioned a moment ago, namely that in the film we have to reckon with energy differences as small as the change in gravitational energy of an atom (and a light one, at that) on a few centimeters height.

It shows also that $\Delta\lambda$ has to be a *positive* quantity. Indeed one can see an almost obious reason why the latent heat should be larger for the film than for the bulk liquid, if one remarks that any atom leaving the surface of the film has to perform an additional work against the very forces that have built the film, which are necessarily mainly attractive. Let it be mentioned that in a more rigorous way, $\Delta\lambda$ measures the decrease $-\Delta\mu$ of the chemical potential caused by the action of the wall on the film. Equation (3) means therefore that this decrease has to compensate for the gravitational increase, thus leaving the chemical potential constant throughout the liquid.

As the simplest approximation we can take into account solely the van der Waals forces between the wall and the atoms of the liquid. It is well known that the usual r^{-6} potential between atoms gives rise to an energy $-Ax^{-3}$ between an atom and a body bounded by a plane, if x is the distance of the atom from the plane. Since what counts is the work done when an atom leaves the surface of the film, we are led to put

$$\Delta\lambda = \frac{A}{t^3}, \tag{4}$$

with t the thickness of the film. Inserting this in (3) one has

$$mgh = \frac{A}{t^3}, \tag{5}$$

that is the Schiff-Fränkel law, which usually is written in the form

$$t = \text{const}\ h^{-1/3}\,. \tag{5}$$

The value of the constant, whose knowledge involves that of the parameter A in (4), can only be given as an order of magnitude, and this turns out to be right. However, the precise prediction about the exponent of h in (5) is not confirmed. The most reliable experimental values seem to favour an exponent $-1/n$ with n near 2.4 or so for h running in the centimeters and approaching the value 3 only for films some tens of cm high [1]. One is thus led to guess that $\Delta\lambda$ should be expressed by a formula of the type

$$\Delta\lambda = \frac{A}{t^3} + \frac{B}{t^2}\,, \tag{6}$$

leading to an equation

$$\frac{A}{t^3} + \frac{B}{t^2} = mgh\,, \tag{7}$$

for the relation between t and h, which we shall refer to as the « equation of the film ».

Two possible terms quadratic in $1/t$ have been pointed out, one by ATKINS [2] and one by me [3], the former being the larger and temperature-independent, while the latter is smaller and vanishes as T tends to zero.

The Atkins term is an effect on the zero-point energy of the liquid due to its confinement to a very small thickness. There is no question of calculating it rigorously, the physics of a liquid being difficult enough even in the absence of special complications.

ATKINS evaluated it by adapting to the liquid the Debye treatment of the thermal motion in solids.

This method requires, as well known, to decompose the thermal motion in elementary monochromatic waves, each characterized by three integers n_x, n_y, n_z determining its wave vector, and each behaving — as far as the energy is concerned — as a quantized harmonic oscillator. The zero point energy, of course, is given by the sum of the fundamental state energies (the « half quanta ») of these oscillators. In a liquid where only longitudinal waves can propagate, *one third* of the degrees of freedom fall into a picture of this kind.

An essential point is that the number of degrees of freedom is finite. That means that you have to introduce only N elementary waves, with N the number of atoms of the liquid and to choose them properly.

Debye's recipe for choosing the elementary waves for a monatomic solid — which are $3N$ in that case — is to choose the $3N$ ones having the lowest wave numbers, that is wave numbers comprised between zero and an upper limit K_D. That makes E_0 a minimum, which seems fit for the fundamental state of a quantum system, and is an important requirement.

To see how a similar procedure can be carried out in our case, suppose that the body of liquid under consideration has the form of a rectangular prism with edges L_1, L_2, L_3. Then note that the wave number of the wave characterized by the integers n_x, n_y, n_z is given by

$$K_n = \frac{1}{2}\left[\left(\frac{n_x}{L_1}\right)^2 + \left(\frac{n_y}{L_2}\right)^2 + \left(\frac{n_z}{L_3}\right)^2\right]^{\frac{1}{2}}, \tag{8}$$

in the assumption — which seems the simplest and most natural one — that we have *nodes* on the faces of the prism.

In a space where we take n_x/L_1, n_y/L_2, n_z/L_3 as co-ordinates, we see from this formula that the frequency $\nu_n = cK_n$ is proportional to the distance of the representative point from the origin. The *density* of the representative points on the other hand is given by $L_1\, L_2\, L_3 = V$. The N points having the lowest frequencies will therefore fill the octant formed by the positive axes up to a radius

$$R = 2K_{\mathrm{D}} \tag{8'}$$

such that

$$V\int_0^{\pi/2} \mathrm{d}\varphi \int_0^{\pi/2} \sin\theta\, \mathrm{d}\theta \int_0^{R} \varrho^2\, \mathrm{d}\varrho = \frac{4}{3}\,\pi\, VK_{\mathrm{D}}^3 = N\,, \tag{9}$$

where ϱ, the distance from the origin, is just $2K_n$, according to (8). For the zero point energy E_0 we shall have (as $\nu = cK_n = (c/2)\varrho$)

$$E_0 = \frac{hcV}{4}\int_0^{\pi/2} \mathrm{d}\varphi \int_0^{\pi/2} \sin\theta\, \mathrm{d}\theta \int_0^{R} \varrho^3\, \mathrm{d}\varrho = \frac{\pi}{2}\, hcVK_{\mathrm{D}}^4\,. \tag{10}$$

In the above formulae, L_1, L_2, L_3 seem to be arbitrary and they disappear altogether from the final result (eq. (10)), showing that the form of the body is immaterial. There is however a feature in the above procedure which is no longer justified if one (or two) of the lenghts L_1, L_2, L_3 is made sufficiently small (by constant $L_1\, L_2\, L_3 = V$), and this is the neglect of the discontinuous character of the distribution of the representative points.

To see this, one has just to note that the interdistance between points along — say — the n_z-axis is $1/L_3$. If L_3 is made very small, the points along the n_z direction fall farther and farther apart until it is no longer sufficiently accurate to consider them as continuously distributed as implicit in deriving formulae (9) and (10).

For convenience, the corrections to be made to (9) and (10) may be thought as arising from two sources:

a) the change in the *boundary* of the integral in (9) and (10), which can no longer be assumed spherical, but has to be replaced by a revolution surface having a staircaselike intersection with any meridian plane;

b) the replacement of the sums over the integers n_z by integrals in dn_z. (This effect can be dealt with conveniently by means of the Euler-Maclaurin formula relating sums to integrals.)

When due account is taken of the above facts one finds corrections to E_0 which depend on powers of the small quantity $(K_D L_3)^{-1}$. Since in our case the small dimension is the thickness t of the film and $K_D = 1/\lambda_D$, the small quantity is λ_D/t. In other words we have:

$$(11) \qquad E_0 = E_{0\infty}\left(1 + a\frac{\lambda D}{t} + b\frac{\lambda D^2}{t^2} \dots\right)$$

At this point a controversial question has arisen about the existence of a 1-st order term. While Atkins found a negative one, I thought there were none. Recently MATSUDA and VAN DER MEIJDENBERG [4] have re-examined the problem, coming to the conclusion that there should be a positive 1-st order term, at least under the most natural assumption that the elementary elastic waves have nodes at the boundary surfaces (as already said).

I must acknowledge that my calculation was indeed wrong, as I simply forgot about correction *b*), misled by the fact that with the boundary I choose to employ it vanishes for integral (9). Unfortunately this is not so with integral (10) and I missed this point. When this was brought to my attention by the work of MATSUDA and VAN DER MEIJDENBERG it convinced me for a while about the existence of the 1-st order term. However, on better thoughts, I have come again to believe that the idea underlying my procedure was basically sound. This is the idea that by suitably choosing the boundary for integrals (9) and (10), that is by changing from $R = \text{const}$ to a suitable $R(\theta)$ (the dependence on θ alone is imposed by the symmetry of the problem) one can get rid of the 1-st order term in E_0.

I think this should be possible even when point *b*) is taken into consideration, although I have not verified that in detail. And if this *can* be done, it *should*, because of the *minimum* requirement for E_0. (Unless of course the 1-st order term should turn negative, which looks very unlikely.) Indeed $R(\theta)$ is subject only to condition (9) which is obviously unable to determine it. Additional conditions can therefore be imposed on $R(\theta)$, like the vanishing of the main term in the correction to E_0.

This degree of arbitrariness concerning the boundary, together with the condition that E_0 be a minimum, have not received enough attention by MATSUDA and VAN DEN MEIJDENBERG. This is all the more regrettable because with the skill and patience they have displayed in their work they certainly would have reached interesting results.

Note in addition that while the 2-nd order increase in E_0 can be explained as a quantum effect akin to the increase of the kinetic energy of a quantized system when confined to smaller and smaller spaces, no such simple explanation could be given for a 1-st order term.

Moreover, the analogous temperature-dependent terms arising from excitations, although of a more intricate thermodynamic character, were studied by me rather carefully and they too do not show any 1-st order contribution. (A repetition of the old error is out of question because the methods were altogether different and more rigorous.)

These latter considerations suggest that changes proportional to t^{-2} should probably be expected for the *entire* zero point energy and not only for the phonon fraction of it, which alone has been considered up to this point and which affects only one third of the degrees of freedom of the liquid. Since we do not know how to deal with the remaining two thirds, this contributes an additional incertitude as regards the magnitude of the t^{-2} term in the equation of the film.

It is worth while mentioning that 1-st oder terms in the expression of the energy per particle (or per gram), if at all present, would have no *direct* bearing on the form of the film. Indeed, if the density ϱ of the liquid portion of the film were independent of its thickness (and as such it would be the same as for the bulk liquid) the t^{-1} terms would contribute to the energy per cm^2 of the film a quantity $\varrho t(\text{const}/t)$ *independent* of the thickness t and therefore unable to have any influence on it. Only through the change of the (average) density, the t^{-1} terms would have an effect, giving rise to 2-nd and higher order terms in the equation of the film, which however would be very difficult to evaluate.

Here is briefly how that comes about. The important point is the change of the energy per cm^2 of the film when t is varied, be it by evaporation or otherwise. If $\Delta\varepsilon$ is the energy difference per gram with respect to the bulk liquid and ϱ is the average density, this means calculating

$$\Delta\mu = \frac{\partial(\varrho t \cdot \Delta\varepsilon)}{\partial(\varrho t)}\,.$$

Singling out the contribution of $E_0 - E_{0\infty}$ to $\Delta\varepsilon$, one has

$$\frac{\partial}{\partial(\varrho t)}\left[\varrho t E_{0\infty}\left(1 + a\frac{\lambda_{\mathrm{D}}}{t} + b\frac{\lambda_{\mathrm{D}}^2}{t^2}\,\ldots\right)\right],$$

If we put ϱ in the form

$$\varrho = \varrho_0 \left(1 + \alpha \frac{\lambda_D}{t} + \beta \frac{\lambda_D^2}{t^2}\right),$$

it is easy to find that the contribution to $\Delta\mu$ arising from the zero point energy change $\Delta E_0 = E_0 - E_{0\infty}$, becomes up to 2-nd order terms, according to eq. (11),

$$- E_{0\infty} (a\alpha + b) \frac{\lambda_D^2}{t^2},$$

showing that a change of 1-st order in E_0 (meaning $a \neq 0$) gives a 2-nd order term in the equation of the film if $\alpha \neq 0$, that is if there is a change in the density.

Note, by the way, the *minus* sign, which, as far as $a\alpha + b$ remains a positive quantity, gives this term a function analogous to the van der Waals term in the equation for t, in spite of the fact that the Atkins term means an *increase* of energy in going from the bulk liquid to the film, while surely the van der Waals field causes the energy to *lower*. What makes the difference is of course that the extra zero point energy is due to *no potential.*

The problem of the *density change* in the liquid region of the film, although no longer critical if the 1-st order term in E_0 is absent, is another difficult question.

Of course, the van der Waals forces exerted by the wall tend to compress the film, with the well known result that the innermost layer for a thickness of some Å units must be altogether a solid. In the liquid phase, the density change is practically confined to a thickness of the order of, say, 10 Å, where the effect is large and can therefore be calculated with some confidence from the compressibility of the liquid. The result turns out to be a contraction of the liquid layer thickness by an amount of the order of one Å, practically independent of t, at least for not too thin films.

Another effect on the density has been considered first by ATKINS and later, in greater detail, by MATSUDA and VAN DER MEIJDENBERG. This should consist in a practically uniform dilatation of the liquid, due to the increase of the zero point energy beyond its normal value. It has been argued that the liquid should lower its density against the elastic forces — as calculated from its compressibility — in order to reduce the zero point energy increase. According to MATSUDA and VAN DEN MEIJDENBERG this effect would be *larger* than the Van der Waals compression, as far as the *average* density is concerned, with the result that the liquid should have in the film an average density slightly *lower* than in the bulk.

The calculation of this density effect requires the knowledge of

$(\partial E_0/\partial V)_{t=\text{const}}$ that is of the dependence of the zero point energy on the specific volume (by constant thickness). This dependence is known only for the phonon portion, but not for that connected to the nonphonon degrees of freedom. This casts a serious incertitude on the results obtained from the consideration of the phonon zero-point energy alone.

If however it should turn out that the 1-st order terms in (11) are absent, the above problem would loose its importance, since the change in the order of magnitude would render the density effect negligible anyway.

We may conclude this discussion about the equation of the saturated static film by the statement that we see reasons for terms in t^{-2} as well as in t^{-3}, but concerning their coefficients nothing more than rough orders of magnitude can be given.

Not long ago the 2-nd order terms in the equation of the film came under criticism from a group of Russian authors who denied their existence altogether [5]. However their argument does not look as particularly convincing. Indeed we have seen the 2-nd order terms (or at least the main Atkins term) arise from the *atomic structure* of the liquid, as evidenced by the very fact that they are terms in $(\lambda_D)^2/t^2$. In a *continuum*, there could be no finite lower limit λ_D for the wave lenght and with $\lambda_D \to 0$ the terms in question would automatically vanish. Since the Russian Authors treated the liquid just as a continuum, their theory could not possibly yield the t^{-2} terms anyway, good or not as they may be [6].

The same authors suggest the existence of t^{-4} terms from acoustical fluctuations in the liquid. This however does not seem consistent with any experimental finding so far.

To end this talk, let us see very briefly what can be said about the static *unsaturated* film.

From considerations perfectly similar to those I made at the beginning, one can find the equation of the unsaturated film in the form:

$$\Delta\lambda - mgh = kT \ln \frac{P_0}{p},$$

where p is the actual pressure and P_0 the vapour pressure of the bulk liquid.

Since we have seen (eq. (6)) that [7] $\Delta\lambda = A/t^3 + B/t^2$, we shall have

$$\frac{A}{t^3} + \frac{B}{t^2} - mgh = kT \ln \frac{P_0}{p}.$$

The gravitational term as well as the t^{-2} term can be neglected against the others for any degree of saturation not too close to unity and temperatures not too close to 0 °K, leaving

$$\frac{A}{t^3} = kT \ln \frac{P_0}{p},$$

which is a well known equation and has been tested to some extent by K. BOWERS [8].

This equation allows an order of magnitude of t to be calculated and it turns out that t is much smaller than for the saturated film. The *solid layer* is also substantially reduced.

I employed the above equation to explore a well known problem: the lowering of the λ-point in the unsaturated film. The idea was as follows. In the unsaturated film, a substantial fraction of the liquid is subjected to an intense van der Waals field from the wall which — in addition — causes the potential energy of an atom of the liquid to show a doubly periodic « peak and valley » character in the region near the solidified layer, due to the fact that the moving atoms have to glide « up and down » on top of those belonging to the crystalline layer underneath. In these conditions it may be conceived that the atoms near the solid layer though belonging to the liquid are not sufficiently free to become superfluid. The superfluid state would therefore not coincide with the lowest energy state. Indeed, this is what is happening with the atoms in the solidified layer as well, whose energy undoubtedly lies below that of the superfluid. However the solidified layer is too little temperature-sensitive to influence to any appreciable extent the condensation process which is generally recognized as being at the root of the λ-transition. On the contrary, the more loose atoms in the quasi-bound liquid states — if they exist at all — might be sufficiently temperature-sensitive.

Of course any theory based on *individual* states for the atoms cannot be rigorous, but it should not be forgotten that the London theory which is of this type, predicts after all the correct qualitative features of helium behaviour. What I am proposing here is a modified London theory capable of dealing with the unsaturated film.

To this aim, I assume there is one or more quasi-bound states below the superfluid state. These states can accomodate only a *fraction* $\theta < 1$ of the atoms of the liquid, because these atoms are to belong to the inner layer of the liquid phase, up to a thickness Δ.

In the condensation process, the quasi-bound states would be filled first, to full capacity, by θN atoms. Only from that moment on the superfluid state would begin to be occupied « macroscopically ».

Calling $f(T)$ the fraction of atoms belonging to the excited states, the λ-point comes to be characterized by

$$Nf(T_\lambda) = (1-\theta)N\,.$$

It is well known that the London theory gives $f(T) = (T/T_{\lambda_0})^{\frac{3}{2}}$ with T_{λ_0} the λ-point for the bulk liquid. Instead, it is found empirically, near the

λ-point $=f(T) \approx (T/T_{\lambda_0})^{5.6}$. Supposing, very tentatively, that $f(T)$ remains the same even in the peculiar conditions of the unsaturated film, the above equation gives

$$\left(\frac{T\lambda}{T_{\lambda_0}}\right)^{5.6} = 1 - \theta .$$

From this and the experimental values for T_λ, the fraction θ and therefore the thickness Δ of the bound state layer can be found. One finds that for saturation degrees from 0.7 to 0.95 this layer would be from 5 to 6 Å thick, as compared with about 4 Å for the solid layer [9]. These figures do not look altogether unreasonable. Choosing to be optimistic, this could be viewed as a mild confirmation of the idea put forward.

Note added in proofs.

On closer examination, the Author has found that it is not possible to cancel the 1-st order term in eq. (11) by the procedure proposed in the text, namely, by deforming the boundary surface of the region containing the representative points of the wave vectors in the $(N_x/L_1, N_y/L_2, N_z/L_3)$ space. Indeed one can prove that the boundary has to be an octant of a sphere if the zero-point energy has to be a minimum.

The expression for E_0 as given by MATSUDA and VAN DEN MEIJDENBERG (ref. ([4]), eq. (9)) is therefore correct, including the 1-st order term, and it does not appear possible to get rid of this term at least as long as one employs a Debye-cut procedure.

A kind correspondence on the subject matter with C. J. N. VAN DEN MEIJDENBERG is gratefully acknowlegded.

REFERENCES

[1] See L. C. JACKSON and L. G. GRIMES: *Phil. Mag. Suppl.*, **7**, 435 (1958); O. T. ANDERSON, D. M. LIEBENBERG, J. R. DILLINGER: *Phys. Rev.*, **117**, 39 (1960).

[2] K. R. ATKINS: *Canad. Journ. of Phys.*, **32**, 347 (1954); *Conf. de Phys. des Basses Temp.* (Paris, 1955), p. 100.

[3] S. FRANCHETTI: *Conf. de Phys. des Basses Temp.* (Paris, 1955), p. 124; *Nuovo Cimento*, **5**, 183 (1957).

[4] H. MATSUDA, C. J. N. VAN DEN MEIJDENBERG: *Physica*, **26**, 939 (1960).

[5] I. E. DZYALOSHINSKIJ, E. M. LIFSCHITZ and L. P. PITAEVSKIJ: *Sov. Phys. JETP* **10**, 161 (1960); or else: *Phil. Mag. Suppl.*, **10**, 165 (1961).

[6] S. FRANCHETTI: *Nuovo Cimento*, **16**, 1158 (1960).

[7] As expression (6) for $\Delta\lambda$, obtained for the saturated film, is only approximate, it might well be that it is no longer quite appropriate for the unsaturated film, in which the thickness is generally much smaller. In particular, features such as *superficial waves*, which have been considered by C. G. KUPER (*Physica*, **24**, 1009 (1959)) and by K. R. ATKINS (*Phys. Rev.*, **113**, 962 (1959))

and can be neglected for saturated films, might prove to be of some importance in any accurate treatment of the unsaturated film.

[8] R. Bowers: *Phil. Mag.*, **44**, 467, 485 (1953). Let it be mentioned that if t^{-4} terms in the expression of the chemical potential for the film were really important, they would show particularly clearly in the case of the unsaturated film. This does not seem to fit Bower's results.

[9] Thanks are due to A. Mazza for help in calculating these data.

Equations of Motion for Rotating Helium II.

H. E. Hall

The Physical Laboratories, University of Manchester - Manchester

1. – Introduction.

The equations of motion for helium II originally derived by Landau [1] are now generally accepted as valid when the flow of the superfluid is irrotational, but numerous experiments have shown that the superfluid can in fact be made to rotate quite easily, thereby violating Landau's condition curl $\boldsymbol{v}_s = 0$. Also, a detailed and convincing series of experiments by Vinen [2] has shown that the nonlinear frictional forces frequently observed in practice are very probably associated with turbulence in the superfluid. The question, therefore, arises of how Landau's equations should be modified to take account of superfluid rotation.

A theoretical starting point for this modification is provided by Onsager and Feynman's idea of quantized vortex lines [3] and experimental justification for taking quantized circulation seriously is provided by Vinen's [4] demonstration that the circulation round a fine wire is quantized. In the next section we shall summarize the derivation of equations of motion from the vortex line model [5].

Recently Lin [6] has obtained a different modification of the Landau equations by considering phenomenologically what extra terms in the equations can arise when all restrictions on the superfluid vorticity and circulation are removed. In this context it is interesting to consider a recent derivation of the equations of the vortex line model by Bekarevich and Khalatnikov [7], using a phenomenological approach similar in spirit to that of Lin.

They make an assumption about the dependence of the internal energy of the liquid on absolute superfluid vorticity in agreement with the vortex line model, but otherwise the derivation is a purely phenomenological one, using the conservation laws. We shall outline this derivation after the microscopic one, since the vortex line model gives a more direct physical insight into the processes that are thought to be involved. We shall then discuss the experimental evidence for the existence of the various additional terms in the equations of motion, and conclude with some brief remarks on the applicability of the equations to turbulent flow.

2. – Derivation from the vortex line model.

In deriving the equations we shall consider only velocity fields that are slowly varying on the scale of the separation between vortex lines, so that the lines can for most purposes be considered as a continuum. Thus $\boldsymbol{v}_n$ and $\boldsymbol{v}_s$ denote normal and superfluid velocities averaged over a region containing many vortex lines.

It is also necessary to define two further velocity fields: $\boldsymbol{v}_L$, the velocity of the vortex lines; and $\boldsymbol{v}_R$, the average drift velocity of rotons that collide with a vortex line.

We first consider flows in which the velocities do not vary parallel to the axis of rotation, and derive the mutual friction force that results from collisions between rotons and vortex lines (phonons may be neglected above 1 °K). Simple kinetic theory arguments show that the force $\boldsymbol{f}_L$ on unit length of line is given by

$$\boldsymbol{f}_L = D(\boldsymbol{v}_R - \boldsymbol{v}_L) + D'\hat{\boldsymbol{\omega}} \times (\boldsymbol{v}_R - \boldsymbol{v}_L) ,$$

where $\hat{\boldsymbol{\omega}}$ is a unit vector in the direction of $\boldsymbol{\omega}$, the average vorticity of the superfluid, and

$$D = \varrho_n \bar{\sigma} v_G , \qquad D' = \varrho_n \bar{\sigma}_\perp v_G ,$$

where v_G is the average group velocity of rotons and $\bar{\sigma}, \bar{\sigma}_\perp$ are suitably averaged collision diameters for momentum exchange parallel and perpendicular to $(\boldsymbol{v}_R - \boldsymbol{v}_L)$.

We now have to relate $(\boldsymbol{v}_R - \boldsymbol{v}_L)$ to $(\boldsymbol{v}_s - \boldsymbol{v}_n)$. The difference between $\boldsymbol{v}_R$ and $\boldsymbol{v}_n$ arises from local dragging of the normal fluid by a vortex line and is given by

$$(\boldsymbol{v}_n - \boldsymbol{v}_R) = \boldsymbol{f}_L / E , \tag{2}$$

where

$$E \doteqdot \frac{-4\pi\eta_n}{\ln \frac{1}{2}\lambda L + 1} ,$$

where λ is the reciprocal penetration depth of viscous waves in the normal fluid and L the roton-roton mean free path. Eliminating $\boldsymbol{v}_R$ between eqs. (1) and (2) we find the force $\boldsymbol{F}(= \boldsymbol{f}_L \omega / k$, where k is the circulation round a vortex line) on the lines in unit volume of superfluid is given by

(3) $$\boldsymbol{F} = B_{\mathrm{L}} \frac{\varrho_{\mathrm{s}}\varrho_{\mathrm{n}}}{2\varrho} \hat{\boldsymbol{\omega}} \times [\boldsymbol{\omega} \times (\boldsymbol{v}_{\mathrm{L}} - \boldsymbol{v}_{\mathrm{n}})] + B'_{\mathrm{L}} \frac{\varrho_{\mathrm{s}}\varrho_{\mathrm{n}}}{2\varrho} \boldsymbol{\omega} \times (\boldsymbol{v}_{\mathrm{L}} - \boldsymbol{v}_{\mathrm{n}}) ,$$

where

$$B_{\mathrm{L}} = \frac{2\varrho}{\varrho_{\mathrm{s}}\varrho_{\mathrm{n}}k} \frac{X}{X^2 + Y^2} , \qquad B'_{\mathrm{L}} = \frac{2\varrho}{\varrho_{\mathrm{s}}\varrho_{\mathrm{n}}k} \frac{Y}{X^2 + Y^2} ,$$

in which

$$X = \frac{1}{E} + \frac{D}{D^2 + D'^2} , \qquad Y = \frac{-D'}{D^2 + D'^2} .$$

We now relate $\boldsymbol{v}_{\mathrm{L}}$ to $\boldsymbol{v}_{\mathrm{s}}$ by means of the equation for the Magnus effect,

(4) $$\boldsymbol{f}_{\mathrm{L}} = \varrho_{\mathrm{s}}(\boldsymbol{v}_{\mathrm{L}} - \boldsymbol{v}_{\mathrm{s}}) \times \boldsymbol{k} ,$$

which is the mechanism by which force is transmitted to the superfluid, and obtain

(5) $$\boldsymbol{F} = B \frac{\varrho_{\mathrm{s}}\varrho_{\mathrm{n}}}{2\varrho} \hat{\boldsymbol{\omega}} \times [\boldsymbol{\omega} \times (\boldsymbol{v}_{\mathrm{s}} - \boldsymbol{v}_{\mathrm{n}})] + B' \frac{\varrho_{\mathrm{s}}\varrho_{\mathrm{n}}}{2\varrho} \boldsymbol{\omega} \times (\boldsymbol{v}_{\mathrm{s}} - \boldsymbol{v}_{\mathrm{n}}) ,$$

where

$$B = \frac{2\varrho}{\varrho_{\mathrm{s}}\varrho_{\mathrm{n}}k} \frac{X}{X^2 + (Y + 1/\varrho_{\mathrm{s}} k)^2} ,$$

and

$$B' = \frac{2\varrho}{\varrho_{\mathrm{s}}\varrho_{\mathrm{n}}k} \frac{Y + 1/\varrho_{\mathrm{s}} k}{X^2 + (Y + 1/\varrho_{\mathrm{s}} k)^2} .$$

It is now convenient to consider the effect of curvature of the vortex lines (*i.e.* flows in which the velocity varies parallel to the axis of rotation) in the absence of mutual friction. The equation of motion of the superfluid can then be written

(6) $$\frac{\partial \boldsymbol{v}_{\mathrm{s}}}{\partial t} = \boldsymbol{v}_{\mathrm{L}} \times \boldsymbol{\omega} + \operatorname{grad} \varphi ,$$

where the first term gives the change in $\boldsymbol{v}_{\mathrm{s}}$ at a point due to flow of vortex lines past the point and the second term represents the effects of pressure and temperature gradients. If a vortex line has a radius of curvature R, it will experience a tranverse force ε/R per unit length, due to the tension ε in it (equal to the energy of unit length of line). This force must be balanced by the Magnus effect so that

$$\frac{\varepsilon}{R} = \varepsilon \frac{(\boldsymbol{k}\cdot\nabla)\boldsymbol{k}}{k^2} = \varrho_s(\boldsymbol{v}_L - \boldsymbol{v}_s)\times k\ . \tag{7}$$

Eliminating $\boldsymbol{v}_L$ between eqs. (6) and (7) we obtain

$$\frac{\partial \boldsymbol{v}_s}{\partial t} = \boldsymbol{v}_s \times \boldsymbol{\omega} + \nu(\boldsymbol{\omega}\cdot\nabla)\hat{\boldsymbol{\omega}} + \text{grad}\,\varphi\ , \tag{8}$$

where $\nu = \varepsilon/\varrho_s k$.

If we now include the effect of mutual friction eq. (7) is replaced by

$$\varrho_s(\boldsymbol{v}_L - \boldsymbol{v}_s)\times\boldsymbol{\omega} = \varrho_s\nu(\boldsymbol{\omega}\cdot\nabla)\hat{\boldsymbol{\omega}} + \boldsymbol{F}\ ,$$

i.e. by

$$\boldsymbol{F} = \varrho_s(\boldsymbol{v}_L - \boldsymbol{v}_s - \nu\ \text{curl}\ \hat{\boldsymbol{\omega}})\times\boldsymbol{\omega}\ , \tag{9}$$

where $\boldsymbol{F}$ is given in terms of $(\boldsymbol{v}_L - \boldsymbol{v}_n)$ by eq. (3). Comparison of eqs. (4) and (9) shows that the effect of vortex line curvature is to replace $\boldsymbol{v}_s$ by $(\boldsymbol{v}_s + \nu\ \text{curl}\ \hat{\boldsymbol{\omega}})$, so that when $\boldsymbol{v}_L$ is eliminated from eqs. (3) and (9) eq. (5) is replaced by

(10) (*)
$$\begin{aligned}\boldsymbol{F} = B\frac{\varrho_s\varrho_n}{2\varrho}\hat{\boldsymbol{\omega}}\times[\boldsymbol{\omega}\times(\boldsymbol{v}_s + \nu\ \text{curl}\ \hat{\boldsymbol{\omega}} - \boldsymbol{v}_n)] +{}\\ + B'\frac{\varrho_s\varrho_n}{2\varrho}[\boldsymbol{\omega}\times(\boldsymbol{v}_s + \nu\ \text{curl}\ \hat{\boldsymbol{\omega}} - \boldsymbol{v}_n)] ={}\\ = B\frac{\varrho_s\varrho_n}{2\varrho}\hat{\boldsymbol{\omega}}\times[\boldsymbol{\omega}\times(\boldsymbol{v}_s - \boldsymbol{v}_n)] + B'\frac{\varrho_s\varrho_n}{2\varrho}\boldsymbol{\omega}\times(\boldsymbol{v}_s - \boldsymbol{v}_n) -{}\\ - B\frac{\varrho_s\varrho_n}{2\varrho}\nu\hat{\boldsymbol{\omega}}\times(\boldsymbol{\omega}\cdot\nabla)\hat{\boldsymbol{\omega}} - B'\frac{\varrho_s\varrho_n}{2\varrho}\nu(\boldsymbol{\omega}\cdot\nabla)\hat{\boldsymbol{\omega}}\ .\end{aligned}$$

Combining eqs. (9) and (10) with eq. (6) and writing the potential φ explicity we have for the equation of motion of the superfluid

$$\begin{aligned}\varrho_s\frac{\mathrm{D}\boldsymbol{v}_s}{\mathrm{D}t} = -\frac{\varrho_s}{\varrho}\text{grad}\ p + \varrho_s S\ \text{grad}\ T +{}\\ + B\frac{\varrho_s\varrho_n}{2\varrho}\hat{\boldsymbol{\omega}}\times[\boldsymbol{\omega}\times(\boldsymbol{v}_s - \boldsymbol{v}_n)] + B'\frac{\varrho_s\varrho_n}{2\varrho}\boldsymbol{\omega}\times(\boldsymbol{v}_s - \boldsymbol{v}_n) -{}\\ - B\frac{\varrho_s\varrho_n}{2\varrho}\nu\hat{\boldsymbol{\omega}}\times(\boldsymbol{\omega}\cdot\nabla)\hat{\boldsymbol{\omega}} + \varrho_s\nu(1 - B'\varrho_n/2\varrho)(\boldsymbol{\omega}\cdot\nabla)\hat{\boldsymbol{\omega}}\ ,\end{aligned} \tag{11}$$

(*) As pointed out in [7], the third term in the mutual friction was given incorrectly in the last paper quoted in [5] (eqs. (51)-(53)), since curl $\hat{\boldsymbol{\omega}}$ is not necessarily perpendicular to $\hat{\boldsymbol{\omega}}$, but this error does not affect the linearized equations used to discuss experimental results.

where the term $\boldsymbol{v}_s \times \boldsymbol{\omega}$ in eq. (8) has been absorbed in $\mathrm{D}\boldsymbol{v}_s/\mathrm{D}t$. Addition of $\boldsymbol{F}$ to the equation of motion of the normal fluid gives

$$\text{(12)}\qquad \varrho_n \frac{\mathrm{D}\boldsymbol{v}_n}{\mathrm{D}t} = -\frac{\varrho_n}{\varrho}\operatorname{grad} p - \varrho_s S \operatorname{grad} T + \eta_n\left(\nabla^2 \boldsymbol{v}_n + \frac{1}{3}\operatorname{grad}\operatorname{div}\boldsymbol{v}_n\right) - $$
$$- B\frac{\varrho_s\varrho_n}{2\varrho}\hat{\boldsymbol{\omega}}\times[\boldsymbol{\omega}\times(\boldsymbol{v}_s-\boldsymbol{v}_n)] - B'\frac{\varrho_s\varrho_n}{2\varrho}\boldsymbol{\omega}\times(\boldsymbol{v}_s-\boldsymbol{v}_n) +$$
$$+ B\frac{\varrho_s\varrho_n}{2\varrho}\nu\hat{\boldsymbol{\omega}}\times(\boldsymbol{\omega}\cdot\nabla)\hat{\boldsymbol{\omega}} + B\frac{\varrho_s\varrho_n}{2\varrho}\nu(\boldsymbol{\omega}\cdot\nabla)\hat{\boldsymbol{\omega}}\,.$$

These are the final equations of the vortex line model.

3. – Phenomenological derivation.

BEKAREVICH and KHALATNIKOV [7] start from the assumption that the change in internal energy of unit volume of liquid can be written as

$$\text{(13)}\qquad \mathrm{d}e = T\,\mathrm{d}S + \mu\,\mathrm{d}\varrho + (\boldsymbol{v}_n - \boldsymbol{v}_s)\cdot\mathrm{d}\boldsymbol{p} + \lambda\,\mathrm{d}\omega\,,$$

where e and $\boldsymbol{p}$ are the energy and momentum of the liquid in a co-ordinate system moving with velocity $\boldsymbol{v}_s$ and μ is the chemical potential. The extra term $\lambda\,\mathrm{d}\omega$ represents the effect of vorticity, and on the vortex line model $\lambda = \varepsilon/k$. The laws of conservation of mass, energy, and momentum are written as

$$\text{(14)}\qquad \frac{\partial\varrho}{\partial t} + \operatorname{div}\boldsymbol{j} = 0\,,$$

$$\text{(15)}\qquad \frac{\partial E}{\partial t} + \operatorname{div}(\boldsymbol{Q}_0 + \boldsymbol{q}) = 0\,,$$

$$\text{(16)}\qquad \frac{\partial j_i}{\partial t} + \frac{\partial}{\partial x_k}(\Pi^0_{ik} + \pi_{ik}) = 0\,,$$

where E, the internal energy in a stationary co-ordinate system, is given by

$$\text{(17)}\qquad E = \tfrac{1}{2}\varrho v_s^2 + \boldsymbol{p}\cdot\boldsymbol{v}_s + e\,,$$

and the unperturbed energy flow and pressure tensors are taken as for irrotational flow [8]:

$$\text{(18)}\qquad \boldsymbol{Q}_0 = (\mu + \tfrac{1}{2}v_s^2)j + ST\boldsymbol{v}_n + \boldsymbol{v}_n(\boldsymbol{v}_n\cdot\boldsymbol{p})\,,$$

$$\text{(19)}\qquad \Pi^0_{ik} = \varrho v_{si}\cdot v_{sk} + v_{si}p_k + v_{nk}p_i + p\,\delta_{ik}\,,$$

where p is the pressure:

(20) $$p = -e + TS + \mu\varrho + (\boldsymbol{v}_n - \boldsymbol{v}_s)\cdot\boldsymbol{p}\,.$$

The effects of vorticity are thus represented by the unknown terms $\boldsymbol{q}$ and π_{ik}. In addition to the conservation laws we have the equation of motion of the superfluid

(21) $$\frac{\partial \boldsymbol{v}_s}{\partial t} + (v_s\cdot\nabla)\boldsymbol{v}_s + \operatorname{grad}\mu = \boldsymbol{f}\,,$$

and the equation of entropy production

(22) $$\frac{\partial S}{\partial t} + \operatorname{div} S\boldsymbol{v}_n = R/T\,,$$

Here the unknown force $\boldsymbol{f}$ and the dissipation function R represent the effects of vorticity. In order to determine $\boldsymbol{q}$, π_{ik}, $\boldsymbol{f}$, and R we differentiate eq. (17) with respect to time, substitute for the various time derivatives from the other equations, and gather together divergence terms to obtain

(23) $$\dot{E} + \operatorname{div}\{\boldsymbol{Q}_0 + (\pi\boldsymbol{v}_n) + \lambda\hat{\boldsymbol{\omega}}\times[\boldsymbol{f} + \boldsymbol{\omega}\times(\boldsymbol{v}_n - \boldsymbol{v}_s)]\} =$$
$$= T(\dot{S} + \operatorname{div} S\boldsymbol{v}_n) + (\pi_{ik} - \lambda\omega\delta_{ik} + \lambda\omega_i\omega_k/\omega)\frac{\partial v_{ni}}{\partial x_k} +$$
$$+ [\boldsymbol{f} + \boldsymbol{\omega}\times(\boldsymbol{v}_n - \boldsymbol{v}_s)]\cdot(\boldsymbol{j} - \varrho\boldsymbol{v}_n + \operatorname{curl}\lambda\hat{\boldsymbol{\omega}})\,,$$

By comparing this with eqs. (15) and (22) we can identify

(24) $$\boldsymbol{q} = (\pi\boldsymbol{v}_n) + \lambda\hat{\boldsymbol{\omega}}\times[\boldsymbol{f} + \boldsymbol{\omega}\times(\boldsymbol{v}_n - \boldsymbol{v}_s)]\,,$$

(25) $$R = -(\pi_{ik} - \lambda\omega\delta_{ik} + \lambda\omega_i\omega_k/\omega)\frac{\partial v_{ni}}{\partial x_k} -$$
$$- [\boldsymbol{f} + \boldsymbol{\omega}\times(\boldsymbol{v}_n - \boldsymbol{v}_s)](\boldsymbol{j} - \varrho\boldsymbol{v}_n + \operatorname{curl}\lambda\hat{\boldsymbol{\omega}})\,.$$

The fact that R has to be a positive quadratic form then requires that

(26) $$\boldsymbol{f} = -\boldsymbol{\omega}\times(\boldsymbol{v}_n - \boldsymbol{v}_s) + \alpha\boldsymbol{\omega}\times(\boldsymbol{j} - \varrho\boldsymbol{v}_n + \operatorname{curl}\lambda\hat{\boldsymbol{\omega}}) +$$
$$+ \beta\hat{\boldsymbol{\omega}}\times[\boldsymbol{\omega}\times(\boldsymbol{j} - \varrho\boldsymbol{v}_n + \operatorname{curl}\lambda\hat{\boldsymbol{\omega}})] - \gamma\hat{\boldsymbol{\omega}}[\boldsymbol{\omega}\cdot(\boldsymbol{j} - \varrho\boldsymbol{v}_n + \operatorname{curl}\lambda\hat{\boldsymbol{\omega}}]\,,$$

with $\beta, \gamma \geqslant 0$ and

(28) $$\pi_{ik} = \lambda\omega\delta_{ik} - \lambda\omega_i\omega_k/\omega + \eta_{iklm}\frac{\partial v_{nl}}{\partial x_m}\,.$$

It is assumed that the viscosity tensor η_{iklm} is as for irrotational superflow on the ground that any change would be quadratic in ω, and other possible quadratic effects are also ignored. Then, if we put

$$\beta = \frac{1}{2}B\varrho_n/\varrho\varrho_s\,, \qquad \beta' = \alpha + 1/\varrho_s = \frac{1}{2}B'\varrho_n/\varrho\varrho_s\,, \qquad \gamma = 0\,,$$

and use the relation $\boldsymbol{j} = \varrho_n \boldsymbol{v}_n + \varrho_s \boldsymbol{v}_s$, eqs. (26) and (21) yield eq. (11) for the superfluid and combination with eqs. (16) and (28) yields eq. (12) for the normal fluid. The term in γ in eq. (26) represents a mutual friction parallel to the axis of rotation, and does not exist on a simple vortex line model. However, thermal vibrations of the vortex lines may give rise to a nonzero value of γ; this question is as yet unresolved either experimentally or theoretically.

In a similar way, by requiring the energy dissipation at a boundary to be positive, BEKAREVICH and KHALATNIKOV obtain the boundary condition

$$(\boldsymbol{v}_L - \boldsymbol{u}) = \zeta\hat{\boldsymbol{\omega}}\times[\boldsymbol{N}\times\hat{\boldsymbol{\omega}}] + \zeta'\boldsymbol{N}\times\hat{\boldsymbol{\omega}}\,, \tag{29}$$

where $\boldsymbol{u}$ is the velocity of the boundary and $\boldsymbol{N}$ is a unit vector directed normally out of the liquid. If $\zeta' = 0$ this is equivalent to the boundary condition used to describe empirically the damping of resonance in vortex wave experiments [5]. Equation (29) represents a slip of vortex lines over the surface at a rate proportional to their inclination to the normal. ζ, $\zeta' = 0$ corresponds to an ideally rough surface, and ζ, $\zeta' \to \infty$ corresponds to a perfectly smooth surface (*e.g.* a free liquid surface).

4. – Discussion.

The direct experimental evidence for the validity of eqs. (11) and (12) comes from the second sound attenuation and vortex wave experiments cited previously [5] and the vortex wave experiments of ANDRONIKASHVILI and ZAKADZE [9]. The extra terms due to rotation that we have to consider are the four mutual friction terms given by eq. (10) and the additional term $\nu(\boldsymbol{\omega}\cdot\nabla)\hat{\boldsymbol{\omega}}$ is the superfluid equation.

Second sound propagation is affected by the first two mutual friction terms. It is the first term that produces attenuation, and the experimental evidence for its existence is very strong; even the actual magnitude and temperature-dependence observed are in reasonable agreement with detailed calculations based on the vortex line model [10].

But no experiment has yet been done which would show up the second (B') term, which is not dissipative, and experimental evidence concerning a possible axial mutual friction (γ term in eq. (26)) is inconclusive; the axial friction is certainly small, but may not be zero.

It is the extra term $\nu(\boldsymbol{\omega}\cdot\nabla)\hat{\boldsymbol{\omega}}$ in the superfluid equation that leads to the prediction of vortex waves, by giving the superfluid a form of torsional rigidity, and the experimental observation of such waves is therefore strong evidence for its existence. Some details of the experimental results are rather strongly dependent on the boundary conditions (29), but the existence of a series of resonances is independent of the precise boundary conditions, and measurement of the resonance positions gives ν directly, with a result in good agreement with the prediction of the vortex line model. Vortex waves are affected by all the mutual friction terms, but under the experimental conditions actually used the last two terms are dominant.

The third (B) term causes attenuation of the vortex waves, and measured attenuation coefficients, though not very accurate, are in satisfactory agreement with the value of B from the second sound attenuation. The fourth (B') term merely causes a slight temperature-dependence of the vortex wave velocity; there is some evidence that an effect of the expected order of magnitude exists, but the experiments are too inaccurate to be of much value.

Thus the experimental tests, though not complete, have all been satisfactory so far. There is good evidence for the first and the third (B) mutual friction terms and for the vortex wave term, and some rather weaker evidence for the fourth mutual friction term. The form of the boundary condition (29), appears correct but neither the experimental facts nor the microscopic theory concerning the constants ζ, ζ' are clear. Despite the incompleteness of the experimental tests, it would be very difficult to find equations other than (11) and (12) which gave an equally satisfactory description of the experimental facts.

It is perhaps worth-while to try and assess the merits of the two derivations of the equations of motion presented here. The microscopic derivation gives a clearer physical picture of the processes involved and enables us to deduce actual numerical values of the constants B, B' and ν; it further shows that they are not true constants, but have a slight (logarithmic) dependence on some characteristic length, such as the spacing between vortex lines (but there is as yet no really conclusive evidence for such a dependence). In both methods the validity of a continuum approximation is assumed, though the precise meaning of this approximation is clearer in the microscopic approach. The real merit of the phenomenological method is that it requires only the one basic assumption about the dependence of internal energy on absolute vorticity; it thus shows that many of the detailed properties of the vortex line model are not essential for the validity of eqs. (11) and (12). The moral is clear; some sort of vortex line model is probably valid for helium II, but we should be rather cautious about assuming all the details of a semiclassical vortex line model without further evidence.

When we come to consider turbulence the question of the validity of the continuum approximation is crucial. VINEN's analysis [2] of the homogeneous turbulence produced by heat flow shows that in this case the dominant scale of turbulent motion is very probably of the same order as the spacing between vortex lines; the continuum equations derived here therefore have at best a qualitative value, and may well be useless. On the other hand TOWNSEND [11] has recently shown that the continuum equations give a good description of pressure flow at high Reynold's numbers, and has also been able to establish theoretical criteria for the validity of the continuum approximation for this type of flow. It thus seems that each case must be considered on its merits; some types of turbulence can be adequately described by eqs. (11) and (12), but others may require a microscopic approach, and are thus analogous to classical turbulence on a scale of the order of the mean free path.

Finally, we may make a few comments on the relation between the equations discussed here and those of LIN [6]. Lin's equations do not contain any of the additional terms discussed here, apart from the extra Coriolis term, which corresponds to $B'=2$, but instead contain three additional viscosity coefficients $\eta^{(\text{sn})}$, $\eta^{(\text{ns})}$, and $\eta^{(\text{ss})}$. The reason for this is that the two « phenomenological » derivations are based on different assumptions. Bekarevich and Khalatnikov exclude Lin's additional viscosities by assuming that Landau's equations are valid for the non-rotating liquid, and Lin excludes the additional terms discussed here by assuming that the equations do not depend explicitly on the vorticity. A true phenomenological derivation would include *all* these extra terms, and perhaps others, and it must be left to experiment and microscopic theory to decide which terms are actually present. We have seen above that there is good experimental evidence for most of the terms involving superfluid vorticity. On the other hand, Lin is able to explain capillary flow and oscillating disk experiments with his equations. But both these experiments concern situations where tubulence is likely, and indeed the capillary flow results have already been satisfactorily accounted for by means of a turbulent solution of the equations discussed here [11]. There is as yet no reason to suppose that other turbulent flows cannot be equally well explained by the vortex line model, but the difficulties inherent in turbulence have prevented solution of all but the simplest problems so far. There is, however, some positive evidence against Lin's extra viscosities. For consistency with experiment he requires that $\eta^{(\text{ns})}+\eta^{(\text{ss})}\sim 0$ and $\eta^{(\text{sn})}\sim 0$. But VINEN's [4] observation of considerable persistent circulations is very hard to understand unless $\eta^{(\text{ss})}\sim 0$ (independently of the boundary conditions), and we are thus forced to conclude that all Lin's additional viscosities are at any rate very small.

It is also worth remembering that the equations considered here do have a microscopic basis, albeit imperfect, whereas Lin's do not. It therefore seems reasonable to adopt the vortex line model and the equations of motion to which it leads as a working hypothesis, the most fruitful at present available, to be modified or discarded in the light of experimental evidence.

REFERENCES

[1] L. D. Landau: *Journ. Phys. U.S.S.R.*, **5**, 71 (1941).
[2] W. F. Vinen: *Proc. Roy. Soc.*, A **240**, 114, 128 (1957); **242**, 493 (1957); **243**, 400 (1958).
[3] L. Onsager: *Suppl. Nuovo Cimento*, **6**. 249 (1949); R. P. Feynman: *Progr. Low Temp. Phys.*, vol. **1** (1955), chap. II.
[4] W. F. Vinen: *Proc. Roy. Soc.*, A **260**, 218 (1961).
[5] H. E. Hall and W. F. Vinen: *Proc. Roy. Soc.*, A **238**, 204, 215 (1956); H. E. Hall: *Proc. Roy. Soc.*, A **245**, 546 (1958); *Adv. in Phys.*, **9**, 89 (1960).
[6] C. C. Lin: This volume p. 93.
[7] I. L. Bekarevich and I. M. Khalatnikov: *Žurn. Ėksp. Teor. Fiz.*, **40**, 920 (1961).
[8] I. M. Khalatnikov: *Žurn. Ėksp. Teor. Fiz. U.S.S.R.*, **23**, 265 (1952).
[9] E. L. Andronikashvili and D. S. Zakadze: *Žurn. Ėksp. Teor. Fiz. U.S.S.R.*, **37**, 322, 562 (1959).
[10] E. M. Lifshitz and L. P. Pitaevsky: *Žurn. Ėksp. Teor. Fiz.*, **33**, 535 (1957) (*Sov. Phys. J.E.T.P.*, **6**, 418 (1957)).
[11] A. A. Townsend: *Journ. Fluid Mech.*, **10**, 113 (1961).

Critical Velocities in Liquid Helium II.

W. F. VINEN (*)

The Royal Society Mond Laboratory, University of Cambridge - Cambridge

1. – Introduction.

It is well known that below a certain velocity the superfluid component of helium II will flow with little or no friction, but that above this velocity complicated nonlinear frictional forces set in. It is the aim of the present paper to consider the nature of these critical velocities (*), and to survey the extent to which we understand them theoretically. As will become clear later, an unsatisfactory situation exists in relation to both the experimental evidence and the theory; it will be shown that many experiments of a certain type remain to be carried out, and that a number of difficult theoretical problems have still to be tackled. It is hoped that the present paper will be of some value, not only because it deals with an intrinsically interesting phenomenum, but also because it will serve to underline certain fundamental problems in the theory of helium II that remain to be solved.

It will be assumed throughout the paper that the supercritical frictional forces originate from turbulence in the superfluid, the turbulence taking the form of a tangled mass of quantized vortex line [1, 2]. This view would probably not be accepted universally (see, for example, ref. [3-5], and the work of LIN described at this school), but the bulk of the experimental evidence seems to favour it [6, 7]. It will also be assumed that the effective core radius in a vortex line is of order 10^{-8} cm.

It is clear that within the framework of our main assumption there exist in principle two types of critical velocity. In the first, which we shall call the «ideal» type, the subcritical flow of superfluid contains no vortex lines, the flow being purely potential (curl $v_s = 0$). If the superfluid were a classical ideal fluid, a transition from a state of purely potential flow to one containing vorticity would be forbidden by the laws of classical hydrodynamics. Thus, very probably, the transition must involve the atomic structure of the liquid and be essentially quantum mechanical, unless, as explained later,

(*) Now at Department of Physics, University of Birmingham.

vortex line can be generated from the normal fluid by thermal excitation. In the second type, which we shall call « nonideal », the subcritical flow does contain vortex lines, either in the form of a uniform array such as occurs in the uniformly rotating liquid [8], or in the form of a turbulent tangle of low density, and the transition involves a sudden increase in the density of line (and probably the breakdown of any orderly array). Such a transition could in principle be understood in terms of the classical hydrodynamics of an ideal fluid, although in any particular case a complete explanation in these terms might not be possible.

It appears from the experimental evidence that both types of critical velocity can occur in practice. The nonideal type is not of great physical interest, and plausible explanations have been given in the two particular cases which have been investigated in detail; it will be discussed therefore only in outline (Section **2**). The ideal type is of much greater interest, and it is here where further experiments are urgently required and where a satisfactory theory is still lacking; this type will therefore be considered in more detail (Section **3**). Critical velocities in the film will be excluded from the discussion, since the existence of a free surface in the film may give rise to special effects not present in ordinary channels [34].

It should be noted that critical velocities could presumably exist in helium that are associated with transitions from laminar flow to turbulent flow in the *normal* fluid. These will not be discussed here.

2. – Nonideal critical velocities.

As already mentioned, two particular critical velocities that appear to be nonideal have been investigated in some detail, and we shall discuss these in turn.

2·1. *Critical velocities in heat currents.* – Nonideal critical velocities that probably involve transitions from one turbulent state to another have been observed in heat flow experiments by Winkel, Broese van Groenou and Gorter [9] (in narrow channels, of order 1 micron wide) and by Vinen [10] (in wide channels, of order 1 mm wide), and further studies of the latter case have recently been reported by Careri, Scaramuzzi and McCormick [11] and by Chase [12]. In the experiments of Winkel *et al.* the subcritical friction was observed directly; in Vinen's experiments, however, it was too small to be observed directly, but the existence of subcritical turbulence was inferred from its effect on the rate of build-up of supercritical turbulence when the small subcritical heat current was suddenly increased to a larger value.

A simple theory of the wide-channel critical velocity was proposed in ref. [10]. An analysis was made of the mechanisms by which vortex line present in a heat current can build up and decay, and by supposing that only certain effects are important an equation was obtained that accounted surprisingly well (semi-quantitatively) for the existence and characteristics of the observed critical velocity. The analysis was quite crude and aimed only to describe the average rate of change of the length of vortex line in a channel containing a high concentration of line (spacing between lines small compared with channel width); and it was therefore restricted to wide channels and not too small heat currents. The final equation has the form

$$\frac{\mathrm{d}L}{\mathrm{d}t} = \chi_1 \frac{B}{2}\frac{\varrho_n}{\varrho} v L^{\frac{3}{2}} - \chi_2 \frac{\hbar}{m} L^2 - \chi_3 \frac{B}{2}\frac{\varrho_n}{\varrho} v \frac{L}{d} + \gamma v^{\frac{5}{2}}, \tag{1}$$

where L is the length of line per unit volume, $v = v_s - v_n$, d is the channel width, B is the constant appearing in the equation for mutual friction in the uniformly rotating liquid [8], χ_1, χ_2, χ_3 are constants of order unity, and γ is a parameter equal to approximately $1 \cdot 1 T^{11}\ \mathrm{cm}^{-\frac{9}{2}}\ \mathrm{s}^{\frac{3}{2}}$. The meanings of the various terms in this equation are as follows. The first term represents a build-up of turbulence through the action of the relative motion of the two fluids: the normal fluid exerts a force on the lines, which causes them to move owing to the Magnus effect, and this movement, combined with interactions between the lines, leads to a continuous stretching of the lines; the form of the term can be justified by physical argument together with dimensional analysis. The second term represents a homogeneous decay process. It can be argued that this decay proceeds through a tendency of lines to cluster and hence for lines of opposite sense to annihilate one another; this process is in some respects analogous to the decay of homogeneous turbulence in an ordinary liquid, and the form of this second term can be obtained either by analogy to an empirical law known to apply to ordinary liquids or by dimensional analysis. The third term represents a perturbing effect of the walls of the channel, and it may originate from, for example, the effective absence of the mechanism represented by the first term for a distance from the wall of the order of the average line spacing. The final term is purely empirical, and was introduced to obtain agreement between the theoretical and observed values for the time taken for turbulence to build up when a heat current is first switched on. It must presumably represent the effect of the mechanism by which the pure superflow initially breaks down (*i.e.* one of the mechanisms involved in the ideal critical velocities to be discussed in the next section), but it may include other effects.

A typical steady state solution ($\mathrm{d}L/\mathrm{d}t = 0$) of eq. (1) has the form shown in Fig. 1. It is clear that there is a critical velocity, and, as already mentioned, it turns out that this predicted critical velocity has just the characteristics (*e.g.* disappearance at high temperatures and in wide channels) that were observed in the experiments (*). Admittedly, the recent experiments of CHASE [12] have shown that the critical velocity depends to some extent on the geometry of the channel in a way not accounted for by the theory, but there is no reason to believe that some minor refinements in the theory would not account for this. Other experiments of CHASE [12, 13] have on the whole provided evidence in favour of the basic ideas in the theory. For example, he has shown that with a channel of cross-section $0.10 \text{ cm} \times 0.05 \text{ cm}$ a steady rotation of the whole apparatus about an axis perpendicular to the heat flow and to the longer side of the cross-section produced no observable effect on the critical velocity until the angular velocity exceeded 1 radian s^{-1}; even at angular velocities considerably less than 1 radian s^{-1} this rotation must almost certainly have produced quite a large number of vortex lines in the channel before the heat flow was switched on, so that, although the observation is consistent with the theory of ref. [10], it is obviously inconsistent with any theory, such as that applicable to an ideal critical velocity, in which the critical velocity is associated directly with the difficulty of nucleating the formation of vortex line.

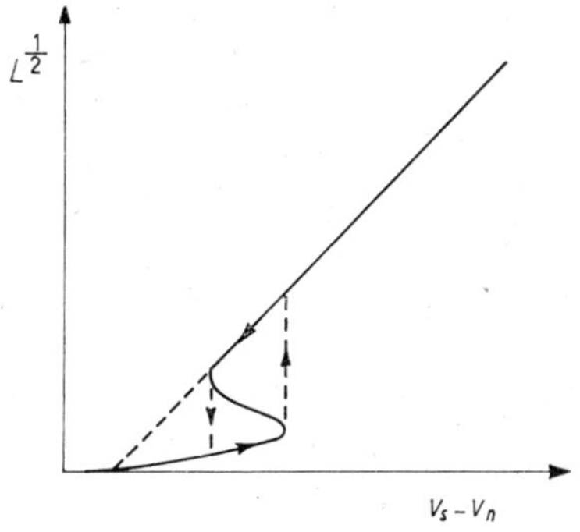

Fig. 1. – Typical steady state solution of eq. (1). The mutual friction between the two fluids, which is the major frictional force, is proportional to L.

The nonideal critical velocity observed by WINKEL *et al.* in a narrow channel has not been considered theoretically. The theory of ref. [10] is not applicable in a narrow channel, although presumably the same fundamental type of critical velocity could still be found.

It should be noted that in the work of ref. [10] it was assumed that the normal fluid flow is not turbulent in a heat current. This seemed a reasonable assumption, since the Reynolds number for the normal fluid in most heat flow experiments has been well below the critical value (and it also seemed reasonable, although not certain, that the mutual friction in a heat flow would have a stabilizing effect on the normal fluid flow). The work of TACONIS and his students described at this school suggests, however, that

(*) Figure 1 also shows that there exists a possible hysteresis associated with the critical velocity (indicated by the broken lines). Hysteresis effects of this type h[illegible] been observed in *some* heat flow experiments [20], but not in cases where the critical velocity is known with certainty to be of the nonideal type. The absence of hysteresis in the experiments of VINEN and CHASE may well have been due to the presence either of too much vibration or of too much disturbance to the flow by end effects to allow the system to remain in a metastable condition.

even in a heat flow, where the two fluids cannot move together, some form of turbulence is set up in the liquid as a whole when the Reynolds number calculated using the normal fluid velocity and the *total* density exceeds the usual critical value. The work of ref. [10] should therefore be re-examined.

2·2. *Critical velocities in Couette motion.* – The flow of helium between rotating concentric cylinders has been studied experimentally by DONNELLY [14] and theoretically by DONNELLY and CHANDRASEKHAR [15]. Two critical velocities are observed; one associated apparently with the normal fluid, the other with the superfluid. The latter critical velocity must involve a transition from a simple rotational flow with a uniform array of vortex lines to a turbulent flow, and it therefore provides an example of another type of non-ideal critical velocity. The problem is analogous to the famous stability problem in classical hydrodynamics solved by G. I. TAYLOR [16], and the theory of DONNELLY and CHANDRASEKHAR is essentially a generalization of TAYLOR'S work. This generalization is based on a simplified model, in which the vortex lines in the superfluid are replaced by a continous vorticity and a mutual friction; the validity of this procedure has been questioned by HALL [17], but the preliminary experimental results obtained by DONNELLY seem to be in fair agreement with the theory.

The case of a heat current in a wide channel and this case of Couette flow are the only ones that have been investigated in detail where the critical velocity is known with fair certainty to be of the non-ideal type. As will be mentioned in Section **3**, many other observed critical velocities may well be of this type, but the evidence is insufficient to decide. It should perhaps be emphasized that these nonideal critical velocities probably involve to a large extent hydrodynamical problems that are not fundamentally different from those encountered in classical hydrodynamics; as we shall see in the next section, ideal critical velocities involve rather different problems, largely quantum-mechanical, which are of greater interest to the physicist.

3. – Ideal critical velocities.

Critical velocities for which the subcritical flow has been frictionless within the experimental error have been observed on many occasions (see review in ATKINS [18]; for examples of more recent work see ref. [19-23]), and it has been generally assumed in effect that these are ideal in the sense defined in Section **1**. This may not be true. The subcritical friction may have been merely too small to measure, as indeed it was in the experiments of ref. [10], where an indirect method, probably inapplicable in other cases, had to be used to detect it. The only experiments where one can be at all sure that there was frictionless flow are the « persistent current » experiments: *i.e.* experiments where the superfluid was observed to rotate indefinitely in a steadily rotating vessel. Unfortunately very few such experiments have been carried out; in fact, only three (HALL [24], VINEN [25], REPPY and LANE [26]), and in no case has there yet been any detailed investigation. Even in these experiments, as will be explained later, the subcritical flow may not have been purely potential, but may have contained some vortex line held in metastable equilibrium, so that the associated critical velocities might not have been strictly ideal. However, even if this proves to be the case, the experiments remain important in that they provide the best evidence that pure frictionless superflow is possible, and that ideal critical velocities must therefore in principle exist.

It should be emphasized that it is not being maintained that practically all observed critical velocities are definitely nonideal: only that this possibility must be borne in mind. Film flow is probably ideally frictionless; and the critical heat currents observed in the very careful experiments of MENDELSSOHN, BREWER and EDWARDS [19-22] with channels of intermediate width (less than about 1 mm) may also have been ideal (*). And it should also be noted of course that the existence of a *nonideal* critical velocity does not preclude the existence of a smaller *ideal* critical velocity in the same experiment. It is merely important that it should be realized that the experimental evidence in relation to ideal critical velocities is as yet inadequate and difficult to interpret, and it is by no means clear what facts have to be explained by a theory. More experiments are required, but the design of suitable experiments is much easier if some guidance can be provided by a framework of theory. The rest of the present paper will therefore be concerned with this theory. A complete theory is not yet possible, but it is believed that enough can be done to give guidance to the experimentalists and to provide the professional theoreticians with some interesting unsolved problems.

3·1. *The Landau critical velocity criterion.* – Throughout the following discussion we shall assume for simplicity that we are dealing with the flow of superfluid through a long straight channel; the same general principles probably apply in other cases.

Suppose to begin with that the flow takes place at absolute zero. Provided the velocity is everywhere the same (or that it is at least irrotational)

(*) For a discussion of the relationship between these experiments and those of ref. [10], see ref. [7]. If the critical velocity observed in the former experiments is ideal, then the mechanism responsible for the last term in eq. (1) must in this case have been absent.

this state can be described (at least approximately) by the wave function

$$\psi(\boldsymbol{R}_1, \boldsymbol{R}_2, ..., \boldsymbol{R}_N) = \left(\exp\left[i \sum_j s(\boldsymbol{R}_j)\right]\right) \Phi(\boldsymbol{R}_1, \boldsymbol{R}_2, ..., \boldsymbol{R}_N)\,, \tag{2}$$

where Φ is the ground state wave function for the N atoms of the *stationary* liquid, and $\hbar s(r)/m$ is the velocity potential for the flow. The function (2) represents a stationary state of the system (it is in fact obtained merely by an adiabatic perturbation of the stationary ground state by the moving boundaries that set up the flow), and in the absence of any perturbation it will not change. However, a perturbation can be provided by interaction with the walls of the channel through which the flow takes place (the interaction arises from the continual oscillation that must be present on an atomic scale in the position of the wall), and as a result of this interaction two types of process could in principle take place: (i) the liquid might slow down as a whole, the velocity remaining irrotational (*i.e.* there might be a direct transition from the state (2) to a similar state with a smaller velocity); or (ii) some form of localized motion, or « excitation », might be generated. Since we are dealing with a liquid, the first process is, however, highly improbable (negligible transition probability), so that only the second need be considered. Now since this second process must be accompanied by the excitation of some motion in the solid, a *necessary* condition for the process to take place is that the liquid as a whole shall thereby lose energy. This condition can easily be shown to be satisfied only if the velocity of flow satisfies the condition

$$v_s > \varepsilon/p\,, \tag{3}$$

where ε is the energy of the « excitation » and p is the component of its momentum parallel to v_s (both referred to co-ordinates moving with the liquid). This is the Landau criterion for the existence of superfluidity; it states that a sufficient condition for the existence of ideal superfluid flow at velocities less than v_c is that there should be no possible « excitation » with $\varepsilon/p < v_c$. In his original theory Landau believed that the only excitations possible were phonons and rotons, and for these the minimum value of ε/p is about $6\cdot10^4$ cm s^{-1}, which is of course much larger than any observed critical velocity. (Strictly speaking the quantity p appearing in eq. (2) should probably be the quantity known to hydrodynamicists as the « impulse » (LAMB [29]); see discussions by LIN and by CHESTER at this Summer School.)

3·2. *The creation of vortex line in flow in a smooth channel at absolute zero.* – This difficulty was removed in principle as soon as it was realized that quantized vortices can exist in the superfluid, since these can provide other types of localized « excitation », for which ε/p can be much smaller. Thus we might consider « excitations » in the form of vortex rings of radius b [18, 28]. It can be shown [29] that such rings have kinetic energy and momentum (or strictly speaking, impulse) given by

$$\varepsilon = \varrho \frac{h^2 b}{2m^2}\left(\ln \frac{8b}{a_0} - \frac{7}{4}\right), \tag{4}$$

and

$$p = \pi\varrho \frac{h}{m} b^2 , \tag{5}$$

(provided the ring is not close to a solid boundary), where a_0 is the radius of the core of the line, so that they will have associated with them a critical velocity

$$v_c = \frac{\hbar}{mb}\left(\ln \frac{8b}{a_0} - \frac{7}{4}\right). \tag{6}$$

In a channel of width d, the radius b will have a maximum value of order d, so that the minimum critical velocity will be given roughly by

$$v_c = \frac{\hbar}{md}\left(\ln \frac{8d}{a_0} - \frac{7}{4}\right). \tag{7}$$

And it can be shown that a similar result is likely to apply to vortex line of more general shape, although the ring probably has the shape having the minimum value of v_c. The critical velocity (7) is always much smaller than that derived for phonons and rotons, and indeed it does agree very roughly, in both its order of magnitude and its dependence on channel width, with some of the experimentally observed critical velocities (see, for example, ref. [18], p. 200). This appears to be very satisfactory, because the small length of vortex line produced as the critical velocity is exceeded will act as a nucleus for the growth of more line, so that, as appears to be observed, a turbulent tangle of vortex line will eventually be produced.

However, the difficulties have not really been removed, as we shall now show. The trouble is that, although the condition (3) is a *necessary* one for the breakdown of ideal superflow, it is not a *sufficient* one; it is still necessary that the breakdown process envisaged, in this case the creation of, say, a vortex ring, can occur with an appreciable probability. The fundamental process involved is a quantum mechanical transition between a state of uniform flow and a state of flow that includes the ring, the transition being induced by interaction with wall and accompanied by the excitation of some motion in the wall. Now we know from experiment (*e.g.* ref. [10]) that ideal superflow can have broken down at velocities as low as $3 \cdot 10^{-2}$ cm s^{-1}; and this would imply the creation of vortex rings (or something similar) with radii as large as $3 \cdot 10^{-2}$ cm. But the creation of such a large ring would involve a change in the wave function over a large volume of the liquid at considerable distances from the wall, and it is difficult to imagine that such a process could be appreciably more probable that the process of slowing the liquid down as a whole (the possibility of which we have already dismissed), even if the fact is recognized that any real wall will be rough on an atomic scale. It should be emphasized that the ring cannot grow gradually from a small size to a large size; for the creation of a small ring involves a considerable *increase* in energy of the liquid, not a decrease. Thus, in a sense, the difficulty is that the creation of vortex line is opposed by a large potential barrier.

Of course, this assertion that large vortex rings are very unlikely to be created (or indeed that the liquid cannot slow down as a whole) is based on intuition, and an attempt should be made to justify it rigorously. And this justification should include a calculation of the size of the largest ring for which there is an appreciable probability of creation. Unfortunately, the problems encountered here appear at present to be intractable, but it is instructive to consider them. The following analysis is not rigorous or complete, but it serves to illustrate the problems.

Suppose that, at time $t=0$, the liquid is flowing irrotationally with a wave function $\Psi_0(\boldsymbol{R}_1, \boldsymbol{R}_2, ..., \boldsymbol{R}_N)$. This wave function cannot remain strictly unchanged, since the boundaries of the liquid, *i.e.* the channel walls, must be oscillating; at any instant t the function must at least have changed to the eigenfunction $\Psi_0(\boldsymbol{R}_1, \boldsymbol{R}_2, ..., \boldsymbol{R}_N, t)$ that is appropriate to the new instantaneous boundary conditions and has been obtained from $\Psi_0(\boldsymbol{R}_1, \boldsymbol{R}_2, ..., \boldsymbol{R}_N)$ by adiabatic perturbation. This small adiabatic perturbation is of course of no consequence, but because the channel walls are moving there exists the possibility also of a transition to some other state. If this state is described in terms of the complete set of instantaneous eigenfunctions $\Psi_k(\boldsymbol{R}_1, \boldsymbol{R}_2, ..., \boldsymbol{R}_N, t)$ of the system by coefficients a_k, the probability of a transition is described by the following solution of the Schrödinger wave equation

$$\dot{a}_k = \sum_{n \neq k} \frac{a_n}{\hbar\omega_{kn}} \left[\exp\left[i\int_0^t \omega_{kn}\,\mathrm{d}t'\right]\right] \left(\frac{\partial H}{\partial t}\right)_{kn}, \tag{8}$$

where ω_{kn} is the angular frequency associated with the transition $n \to k$, H is the instantaneous Hamiltonian, and

$$\left(\frac{\partial H}{\partial t}\right)_{kn} = \int \dots \int \Psi_k^* \frac{\partial H}{\partial t} \Psi_n \,\mathrm{d}\boldsymbol{R}_1 \dots \mathrm{d}\boldsymbol{R}_N. \tag{9}$$

(SCHIFF'S: *Quantum Mechanics*, p. 209). If we assume that the channel wall can be represented by a step function in the potential, of height V_0, moving at any point with velocity $\boldsymbol{u}$, then

$$(10) \qquad \left(\frac{\partial H}{\partial t}\right)_{kn} = \int\dots\int\left[\Psi_k^*\left\{\int\sum_j V_0\,\delta(\boldsymbol{R}_j-\boldsymbol{R})\,\boldsymbol{u}\cdot\mathrm{d}\boldsymbol{s}\right\}\Psi_n\,\mathrm{d}\boldsymbol{R}_1\dots\mathrm{d}\boldsymbol{R}_N\right.,$$

where the inner integration is carried out over the boundary of the liquid. (It can easily be shown that $(\partial H/\partial t)_{kn}$ remains finite as V_0 tends to infinity.)

It follows therefore that the calculation of the probability of exciting a vortex ring, say, involves a calculation of the matrix element (10), where Ψ_n is the liquid wave function for the initial potential flow and Ψ_k is the liquid wave function for the final state with a vortex ring. If we use the simple FEYNMAN wave functions for these flows [2], both Ψ_n and Ψ_k will be of the form (2), although the velocity potential $\hbar s/m$ will be multivalued when a vortex is present. The matrix element will then contain terms of the form

$$(11) \qquad \int\dots\int\left\{\int\delta(\boldsymbol{R}_j-\boldsymbol{R})\,\boldsymbol{u}\cdot\mathrm{d}\boldsymbol{s}\right\}\left[\exp\left[i\sum_j\{s_n(\boldsymbol{R}_j)-s_k(\boldsymbol{R}_j)\}\right]\right]\Phi^*\Phi\,\mathrm{d}\boldsymbol{R}_1\dots\mathrm{d}\boldsymbol{R}_N\,.$$

Now it is easily seen that if Ψ_k represents a state containing a large vortex ring (large compared with an interatomic spacing) the exponential term in this expression will, owing to the disordered nature of the liquid, contribute positively and negatively with almost exactly equal probabilities, so that the whole integral must be extremely small. (The integral may in any particular case vanish simply because of the form of the perturbation provided by the boundary, but we must assume that this is not generally the case.) However, we must remember that the overall transition probability will be governed by the product of the square of (11) and the density of states associated with the vortex ring, and, as we shall see in the next section, this density of states may be very large. A more detailed investigation than is possible here shows that as a result it is not obvious from the mathematics that the overall transition probability is negligible. But the detailed investigation also shows that for the present purpose the Feynman wave function for a vortex is not sufficiently accurate (note that even the orthogonality integral

$$\int\dots\int\left[\exp\left[i\sum_j\{s_n(\boldsymbol{R}_j)-s_k(\boldsymbol{R}_j)\}\right]\right]\Phi^*\Phi\,\mathrm{d}\boldsymbol{R}_1\dots\mathrm{d}\boldsymbol{R}_N\,,$$

does not vanish exactly, as it should), and since the improvement of this wave function represents a very difficult problem the prospects of further immediate progress are not good.

For the time being therefore, we must be content with an intuitive view that the creation of vortex rings (or similar configurations of vortex line) with a size much larger than an interatomic spacing is a highly improbable process in a macroscopically smooth channel at absolute zero, although this view may turn out eventually to be wrong. If we do accept the view, and if we therefore suppose that vortex rings larger than, say, 10^{-6} cm cannot be created, then we are still not in a position to account for the breakdown of ideal superflow at velocities less than about 750 cm s^{-1}.

The situation that we have so far envisaged is of course an ideal one, in that we have supposed the flow to take place at absolute zero in a channel that is everywhere straight and smooth, and we must now see whether the creation of vortex line is likely to be any easier under more realistic conditions.

3·3. *Effect of a finite temperature.* – A finite temperature gives rise to two effects: the presence of phonons in the solid walls; and the presence of thermal excitations, constituting the normal fluid, in the helium. It is reasonable to suppose that the superfluid in itself still behaves like helium at absolute zero (see discussion by DE BOER at this Summer School).

The presence of phonons in the walls means that the superfluid, in interacting with the walls, need no longer lose energy; it can now absorb a phonon from the walls. The maximum energy that can be absorbed in this way will be of the order kT, and therefore the Landau criterion (3) becomes roughly

$$v_s > \frac{\varepsilon - kT}{p} . \tag{12}$$

For vortex line configurations of any appreciable size $\varepsilon \gg kT$, so that for these our earlier conclusions are hardly affected. For phonons and rotons, however, our conclusions must be changed. Interactions in which the superfluid absorbs a phonon from the solid lead of course to the creation of another phonon (or possibly roton) in the liquid, and this phonon joins the normal fluid. The superfluid therefore loses momentum in the sense that its density is decreased. However, this process does not go far. The distribution function for excitations constituting the normal fluid is changed (formally because the effective energy of an excitation is changed from ε to $\varepsilon + \boldsymbol{p}\cdot\boldsymbol{v}_s$), and the effects on the superfluid resulting from absorption of phonons from the solid by the liquid and from absorption of phonons (and possibly rotons) by the solid from the liquid then exactly balance. Effectively frictionless flow of the superfluid therefore continues as a state of metastable equilibrium.

Apart from this effect connected with interaction with the wall, the presence of normal fluid can have two other consequences. First, it can lead to a mutual friction force on any moving vortex line that is produced. Consider again the example of a vortex ring of radius b. Such a ring will always move (under its own velocity field) with a velocity [29]

$$(13) \qquad v_t = \frac{\hbar}{2mb}\left(\ln\frac{8b}{a_0} - \frac{1}{4}\right),$$

in a direction perpendicular to its plane, and unless this velocity is less than the relative velocity $(v_s - v_n)$ the resulting mutual friction will, through the Magnus effect, cause the ring to contract. Thus the production of vortex line will tend to be impeded unless rings of a size greater than that given by

$$(14) \qquad v_s - v_n = \frac{\hbar}{2mb}\left(\ln\frac{8b}{a_0} - \frac{1}{4}\right),$$

can be produced directly, and as we have already seen (compare with eq. (7)) this appears to be very unlikely unless b is not much larger than an interatomic spacing, *i.e.* unless $(v_s - v_n)$ exceeds about 400 cm s^{-1}.

Secondly, there exists the possibility that the interactions among the phonons and rotons constituting the normal fluid may occasionally *produce* large lengths of vortex line (for example, large rings of a size greater than that given by eq. (14)); *i.e.* that large lengths of line may occasionally be produced by thermal excitation. This is an important possibility and must be considered carefully.

Suppose first that the helium is at rest in equilibrium. The probability that a given length, l, of vortex line will be present in thermal equilibrium will be given by

$$(15) \qquad w(l) = g(l) \exp\left[-\frac{\varepsilon(l)}{kT}\right],$$

where $\varepsilon(l)$ is the energy of the line, and $g(l)$ is the density of states. The estimation of $g(l)$ presents us with another fundamental difficulty, which reflects again the inadequacy of the simple Feynman wave function for a vortex line. The function $g(l)$ represents the number of configurations possible for a vortex line of given length l. At first sight one is tempted to think that this is infinite, but oviously it cannot be so. The correct wave function for a line would tell us the answer, but the simple Feynman wave function gives us no indication. It seems that the best we can do at present is to follow up an idea mentioned by FEYNMAN (ref. [2], p. 52), which appears reasonable. We assume that the line is like a chain, with links of length equal to an interatomic spacing a. The chain can begin at only a finite number of points in the liquid, equal to the total number of atoms, and two adjacent links in the chain can be oriented with respect to one another in only a finite number, α, of ways. For this model, the value of g is easily seen to be given roughly by

$$g(l) = \frac{V}{a^3}\,\alpha^{l/a} \tag{16}$$

where V is the total volume of the liquid. Using this value, we get for the probability:

$$w(l) = \frac{V}{a^3}\exp\left[-\frac{G(l)}{kT}\right], \tag{17}$$

where

$$G(l) = \varepsilon(l) - \frac{lkT\ln\alpha}{a}. \tag{18}$$

We can now obtain an upper limit for α by noting that, since vortex line is not present in large quantities in thermal equilibrium in helium II, $G(l)$ must be greater than zero for all temperatures below the λ-point [it will be equal to zero at the λ-point if Feynman's suggestion is correct that the λ-point is characterized by a sudden appearance of a great length of line in equilibrium). Therefore, if we put

$$\varepsilon \approx \frac{10\varrho\varkappa^2 l}{4\pi}, \qquad \left(\varkappa = \frac{h}{m}\right), \tag{19}$$

(strictly speaking this energy varies somewhat with the particular configuration of line, but this value will not be far wrong), we obtain the condition

$$\ln\alpha \leqslant \frac{10\varrho\varkappa^2 a}{4\pi k T_\lambda}; \tag{20}$$

i.e.

$$\ln\alpha \leqslant 15. \tag{21}$$

Therefore we have

$$w(l) \leqslant \frac{V}{a^3}\exp\left[-\frac{15l}{a}\left(\frac{T_\lambda - T}{T}\right)\right], \tag{22}$$

from which it is evident that, except perhaps very close to the λ-point, the probability of finding a length of line appreciably greater than an interatomic spacing is negligible.

We now turn to the situation that obtains if the superfluid is flowing with velocity v (normal fluid at rest). The function $G(l)$ is now modified and becomes

(23) $$G_v(l) = \varepsilon(l) + \boldsymbol{p}(l)\cdot\boldsymbol{v} - \frac{lkT\ln\alpha}{a},$$

where $p(l)$ is the momentum associated with the vortex line of length l. Strictly speaking, $p(l)$ is not a constant for a given l, as has been implied, but varies with the particular configuration of line; however, it will turn out to be sufficient if we have an upper limit for the probability of creating vortex line by thermal excitation, and we can therefore assume that $p(l)$ always has its maximum value, which it has when the line takes the form of circular ring:

(24) $$p(l) = \frac{\varrho\varkappa l^2}{4\pi}.$$

Thus we see that, if v is in the opposite direction to p, the function $G_v(l)$, unlike $G(l)$, passes through a maximum at a value of l given by

$$l^* = \frac{5\varrho\varkappa^2 a^2 - 2\pi akT\ln\alpha}{\varrho\varkappa a^2 v},$$

and becomes negative for l greater than $2l^*$. Therefore, if the system were in strict equilibrium, large numbers of large lengths ($> 2l^*$) of line should be present (in fact, an impossible situation, since the superfluid would then be immediately slowed down). However, it is reasonable to assume that these large lengths of vortex line would have to be formed by gradual lengthening, through thermal agitation, of a short length, and we have to discover whether this process is likely to occur.

We notice that the situation here is closely analogous to that occurring in a supersaturated vapour, and it can be shown that, if some plausible assumptions are made about the kinetics of vortex line formation, the rate of production of vortex line for a given value of v can be calculated by a method similar to that given by FRENKEL for the rate of condensation of a supersaturated vapour [31]. The result shows that the characteristic time for the process, *i.e.* the average time taken for the production of one large length of line, will be not less than that given by

(25) $$\frac{1}{\tau} = \frac{B\varrho_n(kT)^{\frac{3}{2}}V}{2\varrho\varrho_s^{\frac{3}{2}}\varkappa^{\frac{3}{2}}a^7v^{\frac{1}{2}}}\exp\left[-\frac{(5\varrho_s k^2 a - 2\pi kT\ln\alpha)^2}{4\pi\varrho\,\varkappa a^2 vkT}\right],$$

so that for temperatures near 1 °K:

(26) $$\tau \approx 10^{-37}v^{\frac{1}{2}}\exp\left[\frac{10^6}{v}\right].$$

This is already equal to about a year when v is still as large as 10^4 cm s^{-1}.

this discontinuity will involve a drastic change in the wave function and hence a large surface energy. This situation he regards as unrealistic (there is indeed experimental evidence that this large surface energy does not exist [30]), and he proposes a model in which pure superflow can take place only if there is a row of quantized vortices close to the wall. There seem to be difficulties with this model: for example, it does not completely eliminate the velocity discontinuity at the wall, and at a finite temperature it involves continuous dissipation through motion of the vortices relative to the normal fluid. The present author believes that the idea behind this model is probably wrong. A velocity discontinuity, or vortex sheet, *within* the liquid does imply a large change in the wave function: in fact, there must be a node in the wave function (see, *e.g.*, ref. [7]), and it is this that is responsible for the surface tension σ introduced above. But it would seem very probable that such a node exists at the boundary of the liquid at all times, whether or not there is any flow, so that there is no *extra* surface energy associated with the flow. But it should perhaps be admitted that it is not certain that this view is correct, and that the problems involved should therefore be subjected to more rigorous investigation than has been possible so far.

3·6. *Theoretical conclusions and comparison with experiment.* – Our theoretical conclusions may be summarized as follows. In a smooth straight channel ideal superflow is unlikely to break down at velocities much less than 750 cm s^{-1}, provided that the temperature is not very close to the λ-point. The presence of small protuberances or sharp corners in the channel can reduce this value considerably, but in practical cases only to about 10 cm s^{-1}. We must emphasize that these conclusions rely on a number of intuitive guesses, which may turn out to be wrong, and also on a very simple model of a vortex line in which the effective core radius is taken to be of order 1 Å, which again may not be adequate [25]. But the conclusions are probably sufficiently reliable to use in a preliminary assessment of the experimental situation and in planning further experiments.

The striking feature of the experimental results is that, especially in the wider channels, ideal superflow is often observed to have broken down at much smaller velocities than those predicted by the present theory. Thus it seems that the difficulty in creating the first length of vortex line associated with the kind of potential barrier discussed in Section 3·1, does not always in practice exist. It is difficult to see how this can be so unless helium that is apparently undisturbed does always contain a few lengths of vortex line, *not* in thermal equilibrium, and that, as is easily seen to be possible, these lengths act as nuclei for the growth of more line as soon as the superfluid velocity exceeds a certain value, which may be quite small.

There is some experimental evidence that this might indeed be the case

During experiments on the detection of single quanta of circulation [25], it was observed that lengths of line do usually occur in apparently undisturbed helium; it is easy to see that they can be held in metastable equilibrium by having their ends tied to small protuberances in the apparatus. Furthermore, some of the observed characteristics of the breakdown process seem to be in accord with this picture. For example, both in the heat flow experiments of MENDELSSOHN, BREWER and EDWARDS [19-22], and in the persistent current experiments of REPPY and LANE [26], the velocity at which clear breakdown took place depended markedly on the previous history of the helium, even though the helium may have remained undisturbed for a long period of time.

It is therefore tentatively suggested that if pure superflow breaks down at a very low velocity ($<\sim 10$ cm s^{-1}) it does so because of a few lengths of vortex line that happen to be present initially in metastable equilibrium, although at higher velocities there is the possibility of creating new line in the neighbourhood of a sufficiently large and sharp protuberance. It should perhaps be emphasized that if this picture is true the precise value of any critical velocity will depend on more complicated considerations than those used to obtain eq. (7), even if the critical velocity is of the ideal type being discussed here. Thus the fact that eq. (7) is sometimes obeyed in practice may well be to some extent fortuitous; or, if the equation does have significance, it may originate in a way different from that indicated in Section 3·2 (for example, it can easily be seen that it might originate in some circumstances if a small length of line held in metastable equilibrium across a channel acts as a nucleus for the growth of more line). Furthermore, the critical velocity will depend on the precise experimental conditions, and this may provide an explanation of the fact that the measurements of critical velocities by different authors in channels of the same width have sometimes yielded widely differing results.

It should be emphasized that in the discussion here it has been assumed that the critical velocity should be taken as the velocity at which ideal superflow first breaks down as the velocity of flow is *increased*. Dr. D. F. BREWER has raised in discussion the possibility that it should be taken as the velocity at which ideal superflow is re-established as the velocity of flow is *reduced*. At present it appears to the author that the processes associated with this second type of critical velocity are likely to be even less easy to understand, and less fundamental, than those associated with the first.

Brief mention may be made of the directions in which future work should apparently be concentrated. On the experimental side an effort should be made to carry out measurements in which the possibility is eliminated that any vortex line is present initially, difficult though this may be; these experiments should preferably be of the persistent current type, partly because this type is more sensitive to small frictional forces, and partly because the ends of the channel, which are always a source of uncertainty, are thereby eliminated. On the theoretical side, effort should be concentrated on improving the Feynman wave function for a vortex line, in the hope that this will lead to satisfactory values for the matrix elements of the type (11) and to values for the density of states associated with vortex lines.

* * *

I am grateful to Dr. D. F. Brewer and Dr. C. G. Kuper for helpful discussions.

REFERENCES

[1] L. Onsager: *Suppl. Nuovo Cimento*, **6**, 249 (1949).
[2] R. P. Feynman: *Prog. in Low Temp. Phys.*, edited by C. J. Gorter, vol. **1** (Amsterdam, 1955), chap. II, p. 36.
[3] J. F. Allen: *Nature*, **185**, 831 (1960); *Proc. 7-th Int. Conf. Low Temp. Phys.* (Toronto, 1960), p. 515.
[4] C. G. Kuper: *Nature*, **185**, 832 (1960); *Proc. 7-th Int. Conf. Low Temp. Phys.* (Toronto, 1960), p. 516.
[5] J. R. Pellam: *Proc. 7-th Int. Conf. Low Temp. Phys.* (Toronto, 1960), p. 446; *Phys. Rev. Lett.*, **5**, 189 (1960).
[6] H. E. Hall: *Adv. in Phys.* (*Phil. Mag. Suppl.*), **9**, 89 (1960).
[7] W. F. Vinen: *Prog. in Low Temp. Phys.*, edited by C. J. Gorter, vol. **3** (Amsterdam, 1961), chap. I.
[8] H. E. Hall and W. F. Vinen: *Proc. Roy. Soc.*, A **238**, 215 (1956).
[9] P. Winkel, A. Broese van Groenou and C. J. Gorter: *Physica*, **21**, 345 (1955).
[10] W. F. Vinen: *Proc. Roy. Soc.*, A **243**, 400 (1957).
[11] G. Careri, F. Scaramuzzi and W. McCormick: *Proc. 7-th Int. Conf. Low Temp. Phys.* (Toronto, 1960), p. 502.
[12] C. E. Chase: *Phys. Rev. Lett.*, **4**, 220 (1960); *Proc. 7-th Int. Conf. Low Temp. Phys.* (Toronto, 1960), p. 438.
[13] C. E. Chase: *Phys. Rev.*, **120**, 688 (1960).
[14] R. J. Donnelly: *Phys. Rev. Lett.*, **3**, 507 (1959).
[15] S. S. Chandrasekhar and R. J. Donnelly: *Proc. Roy. Soc.*, A **241**, 9 (1957).
[16] G. I. Taylor: *Phil. Trans.*, A **223**, 289 (1923).
[17] H. E. Hall: *Proc. Roy. Soc.*, A **245**, 546 (1958).
[18] K. R. Atkins: *Liquid Helium* (Cambridge, 1959).
[19] D. F. Brewer, D. O. Edwards and K. Mendelssohn: *Phil. Mag.*, (8) **1**, 1130 (1958).
[20] D. F. Brewer and D. O. Edwards: *Proc. 5-th Int. Conf. Low Temp. Phys.* (Madison, Wis., 1957).

[21] K. Mendelssohn: *Suppl. Nuovo Cimento*, **9**, 228 (1958).
[22] K. Mendelssohn and W. A. Steel: *Proc. Phys. Soc.*, A **73**, 144 (1959).
[23] J. N. Kidder and W. M. Fairbank: *Proc. 5-th Int. Conf. Low Temp. Phys.* (Toronto, 1960), p. 560.
[24] H. E. Hall: *Phil. Trans.*, A **250**, 359 (1957).
[25] W. F. Vinen: *Proc. Roy. Soc.*, A **260**, 218 (1961).
[26] J. Reppy and C. T. Lane: *Proc. 7-th Int. Conf. Low Temp. Phys.* (Toronto, 1960), p. 443; *Phys. Rev. Lett.*, **5**, 541 (1960).
[27] L. D. Landau: *J. Phys. U.S.S.R.* **5**, 71 (1941); **11**, 91 (1947).
[28] B. T. Geilikman: *Žurn. Ėksp. Teor. Fiz.*, **37**, 891 (1959).
[29] H. Lamb: *Hydrodynamics*, 6-th ed. (Cambridge, 1952), page 239.
[30] G. A. Gamtsemlidze: *Žurn. Ėksp. Teor. Fiz.*, **34**, 1434 (1958); (*Sov. Phys. J.E.T.P.*, **7**, 992 (1958)).
[31] J. Frenkel: *Kinetic Theory of Liquids* (Dover, 1955).
[32] N. F. Mott: *Phil. Mag.*, **40**, 61 (1949).
[33] V. L. Ginsburg: *Žurn. Ėksp. Teor. Fiz.*, **29**, 254 (1955).
[34] C. G. Kuper: *Physica*, **22**, 1291 (1956).

Liquid Helium II Heat Conductivity and Fountain Pressure: Measurements and Calculations for Narrow Slits (*).

E. F. HAMMEL, W. E. KELLER and P. P. CRAIG

University of California, Los Alamos Scientific Laboratory - Los Alamos, N. Mex.

Introduction.

The objective of this work was to measure the heat conductivity and the fountain pressure in liquid helium II using narrow slits of well-defined geometry. In addition to obtaining the usual limiting values of these quantities as $\Delta T \to 0$, data were obtained over the maximum possible temperature difference for each reference temperature. The results were then compared with theoretical descriptions of these properties. A portion of this work has already been published [1].

Many previous measurements of either the fountain pressure, the heat conductivity or both in narrow slits or channels have yielded results in disagreement with each other and with theory [2, 3]. Recently several additional studies, notably those by BROESE VAN GROENOU, POLL, DELSING and GORTER [4], by BURNHAM, REPPY, PEARSON, SPEES and REYNOLDS [5], and the work with larger channels by BREWER and EDWARDS [6] have been completed and have yielded results more consistent with each other, with related work, and with theory. But even in these experiments the geometry of the smallest channels has not been well defined. Neither has it been possible with some of these channels to measure the heat flow or fountain pressures over wide temperature differences, and with the smallest slits, systematic discrepancies between theory and experiment were still observed. Since the slits used in the present work have well-defined geometries and can also tolerate high

---

(*) Work done under the auspices of the U.S. Atomic Energy Commission. Seminar delivered by E. F. HAMMEL.

pressures across them, it was considered worth-while to repeat and extend previous measurements.

1. – The apparatus.

The apparatus used in the heat conduction measurements is shown in Fig. 1; further details on its construction, etc., are available in ref. [1]. During heat flow experiments the associated fountain pressure was simultaneously obtained by measuring the decrease in volume of the liquid in the hot cell

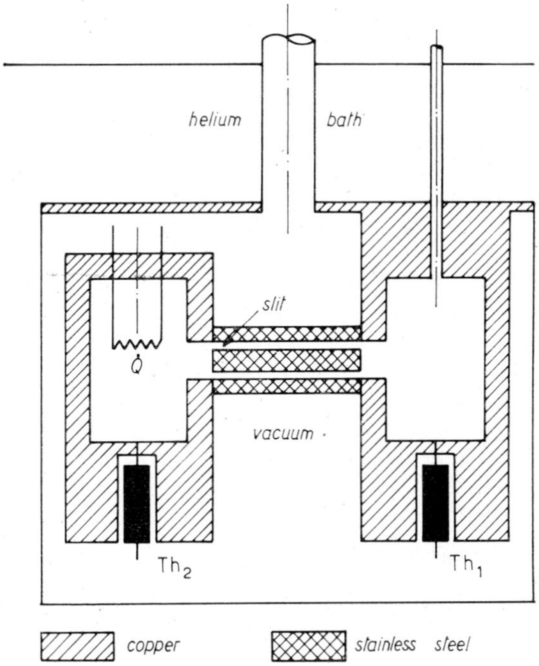

Fig. 1. – Schematic of heat conduction apparatus

caused by the fountain pressure, and computing this pressure from knowledge of the liquid compressibility and density. The data so obtained gave in certain instances fountain pressures slightly larger than the ideal, and these results suggested that the measurements be repeated, preferably using a different pressure-sensing method. A second set of measurements was therefore carried out in which the fountain pressure was measured directly by attaching a commercial [7] pressure transducer to the hot cell. It has subsequently been shown that these direct measurements are more reliable than the previously reported ones.

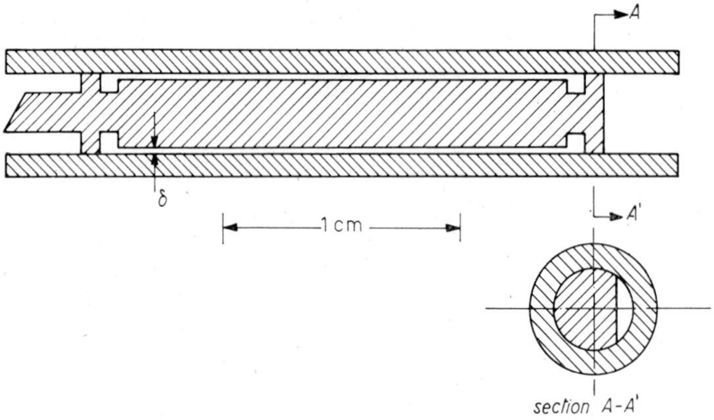

Fig. 2. – Details of slit geometry. δ_1 average gap width from viscous flow of air at 300 °K; δ_2 average gap width from viscous flow of He at 4 °K: slit I: $\delta_1=2.12\ \mu m$, $\delta_2=2.15\ \mu m$; slit II: $\delta_1=$ —, $\delta_2=0.276\ \mu m$; slit III: $\delta_1=3.36\ \mu m$, $\delta_2=$ —.

The slit assembly and the results of the slit calibrations are shown in Fig. 2. Further details on the slit construction and calibration may also be found in ref. [1]. All the slits were shown to be uniform in diameter and accurately circular to at least $\pm 0.1\ \mu m$, which was the limit of accuracy of the gauges used for direct measurement.

2. – Correction for heat flow through stainless steel of slit.

In order to measure accurately the thermal conduction of the helium, it is essential to minimize the conduction through the slit itself and to make the remaining corrections accurately. This was done by constructing the slit of stainless steel, carrying out an independent study of the thermal conductivity of an identical specimen of the steel and by measuring the thermal conductivity of the slit itself *in situ*. The possibility still existed, however, that the stainless steel (with a thermal conductivity varying linearly with the temperature) in contact with the liquid helium (with an effective thermal conductivity varying approximately as the 12-th power of the temperature) would seriously perturb the flow of heat in the helium. In this connection R. LAZARUS [8], J. E. KILPATRICK [9] and R. GARWIN [10] have independently investigated the heat flow in an annular right cylinder composite system of parallel thermal conductors under the following conditions:

a) the usual diffusion equation holds with boundary conditions of flux matching;

b) the external walls are adiabatic;

c) the two plane ends of the cylinder are maintained at uniform temperatures T_0 and T_1.

d) the thermal conductivity of each material is described by a well behaved function of the temperature.

Under these assumptions it was found that the total heat flow from T_1 to T_0 is independent of whether or not heat flows across the interface from one material to the other, and that therefore the total heat flow may be calculated correctly by superposing the independent contributions of the various parallel conductors. The boundary condition of temperature equality in each material at every point along the interface was not required to obtain this result.

At low heat currents the transport of heat in He II is described by linear equations of motion. Thus for small values of ΔT, the heat flow along the helium II-stainless steel system may be calculated by superposition of the heat flow in the helium and in the stainless steel calculated separately. The calculations for small ΔT are found to be in excellent agreement with the experimental results.

For large values of ΔT the linear He II equations of motion no longer obtain, and it is necessary to introduce nonlinear terms. Our experiments have been analyzed in terms of the GORTER-MELLINK [11] mutual friction, using VINEN's [12] recent wide slit values for the Gorter-Mellink constant. With this equation it is no longer rigorous to superpose separate solutions for the He II and for the stainless steel. We have been unable to solve these combined heat transport equations in two dimensions using realistic boundary conditions which permit heat to flow across the He-stainless steel interface. However, if one does calculate the heat flow in the stainless steel and in the helium separately at the high heat currents where the nonlinear terms are admittedly important, one finds that the stainless steel contribution to the total flow is small compared to that of the helium. We therefore argue that the stainless steel in parallel with the He II probably exerts only a small perturbation on the total heat flow, and have used superposition in all of our calculations.

3. – Method of measurement and results.

For both the heat conduction and fountain pressure measurements the « cold » cell was maintained at a constant temperature designated as T_0. The

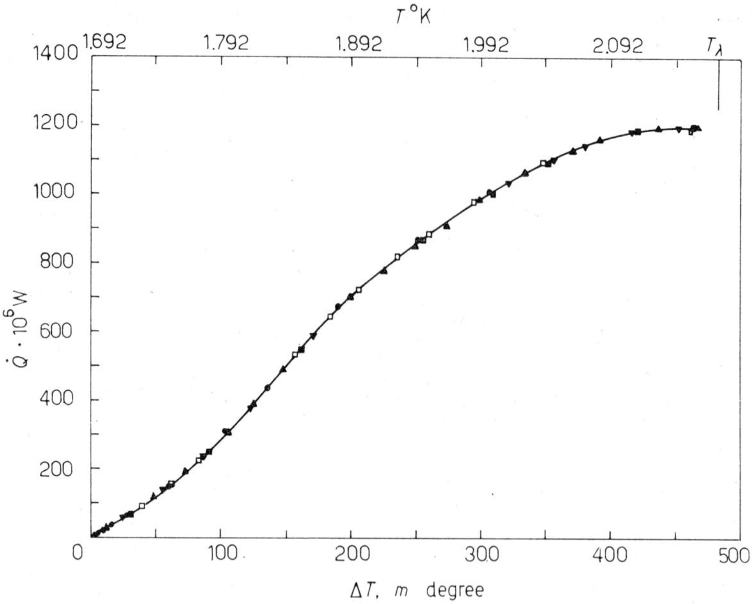

Fig. 3. – $\dot{Q}$ versus ΔT curve for $T_0 = 1.6916$ °K showing reproducibility of results: ● 18 Oct. 1957, □ 19 Oct. 1957, ▲ 14 Nov. 1957, ▼ 15 Nov. 1957, ■ 20 Nov. 1957.

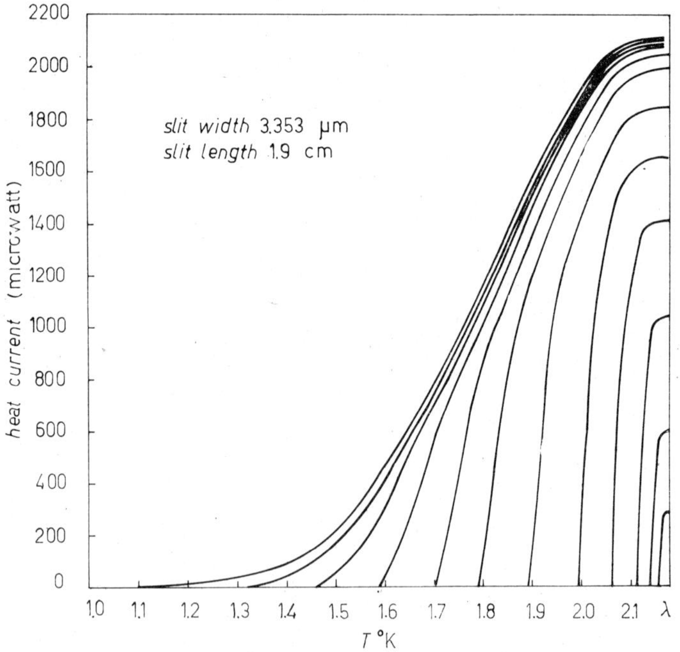

Fig. 4. – Family of heating curves for 3.36 μm slit.

power input to the hot cell was incrementally increased, ΔT, P_f, and $\dot{Q}$ data being recorded for each new power input after equilibrium had been achieved. Then a new cold cell temperature was selected and the sequence repeated. From Fig. 3, in which the reproducibility of the data is displayed, it is clear that representation of each experimental point for all the data would be superfluous. Hence, the data are reported as a family of experimental curves (shown in Fig. 4 for the heat flow measurements using the 3.36 μm slit) one for each separate cold cell temperature. The cold cell temperature T_0 for each run is the intersection of each curve with the abscissa. The associated hot cell temperature and corresponding value of the heat flux is then given by the

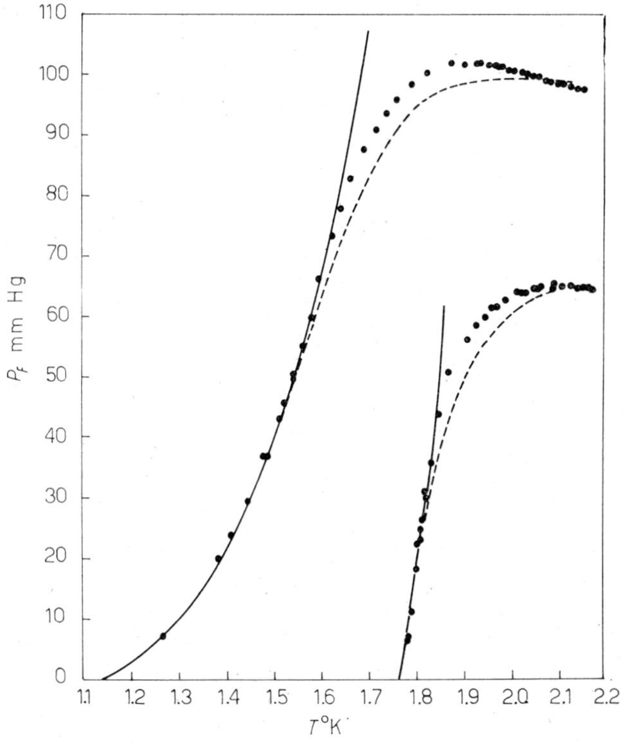

Fig. 5. – Comparison of experimental fountain pressure curves (3.36 μm slit) with the integrated ideal fountain pressure equation (solid curves) and with the integrated Gorter-Mellink equation (dashed curves) using Vinen's A values and no critical velocity. T_0's were 1.143 °K and 1.767 °K.

remainder of each curve from T_0 to T_λ. In Fig. 5 corresponding data are presented for the fountain pressure measurements using the 3.36 μm slit.

4. – Discussion.

In order to systematize these results, Table I summarizes data obtained with the different slits:

TABLE I.

Slit	Gap width	Heat flow measurements	Direct fountain pressure measurements
II	0.276 μm	no (*)	yes
I	2.12 μm	yes	no (**)
III	3.36 μm	yes	yes

(*) Heat flow through stainless steel too large,
(**) Fountain pressure measurements made previously using indirect method.

We now discuss these results in order.

A) Direct fountain pressure measurements from T_0 to T_λ using the smallest slit (slit II) corresponded with the results obtainable from the integrated *H.* London equation $\left(P_f=\int_{T_0}^{T}\varrho s\,\mathrm{d}T\right)$ to within a few percent over the entire range of T_0's to be investigated. Results previously obtained at Leiden both directly by BOTS and GORTER [13] and indirectly by VAN DEN MEIJDENBERG *et al.* [14] as well as measurements by BREWER and EDWARDS [15] are in agreement with these findings except that the earlier data deviated from the theoretical expression at temperatures near T_λ. By combining the previous results and those of the present work, it may be concluded that the validity of the London fountain pressure equation for very small slits, from temperatures of the order of a few tenths of a degree Kelvin to the λ-point, and for temperature differences from one to of the order of a thousand millidegrees, is now satisfactorily established. These considerations are discussed in more detail elsewhere by two of the present authors (EFH and WEK) [16].

B) In order to discuss the 2 and 3 μm slit results systematically it will be advantageous to consider separately for each T_0 the region over which the slit behavior is accurately described by the linear hydrodynamical equations of motion and the region over which dissipation effects other than normal viscosity must be taken into account.

4·1. *The linear region.* – It is well known that several approximations are usually made in describing the flow of heat through narrow slits. Since noninear terms will be added subsequently to describe the nonlinear effects it is important to specify those approximations which have already been made in using the linear theory. From the London-Zilsel hydrodynamic equations for liquid helium II, for the conditions of no mass transfer, stationary flow and no dissipation, one obtains:

$$0 = \varrho s \nabla T - (\eta_n + \eta_n') \nabla(\nabla \cdot \boldsymbol{q}) - \eta_n \nabla^2 \boldsymbol{q} \,. \tag{1}$$

In applying this equation to the flow of heat through narrow slits, the second term on the right-hand side is assumed to be equal to zero on the basis $\nabla \cdot \boldsymbol{q} = 0$. It is then further assumed that $T = T(z)$ only, where the variable z measures distance along the slit. Hence T is considered uniform across a plane perpendicular to the direction of flow. Finally it is assumed that $\boldsymbol{q} = \boldsymbol{q}(x)$ only; that is, $\boldsymbol{q}$ is considered independent of z. With these assumptions the above equation is immediately separable and soluble, after introducing the boundary conditions $\boldsymbol{q}_{\parallel} = 0$, to give the well-known relationship:

$$\langle \boldsymbol{q} \rangle = -\frac{(\varrho s d)^2 T}{12\, \eta_n} \nabla T \,. \tag{2}$$

which has the form of the usual heat conduction equation. To check the validity of this equation in the limit $\Delta T \to 0$, it was used to compute values

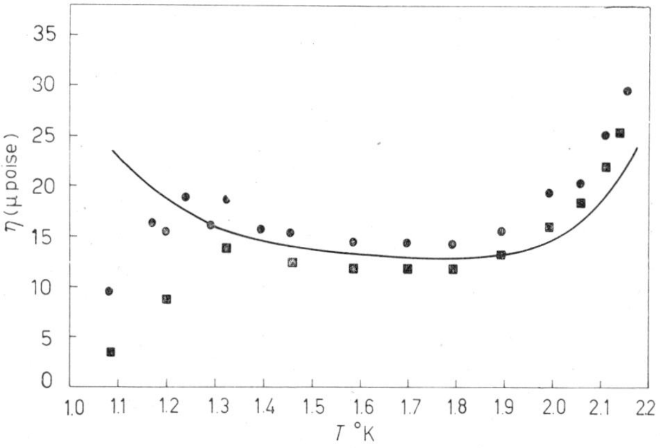

Fig. 6. – Comparison of η_n values from data on 2.13 μm, 2.24 μm, and 3.36 μm slits with bulk liquid data of HEIKKILA and HOLLIS HALLETT [21]: ■ 2.1 μm slit, ● 3.4 μm slit, —— HOLLIS HALLETT. η_n computed from the following equation:

$$\eta_n = \frac{(\varrho S)^2 d^3 \omega T}{12 L \lim\limits_{\Delta T \to 0} (\dot{Q}/\Delta T)} \,.$$

of η_n. These are shown in Fig. 6. Although the agreement is satisfactory, errors of a few percent in the determination of slit width cause large errors in the calculated value of η_n. Hence it is not suggested that these data be considered other than as general confirmation of the linear equations at low ΔT's.

For large ΔT's the above equation [eq. (2)] must be integrated. Rewriting eq. (2)

$$\langle \boldsymbol{q} \rangle = \alpha(T)\frac{\mathrm{d}T}{\mathrm{d}z},$$

where

$$\alpha(T) = (\varrho s d)^2 T / 12\eta_n$$

and integrating we obtain

$$\langle \boldsymbol{q}(T) \rangle = \frac{1}{L}\int_{T_0}^{T} \alpha(T)\,\mathrm{d}T\,. \tag{3}$$

The results are shown in Fig. 7 for the 2.12 μm slit from which it may be noted that the agreement between the calculations and the experiment (except

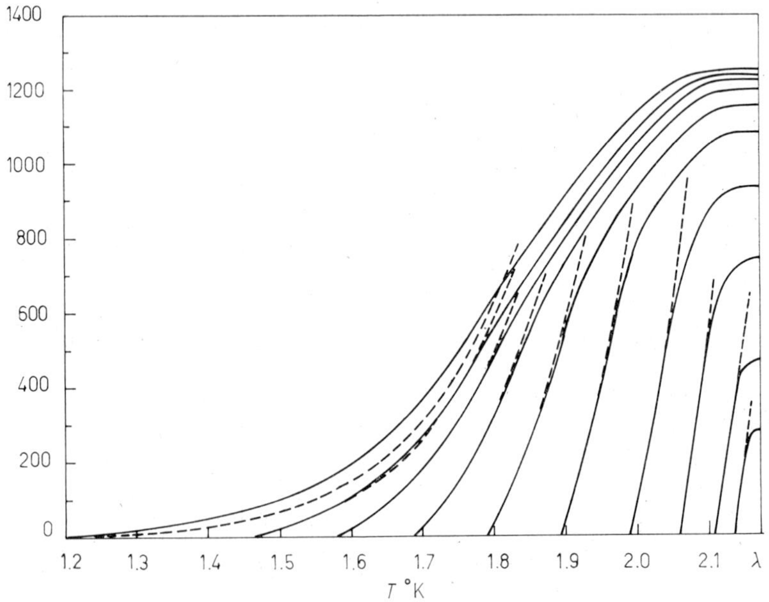

Fig. 7. – Comparison of experimental heating curves for the 2.13 μm slit with calculated results using eq. (3) and assuming additivity of the heat flow through the stainless steel: – – – calculated using

$$\frac{1}{L}\int_{T_0}^{T}\left[\frac{(\varrho S)^2 T\,\mathrm{d}^3\omega}{12\eta} + K_{ss}A_{ss}\right]\mathrm{d}T\,,$$

—— experimental curves.

at the lowest temperatures) over ΔT's as large as 300 mdg is very good. At the lowest temperatures the calculated values lie somewhat below the observed values.

Corresponding data have been obtained for the 3.36 μm slit in the direct fountain pressure measurements, the region of agreement with the integrated fountain pressure equation being approximately the same as was observed in the heat flow experiments as may be seen in Fig. 5.

Critical velocities have been observed in experiments on the flow of He II through narrow channels by BREWER, EDWARDS and MENDELSSOHN [17], MEYER and MELLINK [18], HUNG, HUNT, and WINKEL [19], WINKEL, DELSING and POLL [20], VINEN [12], and many other investigators. The critical velocity denoting the onset of dissipative effects arising from mutual friction might be expected to occur in these slits at very small ΔT's for which the associated superfluid velocities are of the order of a few tenths of a centimeter per second. Since our apparatus was not designed for measurements in this range of ΔT no information on such critical velocities is available from the present work (*).

4·2. *The non-linear region.* – Since the observed results deviate from the prediction of the linear theory for sufficiently large ΔT's it was necessary to select a more detailed description of the heat conductivity and the fountain pressure to compare with the experiments than is afforded by the linear theory. The Gorter-Mellink formulation was chosen as a basis with which the results at these larger ΔT's were to be compared. Again it was necessary to integrate the Gorter-Mellink equation

$$\alpha^{-1}\langle \boldsymbol{q}\rangle + \beta\langle \boldsymbol{q}\rangle^3 = \nabla T$$

where

$$\beta = \frac{A\varrho_n}{\varrho_s^3 s^4 T^3},$$

to give

$$\langle \boldsymbol{q}\rangle = \frac{1}{L}\int_{T_0}^{T} \frac{\alpha\, \mathrm{d}T}{1+\alpha\beta\langle q\rangle^2}.$$

Here A is the coefficient of the Gorter-Mellink mutual friction force

$$F_{sn} = A\varrho_s\varrho_n(v_s - v_n)^3,$$

d is the slit width and L is the slit length.

(*) *Note added in proof.* – For further discussion of these measurements including dissipative effects, see CRAIG, KELLER and HAMMEL: *Annals of Physics*, Jan. 1963.

This equation was solved numerically on the IBM 704 computer by selecting a given value of T_0 and $\langle \boldsymbol{q} \rangle$ and adjusting T until the equality was satisfied. Then $\langle \boldsymbol{q} \rangle$ was increased by an incremental value and the process repeated

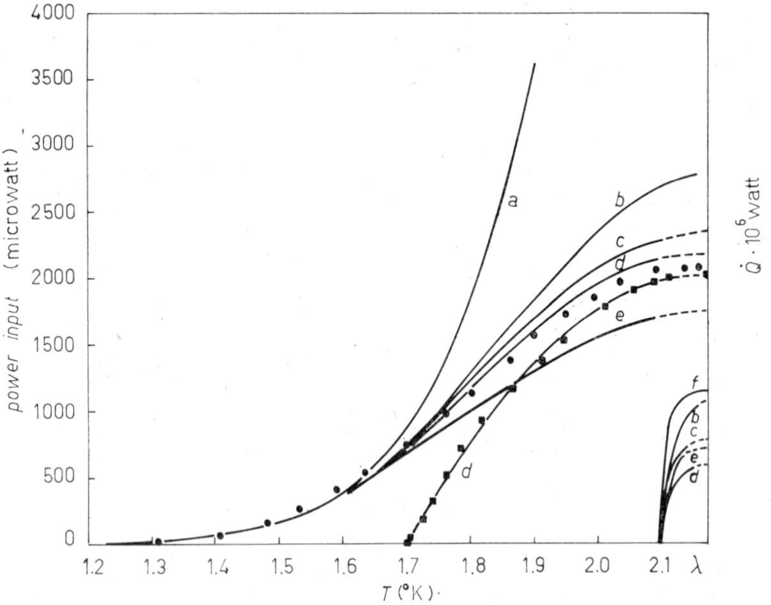

Fig. 8. – Comparison of experimental heating curves for the 3.36 μm slit with calculated curves using different values of the flow parameters: slit width 3.36 μm, slit lengh 1.9 cm. η_n from power measurements, stainless steel conduction added: *a*) integrated heat flow; *b*) A=const=50, exponent=3.0; *c*) VINEN's A, $v_c = 0.09/(\mathrm{d}\varrho_s/\varrho)^{\frac{1}{2}}$ cm/s [22]; *d*) VINEN's A, $v_c=0$; *e*) A=const=50,e xponent=4.0; ● experimental points (T_0=1.2 °K); ■ experimental points (T_0=1.7 °K); *f*) experimental (interpolated) (T_0=2.1 °K).

until the entire heat flow curve was computed. Calculations were made for a variety of parameters as may be seen from Fig. 8 and from which it was concluded that the best overall fit was obtained using VINEN's A values [12] and no critical velocity. Note that there are no adjustable parameters used in this calculation.

In order to compare all of the experimental heat flow results with theory, deviation plots have been prepared. These are shown in Fig. 9. In this figure

$$\Delta T/\Delta T_\lambda = \frac{T - T_0}{T_\lambda - T_0}\,.$$

Upon examining Fig. 9 it is apparent that for T_0's above about 1.9 °K for all $(\Delta T/\Delta T_\lambda)$ and for T_0's below about 1.3 °K for low values of $(\Delta T/\Delta T_\lambda)$

the agreement with the calculated heat flow is poor. This is partly explicable by the fact that in the first mentioned region the Gorter-Mellink correction

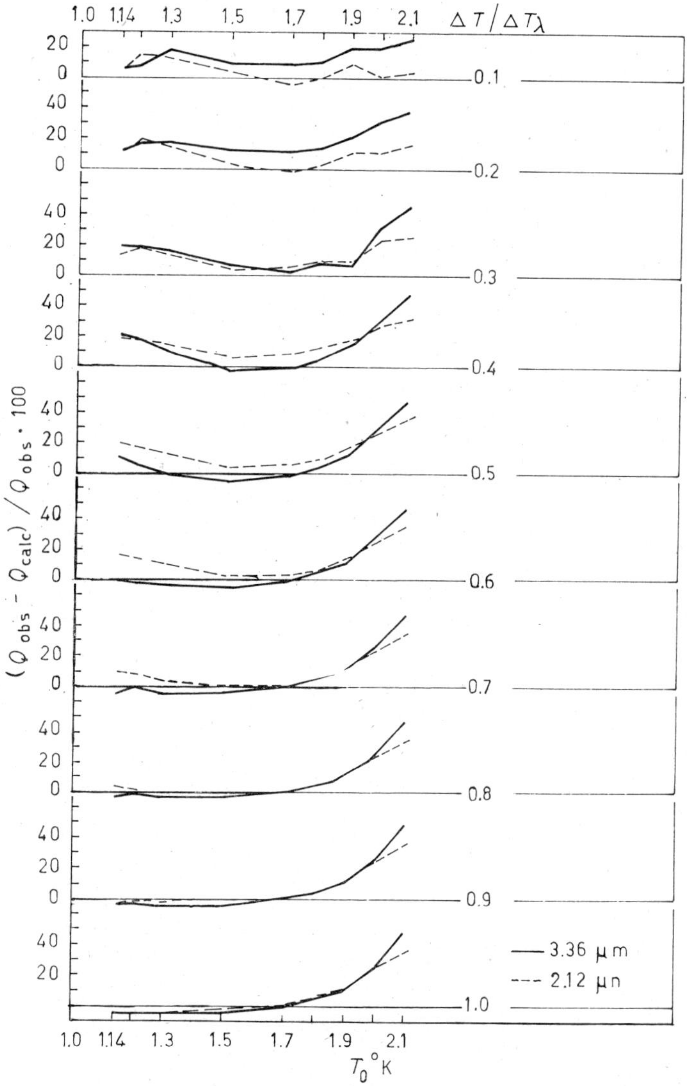

Fig. 9. – Deviation plot for heat conduction measurements using 2.12 μm slit and 3.36 μm slit. Comparison of experimental results with those obtained by integration of the Gorter-Mellink equation, exponent 3, no critical velocity, VINEN's A values.

term becomes large whereas in the last mentioned region the stainless steel correction term becomes large. If the Gorter-Mellink mutual friction is interpreted in terms of the vortex line theory, it may be shown that the region of low $\Delta T/\Delta T_\lambda$ corresponds with a region in which the mean spacing between vortex lines is of the same order of magnitude as the slit width. Under these circumstances statistical equilibrium is no longer maintained among the vortex lines and the form of the mutual friction forces might be expected to change. The results of this analysis are shown in Fig. 10.

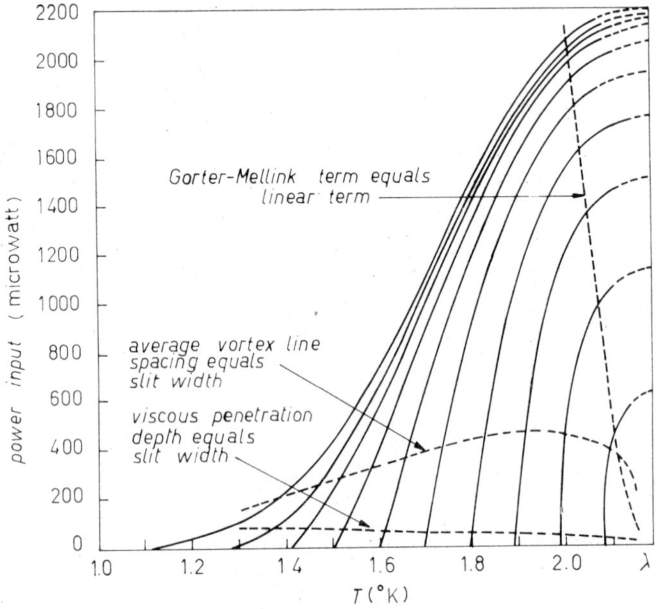

Fig. 10. – Calculated heat conduction curves assuming the Gorter-Mellink dissipative term. The region where the Gorter-Mellink begins to play an important role is indicated, as is the minimum heat current such that the line spacing is less than the slit width (slit width 3.36 μm, slit lenght 1.9 cm. GORTER-MELLINK from VINEN, stainless steel conduction added, no critical velocity).

Fountain pressures were also calculated for the 3.36 μm slit from the equation:

$$P_f = \int_{T_0}^{T} \frac{\varrho s \, \mathrm{d}T}{1 + \alpha\beta \langle q \rangle^2} \, .$$

These calculated fountain pressures are also subject to errors introduced by the use of incorrectly calculated values of $\langle q \rangle^2$ for the ΔT in question because of the non-additivity of the contributions of the He II and the stainless steel to the heat flow when the linear equations are no longer applicable. Again on the assumption that these errors should be small, and that additivity could be assumed, P_f's have been calculated and are compared with the experimental results in Fig. 5.

Thus, subject to the assumption that the computational errors are indeed small, it is concluded that the Gorter-Mellink equations used with Vinen's A values satisfactorily represent the heat flow and fountain pressure data for liquid helium II in narrow slits over extended temperature differences.

REFERENCES

[1] W. E. KELLER and E. F. HAMMEL jr.: *Ann. Phys.*, **10**, 202 (1960).
[2] F. LONDON and P. R. ZILSEL: *Phys. Rev.*, **74**, 1148 (1948).
[3] H. FORSTAT: *Phys. Rev.*, **111**, 1450 (1958).
[4] A. BROESE VAN GROENOU, J. D. POLL, A. M. DELSING and C. J. GORTER: *Physica*, **22**, 905 (1956).
[5] J. BURNHAM, J. REPPY, G. J. PEARSON, A. H. SPEES and G. A. REYNOLDS: *Proc. VII Int. Conf. Low Temp. Phys.* (Toronto, Aug. 1960), p. 564, edited by G. M. GRAHAM and A. C. HOLLIS HALLETT (Toronto, 1961); *Physics of Fluids*, **3**, 735 (1960).
[6] D. F. BREWER and D. O. EDWARDS: *Proc. Roy. Soc.* (*London*), A **251**, 247 (1959).
[7] Consolidated Electrodynamics Corporation, Transducer Division, Monrovia, Cal.
[8] Private communication.
[9] Private communication.
[10] Private communication.
[11] C. J. GORTER and J. H. MELLINK: *Physica*, **15**, 285 (1949).
[12] W. F. VINEN: *Proc. Roy. Soc.*, A **240**, 114, 128 (1957).
[13] G. J. C. BOTS and C. J. GORTER: *Physica*, **22**, 503 (1956); **26**, 337 (1960).
[14] C. J. N. VAN DEN MEIJDENBERG, K. W. TACONIS, J. J. M. BEENAKKER and D. H. N. WANSINK: *Physica*, **20**, 157 (1954).
[15] D. F. BREWER and D. O. EDWARDS: *Proc. Phys. Soc.*, **71**, 117 (1958).
[16] E. F. HAMMEL, jr. and W. E. KELLER: *Phys. Rev.*, **124**, 1641 (1961).
[17] D. F. BREWER, D. O. EDWARDS and K. MENDELSSOHN: *Phil. Mag.*, **1**, 1130 (1956).
[18] L. L. MEYER and J. H. MELLINK: *Physica*, **13**, 197 (1947).
[19] C. S. HUNG, B. HUNT and P. WINKEL: *Physica*, **18**, 629 (1952).
[20] P. WINKEL, A. M. G. DELSING and J. D. POLL: *Physica*, **21**, 331 (1955).
[21] W. J. HEIKKILA and A. C. HOLLIS HALLETT: *Can. Journ. Phys.*, **33**, 420 (1955).
[22] J. G. DASH, *Phys. Rev.* **94**, 1091 (1954).

The Apperance of Friction in Superfluid Helium.

K. Mendelssohn

Clarendon Laboratory - Oxford

1. – Introduction.

Research into the phenomena connected with the first appearance of friction in superfluid helium have been in progress in Oxford for a number of years and some of the earlier results have been reported in here at a previous date [1]. Our work has been confined to flow in tubes of up to about one millimetre diameter and a fairly consistent picture of the onset of dissipation involving the superfluid component has been obtained. Experiments have been carried out of flow initiated by a pressure and by a temperature gradient and these two sets of observations will be treated separately in the following. From all our experiments it appears that turbulence will set in the superfluid when a certain critical velocity is exceeded. This critical velocity appears to be the same, irrespective of whether the flow takes place under a pressure or a thermal gradient. It depends on the tube diameter and the temperature and it is measured relative to the tube and not to the normal component.

2. – Flow under a pressure gradient.

It has been noticed in many of the older experiments on isothermal flow of He II under a pressure gradient that the obtained data showed a rather larger degree of scatter than was to be expected in such a simple kind of observation. Quite erratic behaviour was found by S. M. Bhagat and the author in measurements on a short tube [2] in which the flow, after slowing down for part of the experiment, suddenly re-started with a velocity which, on occasion, was higher than the initial one. This somewhat surprising result led to a systematic investigation in which several thousand individual flow experiments were carried out. Tube diameters of roughly 100, 300, 500 and 700 μm with lengths of 150 cm for the three widest and 100 cm for the finest tube were used. A length of 30 cm of the 100 μm tube was also measured. Care was taken that the flow took place under strictly isothermal conditions and readings were taken for inflow as well as for outflow.

The long 100 μm tube showed almost pure superflow under all conditions, the flow rate rising by only 20% as the level difference was raised from 2 mm to 24 mm. In other tubes it was found that the mean velocity not only depended on the pressure gradient but also on the history of the experiment and it was noticed in particular that the initial level difference, h_i, at which the flow experiment was started, determined the subsequent events. Figure 1 shows a typical set of graphs of the mean velocity, $\bar{v}$, plotted against the pressure gradient, yielding a number of straight lines for different initial heights. Moreover, it should be noted all the lines have finite intercepts in the neighbourhood of 1.5 cm/s for zero pressure gradient. The values given here are those for the short 100 μm tube and they indicate the existence of a critical velocity, v_0, at this value for the temperature of 1.31 °K. The results for all the other tubes follow the same pattern and can be expressed as

$$\bar{v} = v_0 - k \text{ grad } p \tag{1}$$

where k, and to some extent v_0, are functions of h_i. The form of eq. (1) suggests that the flow phenomena can in each case be described by an « effective viscosity », η_{eff}, given by the relation

$$k(h_i) = a^2/8\eta_{\text{eff}} \tag{2}$$

where a is the radius of the tube. Thus the effective viscosity varies with the initial pressure difference and is not an intrinsic property of the liquid. This immediately suggests in analogy with classical turbulence [3] the existence of an eddy viscosity in superfluid helium [4]. In order to determine the seat of this turbulence in our experimental arrangement, the 300 μm tube was cut in half and the two lengths then joined up by a short, wide cylinder. The results obtained are shown in the inset to Fig. 1 and it will first be noticed that the general pattern is identical with that of the 100 μm tube. Secondly, for both values of h_i given here the data are the same for the uncut tube and for that made up of the two halves. Since in the latter there are four ends of the tube involved against the two ends in the former, we must conclude that the turbulence is not an end effect but it is established homogeneously throughout the whole length of the tube. This conclusion is confirmed by the fact that, as is shown in Fig. 2, the effective viscosity plotted against h_i yields the same values for the uncut tube, the two halves in series and the half length of the 300 μm tube.

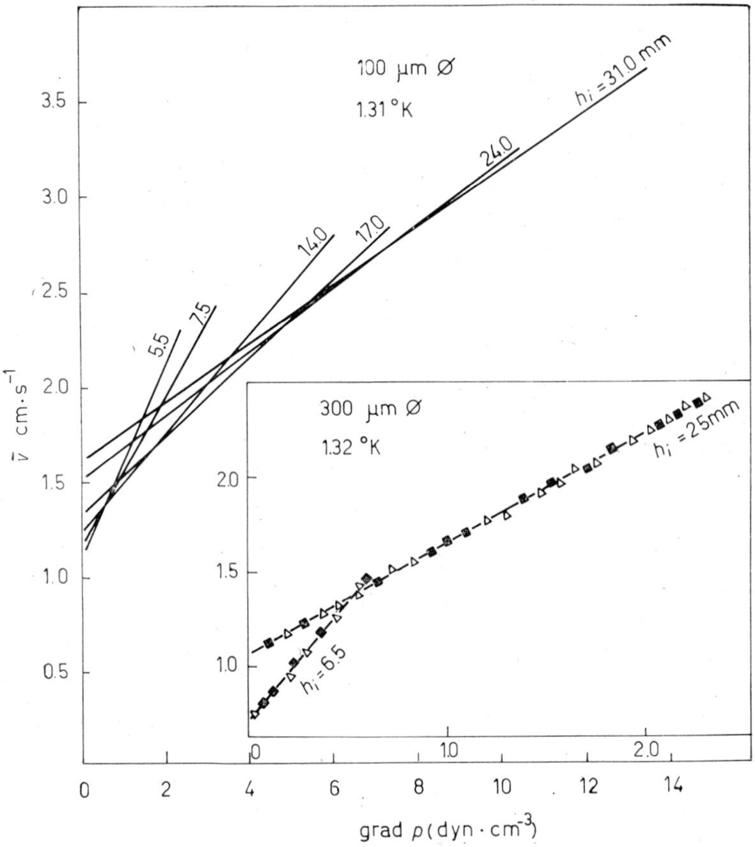

Fig. 1. – Typical graphs for the mean velocity against the pressure gradient, showing the influence of the initial level difference h_i: Inset: △ uncut tube, ■ two halves in series.

Since the magnitude of η_{eff} is always much larger than the value $(\varrho_n\eta_n/\varrho)$ to be expected if the normal fluid alone was contributing by dissipation in laminar flow, the main contribution of the effective viscosity must come from another source. Turbulence in the normal fluid cannot account for it either as the relevant Reynolds number $(\varrho_n v_n a/\eta_n)$ was in our experiments rarely larger than 500. Friction between the superfluid and normal components, as postulated by GORTER and MELLINK [5] must make its appearance if their relative velocity $(v_s - v_n)$ becomes appreciable. While this effect has to be considered in the three wider tubes, it is negligible in the short 100 μm tube. Moreover, for values of h_i which are not too small and at temperatures above 1.6 °K it appears justified to approximate

$$\eta_{\text{eff}} = \eta_s + \eta_n \,. \tag{3}$$

The fact that for each individual flow experiment the eddy viscosity remains constant indicates that at its beginning turbulence up to a certain level is rapidly created and is then maintained during the whole experiment.

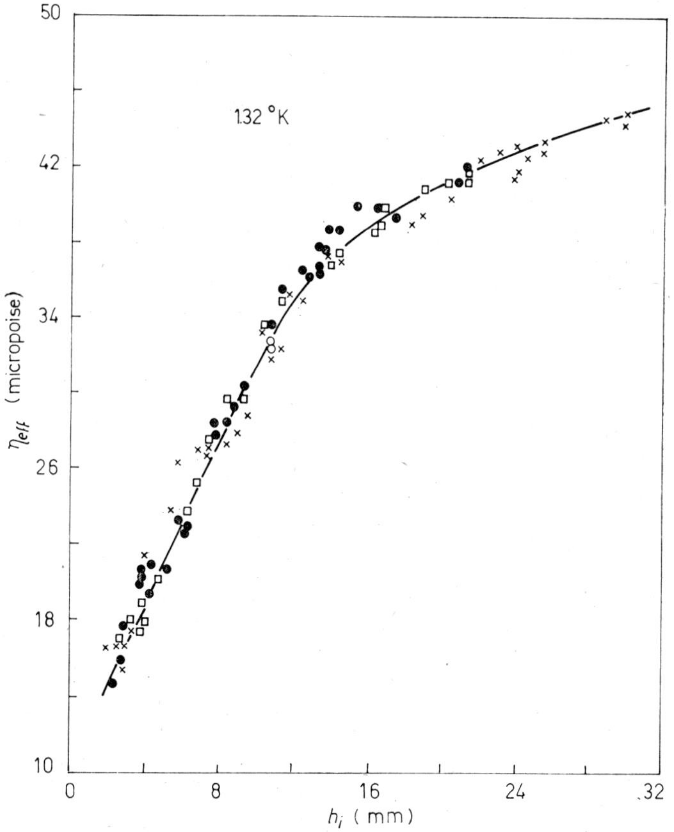

Fig. 2. – The effective viscosity in dependence on the initial level difference, h_i, for the 300 μm tube: ● uncut tube, □ two halves in series, × half tube.

The liquid is probably accelerated initially to a supercritical velocity corresponding to a certain level of turbulence and, assuming such a process, it seems reasonable that this degree of turbulence should depend on h_i.

Summarizing these results, we conclude that the dominant process in the flow of He II under a pressure gradient is turbulence in the superfluid component. In wider tubes this turbulence is responsible for the dependence of the flow on the initial conditions. Dissipation due to the normal component or due to mutual friction between the two components is, generally speaking, of less importance.

3 – Flow under a thermal gradient.

3·1. *The thermo-mechanical pressure.* – Determinations of the heat conductivity of a tube filled with liquid He II carried out several years ago [1, 6] revealed two types of flow mechanism which occurred at low and at high heat currents respectively and which change over into each other at a critical heat current, Q_c. Plotting the thermal resistence, $\Delta T/Q$, against the heat current, Q, one obtains (Fig. 3, inset) at low values of Q a constant value which corresponds to dissipation caused by the laminar stream of normal fluid away from the heater. This is followed, after a narrow transition range to be discussed further below, by a thermal resistence which rises rapidly with Q in a manner as is suggested by the GORTER-MELLINK force [5]. The latter is the dominant process in a heat conductivity experiment but the existence of an eddy viscosity due to turbulence in the superfluid should be noticed when in addition to the thermal gradient the pressure gradient is determined simultaneously. In their early work on the thermo-mechanical effect ALLEN and REEKIE [7] observed that in first approximation the fountain pressure, ΔH, remained proportional to Q over the whole range of their experiment. The reason why this rule should hold even when there is a large degree of mutual friction between the liquids, is due to the fact that the force acts equally on both fluid components. This becomes clear when the two-fluid equations of motion for steady flow

$$(4) \qquad 0 = -\frac{\varrho_s}{\varrho}\,\mathrm{grad}\,p + \varrho_s S\,\mathrm{grad}\,T + A\varrho_s\varrho_n|\boldsymbol{v}_s - \boldsymbol{v}_n|^2\,(v_s - v_n)\,;$$

and

$$(5) \qquad 0 = -\frac{\varrho_n}{\varrho}\,\mathrm{grad}\,p - \varrho_s S\,\mathrm{grad}\,T - A\varrho_s\varrho_n|\boldsymbol{v}_s - \boldsymbol{v}_n|^2\boldsymbol{v}_s - \boldsymbol{v}_n) + \eta_n\nabla^2 v_n$$

are added up to give

$$(6) \qquad \mathrm{grad}\,p = \frac{8\eta_n}{\pi a^4(\varrho S T)}\,Q\,.$$

However, if, as was suggested by the experiments of the preceding section turbulence in the superfluid gives rise to an eddy viscosity, η_s, then the term $(+\eta_s \nabla^2 v_s)$ has to be added to eq. (4) and proportionality of grad p to Q is lost.

There are indeed some indications for such deviations from ALLEN and REEKIE's rule and experiments by BREWER and EDWARDS [8] have made the existence of the effect probable. The work has now been repeated with greater accuracy by BHAGAT and CRITCHLOW [9], using tubes of 100, 300 and 800 μm diameter. A typical result, on the 300 μm tube, is shown in Fig. 3. Up to a certain critical heat current $\Delta H/Q$ is constant, *i.e.* ALLEN

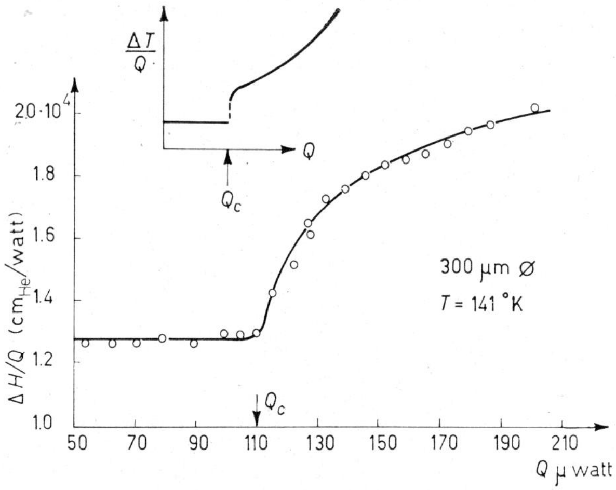

Fig. 3. – Deviation from ALLEN and REEKIE's rule. Inset: plot of thermal resistance against heat current.

and REEKIE's rule is obeyed. At Q_c, which incidentally is identical with the critical heat current obtained in determinations of $\Delta T/Q$, deviations begin to occur, $\Delta H/Q$ rising first steeply and then levelling off, evidently to saturation. Thus these observations furnish direct evidence for turbulence in the superfluid in the case of counterflow experiments.

Summarizing these results together with the earlier work it appears that the onset of friction in the superfluid is primarily due to the appearance of turbulence in it and that this will set in when the superfluid exceeds a certain critical velocity relative to the walls of the tube. The sudden increase in the thermal resistance is due to the collision of the thermal excitations with the vortex motion in the superfluid. This mechanism for the mutual friction process has also been postulated by VINEN [10] from other types of experiment. As reported earlier [1, 11], the onset of turbulence is sometimes delayed by hysteresis and it is significant that the increase in thermal resistance due to mutual friction will not occur under these circumstances either.

3·2. *Growth of turbulence.* – The rise in the thermal resistance due to the interaction of the thermal excitations with vorticity presents a very sensitive method for studying the onset of the latter. A careful study of the transition region in the $(\Delta T/Q)-Q$ diagram was made by W. A. STEELE and the author [12] and has recently been continued in collaboration with P. R. CRITCHLOW [13]. It is found that, as the transition region is gradually approached by increasing Q, a value of the heat current is reached at which the thermal resistance rises with time until a new value of $\Delta T/Q$ is reached. Even if Q as well as the ambient temperature are kept rigorously constant, no intermediate value of the thermal resistance is stable. Since the increase of heat resistance with time was found to be linear it was concluded that a state of higher thermal resistance originated at one end of the tube and that it was propagated through the length of the tube at a steady rate. By placing a number of thermometers at various points along the tube, the progress of this state of higher thermal resistance could indeed be observed directly, as is indicated in Fig. 4. In the most simple case the propagation would start at the cold end of the tube and proceed undisturbed until the whole tube had been filled. All the available evidence suggests that we observe here the first growth of vorticity in the superfluid, possibly in the form of a continuous creation of small vortex rings. A slightly more complex pattern is presented by two propagations starting simultaneously at both ends of the tube. Their speeds are then different which is not surprising since the superfluid in which the propagation proceeds is itself in motion relative to the tube. It is, however, significant that, as the two propagations meet inside the tube, no further changes will occur in the value of $\Delta T/Q$. This indicates that the density of vorticity in both propagations was the same.

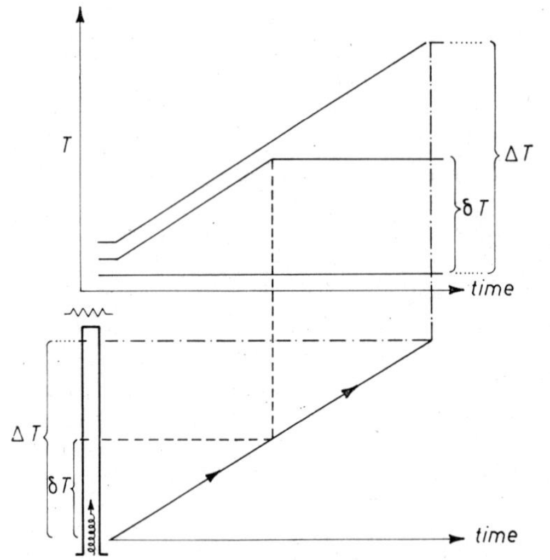

Fig. 4. – Propagation of vorticity, originating at the cold end, along a tube which carries 3 thermometers.

In addition to these flow patterns a great number of more complicated ones have been observed, all of them, however, being made up of a number of linear portions. Recording these various rates of progression it was noticed that they can always be expressed as multiples of the two basic rates originating at the warm and cold ends of the tube. These multiples may be due either to the simultaneous appearance of a number of sources inside the tube, all of equal strength or to sources whose strength varies in simple multiples. Although the latest experiments seem, on balance, to favour the first explanation, a final decision will have to wait until more results are available.

It must seem reasonable that vorticity will make its first appearance at the end of the tubes since it is here that discontinuities in the superfluid velocity must occur. It was also noticed that when a source appeared inside the tube, its place was the same in a number of subsequent experiments which suggests that it was most likely caused by an irregularity or solid impurity. In some cases an additional heater was placed half way along the tube. If then the end heater was first energized and vorticity began to propagate from the warm end, the heat current could be switched subsequently to the second heater. If, at the time of switching over, the turbulent front had already passed the second heater, it would progress further towards the cold end but turbulence would die out in the section between the heaters. On the other hand, if the front had not yet reached the second heater at the switch-over time, it would die out altogether. In this case no new front would originate at the second heater, presumably because the superfluid velocity suffers no discontinuity in the middle of the tube.

In another set of experiments a Y-shaped tube was used whose leg communicated with the bath while the two arms carried heaters and thermometers. Additional thermometers were placed at the junction and at the bath end. Turbulence created at the end of either arm would propagate through it to the bath but the other arm would remain free of vorticity. When heat currents, each smaller than half the critical value were set up in both arms and one of these currents was gradually raised, vorticity would propagate along the leg of the Y as soon as the sum of the currents had reached the critical value in this section. Further increase of the current in the same arm would ultimately lead to propagation of vorticity along it when here, too, the critical heat current had been reached. On the other hand, no vorticity would spread into the arm carrying a heat current lower than the critical although it was in direct contact with the rest of the tube filled with turbulent fluid From these observations is seems clear that although the onset of vorticity may be delayed by hysteresis, its growth will not take place below the critical velocity.

Most of these experiments were carried out on tubes of 800 μm diameter but similar results were obtained on tubes of 500, 670 and 1700 μm diameter. A detailed account of this work is in preparation.

REFERENCES

[1] K. Mendelssohn: *Suppl. Nuovo Cimento*, **9**, 228 (1958).
[2] S. M. Bhagat and K. Mendelssohn: *Physica*, **24**, 147 (1958).
[3] H. Lamb: *Hydrodynamics*, 6-th ed. (C.U.P., 1959), p. 668.
[4] S. M. Bhagat: *Proc. Phys. Soc.*, **75**, 303 (1960).
[5] C. J. Gorter and J. H. Mellink: *Physica*, **15**, 285 (1949).
[6] D. F. Brewer, D. O. Edwards and K. Mendelssohn: *Phil. Mag.*, **1**, 1130 (1956).
[7] J. F. Allen and J. Reekie: *Proc. Camb. Phil. Soc.*, **35**, 114 (1939).
[8] D. F. Brewer and D. O. Edwards: *Phil. Mag.*, **6**, 1173 (1961).
[9] S. M. Bhagat and P. R. Critchlow: *Cryogenics*, **2**, 34 (1961).
[10] W. F. Vinen: *Progr. in Low Temp. Phys.*, vol. III (Amsterdam, 1961).
[11] D. F. Brewer and D. O. Edwards: *Phil. Mag.*, **6**, 775 (1961).
[12] K. Mendelssohn and W. A. Steele: *Proc. Phys. Soc.*, **73**, 144 (1959).
[13] P. R. Critchlow: *Thesis* (Oxford, 1960).

Laminar and Turbulent Flow of He II in Wide Capillaries (*).

K. W. TACONIS and F. A. STAAS

Kamerlingh Onnes Laboratorium - Leiden

1. – Introduction.

It is well known that the flow properties of He II are complicated by the appearance of mutual friction between the normal and superfluid. Five different interesting types of flow can be distinguished in wide capillaries:

1) isothermal gravitational flow in which the superfluid velocity in general exceeds the normal flow velocity. The excess flow may induce a temperature difference and very rigorous precautions should be taken to keep the flow isothermal;

2) superfluid flow only, whereas the normal fluid is kept at rest, for instance between two superleaks. Experiments based on this principle are indicated to study the mutual friction;

3) pure heat conduction through capillaries so that there is no net mass transport. The superfluid and normal fluid velocities v_s and v_n are connected to the densities ϱ_s and ϱ_n by $v = 0 = v_s \varrho_s + v_n \varrho_n$. In wide capillaries already at very small heat flow the critical velocity is surpassed and the temperature difference rises steeply with the heat input. The decreasing heat conduction is due to the mutual friction. It is very easy to measure the temperature difference accurately, but very difficult to measure the pressure difference over the capillary accurately enough in order to analyse the flow behaviour;

4) heat flow in such a way that both fluids have artificially got the same velocity. This can be attained by placing a wide superleak parallel to the capillary. It is one of our main objects to explain this method here beneath;

5) heat flow as under 4 but now regulating the flow through the superleak in order to obtain all different relative velocities between the normal and superfluid. The procedure will be called further-on the overflow method.

The latter three flow methods are used to study the flow behaviour at low and high velocities especially extended into the turbulent region.

(*) Seminar delivered by K. W. TACONIS.

2. – Apparatus, method and results.

A heater H and a thermometer T_2 are attache to a small vessel in a vacuum chamber which is connected with a second vessel in contact with the sur-

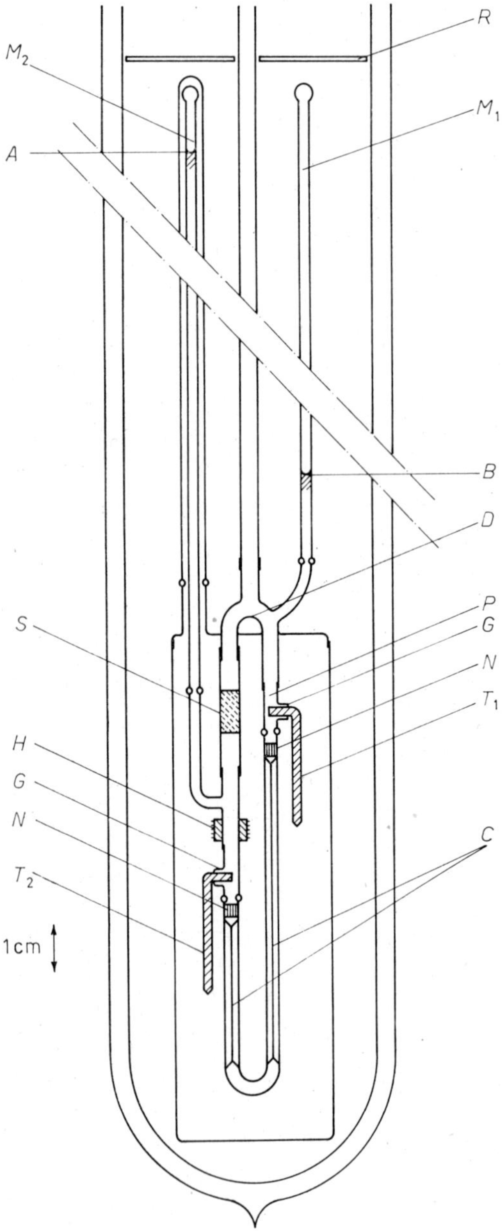

Fig. 1. – Apparatus.

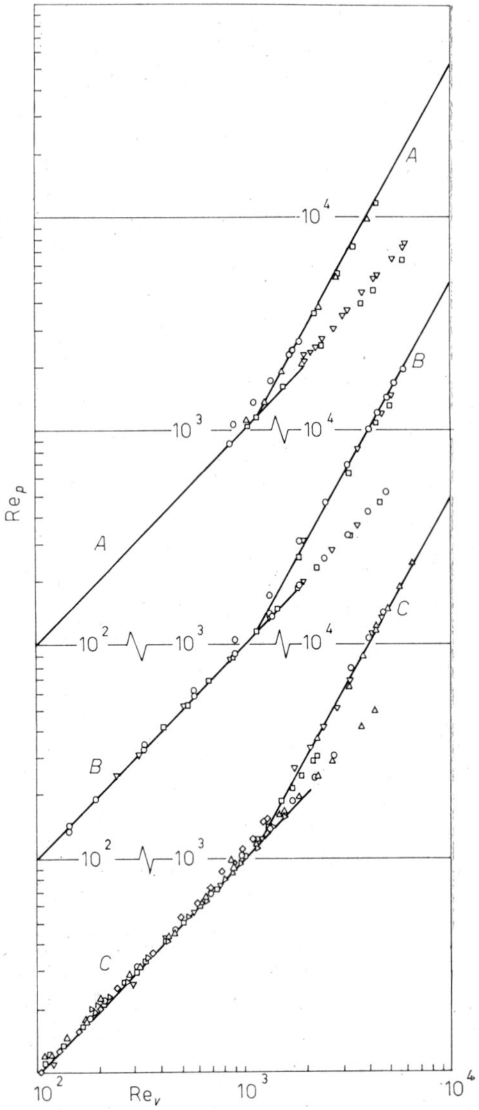

Fig. 2. – Results for the 173.5 μm capillary: graph A: ○ $T=1.019$ °K, △ $T=1.136$ °K, ▽ $T=1.0825$ °K, □ $T=1.190$ °K; graph B: ○ $T=1.496$ °K, □ $T=1.280$ °K, ▽ $T=1.404$ °K; graph C: □ $T=1.999$ °K, △ $T=1.703$ °K, ▽ $T=1.803$ °K, ◇ $T=2.104$ °K, ○ $T=1.598$, ▷ $T=1.899$.

rounding helium bath by means of a capillary C and a superleak S. Both vessels have got manometer tubes M_2 and M_1 in which the levels A and B of the liquid helium can be observed. A certain heat input in H gives a temperature difference that results in a level difference in the manometer tubes due to the fountain effect over the superleak. This superleak has a large flow capacity so that, independent of the heat flow through the tube C, the fountain pressure difference at a certain temperature difference has always its maximum London value (ϱS).

The normal fluid flow through C is therefore always determined by this London pressure difference and as the normal fluid mean velocity $\bar{v}_n$ is known from the heat input the viscosity η_n can be calculated accurately, although the capillary is wide, because the mutual friction is eliminated here. Indeed the superfluid will stream together with the normal fluid, the superleak having enough capacity to supply plenty superfluid to avoid mutual friction in the capillary. Increasing the heat input gradually at a certain critical Reynolds number turbulence appears just like in an ordinary fluid like water. A result of the research is that always the total density has to be taken into account. An elegant representation of the results is obtained in a special Δp-v diagram using along the logarithmic abcissa $\mathrm{Re}_v=(2\varrho r\bar{v}_n)/\eta_n$ the Reynolds number as co-ordinate for the velocity and using along the logarithmic ordinate $\mathrm{Re}_p=(\varrho r^3\Delta p)/4\eta_n^2 l$ as co-ordinate for the pressure difference Δp. Re_p is obtained from Re_v by eliminating $\bar{v}_n$ using Poiseuille's law $\Delta p=(8\eta_n l/r^2)\bar{v}_n$.

In such a diagram laminar flow gives a straight line under 45° and turbulent flow obeying the empiric equation of Blasius written here as $\mathrm{Re}_p=4.94\cdot10^{-3}\,\mathrm{Re}_v^{1.75}$ a straight line with a tangent 1.75.

Some results are shown in Figs. 2 and 3 for the 173.5 and 255 μm capillaries. They cover indeed the two straight lines of each graph (A, B, C, D and E). Laminar flow can, just like in an ordinary liquid, with increasing velocity persist up till rather high Reynolds numbers (6 000) and suddenly jump over to turbulence. For He II, however, it remains turbulent in decreasing the velocity down to the intersection point of the two straight lines; there is no lower limit for the Reynolds number.

3. – Overflow method.

In order to vary the superfluid flow velocity relative to the normal flow velocity one may proceed as follows: the liquid helium in the flow apparatus in Fig. 1 is filled only up to somewhere in the neighbourhood of bend D in the outer vessel. If now enough heat is supplied the fountain effect over the superleak will empty part of the helium out of the outer vessel (left of D)

and this portion of the liquid helium will give a rise of the level in the manometer tube M_2. A stationary state will be reached because the heat flow

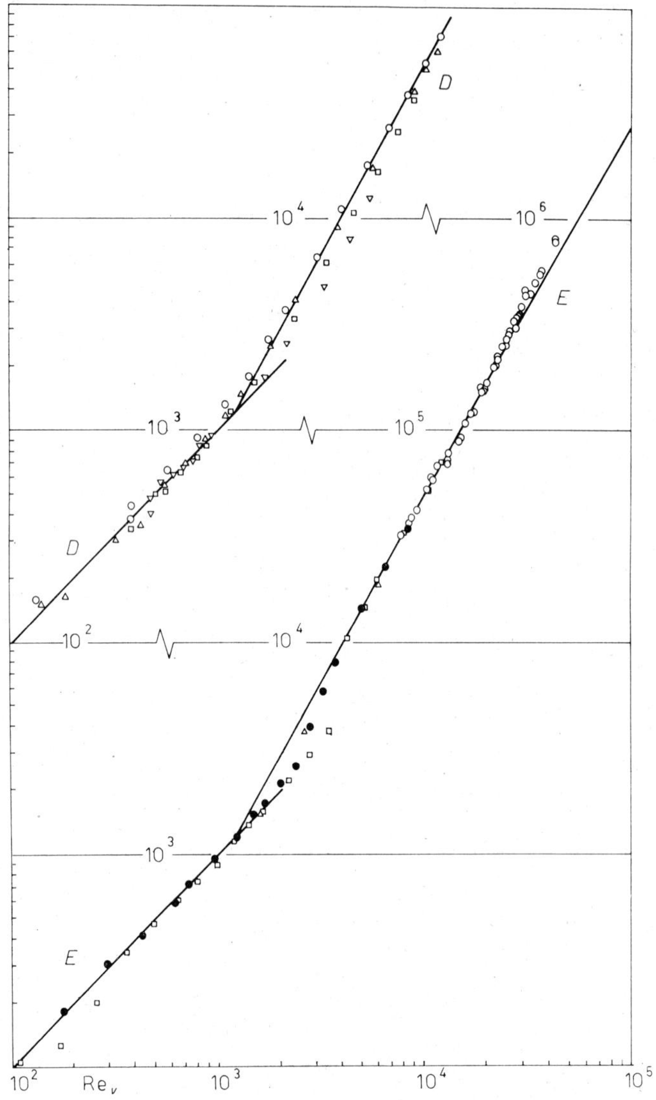

Fig. 3. – Results for the 255 μm capillary: graph D: ○ $T=1.604$ °K, □ $T=1.189$ °K, △ $T=1.405$ °K, ▽ $T=1.022$ °K; graph E: ● $T=2.008$ °K, □ $T=1.807$ °K, △ $T=1.900$ °K, ○ overflow data.

through C will give a helium flow to the outer vessel and the same amount will flow over the rim at D and than through the superleak back to the warm vessel. In this way very high Reynolds numbers (as high as 35 000) can be reached as still more heating will keep all levels unchanged only the temperature difference between the thermometer T_2 and T_1 will increase and subsequently the pressure difference increases too because of a higher vapour pressure above the level A in the manometer tube M_2. The data on graph E in Fig. 3 are obtained by this method.

The work was completed by studying also the classical heat conduction by taking away the superleak connection. Evidently not very high heat inputs can be used here as due to the mutual friction only small pressure heads can be reached. The vapour pressure in M_2 soon pushes the level into the vacuum can. The results are given in Fig. 4. A more complete description of the work will appear in the Communication of the Kamerlingh Onnes Laboratory No. 328 d and in *Physica* (in press).

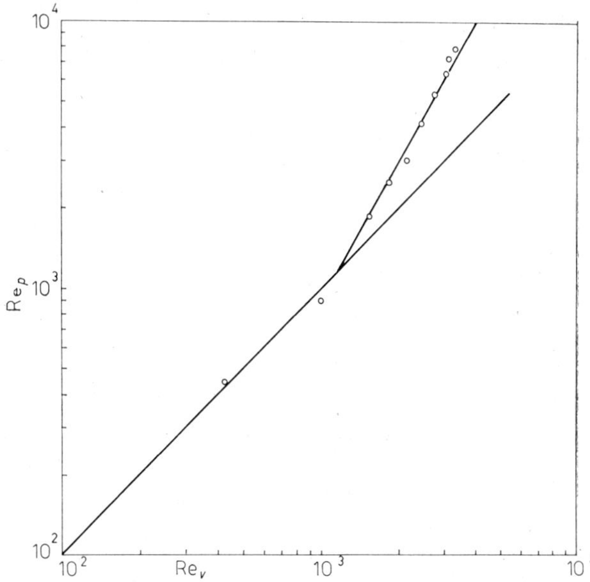

Fig. 4. – Pure heat conduction at 1.7 °K.

Dilute Mixtures of ^3He in Superfluid ^4He.

J. J. M. Beenakker and R. de Bruyn Ouboter

Kamerlingh Onnes Lab. - Leiden

1. – Introduction.

Measurements on the specific heat of ^3He-^4He mixtures at concentrations of 4.66, 9.4 and 15% of ^3He in liquid ^4He show, between 0.4 and 1 °K, an almost constant contribution of the ^3He to the specific heat which is close to $\frac{3}{2}RX$, cf. Fig. 1. The contribution of the ^4He to the specific heat is

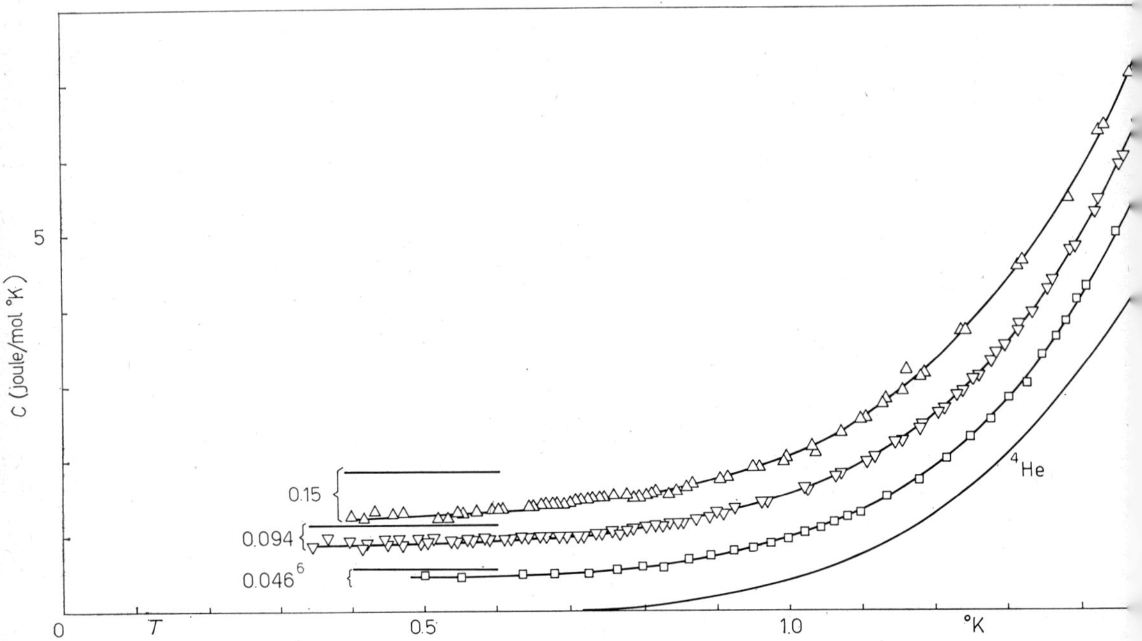

Fig. 1. – The specific heat of ^3He-^4He mixtures as a function of the temperature T at different concentrations: $X = 0.0466$, 0.094 and 0.15. The horizontal lines at low temperatures just above the curves are the theoretical values according to the theory of the ^3He gaslike spectrum ($C_3 = \frac{3}{2}RX$).

practically negligible at these temperatures. This behaviour can be explained by means of the ideas put forward by POMERANCHUK, in which he assumes that in a dilute solution the ^{3}He atoms do not interact with one another and that the assembly of ^{3}He atoms is nondegenerate. The ^{3}He atoms can be treated as practically free particles which move through the superfluid with an effective mass m_3^*. Hence the specific heat will be that of an ideal monoatomic gas. To describe the properties in more detail the energy spectrum associated with the ^{3}He atom is taken to be

$$E = E_{03} + \frac{p^2}{2m_3^*}\,, \tag{1}$$

where E_{03} is the effective potential and $p^2/2m_3^*$ is the kinetic energy associated with the translational motion through the superfluid of the ^{3}He particle, which has an effective mass m_3^*.

Since the ^{3}He particle can readily collide with the rotons and phonons, it will nearly always participate in the motion of the normal component and the density of the normal component ϱ_n is equal to:

$$\varrho_n = \varrho_{n,4} + \frac{m_3^*}{m_4}\varrho X\,, \tag{2}$$

Below 1 °K one has $\varrho_{n,4} \ll (m_3^*/m_4)\varrho X$ since the phonons and rotons make a negligible contribution, so that the normal fraction is almost equal to

$$x \equiv \frac{\varrho_n}{\varrho} = \left(\frac{m_3^*}{m_4}\right) X\,.$$

From the Andronikashvilli type experiments by PELLAM, BEREZNIAK and ESEL'SON, and DASH and TAYLOR one finds $x \approx 2X$ or $m_3^*/m_3 \approx 2.7$.

This rather large effective mass has as a consequence that the Fermi-degeneracy temperature,

$$T_0 = \frac{\hbar^2}{2km_3^*}\left(\frac{3\pi^2 XN}{V^m}\right)^{\frac{2}{3}}, \tag{3}$$

is so low that the specific heat is not influenced by degeneracy. At not too low concentrations T_0 is of the order of the phase separation temperature. Another confirmation of these ideas comes from the velocity of second sound in dilute solutions. At low enough temperatures one can neglect the ^{4}He contribution and one has the expression for the sound velocity of an ideal monoatomic gas with mass m_3^*, which is given by

$$V_{\text{II}} = \sqrt{\frac{c_p}{c_v}\frac{\partial P}{\partial \varrho}} = \sqrt{\frac{5}{3}\frac{kT}{m_3^*}} \tag{4}$$

From this expression one sees that the velocity of second sound becomes at lower temperatures independent of the ³He concentration and proportional to $\sqrt{T}$. This is in agreement with the results of KING and FAIRBANK, who found a value for $m^*/m_3 \approx 2.8$ and with those of KRAMERS and NIELS that give $m_3^*/m_3 \approx 2.5$.

2. – The thermodynamic properties.

The statistical mechanics of the ³He particles is nearly the same as for an ideal gas and the internal energy U, the entropy S, the specific heat C and the partial chemical potentials μ_i are equal to

$$U = (1-X)\,U_4^0 + X\,[NE_{03} + \tfrac{3}{2}RT]\,, \tag{5}$$

$$S = (1-X)\,S_4^0 + XR\left[\ln\left\{\frac{g_3 V_4^0}{XN}\left(\frac{m_3^* kT}{2\pi\hbar^2}\right)^{\frac{3}{2}}\right\} + \frac{5}{2}\right], \tag{6}$$

$$C = (1-X)\,C_4^0 + X\,\tfrac{3}{2}R\,, \tag{7}$$

$$\mu_4 = \mu_4^0 + RT\ln(1-X)\,, \tag{8}$$

$$\mu_3 = NE_{03} + RT\ln\left[\frac{XN}{g_3 V_4^0}\left(\frac{2\pi\hbar^2}{m_3^* kT}\right)^{\frac{3}{2}}\right]. \tag{9}$$

All quantities are expressed per mole; g_3 is the statistical weight. To obtain the constant E_{03} we analysed the vapour-liquid equilibrium data. The equilibrium condition between the liquid and vapour phase $\mu_{i\text{L}} = \mu_{i\text{V}}$ gives the following equation:

$$\mu_{3\text{L}} = NE_{03} + RT\ln\left[\frac{XN}{g_3 V_4^0}\left(\frac{2\pi\hbar^2}{m_3^* kT}\right)^{\frac{3}{2}}\right] = \mu_{3\text{V}} = RT\ln\left[\frac{P_3}{g_3 (kT)^{\frac{5}{2}}}\left(\frac{2\pi\hbar^2}{m_3}\right)^{\frac{3}{2}}\right], \tag{10}$$

or

$$NE_{03} = RT\ln\left[\frac{P_3 V_4^0}{XRT}\left(\frac{m_3^*}{m_3}\right)^{\frac{3}{2}}\right], \tag{11}$$

if we assume the vapour phase to be ideal. From the smoothed vapour pressure measurements of ROBERTS and SYDORIAK we calculated the partial vapour pressure P_3 which at temperatures below 1 °K is nearly equal to the total vapour pressure P_{tot}. With formula (11) we calculated NE_{03} and found its value between 0.6 and 1 °K to be independent of temperature.

For m_3^* we used the experimentally determined value 2.7 m_3, the value of NE_{03} is rather insensitive to this choice.

NE_{03} has to be interpreted as the depth of the effective potential well, relative to the gas phase, in which the ^{3}He particle moves.

Comparing the value of NE_{03} of — 23.5 J/mole, with the value for pure ^{3}He, obtained from the heat of vaporization at absolute zero L_{03}^0, of — 21.2 J/mole, we see that the potential well of ^{3}He in a ^{4}He surrounding is only slightly larger, as one should expect in a « cell model » taking into account the zero point energy of the ^{3}He and the large compressibility of ^{4}He.

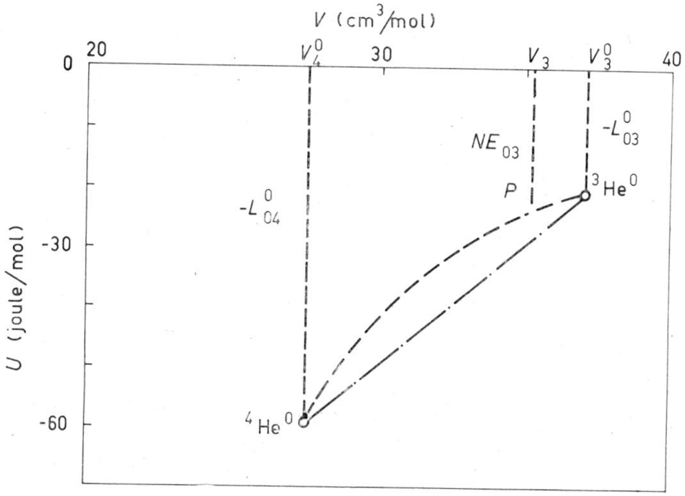

Fig. 2. – $U_{03}^0 = -L_{03}^0$, $U_{04}^0 = -L_{04}^0$, NE_{03} as a function of the molar volume V. V_3 is the partial molar volume for a dilute mixture of ^{3}He in liquid ^{4}He.

NE_{03}, $U_{03}^0 = -L_{03}^0$, $U_{04}^0 = -L_{04}^0$ as functions of the molar volume are plotted in Fig. 2. V_3 is the partial molar volume for a dilute solution of ^{3}He in liquid ^{4}He derived from the molar volume experiments by KERR.

We will now compare the heat of mixing H^E and the excess Gibbs function G^E with the experimental data.

H^E and G^E are given by:

$$\frac{G^E}{X} = \mu_3^E = \left[NE_{03} + RT\ln\left\{\frac{RT}{P_3^0 V_4^0}\left(\frac{m_3}{m_3^*}\right)^{\frac{3}{2}}\right\}\right] = RT\ln\left(\frac{P_3}{XP_3^0}\right) = RT\ln\left(\frac{X_u}{X_1}\right), \tag{12}$$

$$\frac{H^E}{X} = [H_3 - H_3^0] = [NE_{03} - H_3^0(T) + \tfrac{3}{2}RT], \tag{13}$$

where X_u and X_l are the concentrations of the upper and lower phase that are in equilibrium with each other in the phase separation region.

From the specific heat measurements in the phase separation region we derived values for H^E, while G^E was derived from vapour pressure data and the phase separation curve.

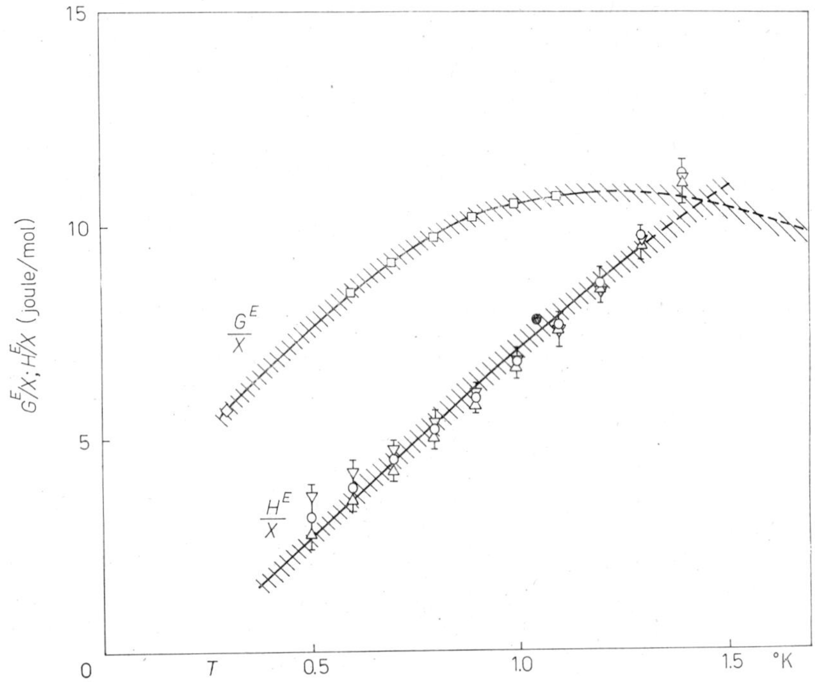

Fig. 3. – G^{E^F}/X and H^{E^P}/X as a function of the temperature T for dilute mixtures of ^{3}He in liquid ^{4}He. The excess Gibbs function is defined by means of the relation:

$$G^E = X\mu_3 + (1-X)\mu_4 - \left[X\mu_3^0 + (1-X)\mu_4^0 + RT\{X \ln X + (1-X)\ln(1-X)\}\right].$$

The heat of mixing $H^E(X, T)$ is defined as the heat required to keep the temperature constant when X mole of the pure liquid ^{3}He is added to $(1-X)$ mole pure liquid ^{4}He. □ G^E/X derived from the vapour pressure measurements of ROBERTS and SYDORIAK for a 10% mixture with formula (12); ◇ G^E/X derived from the phase separation curve with formula (12); ⊗ H^E/X direct experimental point of SOMMERS, KELLER and DASH; △ $X=0.0466$, ⊙ $X=0.094$, ▽ $X=0.15$. Our experimental values of H_Q/X.

These data are in Fig. 3 compared with the results obtained from theory. The agreement is again rather good.

3. – Nernst's heat theorem.

A combination of G^E and H^E gives the excess entropy S^E

$$G^E = H^E - TS^E \,. \tag{14}$$

Here S^E is the deviation of the entropy of mixing ΔS from the value for an ideal classical solution:

$$-R[X \ln X + (1-X)\ln(1-X)] \,,$$

the so called Gibbs paradox value. Nernst's heat theorem states that the entropy at absolute zero should be equal to zero. In classical thermodynamics this is not the case for an ideal mixture, where the Gibbs paradox value (positive) remains. In quantum statistics this difficulty does not arise as here this term goes to zero with decreasing temperature, because of the degeneracy. Hence if one describes the thermodynamic properties of a mixture in terms of the classical ideal solution expression one will obtain at $T = 0\,^\circ$K a deviation from such a behaviour given by:

$$S^E = +R[X \ln X + (1-X)\ln(1-X)] \quad \text{(negative)} \,.$$

For comparison with experiment we have plotted in Fig. 4 the excess entropy *vs.* temperature for a mixture of 10% ^{3}He, where the phase separation occurs at low enough temperature (0.31 °K) that it does not confuse the situation too much. We see that the total entropy of mixing goes to zero with decreasing temperature, in agreement with what Nernst's heat theorem suggests.

Another consequence of the quantum degeneracy of the entropy of mixing is the finite slope of the phase separation curve at absolute zero. The logarithmic behaviour of the classical entropy of mixing would give rise to a vertical slope in the T-X diagram at absolute zero.

In FEYNMAN's theory the motion of the ^{3}He atom through the superfluid is treated as a microscopic hydrodynamical problem and the effective mass m_3^* is the sum of the true mass of the ^{3}He atom and one half of the mass of the displaced fluid. Thus $m_3^* = m_3 + \frac{1}{2}(V_3/V_4)m_4$, which gives $m_3^*/m_3 \approx 1.9$. In this model the backflow of the superfluid around the ^{3}He atom gives rise to the large effective mass m_3^*. We like to stress the fact that in this classical picture the large volume occupied by the ^{3}He in the solution rises the effective mass considerably, ($V_3 > V_4^0$).

We like to point out that the ideal gas spectrum can only remain valid as long as the ^{3}He particles do not create ^{4}He excitations in the background fluid, what will be of importance at higher temperatures.

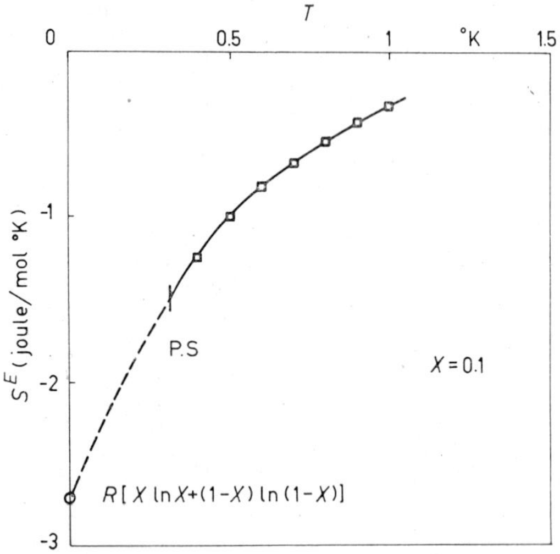

Fig. 4. – The excess entropy S^E as a function of the temperature T for a dilute mixture of ^{3}He in liquid ^{4}He: □ S^E calculated from our experiments ($X=0.1$), ○ $\lim_{T\to 0;\, X=0.1} S^E = + R[X \ln X + (1-X) \ln (1-X)]$ (Nernst's theorem). The excess entropy is defined by means of the relation

$$S(T, X) = XS_3^0(T) + (1-X)S_4^0(T) - R[X \ln X + (1-X) \ln (1-X)] + S^E(T, X) .$$

This might be one of the reasons of the temperature-dependence of m_3^* as found in second sound measurements by LYNTON and FAIRBANK at higher temperatures.

BIBLIOGRAPHY

More details and full references are given in:

R. DE BRUYN OUBOTER, J. J. M. BEENAKKER and K. W. TACONIS: *Commun. Kamerlingh Onnes Lab., Leiden, Suppl.*, **116c**; *Physica*, **25**, 1162 (1959).
R. DE BRUYN OUBOTER, K. W. TACONIS, C. LE PAIR and J. J. M. BEENAKKER: *Commun. Kamerlingh Onnes Lab., Leiden, Suppl.*, **324b**; *Physica*, **26**, 853 (1960).
R. DE BRUYN OUBOTER and J. J. M. BEENAKKER: *Commun. Kamerlingh Onnes Lab., Leiden, Suppl.*, **119a**; *Physica*, **27**, 219 (1961).
R. DE BRUYN OUBOTER: *Thesis* (Leiden, 1961).

Second Sound and Dissipative Processes in Very Dilute ^{3}He-^{4}He Mixtures.

H. C. KRAMERS

Kamerlingh Onnes Laboratorium - Leiden

1. – Introduction.

The present exposé will be concerned with the discussion of experimental results obtained from the observation of heat pulses propagated through the liquid mixture. The experiments were performed in Leiden by Mrs C. G. NIELS-HAKKENBERG and the present author. The discussion will be limited to

a) the temperature range between 0.1 and 1.0 °K;

b) concentrations below roughly 1% ^{3}He.

It will be apparent from the foregoing talk that these restrictions put the discussion within the limits of the Pomeranchuk picture of a dilute mixture. Phase separation effects can be disregarded completely and the addition of ^{3}He atoms may essentially be interpreted as the appearance of a new type of excitation in addition to the already existing phonons and rotons.

Just as in the pure liquid the assembly of the excitations may drift through the liquid and carry momentum: the ratio of the total momentum and the drift velocity is just the normal density ϱ_n. All the excitations move together in agreement with the early experimental observation that the impurities move with the normal fluid. The restrictions put forward mean further that all the three partial excitation gases are dilute; the normal density, the specific heat, etc., are therefore built up out of additive contributions. For phonons and rotons these contributions have been calculated by Landau; the gas of impurity excitations behaves essentially as a simple perfect gas. Table I gives some useful properties and numerical data of the excitations.

Just as in pure ^{4}He second sound is the phenomenon determining the propagation of heat pulses, provided the condition of local equilibrium is essentially fulfilled. If this is no more true or in other words: if the mean free paths become large (*i.e.* at low enough temperatures) the experimental results are completely changed and must be interpreted differently. The latter case is considered in the last section.

TABLE I.

	Phonon	Roton	^{3}He excitation
Energy spectrum	$\varepsilon = v_{\text{I}} p$ $v_{\text{I}} = 240$ m/s	$\varepsilon = \Delta_{\text{r}} + (p - p_0)^2/2\mu_{\text{r}}$ $\Delta_{\text{r}}/k = 8.65\ ^\circ\text{K}$ $p_0/\hbar = 2.0 \cdot 10^8\ \text{cm}^{-1}$ $\mu_{\text{r}}/m_4 = 0.16$	$\varepsilon = p^2/2\mu$ $\mu/m_3 = 2.5 - 3$
Number density	$N_{\text{p}} = 9.6\,\pi \left(\frac{kT}{hv_{\text{I}}}\right)^3$ $= 2 \cdot 10^{19} T^3\ \text{cm}^{-3}$	$N_{\text{r}} = 4 \cdot 10^{22}\, T^{\frac{1}{2}} \exp\,[-\Delta_{\text{r}}/kT]\ \text{cm}^{-3}$	$N_{\text{i}} = 2.1 \cdot 10^{22}\, X \cdot \text{cm}^{-3}$
Relative normal density contribution $\varrho_{\text{n}}/\varrho = x$	$x_{\text{p}} = \frac{16}{45}\pi^5 \frac{h}{v_{\text{I}}} \left(\frac{kT}{hv_{\text{I}}}\right)^4$ $= 1.2 \cdot 10^{-4}\, T^4$	$x_{\text{r}} = (p_0^2/3\varrho kT)\, N_{\text{r}}$ $= 4.3\, T^{-\frac{1}{2}} \exp\,[-\Delta_{\text{r}}/kT]$	$x_{\text{i}} = \mu X/m_4$ $\approx 2X$
Specific heat per unit of mass c_v (c.g.s. units)	$c_{v\text{p}} = \frac{16}{15}\pi^5 k \left(\frac{kT}{hv_{\text{I}}}\right)^3$ $= 2.04 \cdot 10^5\, T^3$	$c_{v\text{r}} = kN_{\text{r}} \left\{\left(\frac{\Delta_{\text{r}}}{kT}\right)^2 + \left(\frac{\Delta_{\text{r}}}{kT}\right) + \frac{3}{4}\right\}$ $\approx 4.1 \cdot 10^8 (T^{-\frac{3}{2}} + 0.11\, T^{-\frac{1}{2}}) \cdot$ $\cdot \exp\,[-\Delta_{\text{r}}/kT]$	$c_{v\text{i}} = \frac{3}{2}\frac{kX}{m_4}$ $= 0.33 \cdot 10^8\, X$

If the condition of local equilibrium is only slightly affected second sound is still apparent but shows damping effects. The experimental data on this absorption can be compared with the theoretical deductions of LANDAU, KHALATNIKOV and ZHARKOV on the coefficients of the irreversible processes. The work of these Russian authors is a very extensive investigation of means of a kinetic theory of the gases of the excitations.

KHALATNIKOV gave the first derivation of the absorption coefficient of second sound in terms of viscosity, heat conductivity and diffusion.

2. – Second sound experiments.

The heat pulse is propagated along a cylindrical tube with a heater on one end and a carbon resistor thermometer on the other end. In order to avoid possible heat flush effects one single pulse was emitted and observed. For this same reason it was not possible to use a resonant cavity method though in principle much more accurate results might be obtained from it with regard to the absorption. The pulse method gives only reliable results if the absorption is rather large.

Absorption of second sound shows itself in a broadening of the original narrow pulse to a Gauss shaped signal. Following Khalatnikov one can write

$$v^2 = v_{\text{II}}^2 + i\alpha\omega\,; \tag{1}$$

v is a complex velocity, v_{II} the second sound velocity (dispersion effects will be neglected), ω the angular frequency of a propagated wave and α a frequency-independent quantity determining the absorption. The commonly defined absorption coefficient for such a wave equals $\beta = \frac{1}{2}(\omega^2/v_{\text{II}}^3)\alpha$.

It can easily be shown that α and the half-width of an originally narrow pulse (Δt) are connected by $(\Delta t)^2/z = 1.4(\alpha/v_{\text{II}}^3)$; z being the tube-length. The second sound velocity is determined by the velocity of the top of the bell-like signal. It should be mentioned that in this analysis surface absorption effects are entirely disregarded; this is probably justified, because of the large effective frequency in a narrow pulse.

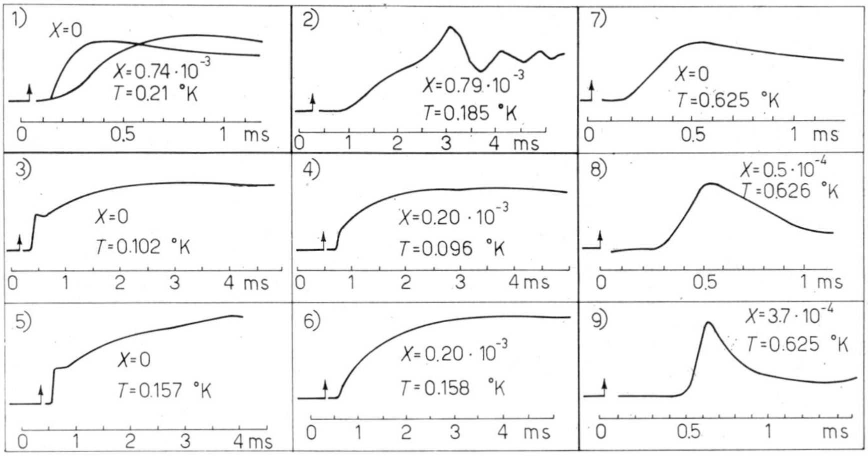

Fig. 1. – Some examples of typical pulse shapes (see text).

The results show that even a small admixture of ^{3}He gives a considerable narrowing of the pulse (see Fig. 1-7, 8, 9) compared with the pure liquid observations. In other words the absorption is greatly reduced.

3. – The second sound velocity and the effective mass of the ^{3}He excitations.

The equations for the two-fluid model together with the equation for conservation of the ^{3}He allow for the derivation of the second sound velocity in a dilute mixture:

$$v_{II}^2 = \frac{\varrho_s}{\varrho_n}\left[\frac{T}{c}\left(S_0 + \frac{kX}{m_4}\right)^2 + \frac{kTX}{m_4}\right]. \tag{2}$$

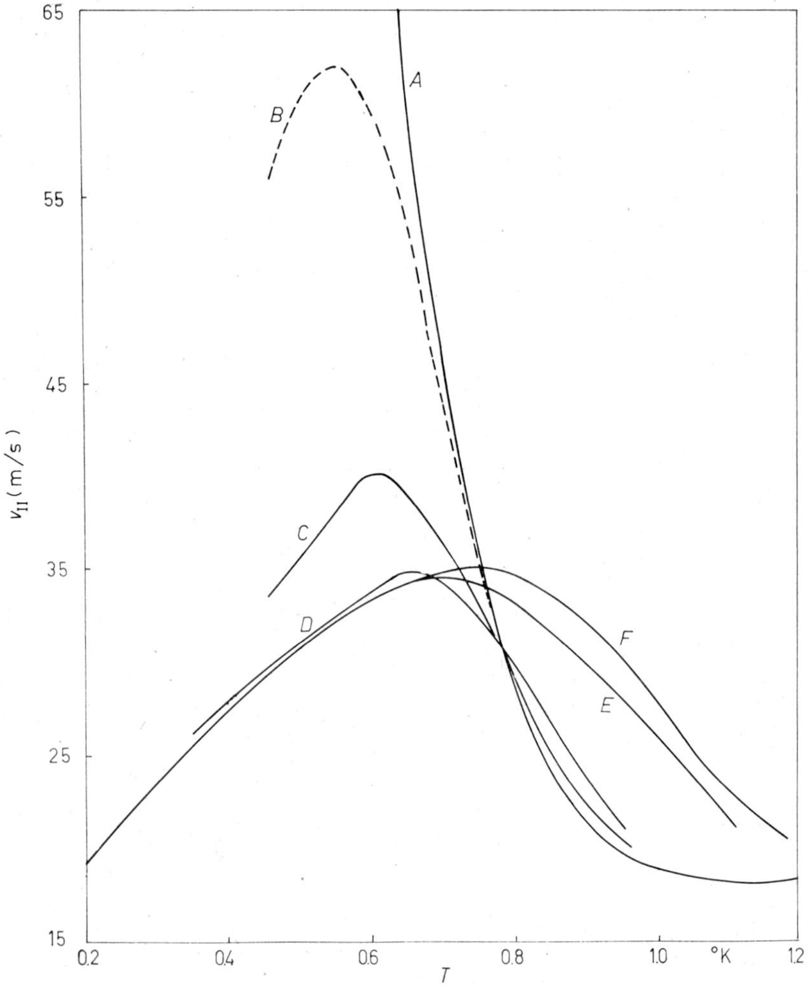

Fig. 2. – Second sound velocities for different concentrations: A) $X=0$; B) $X=3\cdot10^{-3}$; C) $X=1.0\cdot10^{-4}$; D) $X=3.7\cdot10^{-4}$; E) $X=2.0\cdot10^{-3}$; F) $X=3.0\cdot10^{-3}$.

In this expression X is the molar concentration, c and S are the specific heat and the entropy per gram respectively, m_4 is the atomic mass of a ^{4}He atom and k the Boltzmann constant. Symbols with a 0-suffix indicate « pure » values only.

For pure ^{4}He this expression becomes

$$v_{II}^2 = \frac{\varrho_s}{\varrho_n}\frac{TS_0^2}{c_0}\,. \tag{2a}$$

As is well known this expression goes to a finite value equal to $v_I^2/3$ at absolute zero (140 m/s).

In a dilute mixture, the pure contributions can be neglected at sufficiently low temperatures and the expression (2) reduces to

$$v_{II}^2 = \frac{5}{3}\frac{kT}{\mu}\,, \tag{2b}$$

independent of the concentration; μ is the effective mass. It is seen that for any concentration v_{II} goes to zero as $T^{\frac{1}{2}}$ for $T \to 0$.

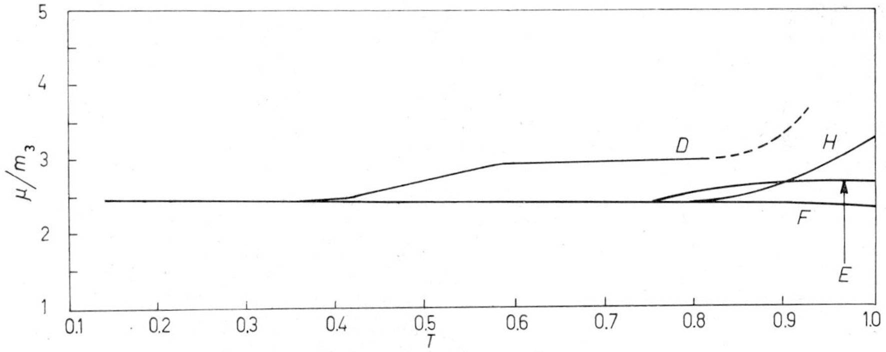

Fig. 3. – Values of the effective mass of ^{3}He excitations: D) $X=3.7\cdot10^{-4}$; E) $X=2.0\cdot10^{-3}$; F) $X=3.0\cdot10^{-3}$; H) $X=1.5\cdot10^{-3}$.

The value of the effective mass μ can be deduced from experiments by means of (2) or (2b). Results for the velocity and μ are given in Figs. 2 and 3. It is seen that for concentrations of the order 10^{-3} μ equals 2.5 m_3 and is independent of temperature. For the lower concentrations μ appears to increase, starting from the same value, if the temperature is raised. These latter values are rather more uncertain, however, but the present author's opinion is that it may be a real effect.

4. – Absorption of second sound.

In order to get a clear picture it is worthwhile to recapitulate first the situation in pure ^{4}He. Formally the four equations for the two-fluid model, *i.e.* the two equations of motion, and the continuity equations for mass and entropy can be supplemented to include terms arising from first order deviations from equilibrium. In this way terms appear designating heat conductivity and viscosity. A simple derivation then leads to an expression for the absorption parameter α for second sound

$$\alpha_{\text{II}} = \frac{\varrho_s}{\varrho\varrho_n}\left(\frac{3}{4}\eta_n + \zeta_{22}\right) + \frac{\lambda}{\varrho c_v}, \tag{3}$$

η_n is the well-known normal viscosity, ζ_{22} a second viscosity coefficient, and λ the coefficient of thermal conductivity of the normal fluid. This heat conductivity is a rather peculiar phenomenon: it applies to a transport of energy due to a gradient of temperature while the normal fluid is fixed. Since ordinarily the normal fluid starts to move if one tries to set up a temperature gradient it cannot be measured in the conventional way. It proves, however, to be the main source for attenuation of second sound as might already be suggested by the very nature of the latter. The viscosity terms in α_{II} can be estimated to contribute for about 10 % only. They will be disregarded in what follows.

The kinetic theory of the excitations allows for a microscopic picture and an estimate of the magnitude and the behaviour as a function of temperature of the coefficient λ. In general both types of excitations may contribute to the energy flow and λ may be written as the sum of a phonon part (λ_p) and a roton part (λ_r). In each of these the resistance to the flow arises from collisions with both types of excitations.

It is not possible to go here into the detailed theory, which is worked out along lines described in the book of CHAPMAN and COWLING. The picture is complicated, however, because the number of excitations need not be conserved. Moreover the interaction processes have either to be guessed at or the calculation of cross-sections involves the introduction of quantities only very roughly obtainable from experiment.

Nevertheless this theory has succeded surprisingly well in explaining the dissipation processes occurring.

It turns out that below 1 °K λ_r can be neglected in comparison to λ_p. Moreover the resistance arises solely from phonon-roton collisions. One can write λ_p in a form analogous to a well-known formula in elementary kinetic theory:

$$\lambda_p = \alpha_p (\varrho c_v)_p v_p^2 \theta \left(1 - \frac{\varrho S T}{\varrho_n v_p^2}\right), \tag{4}$$

in which α_p is a factor of order $\frac{1}{2}$, $(\varrho c_v)_p$ the phonon part of the specific heat, v_p the phonon velocity ($v_p = v_I$) and θ a time characterizing the process. θ may be evaluated from

$$\frac{1}{\theta} = \frac{1}{t_{pr}} = \overline{N_r \sigma_{rp} v_p}, \tag{5}$$

in which N_r is the number of rotons per unit of volume and σ_{rp} is the cross-section for phonon-roton collisions. The corresponding m.f.p. equals, of course, θv_p.

The last factor in expression (4) is due to the peculiar character of λ. It arises because the energy transport has to be corrected for motion of the normal fluid as a whole.

Mainly due to the appearance of N_r in eq. (5) λ rises rapidly with decreasing temperature, just as observed in the experiment.

Turning now to the dilute mixtures a similar picture may be expected. From a theoretical point of view the situation may be summarized as follows:

a) a new type of irreversible process is possible *i.e.* diffusion of the ^{3}He atoms;

b) there are now three kinds of excitations and consequently a new set of collision processes have to be considered in the calculation of the kinetic coefficients;

c) the attenuation of second sound is determined by thermal conductivity as in the pure liquid, but in particular at low temperatures a (thermo-) diffusion term has to be added and may become predominant. Viscosity terms give again only a minor contribution.

Considering first the thermal conductivity. Below 1 °K as before only λ_p is of importance. Formally expression (4) is not changed; only a temperature-dependent factor has to be added which will be left out of the present discussion. It may change from 1 to about 10 in extreme cases.

A considerable reduction of λ_p occurs because the characteristic time θ is appreciably shortened due to the occurrence of phonon-^{3}He collisions. One may write

$$\theta^{-1} = t_{pr}^{-1} + t_{pi}^{-1}, \tag{6}$$

t_{pr} having the same meaning as before and t_{pi} being determined by

$$t_{pi}^{-1} = \overline{N_i \sigma_{pi} v_p}\ , \tag{7}$$

N_i is here the number density of the impurities and σ_{pi} the cross-section for phonon-impurity collisions.

It should be remarked that the factor between brackets occurring in eq. (4) still contains only « pure » values of S and ϱ_n. The reason is that motion of the ^{3}He part of the normal fluid has already been accounted for by the definition of the thermal conductivity which excludes the net flow of mass of the ^{3}He atoms; consequently the conduction of heat must not be corrected for this part of the normal flow. A direct outcome of this term is that λ_p goes rapidly to zero if one reduces the temperature below 0.5 °K. This does not happen in the « pure » liquid because the effect is compensated by the simultaneous rapid rise of t_{pr} (due to the occurrence of N_r in eq. (5)).

The diffusion contribution to second sound absorption provides the same kind of « leak » in a second sound wave with respect to the ^{3}He excitations as the heat conductivity does provide for the ordinary excitations. The contribution turns out to be negligible from 0.6 to 1.0 °K but must be considered seriously below this temperature range. Khalatnikov and Zharkov's calculations indicate that both impurity-phonon and impurity-roton collisions play a part in resisting the transport of impurities. The result of this calculation is a rather rapid rise of the second sound attenuation towards lower temperatures.

The experimental results are represented in Fig. 4 in terms of an effective coefficient of thermal conductivity. Leaving out first the concentrations of 10^{-3} and larger the following deductions can be made:

a) for temperatures above 0.5 °K the results are in reasonably good agreement with theory taking into account the heat conductivity only. A numerical factor « of order 1 » as Zharkov states it, has to be adapted, however;

b) for lower temperatures the expected rise in attenuation due to diffusion is not observed.

It must be concluded that the time characterizing diffusion (or the corresponding m.f.p.) does not increase with decreasing temperature as fast as the mentioned theory predicts. A logical conclusion might be that diffusion of the ^{3}He is mainly opposed now by the impurities themselves. The situation should be very much similar to a monatomic gas (the other excitations can be neglected), in which the temperature-dependence of D is very weak ($T^{\frac{1}{2}}$). This conclusion is not inconsistent with the experimental data, though it has to be investigated further.

For concentrations from 10^{-3} upward the absorption measured is in the whole temperature region smaller than the theory indicates (the absorption

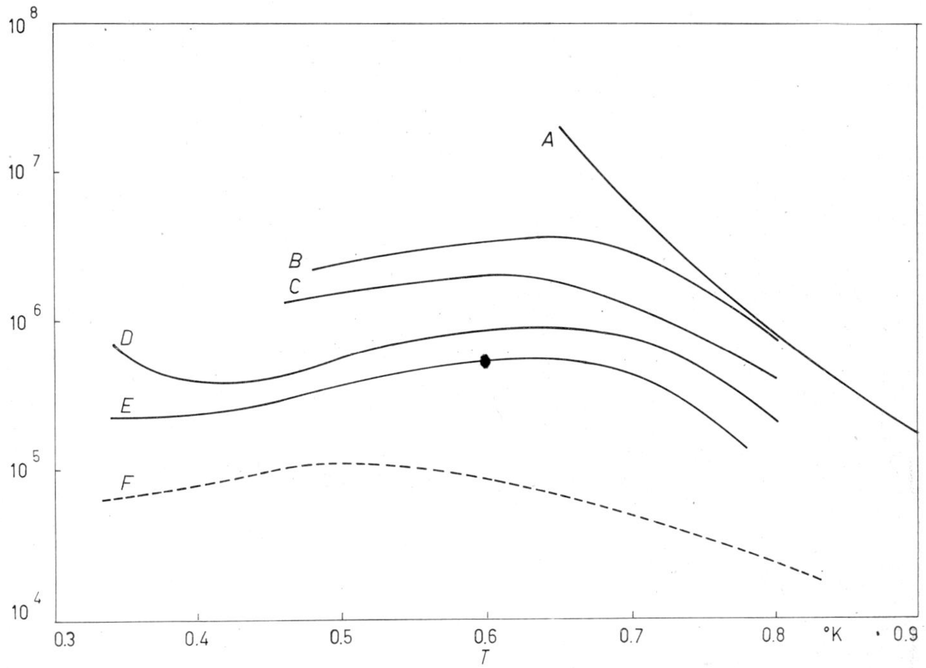

Fig. 4. – Effective thermal conductivity deduced from second sound attenuation: *A*) $X=0$; *B*) $X=0.5\cdot10^{-4}$; *C*) $X=1.0\cdot10^{-4}$; *D*) $X=2.6\cdot10^{-4}$; *E*) $X=3.7\cdot10^{-4}$; *F*) $X=2.0\cdot10^{-3}$.

should be roughly inversely proportional to the concentration). This is as yet not understood; it appears to be a real effect even though the experimental data are rather inaccurate for small attenuation.

5. – The extreme low temperature region $((0.1\div0.3)\ ^{\circ}K)$.

In pure ^{4}He no second sound can be propagated below 0.5 °K because of the large phonon m.f.p. (l_p). The latter may even easily surpass the tube dimensions. Therefore, at low enough temperatures the detected signal in a heat pulse experiment has to be interpreted as due to phonons being scattered at the walls only. As has been shown, in a dilute mixture the m.f.p. of the phonons is greatly reduced by the presence of the impurities and consequently second sound can be found at much lower temperatures. However, for small concentrations a similar effect as in the pure liquid may occur at sufficiently low temperatures. The curious effect happens that in a certain range of temperature this phonon signal and the second sound signal (here completely determined by the ^{3}He) overlap, as has been observed by FAIRBANK and SANDIFORT (see Fig. 1, 2)). At still lower temperature the second sound signal disappears completely.

An extensive investigation is in progress in Leiden on the behaviour of this phonon signal. A detailed survey shows that the ^{3}He still plays an appreciable role in limiting the m.f.p. as can be seen in Fig. 1, 1) in which a « pure » phonon signal is compared with a mixture phonon signal. For a not too short tube a simple analysis of the pulse shape allows for a determination

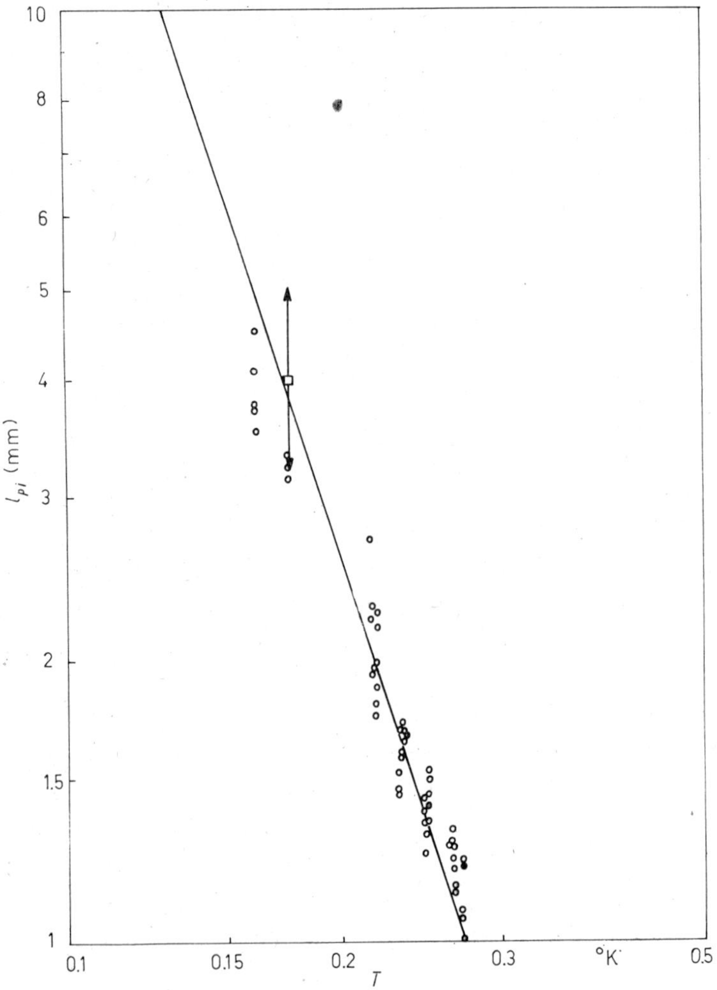

Fig. 5. – Phonon-impurity mean free path deduced from the « phonon » signal for a concentration $X = 0.74 \cdot 10^{-3}$.

of the mean free path. After correction for the limitation of the latter due to the tube dimensions a phonon-impurity m.f.p. can be obtained, which is represented in Fig. 5.

The result gives a much smaller magnitude than one would expect from the Khalatnikov-Zharkov theory and the temperature-dependence appears to be T^{-3} in contrast to T^{-4} or T^{-5} as this theory predicts (which one of the two theoretical possibilities one should expect depends on the choice of an unknown parameter). One may perhaps conclude that the theory can no longer be applied in this very low temperature range but the situation is certainly not yet clear and is still being investigated.

The single square point in Fig. 5 was obtained by a different method; it was deduced from an observation at a lower concentration, supposing l_D to be inversely proportional to the concentration. Earlier measurements in pure ^{4}He showed at sufficiently low temperatures a sharp edge at the front of the pulse which was interpreted as arising from phonons travelling directly from source to thermometer. The same effect proves to occur in extremely dilute mixtures. It disappears, however, at a much lower temperature than in the former case (see Fig. 1, 3-6). At the temperature of disappearance a m.f.p. can again be deduced; it agrees nicely with results of the foregoing method.

It should be mentioned here that Fairbank and Sandifort concluded to much higher values for the mean free path but their method of evaluation is rather crude and might easily give the wrong order of magnitude.

The author must apologize not or hardly to have discussed in this summary the experimental work of others. Very much connected is the work of H. A. FAIRBANK and SANDIFORT on heat pulses and also that of HARDING and WILKS on the attenuation of normal sound. Important are further recent determinations by PTUKHA of the thermal conductivity of dilute mixtures; the latter results roughly agree with the results from the work discussed here.

REFERENCES

For references on the original Landau and Khalatnikov theory see K. R. ATKINS: *Liquid Helium* (Cambridge, 1959).

Theoretical papers:

[1] I. POMERANCHUK: *Žurn. Ėksp. Teor. Fiz.*, **19**, 42 (1949).
[2] I. M. KHALATNIKOV: *Žurn. Ėksp. Teor. Fiz.*, **23**, 265 (1952).
[3] I. M. KHALATNIKOV and V. N. ZHARKOV: *Žurn. Ėksp. Teor. Fiz.*, **32**, 1108 (1957); *Sov. Phys. J.E.T.P.*, **5**, 905 (1957).

Experimental papers:

[4] H. C. KRAMERS: *Proc. Kon. Akad. van Wet.* (*Amsterdam*), **59**, 35, 48 (1956).
[5] K. N. ZINOVIEVA: *Žurn. Ėksp. Teor. Fiz.*, **31**, 31 (1956); *Sov. Phys. J.E.T.P.*, **4**, 36 (1957).
[6] D. J. SANDIFORT and H. A. FAIRBANK: *Proc. Toronto Conf.* (Toronto, 1960), p. 644.
[7] C. G. NIELS-HAKKENBERG and H. C. KRAMERS: *Proc. Toronto Conf.* (Toronto, 1960), p. 644.
[8] G. O. HARDING and J. WILKS: *Proc. Toronto Conf.* (Toronto, 1960), p. 647.
[9] T. P. PTUKHA: *Žurn. Ėksp. Teor. Fiz.*, **39**, 896 (1960); *Sov. Phys. J.E.T.P.*, **11**, 621 (1961).

The Nature of Ions in Liquid Helium.

K. R. ATKINS

Department of Physics, University of Pennsylvania - Philadelphia, Pa.

1. – Brief summary of the experimental results.

The mobility of ions in liquid helium has been investigated by CARERI and his coworkers [1-5] by MEYER and REIF [6] and by DAHM, LEVINE, PENLEY and SANDERS [7].

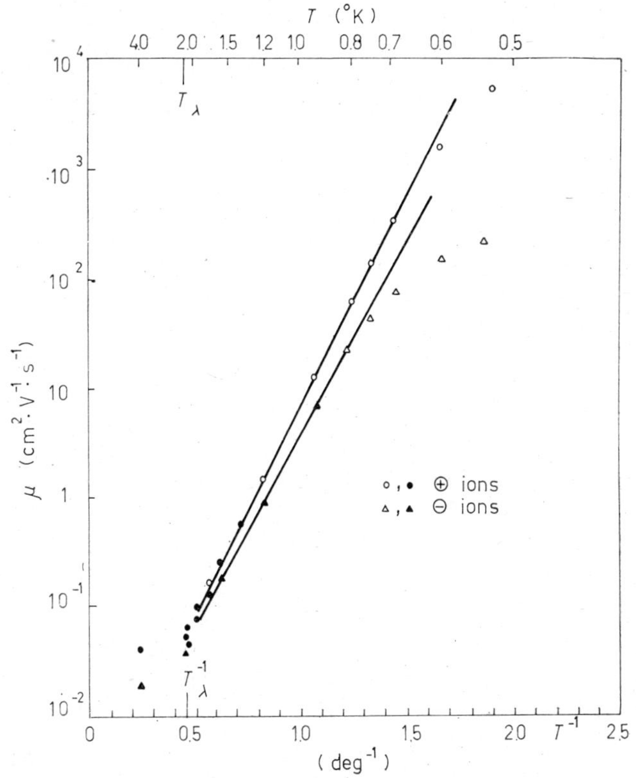

Fig. 1. – The zero-field mobility μ on a logarithmic scale as a function of the reciprocal absolute temperature. (After MEYER and REIF, ref. [6]).

The mobility as a function of temperature at low pressures near the vapour pressure is shown in Fig. 1. Between 2 °K and 1 °K the mobility varies as $\exp[\Delta'/kT]$, where Δ' is very nearly equal to the roton parameter normally represented by this symbol. It is clear then that the major process determining the mobility in this temperature range is scattering of the ions by rotons. Below 1 °K there is a departure from the exponential law, suggesting that scattering by phonons is becoming important. At these lowest temperatures, the mobility varies approximately as $1/T^{3.3}$.

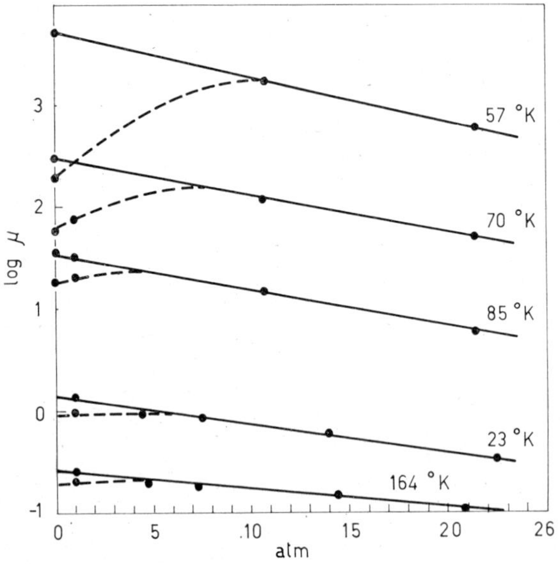

Fig. 2. – Pressure dependence of the mobilities at various temperatures. The solidcurves refer to the positive ion, the dashed curves to the negative ion. The point at 0.57 °K and zero pressure was obtained by extrapolating the curve of Fig. 1 to this temperature, neglecting its bending over due to phonon scattering. (After MEYER and REIF, ref. [6], but see also CUNSOLO and MAZZOLDI, ref. [5]).

The mobility of negative ions is less than that of positive ions by a factor which lies between 1 and 2. However, this difference is present only at low pressures. As the pressure is raised (Fig. 2), the mobility of the positive ions decreases whereas the mobility of the negative ions initially increases until it is equal to that of the positive ions and then at higher pressures the two mobilities are indistinguishable.

The mobility is independent of field strength in weak fields and the drift velocity u is then proportioned to the field E:

$$u = \mu E \, . \tag{1}$$

However, as the field strength increases, this linear relationship breaks down and the mobility decreases (Fig. 3). At high field strengths the drift velocity increases very little with the field and may be approaching a saturation value.

Assuming that the ions obey classical Boltzmann statistics, their diffusion coefficient D can be deduced from their mobility μ by the Nernst-Einstein relation,

$$D = kT\mu/e \, . \tag{2}$$

In this way the diffusion coefficient of the positive ion at 1.2 °K is found to be $1.5 \cdot 10^{-4}$ cm^2 s^{-1}, which should be compared with a value of about

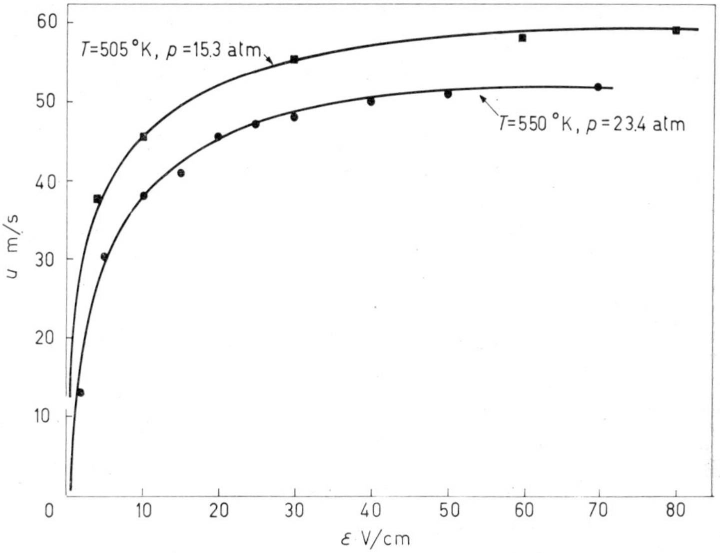

Fig. 3. – Dependence of the drift velocity of the negative ion on electric field at low temperatures and high pressures.

$2 \cdot 10^{-2}$ cm^2 s^{-1} for the diffusion coefficient of ^{3}He in dilute ^{3}He-^{4}He mixtures [8]. Clearly an ion in liquid helium is a very different entity from a ^{3}He atom in solution.

2. – The electrostriction model.

Let us first imagine the ion to be a point charge fixed at a definite point in the liquid, which will be treated as a classical continuum [9]. When a polarizable fluid is distributed throughout a nonuniform electric field, its pressure and density increase with increasing field. The strong electric field due to the ion therefore increases the density of the liquid in the immediate vicinity of the ion.

At a distance r from the ion the electric field is

$$E = \frac{e}{\varepsilon r^2}, \qquad (3)$$

where ε is the local value of the dielectric constant. A straightforward classical thermodynamic treatment gives

$$\int_{p_0}^{p} V\,\mathrm{d}p = \frac{1}{2} N\alpha_0 e^2/r^4 \left(1 + \frac{4\pi\varrho}{M} N\alpha_0\right), \tag{4}$$

p is the pressure in the liquid at a distance r from the ion; p_0 is the external pressure on the liquid ($p = p_0$ when $r = \infty$); V is the molar volume of the liquid at pressure p. N is Avogadro's number; α_0 is the electric polarizability of a helium atom; ϱ is the density of the liquid and M its atomic weight.

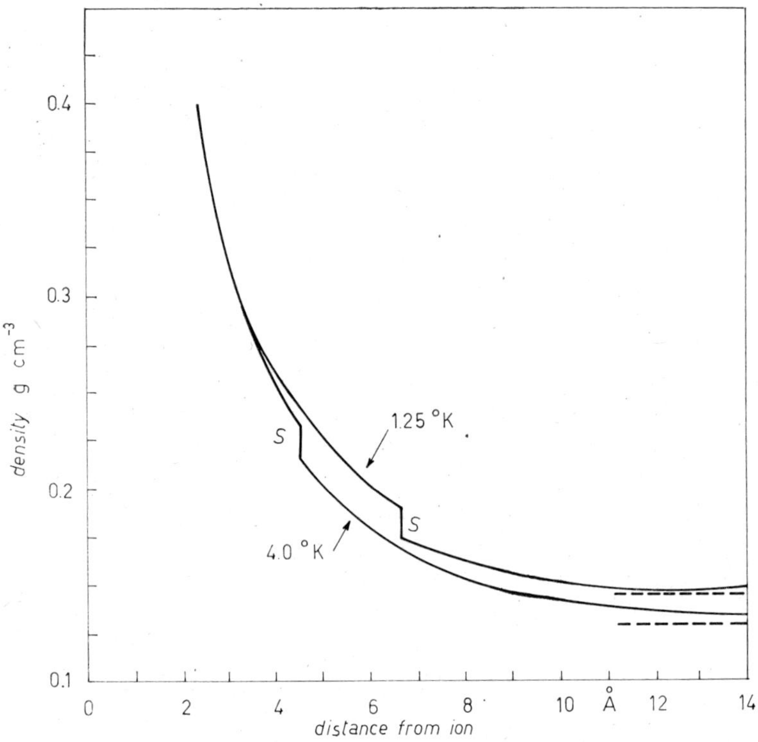

Fig. 4. – Variation in density of liquid near a localized point charge. These curves are based on a classical calculation for a continuum and the surface energy at the solid-liquid boundary has been ignored.

Since the equation of state of liquid helium is known, eq. (4) enables the pressure p and density ϱ to be calculated as a function of the distance r from the ion.

Figure 4 shows the results of such calculations when the external pressure

p_0 is the vapour pressure. It will be seen that there is a very large increase in density in the immediate vicinity of the ion.

3. – The helium snowball.

In Fig. 4 it will be seen that there is an abrupt discontinuity in the density at the point marked S. Here the pressure in the liquid has risen to melting pressure and the liquid has solidified. This suggests that the ion might be the center of a small solid helium « snowball » with a diameter of about 7 Å at 1.25 °K and containing about 30 helium atoms. However, we cannot adopt this picture without taking into account the surface tension σ_{sl} between solid and liquid helium. The pressure $p_m(r)$ in the liquid just outside a solid sphere of radius r differs from the melting pressure $p_m(\infty)$ for a plane solid surface in accordance with the equation

$$p_m(r) - p_m(\infty) \simeq \frac{V_s}{V_l - V_s} \frac{2\sigma_{sl}}{r}, \tag{5}$$

where V_s and V_l are the molar volumes of solid and liquid at the melting pressure. The effect of surface tension is to reduce the radius of the « snowball », but its importance cannot be estimated because σ_{sl} is not known. In any case, the dimensions are so small that any theory based on the properties of bulk liquid and solid is likely to break down. The existence of the « snowball » must therefore remain in doubt until a more rigorous atomistic approach is developed.

If the pressure of the liquid were held at a value just below the melting pressure, it would be very easy for the electrostriction effect to upset the balance and quite a large snowball might form. It should be emphasized, though, that it would be necessary to approach the melting pressure very closely. To obtain a snowball of radius 50 Å, it would be necessary to approach within one hundredth of an atmosphere of the melting pressure, even if surface tension effects are ignored.

4. – The effective mass on the electrostriction model.

The *extra mass* associated with the presence of the ion is readily calculated from the data of Fig. 4, by integrating over all space the difference between the density and the density in the absence of the ion. The result is equal to about 40 times the mass of a helium atom. This suggests that the ion carries around with it a large number of helium atoms and therefore has a large effective mass.

However, the effective mass relevant to transport phenomena must not be equated exactly to the extra mass as calculated above. If, for example, the ion is really surrounded by a solid sphere, then inside this sphere we must integrate the *total* density, not just the excess density. Moreover, in classical hydrodynamics, it is well known that when a solid sphere moves through a liquid its mass must be increased by one half of the mass of liquid it displaces. This allows for the kinetic energy of the liquid flowing round the sphere.

In our case the problem can be formulated as follows. A point charge is located in the liquid and by electrostriction increases the density of the liquid in its vicinity.

The charge is now given a velocity v. What is the flow pattern of the liquid in its vicinity? Is the density field influenced by the motion? What is the total kinetic energy of the flow? The total kinetic energy could be equated to $\frac{1}{2}M_i v^2$ to provide at least one definition of the effective mass M_i, but if it should turn out that the kinetic energy E is not related to the total momentum P by the equation

$$E = E_0 + \frac{P^2}{2M_i}, \tag{6}$$

then the situation might be as complicated as in the case of electrons in a metal.

As the ion moves through the liquid accompanied by its density field, the atoms of the surrounding liquid must adjust their positions. These atoms have an average velocity comparable with the velocity of sound in the liquid, u_1, and density variations are able to propagate with the velocity of sound. As long as the velocity of the ion is less than the velocity of sound, there is no problem, but clearly the situation becomes very complicated when the velocity approaches the velocity of sound.

5. – The positive ion.

The above classical treatment applies to a localized point charge. The actual ion, whatever its nature, must be described by a wave function extending over a finite region of space. If the average velocity of the ion inside this region were appreciably greater than u_1, the velocity of sound in the liquid, the electrostriction effects in the surrounding liquid would not be able to follow the motion of the ion, which would therefore be more nearly equivalent to a charge distribution of finite diameter. Outside the effective diameter of this cloud of charge the electric field would be almost the same as for a point charge and the electrostriction effects would be unaltered. Inside the cloud, however, the electric field would be reduced and the electrostriction effects would be smaller. We must, therefore, consider this point carefully.

The velocity of sound u_1 and the zero-point velocity of a ^{4}He atom in the liquid both have an order of magnitude $\hbar/m_4\delta$, where m_4 is the mass of the atom and δ is the average interatomic distance. This is a simple consequence of the fact that the gradient of the wave function is of order $1/\delta$; it must change from zero to a maximum and back to zero again as an atom is moved through a distance of the order of δ [10]. A similar argument applies to an ion of a foreign atom. If this ion is heavier than a ^{4}He atom, its zero-point velocity can exceed the velocity of sound only if it is trapped within a cell of dimensions less than δ. Our previous approximation of a localized point charge is then adequate. Of course the ion can drift through the liquid with a thermal velocity $v_t = (3kT/M_i)^{\frac{1}{2}}$, but since its effective mass M_i must be greater than m_4, v_1 is always appreciably less than u_1.

If the positive monatomic helium ion, He^+, can exist in the liquid, the following interesting possibility arises. The above argument can be applied to the motion of the He^+ nucleus, but there is the additional possibility that the positive hole might jump to one of the neighboring neutral He atoms. This would occur after an average time interval $\tau \sim h/2z\Delta E$ where z is the number of neighboring atoms (the equivalent of a co-ordination number) and ΔE is the energy separation between the nuclear-symmetric and nuclear-antisymmetric wave functions for a He^+ ion at the appropriate distance from a neutral He atom. Taking $z \sim 5$ and $\Delta E \sim 0.1$ eV we find $\tau \sim 3 \cdot 10^{-5}$ s and the « velocity » for a single jump $\delta/\tau \sim 10^7 \gg u_1$. The subsequent motion is a random walk and the distance travelled after n jumps is on the average $n^{\frac{1}{2}}\delta$. The effective velocity is therefore $n^{\frac{1}{2}}\delta/n\tau = \delta/n^{\frac{1}{2}}\tau$. When this velocity falls to a value comparable with u_1, the electrostriction effects can begin to follow the motion. This happens when

$$n \sim (\delta/u_1\tau)^2 \sim 10^5 . \tag{7}$$

The effective radius of the charge cloud would then be

$$R \sim n^{\frac{1}{2}}\delta \sim \delta^2/u_1\tau \sim 300\,\delta \sim 10^{-5}\ \text{cm} . \tag{8}$$

Distribution of the charge over such a large volume would almost completely obliterate the electrostriction effects.

However, it is unlikely that the positive ion is monatomic. The He_2^+ molecule-ion is known to have a dissociation energy of 2.5 eV [11] which is very large compared with all the other forms of energy involved in the problem. Moreover, He_2^+ ions are more numerous than He^+ ions in the gas at high pressures [12]. It therefore seems probable that the positive ion is He_n^+ where

n is a small integer. It is probably a compact unit; the internuclear distance in He_2^+ is 1.1 Å, as compared with $\sigma = 3.6$ Å for the liquid. We have already seen that such a massive entity is highly localized from our point of view. Moreover, rapid jumping of the hole is no longer possible since it would involve rearrangement of the massive nuclei and this obviously cannot occur with a velocity greater than the zero point velocity of the He atoms. We conclude that the electrostriction theory given in the previous section may be a reasonable approximation for the positive ion.

6. – The negative ion.

The negative monatomic helium ion, He^-, is probably unstable [13, 14]. The neutral diatomic molecule, He_2, is on the verge of stability. The He_2^- ion, if formed, must be very loosely bound. It is possible that a small compact polyatomic ion He_n^- is the stable form in the liquid, and, if this is so, the electrostriction theory might be valid as in the case of the positive ion. It would then be difficult to understand the difference in mobility between positive and negative ions at low pressures, so we shall consider two alternative theories.

KUPER [15] has developed the idea [1] that the negative ion is a free electron contained inside a large bubble. The large size of the bubble lowers the zero-point energy of the electron and this offsets the energy needed to form the surface of the bubble. Dr. KUPER will no doubt explain this theory in greater detail himself.

Another possibility [16] is that the electron moves almost freely in the regions between the widely spaced atoms and is rather analogous to an electron at the bottom of the conduction band in an insulator. The electronic wave function is spread out over a large region of liquid, but inside this region it is energetically favourable to increase the density of the liquid slightly, because this lowers the energy of the bottom of the conduction band.

Suppose that the electron is confined within a region of radius R inside which the density of the liquid has increased in order to bind the electron more strongly. Let the binding energy, which is analogous to the energy at the bottom of the conduction band, be $-B(\delta)$, a function of the interatomic spacing δ. Because the electron is confined within a small domain we must add to this an energy $h^2/8m^*R^2$, where m^* is an effective mass not very different from the mass of a free electron. Adding also the energy needed to compress the liquid, that is to change δ from δ_0 to $\delta_0 - \Delta$, the total energy is

$$E = -B(\Delta) + \frac{h^2}{8m^*R^2} + \frac{6\pi R^3}{K}\left(\frac{\Delta}{S_0}\right)^2 . \tag{9}$$

(1) Ref. to mobility of photoelectrons.

K is the compressibility. Minimizing with respect to R and Δ

$$R = \frac{K\delta_0^2 m^*}{2\pi h^2}\left(\frac{\mathrm{d}B}{\mathrm{d}\delta}\right)^2, \tag{10}$$

$$\Delta = \frac{2\pi^2 h^6}{3K^2\delta_0^4 m^{*3}}\frac{1}{(\mathrm{d}B/\mathrm{d}\delta)^2}. \tag{11}$$

The extra mass associated with the region is

$$M_e' = 4\pi R^3 \varrho \Delta/\delta_0, \tag{12}$$

$$= -\frac{K\delta_0\varrho}{3}\left(\frac{\mathrm{d}B}{\mathrm{d}\delta}\right), \tag{13}$$

$$= \frac{\varrho h}{3}\left(\frac{2\pi K R}{m^*}\right)^{\frac{1}{2}}. \tag{14}$$

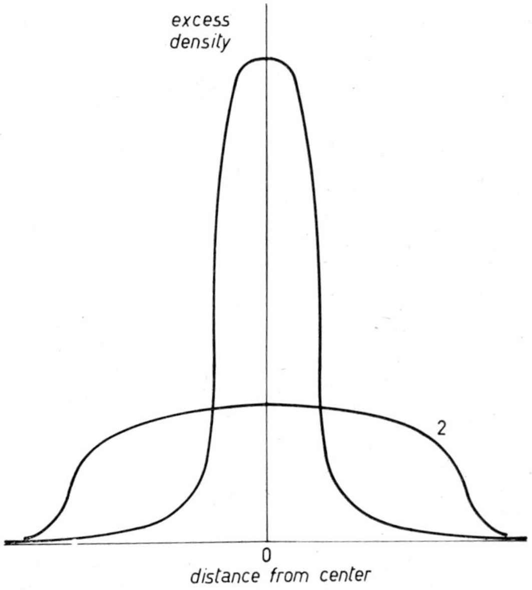

Fig. 5. – Schematic representation of the liquid density in the vicinity of an ion. Curve 1: all ions heavier than helium, positive helium ion. Curve 2: almost free electrons. In both cases it has been assumed that the ion complex moves through the liquid with a velocity less than that of sound.

Whatever the values of B and R, it turns out that this extra mass exceeds the extra mass that would have been present inside the region due to the long range electrostriction effects discussed previously. The present considerations cannot, therefore, decrease the effective mass of the ion, but may appreciably increase it.

Figure 5 illustrates the difference between this picture and the previous picture of the positive ion. In the case of the positive ion the wave function of the ion is highly localized and the density field is sharply peaked. In the case of the negative ion, the wave function of the electron is spread out over a large region and the density is increased slightly inside this region.

The experimental behavior of the mobility (Fig. 2) suggests that as the pressure is raised the negative ion becomes identical with the positive ion. According to Kuper's model, this would be a consequence of the bubble collapsing due to an increasing inwards pressure from the surrounding liquid. According to the alternative model, it would be due to the energy of the bottom of the conduction band initially decreasing with increasing density, but going through a minimum and then increasing. This might happen if at low densities the attractive forces between electrons and helium atoms were to predominate, but at higher densities the repulsive forces took over.

7. – The mobility at high fields strengths.

Assuming that the ions obey classical statistics, they can be treated like an ideal gas and shown to have an average thermal velocity

$$v_t = (3kT/M_i)^{\frac{1}{2}} . \tag{15}$$

At 0.5 °K, if the effective mass M_i is about 40 times the mass of a helium atom, then $v_t \sim 10^3$ cm s^{-1}. Note that this is much less than the velocity of sound in liquid helium, $u_1 = 2.4 \cdot 10^4$ cm s^{-1}.

In the presence of an electric field there is a uniform drift velocity $u = \mu E$ superimposed on these random thermal velocities. As long as $u \ll v_t$ this drift velocity is given by

$$u = elv/M_i v_t \tag{16}$$

where l is the mean free path. Under these circumstances, u is linearly proportioned to E. However, when $u \gg v_t$,

$$u = (2elE/M_i)^{\frac{1}{2}} \tag{17}$$

and u varies as $E^{\frac{1}{2}}$. The data of MEYER and REIF (²), similar to those given in Fig. 3 show that the linear relation between u and E breaks down when $u \sim 10^3$ cm s^{-1}. This strongly suggests a large effective mass M_i, but it would be valuable to have a theory of the mobility in the intermediate region where $u \sim v_t$ and to deduce the effective mass M_i from the shape of the curves in Fig. 3.

(²) Ref. to latest REIF papers.

REFERENCES

[1] G. CARERI, F. SCARAMUZZI and J. O. THOMSON: *Nuovo Cimento*, **13**, 186 (1959).
[2] G. CARERI, U. FASOLI and F. S. GAETA: *Nuovo Cimento*, **15**, 774 (1960).
[3] G. CARERI, F. SCARAMUZZI and J. O. THOMSON: *Nuovo Cimento*, **18**, 957 (1960).

[4] G. CARERI: *Progress in Low Temperature Physics*, edited by C. J. GORTER, Vol. **3** (Amsterdam, 1961), p. 58.
[5] S. CUNSOLO and P. MAZZOLDI: *Nuovo Cimento*, **20**, 949 (1961).
[6] L. MEYER and F. REIF: *Phys. Rev.*, **110**, 279 (1959); **119**, 1164 (1960); *Phys. Rev. Lett.*, **5**, 1 (1960); *Phys. Rev.*, **123**, 727 (1961).
[7] A. DAHM, J. LEVINE, J. PENLEY and T. M. SANDERS: *Proc. of the 7-th International Conference on Low Temperature Physics*, edited by G. M. GRAHAM and A. C. HOLLIS HALLETT (Toronto, 1961), p. 495.
[8] J. J. M. BEENAKKER, K. W. TACONIS, E. A. LYNTON, Z. DOKOUPIL and G. VAN SOEST: *Physica*, **18**, 433 (1952).
[9] K. R. ATKINS: *Phys. Rev.*, **116**, 1339 (1959).
[10] R. P. FEYNMAN: *Progress in Low Temperature Physics*, edited by C. J. GORTER Vol. **1**, (Amsterdam, 1955), chap. II.
[11] L. PAULING: *Journ. Chem. Phys.*, **1**, 56 (1933).
[12] A. V. PHELPS and S. C. BROWN: *Phys. Rev.*, **86**, 102 (1952).
[13] TA-YOU WU: *Phil. Mag.*, **22**, 837 (1936).
[14] E. HALÖIEN and J. MIDTAL: *Proc. Phys. Soc.*, A **68**, 815 (1955).
[15] C. G. KUPER: *Phys. Rev.*, **122**, 1007 (1961).
[16] K. R. ATKINS: Office of Naval Research Technical Report no. 3 on Contract N-onr 551 (28).

The Structure and the Mobility of Ions in Liquid Helium.

C. G. KUPER

Department of Theoretical Physics, St. Salvator's College - St. Andrews

1. – Introduction.

There has been a great deal of interest lately in the « many-body problem » — the derivation of macroscopic properties of matter in bulk from first principles. Much progress has been made, by what are essentially perturbation-theoretical methods, in describing homogeneous systems, *e.g.* the properties of imperfect Bose and Fermi gases, and there has been some progress also in the theory of impurities in such an imperfect gas. These imperfect gas models give some understanding of the properties of liquid ^{3}He and ^{4}He, and can also be applied to mixtures (*e.g.* the behaviour of ^{3}He atoms in liquid ^{4}He). But, of course, it must be remembered that these liquids are sufficiently far from « imperfect gases » to make any quantitative theory impossible using current techniques.

In the case of *ions* in liquid helium the position is even worse, on account of the Coulomb interactions. If perturbation theory were applicable, then the diffusion coefficient for ions, $D = kT\mu/e$ (where μ is the mobility) should be similar to that of ^{3}He in ^{4}He; as Atkins has pointed out, it is smaller, experimentally, by a factor $\sim 10^2$. The problem is thus quite outside the scope of present-day systematic techniques; all that one can really hope to do is to try to get a picture of the « dressed » ion by semi-intuitive means. ABE and AIZU [1] have constructed a theory based on a BOGOLIUBOV [2] model for pure ^{4}He, but their approach seems to me unconvincing: the « effective interaction » has to be postulated to reproduce *either* the correct spectrum of elementary excitations *or* the correct structure factor. Since a Bogoliubov gas satisfies the FEYNMAN [3] relation

$$E(k) = \hbar^2 k^2 / 2m S(k) , \tag{1}$$

the model cannot represent both these features correctly. Even more important, the « effective interaction » between He atoms is not necessarily at all similar to the « effective interaction » between an atom and an ion.

2. – Structure of the positive ion.

I believe that ATKINS' [4] theory of positive ions in liquid helium is essentially correct. However the following alternative derivation of his results (due to Dr. B. DURNEY) helps to clarify the nature of Atkins' approximations.

We will assume that He atoms are hard spheres with van der Waals' attractions outside the core radius, and that the interaction between an electron or a molecular ion and a neutral atom also has a hard core. The system, of N He atoms and one ion, is specified by the Hamiltonian

$$H = \sum_j p_j^2/2m_{\text{He}} + \sum_{i \neq j} \tfrac{1}{2} V_{ij} + \sum_j V_{j,\text{ion}} + p_{\text{ion}}^2/2m_{\text{ion}} \tag{2}$$

where

$$\begin{cases} V_{ij} = V(r_{ij}) = \infty\,, & r_{ij} < a\,, \\ \qquad\quad = -\gamma r_{ij}^{-6}\,, & r_{ij} > a\,, \end{cases} \tag{3}$$

and

$$\begin{cases} V_{j,\text{ion}} = V'(r_{j,\text{ion}}) = \infty\,, & r_{j,\text{ion}} < a\,, \\ \qquad\qquad = -\alpha e^2/\varepsilon r^4, & r_{j,\text{ion}} > a\,. \end{cases} \tag{4}$$

Here ε is the dielectric constant, a is the core diameter (say $a = 2.6$ Å for atom-heavy–ion interactions and $a = 1.3$ Å for atom-electron interactions), and α is the atomic polarizability, $N_0\alpha = 0.125$. If we approximate to the effect of the hard core by a pseudopotential

$$V_{\text{core}} = 2\hbar^2 a m^{-1} \delta(\boldsymbol{r})\,, \tag{5}$$

where m is the reduced mass for the ion-atom system, then we see that if $m \sim m_{\text{He}}$, the Fourier coefficients of V_{core} are small compared with the electrostatic part of $V_{j,\text{ion}}$

$$(i.e.\ V_{\text{e.s.}} = 0, \quad r < a; \qquad V_{\text{e.s.}} = -\alpha e^2/r^4, \quad r > a) \tag{5'}$$

but if $m \sim m_{\text{e}}$, the opposite is true.

Following Atkins, we will assume that the « bare » positive ion is a molecular ion, He_n $(n \geqslant 2)$. Let us assume that the ion is localized at the origin of co-ordinates, and consider two volume elements containing (on the average) the same number of atoms. Let V_0 be the volume of a volume element very far from the ion, and $V_0 - \delta V$ the volume of one less distant. Then, if p is the pressure, ϱ the chemical potential, $\mathscr{E}$ the electric field, Z the (macrocanonical) partition function and Z_g the grand partition function,

$$(6) \quad \begin{cases} p(V_0 - \delta V) = kT \ln Z_g(T, \zeta, V_0, \mathscr{E}), \\ p_0 V_0 = kT \ln Z_g(T, \zeta, V_0, 0), \end{cases}$$

$$(7) \quad \begin{cases} Z_g(T, \varrho, V, \mathscr{E}) = \sum \exp[N\zeta/kT] Z(N, T, V, \mathscr{E}), \\ \qquad = \sum \exp[N\zeta/kT] Z(N, T, V_0, 0) \exp[N\alpha e^2/2r^4 kT\varepsilon^2, \end{cases}$$

$$(8) \quad = Z_g(T, \zeta + \alpha e^2/2\varepsilon^2 r^4, V_0, 0).$$

Hence

$$(9) \quad \begin{cases} (p - p_0) V_0 = kT\{\ln Z_g(T, \zeta + \alpha e^2/2\varepsilon^2 r^4, V_0, 0) - \ln Z_g(T, \varrho, V_0, 0)\}, \\ \qquad = (\alpha e^2/2\varepsilon^2)\{(r + \Delta r)^{-4} - r^{-4}\}\partial \ln Z_g/\partial\zeta. \end{cases}$$

But

$$(10) \quad \bar{N}_0 = kT\, \partial \ln Z_g(\mathscr{E} = 0)/\partial\zeta,$$

so that

$$(p - p_0) V_0 = (\bar{N}_0 \alpha e^2/2\varepsilon^2)\Delta(r^{-4})$$

and a similar expression for $(p' - p)V$, where p' refers to a point nearer to the ion, so that, finally

$$(11) \quad \int_{p_0}^{p} V \mathrm{d}p = \bar{N}_0 \alpha e^2/2\varepsilon^2 r^4,$$

which is Atkins' result.

We see that Atkins' theory is similar in spirit to a Thomas-Fermi approximation — there is assumed to be a large number of particles in each volume element of substantially constant density and potential. The approximation must be good far from the ion, but will not be expected to hold very close to the ion. Nevertheless it will be of interest to see how far Atkins' model can account for the experimental results. Using the empirical equation of state for ^{4}He, Atkins finds, on integration of (11), that there is a central solid sphere, of radius

$$(12) \quad b_+ = 6.3\ \text{Å},$$

outside which there is a density field

$$\varrho - \varrho_0 \sim r^{-4} . \tag{13}$$

The excess mass quoted by Atkins is about 40 atomic masses. For transport properties the effective mass of the ion should be roughly the sum of the following three terms: (*a*) Atkins' excess mass, (*b*) the difference, $\frac{4}{3}\pi\varrho_0 b_+^3$, between the total and the excess mass inside the solid sphere (since within this radius all matter has to be displaced rigidly when the ion moves), and (*c*) the « hydrodynamic » mass, $\frac{1}{2}\cdot\frac{4}{3}\pi\varrho_0 b_+^3$. The total effective mass is 100 atomic masses. The scattering properties will be discussed in Section **4**.

3. – Structure of the negative ion.

Three possible structures for the « bare » negative ion have been advanced: that it is (*a*) a free electron, (*b*) an impurity (*e.g.* O^- [5]), and (*c*) a molecular complex, like the positive ion. Atkins has given some arguments against possibility (*c*). If indeed (*b*) or (*c*) were correct, then Atkins' theory should be applicable, and the negative and positive ionic mobilities should be identical (since the bare mass is only a small fraction of the total mass). It will therefore be assumed hereafter that alternative (*a*) is correct, *i.e. the bare negative ion is a free electron*; this is the only postulate compatible with the observed difference in mobility between positive and negative ions.

With this assumption, the small mass of the bare ion makes V_{core} small compared to $V_{\text{e.s.}}$ (see eqs. (5) and (5')). Hence, at least for *S*-wave scattering, $V_{\text{e.s.}}$ may be neglected. We will, in fact, neglect it entirely for the present.

In order to see how the positions of atoms in the liquid are correlated with the position of the electrons, let us localize the electron within a sphere of radius b_-. Then the electron wave function $\psi_e \sim b_-^{\frac{3}{2}}$ inside the sphere and ~ 0 outside. If now an atom enters the sphere, the surface of the atomic core has to be an additional nodal surface. The energy required is

$$\begin{cases} E_{\text{intrusion}} \sim \displaystyle\int \psi_e^* V_{\text{core}} \psi_e \, d^3 r , \\ \qquad \sim 2\pi\hbar^2 a / m_e b_-^3 . \end{cases} \tag{14}$$

Even if b_- is as large as 15 Å, $E_{\text{intrusion}} \sim 3\cdot 10^{-14}$ erg, which is still large compared to the zero-point kinetic energy of an atom, $E_0 \simeq 40$ cal/mole $\simeq 3\cdot 10^{-15}$ erg/atom. The electron wave packet thus acts as a barrier, nearly impenetrable to helium atoms.

The above arguments strongly suggest that a « bubble » should be a good model for the « dressed » negative ion (compare FERRELL [8], who has proposed a similar structure for a positronium atom in liquid helium) (*). There is clearly an optimum degree of localization of the electron. If b_- is too small, the electron's kinetic energy will be very large, while if b is too large (but still $\leqslant 15$ Å), then the work done against the zero-point pressure in removing atoms from the sphere of radius b_- becomes large. *Outside* the bubble, there will be an Atkins electrostriction region, but because the bubble turns out to be fairly large, the electrostrictive effects are relatively unimportant.

The size of the bubble is to be determined by minimizing the total energy, or, equivalently, by requiring that the pressure exerted by the electron on the liquid be equal to the pressure exerted by the liquid on the electron. If the van der Waals' attractions were of sufficiently long range compared to the bubble radius, then the pressure of the liquid on the bubble wall would be just the zero-point kinetic energy per unit volume,

$$p_0 = E_0\,\varrho_0 = 6\cdot 10^7 \text{ dyne cm}^{-2} = 60 \text{ atm}\,.$$

This corresponds to the assumption made in all perturbation-theoretical calculations. The opposite extreme, of bubble radius large compared to the range of the attractive forces, is nearer to the truth. Here the pressure is reduced by half of the van der Waals' potential of an atom deep in the liquid. A more careful calculation [7] gives (**)

$$p_{\text{v.d.w.}} = -38(1 - c^2/b_-^2) \tag{15}$$

where c is the mean interatomic distance, 3.6 Å, and p is in atmospheres.

The electrostatic interaction gives a pressure term

$$p_{\text{es}} = \varrho e^2\alpha/2b_-^4\, m_{\text{He}} = \frac{5.2\cdot 10^4 \text{ atm Å}^4}{b_-^4}\,. \tag{16}$$

The electron kinetic energy is $\pi\hbar^2/2mb_-^2$ (taking the electron wave function to be $\psi_{\text{e}} \propto r^{-1}\sin(\pi r/b_-)$ for definiteness), so that the outward pressure is

$$p_{\text{kin}} = \pi\hbar^2/4mb_-^5 = \frac{7.9\cdot 10^6 \text{ atm Å}^5}{b_-^5}\,. \tag{17}$$

(*) The first suggestion of a bubble structure in liquid He appears to be due to FERRELL [8] and more specifically, the bubble for the negative ion was first suggested by CARERI [6]. I wish to apologize to these authors for the misattribution quoted in ref. [7].

(**) This calculation is still not good enough to exhibit a surface tension contribution, but an *ad hoc* inclusion of the surface tension does not make much difference.

For equilibrium

$$p_0 + p_{\text{v.d.w.}} = p_{\text{kin}} - p_{\text{es}}\,, \tag{18}$$

which has been solved graphically, to give (Fig. 1)

$$b_- = 12.1\ \text{Å}\,. \tag{19}$$

For the negative ion, the effective mass is the sum of only the hydrodynamic and electrostrictive masses, which, for a sphere of radius 12.1 Å, are respectively $\simeq 82\ m_{\text{He}}$ and $\simeq 17\ m_{\text{He}}$, so that

$$m_{\text{eff}} \simeq 100\ m_{\text{He}}\,, \tag{20}$$

fortuitously close to the positive ion's effective mass.

Atkins has proposed an alternative model for the negative ion, a free electron in the conduction band. If the present assumption of a hard-core interaction is at all reasonable, then the zero-point kinetic energy of such an electron is very high, so that the bubble model is more plausible.

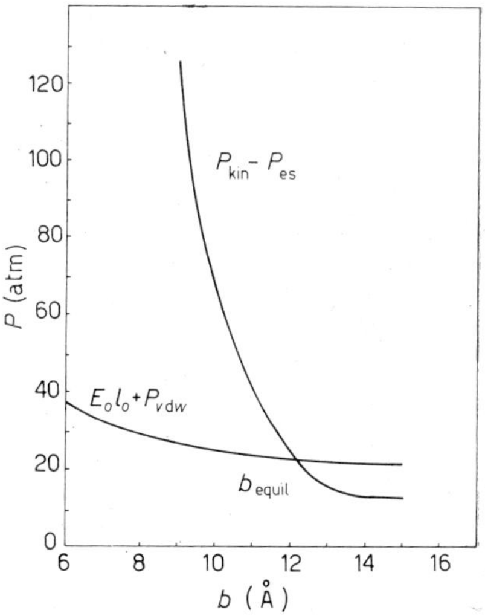

Fig. 1. – Graphical solution of eq. (18).

4. – Scattering and mobility.

The experiments of CARERI *et al.* and of MEYER and REIF [9] show that, at temperatures $\gtrsim 1\ {}^\circ\text{K}$, the predominant process scattering ions is collision with rotons. At lower temperatures both phonons and ^{3}He atoms become more important. I hope to show that the scattering cross-sections predicted on the basis of the electrostriction model for the positive ion and the bubble model for the negative ion are in reasonable accord with experiment.

The experimentally measured quantity is the *mobility*, and there are some complications when we relate the mobility to the cross-section. If we neglect the « persistence of velocity » in a collision, then the relation between mobility and cross-section is

$$\mu = (e/mN\sigma)\big(\langle v_1^2\rangle + \langle v_2^2\rangle\big)^{-1} \tag{21}$$

where $\langle v_1^2\rangle$ and $\langle v_2^2\rangle$ are the average squared velocities of carriers and scattering centres respectively (CARERI [6]). For ion-roton collisions, the thermal momentum of the two colliding particles are comparable, so that the persistence of velocity is unimportant. The « experimental » cross-sections for roton-ion collisions are those quoted by CARERI *et al.*, assuming the effective mass to be 100 m_{He}.

The assumption made in deriving (21) is that the ion loses all « memory » of its previous history in each collision. This is quite untrue when it collides with a « classical » particle of much smaller mass than its own, and Reif has pointed out that under these conditions the « persistence » of velocity cannot be ignored. The experimental ion-^{3}He cross-sections quoted are those of MEYER and REIF, with the correction from centre-of-mass to laboratory coordinates taken into account.

It can be shown that the electrostriction zone outside the solid or bubble core has little effect on the scattering [7], so that the scattering is, in an adequate approximation, elastic scattering by a rigid sphere, where the radius l of the sphere is the closest distance of approach of the ion and the scatterer, *i.e.*

$$l = b + b_0 , \tag{22}$$

where $b = b_+$ or b_-, and b_0 is the effective radius of the roton or ^{3}He atom. (I have not made any calculations for phonon scattering yet.) For the roton, we take the value $b_0 = 4$ Å, an estimate made by LANDAU and KHALATNIKOV [10] for roton-roton scattering. For the ^{3}He atom, we will take $b_0 = 1.3$ Å.

If the momentum of the incident particle is $\hbar k$, then when $kl \gg 1$ (« high-energy » scattering), the cross-section is twice geometrical. This is the case for roton-ion collisions, since $k \simeq k_0 = 2$ Å^{-1}, so that $kl_+ \simeq 20$; $kl_- \simeq 30$. Half of the scattering in this high-energy limit is isotropic, and the other half is forward [11]. The forward scattering plays no role in the gas-kinetic calculation of the mobility, so that we have

$$\sigma_{\text{roton-ion}} = \pi l^2 . \tag{23}$$

On the other hand for ^{3}He atoms, k will be small, $kl < 1$ (the « low-energy » region). Here only S-waves will be scattered, and the cross-section is four times geometrical,

$$\sigma_{^3\text{He-ion}} = 4\pi l^2 . \tag{24}$$

The situation becomes still simpler at higher temperatures, when the mean free path becomes $< l$. The ion can then be thought of as a sphere moving in a viscous medium, and the mobility follows from Stokes' law,

$$\mu = e/6\pi\eta l , \tag{25}$$

TABLE I. – *Table of cross-sections* (cm^2).

	Positive ion		Negative ion	
	Roton	^{3}He atom	Roton	^{3}He atom
Theoretical	$0.33 \cdot 10^{-13}$	$0.73 \cdot 10^{-13}$	$0.82 \cdot 10^{-13}$	$2.20 \cdot 10^{-13}$
Experimental	$0.57 \cdot 10^{-13}$	$0.31 \cdot 10^{-13}$	$0.93 \cdot 10^{-13}$	$2.12 \cdot 10^{-13}$

where η is the viscosity. This expression gives a reasonable description in He I, and in He II very near the λ-point. It should also apply in pure ^{3}He above say 1 °K, but so far as I am aware, no data are yet available to test its validity.

5. – Behaviour of ions under pressure.

The application of high pressures would be expected to have rather different effects on the positive and negative ions. The Atkins structure should grow with application of external pressure, since the distance from the centre at which the solidification pressure is reached will increase. But, as Atkins has pointed out, the surface tension of the liquid-solid interface is not known, and may greatly influence this conclusion. On the other hand the bubble should clearly shrink with application of external pressure, until it is small enough to have a solid zone outside it. Thus increase of pressure should increase the mobility of negative ions, while reducing the mobility of positive ions. Both of these effects are observed [6], [9], but there appear to be some quantitative difficulties. Thus Meyer and Reif find that the variation of mobility of positive ions is explicable entirely in terms of the variation in the number of rotons (since Δ, the roton energy, varies with pressure). They find no trace of growth of the « snowball » with pressure.

At moderate pressures (< 7 atm) the negative ion behaves in the manner predicted, but above 7 atm Meyer and Reif find that the positive and negative ion mobilities are identical. If this is confirmed, it must mean that the bubble collapses, and that the high-pressure structure of the positive and negative ions is identical. If the bubble model continues to be valid at high pressures, but with a « snowball » around it, then its effective mass will be rather smaller than the pure snowball, so that the negative ions should eventually become *more* mobile than the positives. The theoretical discussion of whether the bubble should collapse is difficult, since for this problem a « hard sphere » model for the electron-atom interaction is certainly quite inadequate.

6. – « Hot » ions.

There are indications that some new phenomena appear in the presence of high electric fields. The ionic velocity seems to depart from the (low field) linear dependence on the electric field, and to tend to a limiting value of about 40 to 45 m/s [9]. This is of the order of Landau's critical velocity, so that it looks as if the ions have been accelerated to such a speed that they can emit excitations (possibly more complicated than simple rotons). However at still higher fields, Meyer and Reif observe what they call the « runaway effect »: the ions behave like free particles, insofar as they retain all the energy given them by the applied field. A retarding field has to be at least as large as the acceleration field to stop the hot ions. If the limiting velocity were caused by emission of excitations, a Frank-Hertz effect should be seen, *i.e.* the ions should lose energy in discrete steps; no such effect is observed by MEYER and REIF.

A speculative interpretation is that the moving ion can create *and bind* virtual excitations, in such a way that its effective mass becomes very large as the velocity increases. For example, on account of the pressure dependence of Δ, it is possible to create rotons within the electrostriction zone of an ion, but these rotons will not have enough energy to escape from the neighbourhood of the ion.

REFERENCES

[1] R. ABE and K. AIZU: *Phys. Rev.*, **123**, 10 (1961).
[2] N. N. BOGOLIUBOV: *Journ. Phys.* (*USSR*), **11**, 23 (1947).
[3] R. P. FEYNMAN: *Phys. Rev.*, **94**, 262 (1954).
[4] K. R. ATKINS: *Phys. Rev.*, **116**, 1339 (1959).
[5] D. W. SWAN: *Proc. Phys. Soc.* (*London*), **76**, 36 (1960).
[6] G. CARERI: *Prog. Low Temp. Phys.*, vol. **3** (edited by GORTER, 1961), p. 58; G. CARERI, J. REUSS, F. SCARAMUZZI and J. O. THOMSON: *Proc. 5-th Int. Conf. Low Temp. Phys.* (Madison, 1957), p. 155; G. CARERI, F. SCARAMUZZI and J. O. THOMSON: *Nuovo Cimento*, **13**, 186 (1959); G. CARERI, S. CUNSOLO and F. DUPRÈ: *Proc. 7-th Int. Conf. Low Temp. Phys.* (Toronto, 1960), p. 498; G. CARERI, U. FASOLI and F. S. GAETA: *Nuovo Cimento*, **15**, 774 (1960).
[7] C. G. KUPER: *Phys. Rev.*, **122**, 1007 (1961).
[8] R. A. FERRELL: *Phys. Rev.*, **108**, 167 (1957).
[9] L. MEYER and F. REIF: *Phys. Rev.*, **110**, 279 (1958); **119**, 1164 (1960); *Phys. Rev. Lett.*, **5**, 1 (1960); *Proc. 7-th Int. Conf. Low Temp. Phys.* (Toronto, 1960), pp. 487, 501; *Phys. Rev.* **123**, 727 (1961).
[10] L. D. LANDAU and I. M. KHALATINKOV: *Žurn. Ėksp. Teor. Fiz.*, **19**, 637 (1949).
[11] N. F. MOTT and H. MASSEY: *The Theory of Atomic Collisions* (Oxford, 1949), p. 38.

Ions in Liquid Helium. Some Contemporary Aspects.

G. CARERI

Istituto di Fisica dell'Università - Roma

The behaviour of ions in liquid helium is a comparatively new subject. I am not going to present here a review [1] of what has been done, but I will try to show what the open problems are today in this field. In my opinion progress should be expected along the following main lines.

1. – The nature of the ions themselves.

The theoretical aspect of this problem is considered by ATKINS and KUPER elsewhere in this volume and will not be treated here.

Let me only mention that the experimental evidence is still quite indirect since it is obtained from the surface effect [2] and from the change of the mobility with the pressure [3, 4]. A direct measurement of the effective mass by cyclotron resonance would be most valuable, but experimental difficulties have so far delayed this program.

2. – Properties of the ionic plasma.

Besides the measurements of the recombination coefficient [5], the properties of this plasma have not received attention by theorist or experimentalist. Cooperative effects, like plasma oscillation, could presumibly be detected, because the ionc density is such as to have a Debye distance well below the dimensions of the experimental apparatus and a frequency in the convenient radio range.

3. – The interaction of the ions with the equilibrium excitation gas.

This problem has received considerable attention both by theorists and experimentalists. While there is not doubt about the experimental values of the low field mobility [6-9] in a wide temperature range (from the λ point to 0.6 °K), the pressure dependence is still somewhat controversial [3, 4]. The theory is in good progress [10, 11], but suffers from the lack of knowledge of the experimental value of the effective mass. There is little doubt that a mean free path picture is essentially correct, but one would like to understand why this approximation holds also close to the λ point.

4. – « Hot » ions.

By the term « hot ions » we mean ions that are moved by the electric field with a drift velocity comparable or larger than the average thermal velocity. This aspect of the ionic motion is the most interesting and quite mysterious at the present time. The available information lasts upon the following groups of experiments:

a) REIF and MEYER [6] measured the drift velocity with increasing high fields down to 0.5 °K, and found that a limiting value of about 40 m/s was reached. However these hot ions had some « runaway » properties, namely they no longer obeyed the electric « shutters » and were more sensitive to the actual potential of the electrodes rather than to the field, as if they were free particles. No quantitative data are reported as yet;

b) CARERI, DUPRÈ and MODENA [12] have measured the drift velocity at 0.2 °K by the magnetic deflection method with increasing fields, but they found that a limiting velocity of about 8 m/s was reached and not 40 m/s as one should have expected on the basis of the Reif and Meyer experiment;

c) CARERI, CUNSOLO and MAZZOLDI [13] have measured the drift velocity at moderately high fields with great accuracy in a narrow temperature range ((1.0÷0.9) °K), and found that above a critical field strength E_c the drift velocity did not increase linearly with the increasing field, but suffered a discontinuity. While the value of this critical field E_c depends on the temperature, the corresponding critical drift velocity appears to be a temperature-independent quantity and has the value (4.7 ± 0.15) m/s. A repetition of this phenomenon has been found at drift velocity twice as large. A possible explanation of this strange behaviour can be found in terms of the production of vortex rings behind the ion in motion, but this requires an exceedingly large value for the effective mass of the ions;

d) the behaviour of β-irradiated liquid helium under an applied field has been studied by GAETA [14]. He found a discontinuity in the plot of current density versus temperature at constant field, which occurs at the same values of these parameters as the first discontinuity experiments observed above under *c*). Perhaps the recombination process is inhibited when supercritical velocities are reached, but the reason for this hindrance is not clear at all.

While any conclusion or unified presentation of the hot ions behaviour seems premature, one can certainly say that the sharpness of the transition in *c*) and some other features give evidence for one of the two possibilities:

1) the effective mass of the single ions is indeed much larger than we thought, and then by creation of vortex rings of quantized circulation one can try to explain the above observed phenomena;

2) there is a new kind of cooperative phenomenon which is produced when the drift velocity is comparable to the thermal velocity, and changes the inertial properties of the beam.

At the present time one cannot go any further, but these remarks are sufficient to show the extreme interest of these possibilities.

5. – The interaction with the vorticity.

It has been found [15] that negative ions are extremely sensitive to the vorticity developed in a wide channel, while the positive ones are not. This effect can be easily exploited [16] to detect the threshold of the turbolent flow. But the nature of the phenomenon is still obscure, and it is not yet clear if the negative charges are trapped or only scattered by the vortical lines.

CARERI, MCCORMICK and SCARAMUZZI are at the present investigating the properties of rotating helium, in some buckets of different geometry where a beam of ions can travel. The first results show that a beam of positive ions is practically unaffected by the rotation, while a beam of negative ions is very much attenuated when it moves perpendicular to the rotation axis. The Feynman picture of rotating helium seems therefore substantiated by this experiment, but more quantitative work is still needed to relate the vorticity to the negative ion current attenuation.

Although any subdivision of a lively subject is artificial, and nature usually reserves more surprises than we expect, I believe the above outlined questions to be appealing enough to stimulate our work in this field.

REFERENCES

[1] G. CARERI: *Prog. in Low Temperature Physics*, vol. **3** (Amsterdam, 1961), p. 58.
[2] G. CARERI, U. FASOLI and F. S. GAETA: *Nuovo Cimento*, **15**, 774 (1960).
[3] S. CUNSOLO and P. MAZZOLDI: *Nuovo Cimento*, **20**, 949 (1961).
[4] L. MEYER and F. REIF: *Phys. Rev.*, **123**, 727 (1961).

[5] G. Careri and F. Gaeta: *Nuovo Cimento*, **20**, 152 (1961).
[6] F. Reif and L. Meyer: *Phys. Rev.*, **119**, 1164 (1960).
[7] G. Careri, F. Scaramuzzi and J. O. Thomson: *Nuovo Cimento*, **13**, 186 (1959).
[8] G. Careri, S. Cunsolo and F. Duprè: *Proc. VII Int. Conference Low Temp.* (Toronto, 1961), p. 498.
[9] A. Dahm, J. Levine, J. Penley and T. M. Sanders: *Proc. VII Int. Conference on Low Temperature Physics* (Toronto, 1961), p. 495.
[10] J. De Boer and A.'t Hooft: *Proc. VII Int. Conference on Low Temperature Physics* (Toronto, 1961), p. 510.
[11] G. G. Kuper: *Phys. Rev.*, **122**, 1007 (1961).
[12] G. Careri, F. Duprè and I. Modena: *Nuovo Cimento*, **22**, 318 (1961).
[13] G. Careri, S. Cunsolo and P. Mazzoldi: *Phys. Rev. Lett.*, **7**, 151 (1961).
[14] F. Gaeta: unpublished results.
[15] G. Careri, F. Scaramuzzi and J. O. Thomson: *Nuovo Cimento*, **18**, 957 (1961).
[16] G. Careri, F. Scaramuzzi and W. D. McCormick: *Proc. VII Int. Conference Low Temp.* (Toronto, 1961), p. 502.

Properties of Liquid ^{3}He Below 1 °K (*).

H. A. FAIRBANK

Yale University - New Haven, Conn.

1. – Introduction.

Since this is the first session on liquid ^{3}He it may be appropriate to make a few general introductory remarks. ^{3}He with nuclear spin $\frac{1}{2}$ is a Fermi particle. It's normal boiling points is 3.2 °K; critical temperature = 3.3 °K; critical pressure = 845 mm Hg; it exists as a liquid down to 0.008 °K and probably to 0 °K under saturated vapor pressure and as a solid only above 29 atmospheres pressure. Because of its scarcity (1 part in 10^7 of well helium is ^{3}He) it is only since 1948 when small quantities produced from the decay of ^{4}H became available that investigations of its properties have been possible.

The early experiments on liquid ^{3}He by OSBORNE, ABRAHAM and WEINSTOCK [1] and DAUNT and HEER [2] which showed that liquid ^{3}He was not a superfluid above 0.25 °K strengthened the view that liquid ^{4}He was a quantum Bose liquid whose superfluid properties were a direct consequence of the fact that its excitations obeyed Bose statistics. This left open the very interesting question as to how ordering occurs in liquid ^{3}He near 0 °K and whether it could be described also as a quantum Fermi liquid. The expectation that ^{3}He may prove to be as interesting as ^{4}He near 0 °K is now becoming evident. The lowest temperature range is clearly of greatest interest since it is here that quantum statistics may be expected to have the greatest influence on the properties and it is likewise the temperature region for which fundamental theoretical calculations are at present possible. I will make no attempt at completeness but will mention some of the significant theoretical and experimental results found so far.

(*) Assisted by the Office of Ordnance Research and the National Science Foundation.

2. – Theoretical predictions.

Consider first the behaviour of an ideal gas of non-interacting atoms obeying Fermi statistics. Near 0 °K, well below the Fermi degeneracy temperature, the nuclear susceptibility χ, specific heat C, entropy S, viscosity η, and thermal conductivity K have the following dependence on T:

$$\chi = \text{constant}\,, \qquad C \propto T\,, \qquad S \propto T\,, \qquad \eta \propto T^{-1}, \qquad K \propto T^{-2}\,.$$

In 1956 Landau [3, 4] included interactions between the atoms in his Fermi liquid theory which will be discussed at length in Professor SESSLER's lectures. The theory has been applied to ^{3}He by ABRIKOSOV and KHALATNIKOV [5], HONE [6], BEKAREVICH and KHALATNIKOV [7]. In this approach, the excited state of liquid ^{3}He is described in terms of a gas of quasi-particles, equal in number to the number of atoms. The ideal gas temperature-dependence of the quantities listed above is retained in this theory in the low temperature limit but the magnitudes are different. The theory allows calculation of the quantities in terms of some parameters which can be determined by experiment. Thus, the effective mass m^* of the quasi-particles is given in terms of the specific heat C by

$$m^*/m = C/C_{\text{f}}$$

where m is the mass of the bare ^{3}He atom and C_{f} is the ideal Fermi gas specific heat. The ideal gas relationship between all quantities is not preserved; for example, the exchange forces have a strong influence on the nuclear susceptiblity and little on the specific heat.

A dramatic prediction of the Fermi liquid theory is the existence of a new kind of sound propagation, called by Landau « zero sound ». The mean collision time τ for the quasi-particles in a Fermi liquid is inversely proportional to T^2 and hence the attenuation of orinary sound is also proportional to $1/T^2$. When $\omega\tau > 1$ ordinary sound can no longer be propagated (ω is the angular frequency of the wave). For $\omega\tau \gg 1$ propagation of a new type of wave, zero sound, with a different velocity is possible. Zero sound involves an oscillating distortion of the Fermi surface; in ordinary sound the Fermi surface is displaced as a whole. Since $\tau \approx 10^{-12}\,T^{-2}$ s (°K)^2 experimental detection of zero sound should require measurements at very low temperatures and/or very high frequencies and thus is a formidable project.

BRUECKNER and GAMMEL [8] have developed independently a theory of liquid ^{3}He applicable near 0 °K in which the properties of the quasi-particles are determined in terms of two-body forces between the atoms. The theory gives the following quantitative results:

$$\chi/\chi_f = 12.0\,,$$
$$C/C_f = m^*/m = 1.84\,,$$
$$\text{Compressibility} = 5.3\,\%\ \text{per atm}\,,$$
$$\text{Coefficient of thermal expansion} = -\,0.076\,T\,(°\text{K})^{-1}\,.$$

Probably the most exciting theoretical predictions on the properties of liquid ^{3}He concern the possibility of superfluidity. The conduction electrons in metals constitute another Fermi gas system which, under certain conditions, becomes superconducting. Recently, the possibility has been explored [10-15] that pairing of ^{3}He particles might occur in liquid ^{3}He at a low temperature similar to the BCS theory [16] pairing of electrons of opposite momentum and spin in superconductors. A rather firm prediction has been made that a phase transition to a highly correlated state will occur in this liquid. The pairing of Fermions is a result of *d* state interactions and should result in an energy gap in the excitation spectrum and therefore superfluidity. The predicted behaviour of the specific heat and nuclear magnetic susceptibility above and below the transition is shown in Fig. 1 and Fig. 2. Early estimates

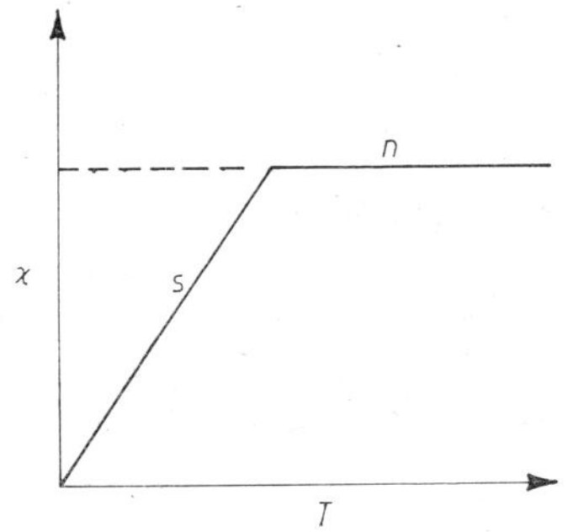

Fig. 1. – The specific heat of a Fermi liquid as a function of temperature, showing the predicted phase transition from the normal state *n* to the superfluid state *s*.

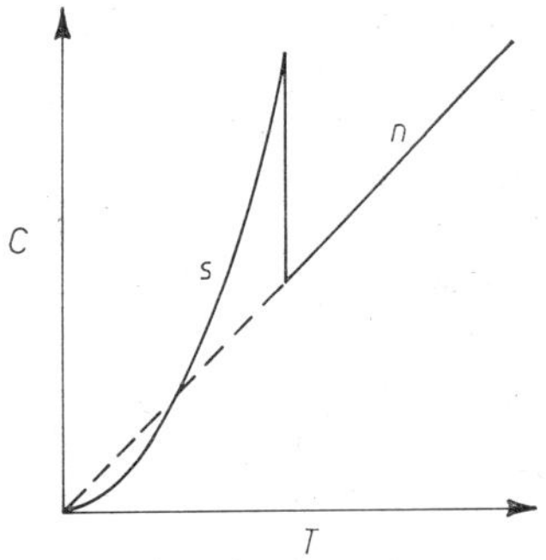

Fig. 2. – The nuclear magnetic susceptibility of a Fermi liquid as a function of temperature, showing the behaviour in the normal state *n* and superfluid state *s*.

gave the transition temperature at 0.08 °K which is a few millidegrees below the lowest temperature at which measurements on ^{3}He had been made up to that time. However, these estimates are not very firm and can be revised downward if required by experimental evidence. In any event, these exciting predictions have supplied ample incentive for further experimental measurements at the lowest possible temperatures.

3. – Experimental results.

3·1. *Specific heat.* – The low temperature behaviour of the specific heat is shown in Fig. 3 and Fig. 4. It is clear that no superfluid transition (as

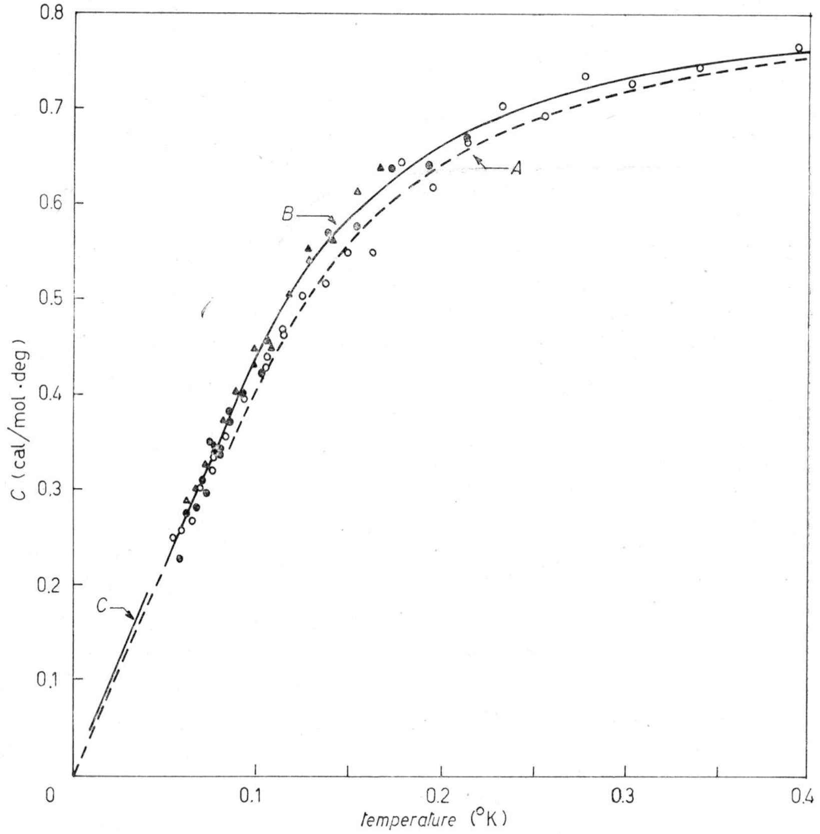

Fig. 3. – The specific heat of liquid ^{3}He as a function of temperature at or near saturated vapor pressure. The curve A is the best fit to the earlier data of BREWER, DAUNT and STREEDHAR [17]. Curve B is for the measurements of STRONGIN, ZIMMERMAN and FAIRBANK [19]. The points shown are for several different runs made by these authors. Curve C is a straight line fit of the data of ANDERSON *et al.* [18] given in Fig. 4.

typified in Fig. 1) occurs above 0.008 °K. $C \propto T$ as predicted by the Fermi liquid theories. Values of $m^*/m = C/C_f$ from each set of data are as follows: BREWER *et al.*, 2.00 ± 0.05; ANDERSON *et al.*, 2.35 ± 0.20; STRONGIN *et al.*,

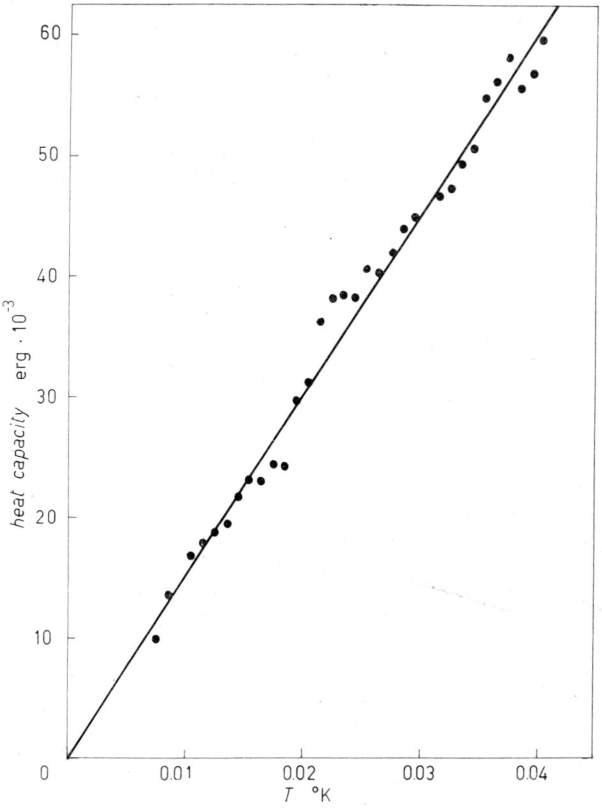

Fig. 4. – Heat capacity of a sample of liquid ^{3}He at a pressure of 14 cm Hg. (Data of ANDERSON, SALINGER, STEYERT and WHEATLEY [18]).

2.19 ± 0.13. From measurements of the specific heat and entropy under pressure BREWER and DAUNT [20] made an estimate of m^*/m for several different densities and obtained the results shown in Fig. 5. The increase in m^* with density is in qualitative agreement with the predictions of BRUECKNER and GAMMEL [8].

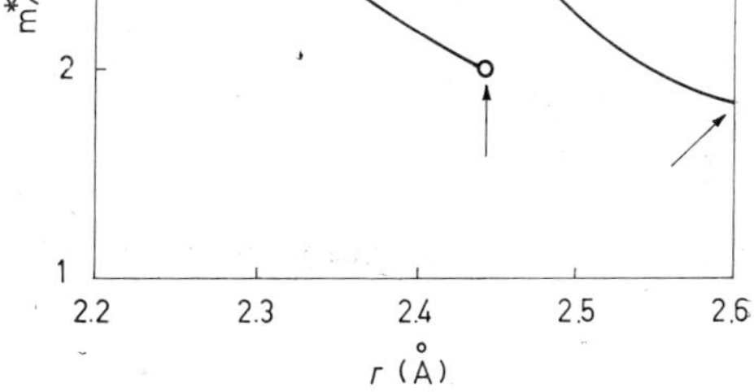

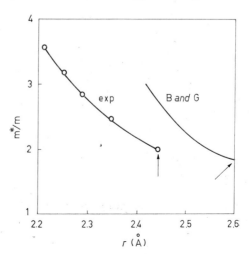

Fig. 5. – Variation of m^*/m with interatomic distance. Experimental curve is calculated from data by BREWER and DAUNT [20]. *B* and *G* curve is from the theory of BRUECKNER and GAMMEL [8].

3·2. *Nuclear magnetic susceptibility.* – The earlier results of FAIRBANK, ARD and WALTERS [21] are shown in Fig. 6. At the lowest temperatures the susceptibility becomes constant in agreement with the Fermi liquid theories. Recent measurements by ADAMS [22] down to 0.07 and ANDERSON, HART

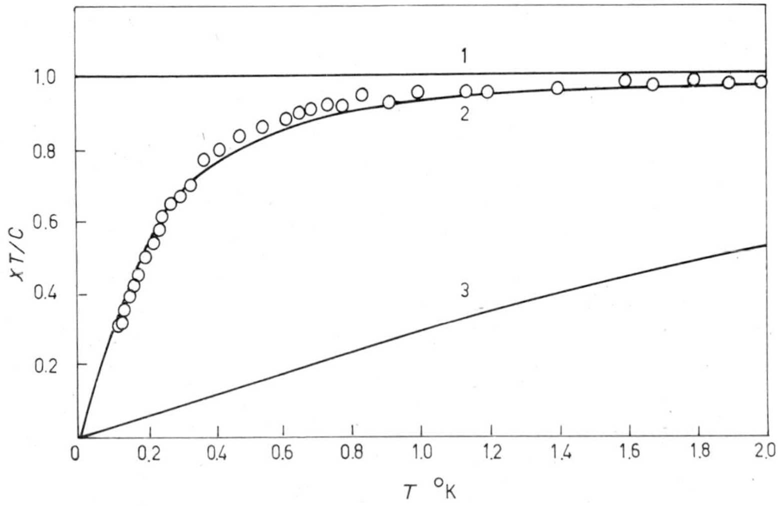

Fig. 6. – Plot of $\chi T/C$ *vs.* T from the data of FAIRBANK, ARD and WALTERS [21]. C is the normalizing Curie constant. Curve 1 represents the classical Curie-law behaviour. Curve 2 represents an ideal Fermi gas with degeneracy temperature of 0.48 °K. The points represent the experimental data normalized to curve 2 at 1.2 °K. Curve 3 represents an ideal Fermi gas with the same density and atomic mass as liquid ^{3}He.

and WHEATLEY [23] down to 0.031 °K confirm that the susceptibility is constant at the lowest temperatures with no suggestion of the behaviour of Fig. 2 expected in a superfluid transition. Note also from Fig. 6 that $\chi/\chi_f = 11$ (to be compared with the BRUECKNER and GAMMEL value of (12)).

3·3. *Thermal conductivity.* – Recent measurements of the thermal conductivity by ANDERSON, SALINGER and WHEATLEY [24] are shown in Fig. 7. Below 0.04 °K the smooth curve on the figure represents the relation

$$K = 48/T \text{ erg/cm s}$$

which is to be compared with the Fermi liquid value of

$$K = 13/T$$

evaluated by ANDERSON *et al.* [24] from the expression of ABRIKOSOV and KHALATNIKOV [5] for a Fermi liquid. Again no evidence for a superfluid phase transition was observed above 0.03 °K.

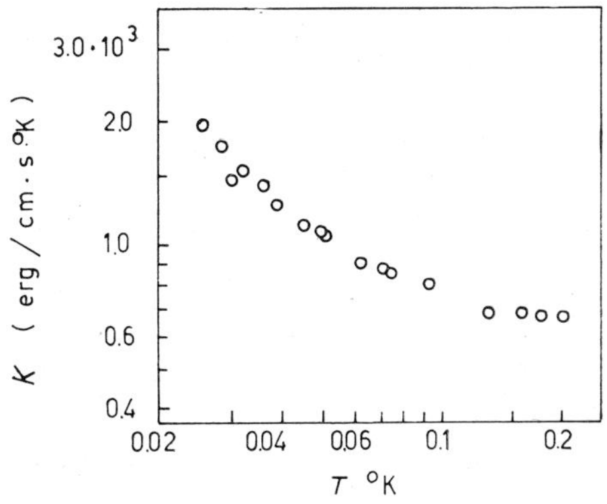

Fig. 7. – Log-plot of the thermal conductivity of ^{3}He (at 8 to 10 cm Hg pressure) *vs.* temperature from data of ANDERSON, SALINGER and WHEATLEY [24]. The values at the high temperature end are in good agreement with the earlier data of LEE and FAIRBANK [25].

3·4. *Coefficient of self diffusion.* – Using the spin-echo technique ANDERSON, HART and WHEATLEY [23] have measured the diffusion coefficient D from 0.15 °K to 0.032 °K. Between 0.1 °K and 0.03 °K they find D to be proportional to $T^{-1.55}$ with no indication of a phase transition. The Fermi liquid behaviour would require D to be proportional to T^{-2} in the low temperature limit. [Note: In a later paper which appeared after the summer school ANDERSON, REESE, SARWINSKI and WHEATLEY find that $D=1.54\cdot 10^{-6}\ T^{-2}\ \text{cm}^2\ {}^\circ\text{K}^2/\text{s}$ below $(0.04\div 0.05)$ °K in agreement with the temperature-dependence predicted by the theory; *Phys. Rev. Lett.*, **7**, 220 (1961)].

3·5. *Viscosity.* – ZINOV'EVA [26] has measured the viscosity of liquid ^{3}He down to 0.35 °K and found it to increase with decreasing temperature but not as rapidly as the predicted $1/T^2$ of a Fermi liquid. In view of the behaviour of the thermal conductivity, which does not show the Fermi liquid temperature-dependence above $(0.04\div 0.05)$ °K, this is not surprising. Measurements another order of magnitude lower in temperature are needed.

3·6. *Thermal boundary resistance.* – ANDERSON *et al.* [18] also report measurements of the thermal boundary resistance between liquid ^{3}He and the walls of the container (Epibond 100 A). They find that the boundary resistance (the temperature difference per unit heat flux) to be proportional to T^3 in agreement with the temperature-dependence predicted by BEKAREVICH and KHALATNIKOV treating the ^{3}He as a Fermi liquid. The magnitude of the thermal resistance is found to be 3 to 6 times lower than the theoretical value which, the authors point out, is reasonable agreement in view of the uncertainty in the velocity of sound in Epibond 100 A.

3·7. *Coefficient of thermal expansion* α_p. – Measurements of the coefficient of thermal expansion have been reported by BREWER and DAUNT [20] and by LEE, WALKER and FAIRBANK [27] down to 0.15 °K and by KERR and TAYLOR [28] down to 0.3 °K. α_p is negative below about 0.5 °K and appears to reach a minimum near 0.15 to 0.2 °K (at sat. vapor pressure). RIVES and MEYER [29] have taken measurements down to 0.045 °K and find a linear dependence on temperature below 0.1 °K given by $\alpha_{0.18\text{ atm}} = -(0.12 \pm 0.02)\ (°\text{K})^{-1} T$ compared with $\alpha_p = -0.76\, T\ (°\text{K})^{-1}$ predicted by BRUECKNER and GAMMEL [8]. They also find an unexpected minimum in the density in the neighborhood of 0.1 °K at 20 and 28 atm pressure.

In conclusion, there appears to be good evidence from both the thermal and transport properties that ^{3}He is behaving as a Fermi liquid below about 0.04 to 0.05 °K. Detection of zero sound would provide strong additional proof. It is clear also that no correlated superfluid phase exists above 0.008 °K at saturated vapor pressure. There are strong incentives to carry measurements to lower temperatures and higher pressures since the transition temperature may occur at a higher temperature at elevated pressure. The most interesting experiments on liquid ^{3}He may yet be ahead.

REFERENCES

[1] D. W. OSBORNE, B. WEINSTOCK and B. M. ABRAHAM: *Phys. Rev.*, **75**, 303 (1949).
[2] J. G. DAUNT and C. V. HEER: *Phys. Rev.*, **79**, 46 (1950).
[3] L. D. LANDAU: *Žurn. Ėksp. Teor. Fiz.*, **30**, 1058 (1956). [Translation, *Sov. Phys. J.E.T.P.*, **3**, 920 (1957)].
[4] L. D. LANDAU: *Žurn. Ėksp. Teor. Fiz.*, **32**, 59 (1957). [Translation, *Sov. Phys. J..E.T.P.*, **5**, 101 (1957)].
[5] A. A ABRIKOSOV and I. M. KHALATNIKOV: *Reports on progress* in *Physics*, vol. **22**, (London, 1959), p. 329.
[6] D. HONE: *Phys. Rev.*, **121**, 669 (1960).
[7] I. L. BEKARAVICH and I. M. KHALATNIKOV: *Žurn. Ėksp. Teor. Fiz.*, **39**, 1699 (1960). [Translation, *Sov. Phys. J.E.T.P.*, **12**, 1187 (1961)].
[8] K. A. BRUECKNER and J. L. GAMMEL: *Phys. Rev.*, **109**, 1040 (1958).
[9] K. A. BRUECKNER and K. R. ATKINS: *Phys. Rev. Lett.*, **1**, 315 (1958).
[10] L. N. COOPER, R. L. MILLS and A. M. SESSLER: *Phys. Rev.*, **114**, 1377 (1959).
[11] K. A. BRUECKNER, T. SODA, P. W. ANDERSON and P. MOREL: *Phys. Rev.*, **118**, 1442 (1960).
[12] V. J. EMERY and A. M. SESSLER: *Phys. Rev.*, **119**, 43 (1960).
[13] P. W. ANDERSON and P. MOREL: *Phys. Rev. Lett.*, **5**, 136 (1960).
[14] A. E. GLASSGOLD and A. M. SESSLER: *Nuovo Cimento*, **19**, 723 (1961).
[15] L. H. NOSANOV and R. VASUDEVAN: *Phys. Rev. Lett.*, **6**, 1 (1961).
[16] J. BARDEEN, L. N. COOPER and J. R. SCHRIEFFER: *Phys. Rev.*, **108**, 1175 (1957).
[17] D. F. BREWER, J. G. DAUNT and A. K. STREEDHAR: *Phys. Rev.*, **115**, 836 (1959).

[18] A. C. ANDERSON, G. L. SALINGER, W. A. STEYERT and J. C. WHEATLEY: *Phys. Rev. Lett.*, **6**, 331 (1961).
[19] M. STRONGIN, G. O. ZIMMERMAN and H. A. FAIRBANK: *Phys. Rev. Lett.*, **6**, 404 (1961).
[20] D. F. BREWER and J. G. DAUNT: *Phys. Rev.*, **115**, 843 (1959).
[21] W. M. FAIRBANK, W. B. ARD and G. K. WALTERS: *Phys. Rev.*, **95**, 566 (1954); *Proc. of the Symposium on Liquid and Solid* ^{3}He (Columbus, O., 1957), p. 205; *Suppl. Nuovo Cimento*, **9**, 297 (1958).
[22] E. D. ADAMS: *Ph. D. Thesis* (Durham, N.C., 1960).
[23] A. C. ANDERSON, H. R. HART jr. and J. C. WHEATLEY: *Phys. Rev. Lett.*, **5**, 133 (1960).
[24] A. C. ANDERSON, G. L. SALINGER and J. C. WHEATLEY: *Phys. Rev. Lett.*, **6**, 443 (1961).
[25] D. M. LEE and H. A. FAIRBANK: *Phys. Rev.*, **116**, 1359 (1959).
[26] K. N. ZINOV'EVA: *Žurn. Ėksp. Teor. Fiz.*, **34**, 609 (1958). [Translation, *Sov. Phys. J.E.T.P.*, **7**, 421 (1958)].
[27] D. M. LEE, H. A. FAIRBANK and E. J. WALKER: *Phys. Rev.*, **121**, 1258 (1961).
[28] E. C. KERR and R. D. TAYLOR: in *Proc. of the VI International Conference on Low Temperature Physics*, edited by GRAHAM and HALLETT (Toronto, 1961), p. 605.
[29] J. E. RIVES and H. MEYER: *Phys. Rev. Lett.*, **7**, 217 (1961).

Properties of ^{3}He Near the Melting Curve.

D. F. BREWER

Clarendon Laboratory - Oxford

The topics to be covered in this seminar are 1) the melting curve at low temperatures, between 0.06 °K and 1 °K, involving moderate pressures of 30 to 40 atm.: this is the region of main interest, and has been investigated in several different laboratories; 2) the melting curve up to 31 °K, corresponding to a melting pressure of around 3 400 atm; and (3) some of the solid and fluid properties near the melting curve. We shall also, where appropriate, make comparison with the corresponding properties in ^{4}He.

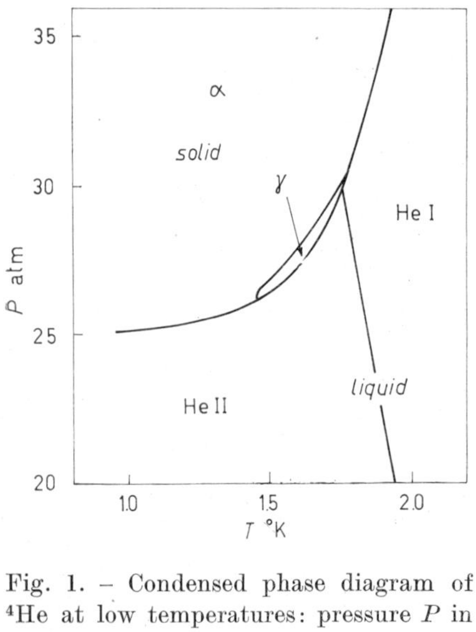

Fig. 1. – Condensed phase diagram of ^{4}He at low temperatures: pressure P in atmospheres plotted against temperature T in °K, showing the α and γ solid phases, and liquid helium I and II.

The slope of the melting curve, dP_m/dT, is given by the Clausius-Clapeyron equation

$$\frac{dP_m}{dT} = \frac{S_l - S_s}{V_l - V_s} = \frac{\Delta S_m}{\Delta V_m}, \qquad (1)$$

where S_l, S_s are the liquid and solid molar entropies, V_l and V_s the corresponding molar volumes, all at the melting curve. From Nernst's heat theorem, $\Delta S_m \to 0$ as $T \to 0$, so that the slope of the melting curve must be zero at the absolute zero (provided $V_l \neq V_s$). Since usually solids are denser than liquids ($V_s < V_l$) and more ordered ($S_s < S_l$) it also follows from eq. (1) that the slope of the melting curve at higher temperatures is generally positive. These two predictions are confirmed in ^{4}He, as shown in Fig. 1. Note that the slope is zero at the comparatively high temperature of 1 °K, because of the lambda phenomenon which results in a high degree of ordering in liquid He II. Also notable is the existence of three solid phases (only two of which are shown in the diagram): above about 15 °K and 1100 atm there is a f.c.c. structure [1] (β-phase); at intermediate pressures and temperatures, an h.c.p. structure (α-phase, shown in Fig. 1); and finally the γ-phase, recently discovered by VIGNOS and FAIRBANK [2] by the observation of a discontinuity in the velocity of sound across the α-γ phase boundary. The structure of the γ-phase is not known, but by analogy with ^{3}He, which as we shall see also has three solid phases, it seems likely to be b.c.c.

In ^{4}He, the entropies of both solid and liquid at the melting curve below 1 °K are extremely small. In ^{3}He, the situation is quite different, because the atom contains an uneven number of particles and has a nuclear spin $\frac{1}{2}$. Solid ^{3}He below 1 °K (where the lattice entropy is very small) will therefore have an entropy of $R \ln 2$ due to spin disorder, until such a low temperature is reached that the energy of interaction tending to align the spins becomes comparable with the thermal energy kT. The crucial question in solid ^{3}He is, what is the size and nature of this interaction energy. POMERANCHUK [3] in 1950 suggested that it would be the magnetic spin-spin interaction, $\sim \mu^2/r^3$ (μ is the magnetic moment, r the separation between atoms). Putting this equal to kT_c, one gets $T_c \sim 10^{-7}$ °K as the temperature where the solid entropy would drop appreciably below $R \ln 2$. Liquid ^{3}He also has a high entropy: if it were to behave as a Fermi-Dirac gas with the same particle mass and number density, its degeneracy temperature T_D at the saturated vapour pressure would be about 5 °K, increasing to over 6 °K at the melting curve. At about these temperatures, antiparallel spin alignment would begin, and below $T/T_D \sim 0.1$ the entropy would decrease nearly linearly with temperature; below $T/T_D \sim 1/7$, the entropy would be less than $R \ln 2$. In fact, liquid ^{3}He does not behave as a Fermi gas, but its entropy *is* less than $R \ln 2$ at about 0.3 °K. Thus if Pomeranchuk's prediction about the solid entropy is correct, the melting curve should have a minimum at this temperature (see eq. (1) with $S_l = S_s$) and then a negative slope, where the solid is more ordered than the liquid, down to $\sim 10^{-7}$ °K.

This is not the only theory of solid ^{3}He, however. BERNARDES and PRIMAKOFF [4] suggested that the amplitude of zero-point motion with these light atoms is so large that they cannot be regarded as rigidly fixed in the lattice. They calculated that the overlapping of the atomic wave functions would give an exchange interaction leading to spin ordering in the solid (antiparallel at low pressures, parallel at high pressures) at a few tenths of a degree. The melting curve would still have a minimum, but would clearly deviate from Pomeranchuk's at much lower temperatures.

The first measurements of the melting curve were made by WEINSTOCK, ABRAHAM and OSBORNE [5], using the blocked capillary technique. Instead of having a minimum, the melting curve became horizontal below 0.32 °K (see Fig. 2, curve *A*). As pointed out by Sydoriak (see, for example, ref. [6]), this result is in error and the blocked capillary method will not work below the temperature of the minimum, T_{min}: for at any lower cryostat tempera-

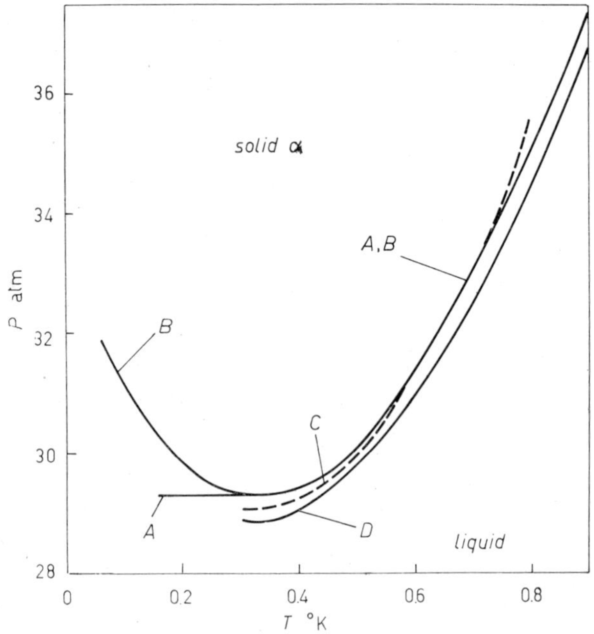

Fig. 2. – Condensed phase diagram of ^{3}He at low temperatures: pressure *P* in atmospheres plotted against temperature *T* in °K. Curve *A*: Weinstock *et al.* [5]; *B*: Baum *et al.* [8]; *C*, Lee *et al.* [10]; *D*: Sydoriak *et al.* [9].

ture *T*, there is always some point along the capillary tube where $T=T_{min}$, and a solid block will always form there at the minimum melting pressure, even though the cryostat pressure is higher. There were subsequently several indirect confirmations of the minimum [7], but direct measurements were not made until quite recently, when new methods were devised to overcome the difficulty of the blocked capillary. The first of these, by Baum *et al.* [8], made use of a resistance strain gauge as a pressure transducer, cemented to a thin-walled vessel containing the ^{3}He. Pressure changes in the vessel as its temperature was reduced below the minimum could then be deduced from the resistance changes, even though the vessel was sealed off from the outside world by the solid block in the capillary tube leading to it. The results (Fig. 2, curve *B*) showed a minimum at 0.32 °K and 29.3 atm, and at lower temperatures (down to 0.06 °K) a steady increase in P_m. Sydoriak *et al.* [9] solved the difficulty by using as their ^{3}He vessel a spring-loaded bellows; when pressures less than P_{min} were applied to liquid ^{3}He surrounding the bellows, pressures greater than P_{min} could be reached inside. They found a minimum at 0.33 °K and 28.9 atm (Fig. 2, curve *D*). LEE *et al.* [10], by dielectric constant measurements, placed the minimum at 29.1 atm and 0.32 °K (curve *C*). These differences in experimental values of P_{min} are outside the estimated error, and are so far unexplained.

The solid entropy along the melting curve can now be obtained by use of eq. (1), together with extrapolated values [9] of ΔV_m, the liquid entropy at the saturated vapour pressure, and the entropy of compression of the liquid up to the melting curve [10, 11]. Although, as discussed by Dr. H. A. FAIRBANK in the preceding seminar, the experimental entropy values at the saturated vapour pressure disagree slightly, the solid entropy comes out to be constant at $R \ln 2$ within a few percent down to 0.06 °K. This result places the antiferromagnetic transition temperature well below the value of 0.1 °K predicted by BERNARDES and PRIMAKOFF [4, 12].

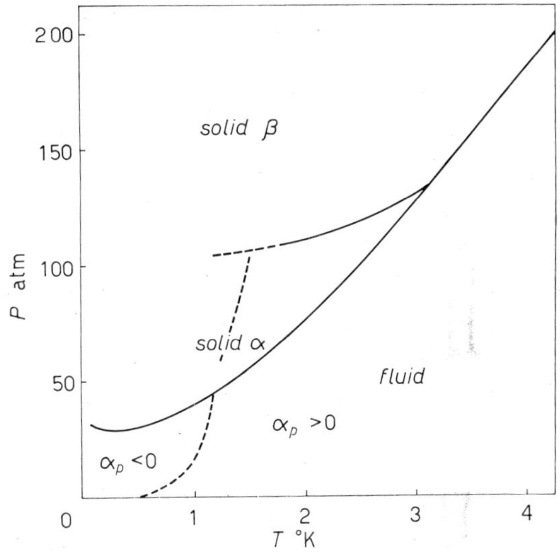

Fig. 3. – Condensed phase diagram of ^{3}He: pressure P in atmospheres plotted against temperature T in °K, showing the α and β solid phases, and the line dividing the liquid into regions of positive and negative expansion coefficient (α_p).

Experiments on the melting curve at much higher temperatures require special high-pressure techniques, and have been pursued exclusively at the Los Alamos Scientific Laboratory [13]. Figure 3 shows a part of this high pressure region, including the α phase (b.c.c [14]) and β phase (h.c.p [14]), the latter discovered by GRILLY and MILLS [13] in the course of measurements of ΔV_m. SCHUCH and MILLS [15] have recently found a new high pressure γ phase (f.c.c, not shown in Fig. 3), which is of great interest since both ^{3}He and ^{4}He are now known to have these three solid phases, depending in a qualitatively similar way on the pressure (unfortunately for historical reasons the labelling of the phases is not consistent in ^{3}He and ^{4}He). It does not necessarily follow from this analogous behaviour that the phase transitions in solid ^{3}He have no connection with its nuclear spin, although this does seem probable. Figure 3 also shows the line of demarcation between positive and negative expansion coefficients in the liquid; it appears from the thermal conductivity data of WALKER and FAIRBANK [16] that this line probably continues up through the solid phase, as indicated in the diagram (see also Fig. 5).

The melting pressure continues to increase with temperature up to measured values of around 3400 atm at 31 °K. A point of interest here, as with all melting curves, is whether there is a critical point or termination of the curve, above which solid and liquid are indistinguishable, indicated by ΔS_m becoming zero. No such termination has ever been found, but it does generally happen that ΔS_m decreases as P_m increases, and could well extrapolate to zero at finite P_m and T_m. With ^{3}He and ^{4}He, GRILLY and MILLS [13] have found that ΔS_m *increases* with P_m in the measured region, so that they are anomalous in this respect. However, the rate of rise of ΔS_m with P_m decreases, and by using an analytical expression for ΔS_m as a function of P_m, the data have been extrapolated to show that ΔS_m may go through a maximum at 4080 atm (^{3}He) and 3500 atm (^{4}He) and become zero at $P_m = 77\,000$ atm, $T_m = 235$ °K for ^{3}He, and $P_m = 62\,000$ atm, $T_m = 197$ °K for ^{4}He. Although these extrapolated values where $\Delta S_m = 0$ lie an order of magnitude beyond the measurements, they do show that, as in other cases, the data are not incompatible with the existence of critical points in the melting curves of ^{3}He and ^{4}He.

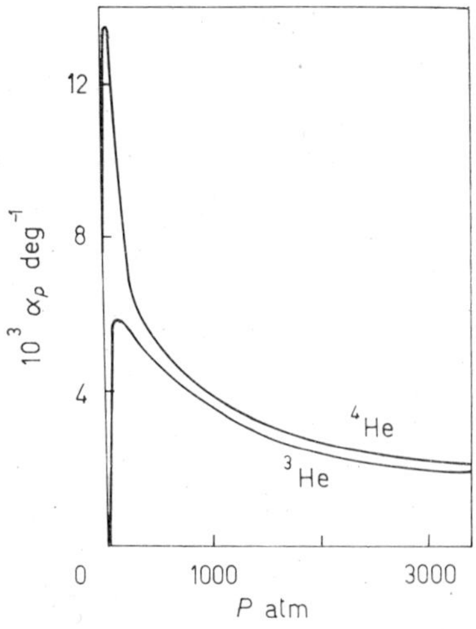

Fig. 4. – Isobaric expansion coefficient (α_p) of liquid ^{3}He and ^{4}He along their melting curves, plotted against pressure P in atmospheres.

Some of the interesting solid and liquid properties along the melting curve, also measured [9, 13] at Los Alamos, are the compressibility, expansion coefficient, and change in volume on melting. The compressibility behaves as expected, that is, it decreases smoothly with increasing temperature and pressure. The isobaric expansion coefficient, α_p, is shown in Fig. 4 as a function of pressure for ^{3}He and ^{4}He, in the high-temperature region. The behaviour is closely parallel in the two isotopes: as the temperature and pressure are reduced, α_p rises to a maximum, then drops sharply through zero to negative values. The measurements on ^{3}He have been taken down to 0.4 °K by SYDORIAK, MILLS and GRILLY [9], and are given in Fig. 5. In the liquid, α_p becomes negative below about 1.2 °K, which is consistent with the lower-pressure behaviour discussed in Prof. H. A. FAIRBANK's seminar and indicated in Fig. 3. The solid expansion coefficient was calculated from α_p in the liquid, with the assumption that $\beta_l \gg \beta_s$, where β is the compressibility, which is not known for the solid phase. The upper broken curve in Fig. 5 corresponds to $\beta_l = \beta_s$, the lower one to $\beta_l = 0.99\beta_s$, and they go through zero at 1.0 to 1.1 °K, in qualitative agreement with the calculations of BERNARDES and PRIMAKOFF [4] and of GOLDSTEIN [17], as well as with the observations of

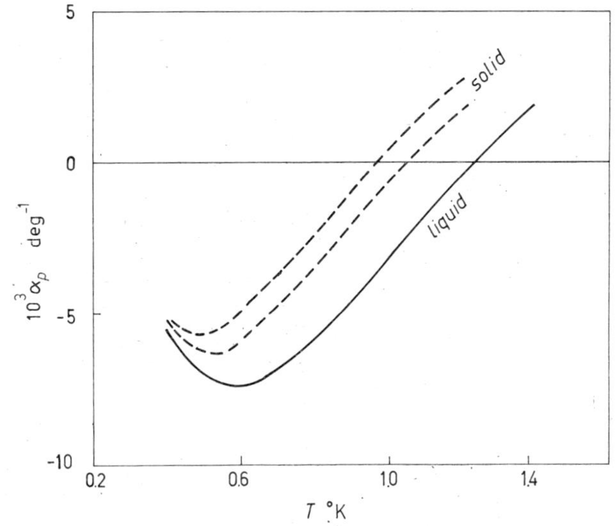

Fig. 5. – Isobaric expansion coefficient (α_p) of liquid and solid ^{3}He at the melting curve, plotted against temperature T in °K. The upper broken curve is calculated with the assumption that $\beta_s = \beta_l$, the lower broken curve with $\beta_s = 0.99\ \beta_l$ (see text).

WALKER and FAIRBANK already mentioned [16]. SYDORIAK *et al.* further concluded from their observations, by thermodynamic reasoning, that at the minimum in the melting curve, both α_p and $d\alpha_p/dT$ in the liquid become zero. If this is true, the behaviour of the liquid must change quite sharply between the melting curve and lower pressures, where α_p was found [10, 11] to be negative down to 0.15 °K (*).

Most of the useful information about ^{3}He which can be obtained from its properties at the melting curve, of which we have given a brief summary, are now known. Experiments at much higher pressures may give some interesting results on the variation of ΔS_m with increasing pressure. Measurements at temperatures below 0.06 °K will be rather difficult, and the information they give can be found more directly and with less trouble from separate measurements of the solid and liquid entropy by specific heat experiments

(*) Recent measurements by RIVES and MEYER and by BREWER and KEYSTON [18] also give conflicting evidence as to whether α_p under pressure becomes positive below 0.1 °K.

(see ref. [12]). The observations which do need clarification are those concerning the expansion coefficient of the liquid under pressure below 0.1 °K (see ref. [18]).

REFERENCES

[1] J. S. Dugdale and F. E. Simon: *Proc. Roy. Soc.*, A **218**, 291 (1953); R. L. Mills and A. F. Schuch: *Phys. Rev. Lett.*, **6**, 263 (1961).

[2] J. H. Vignos and H. A. Fairbank: *Phys. Rev. Lett.*, **6**, 265 (1961).

[3] I. Ia. Pomeranchuk: *Žurn. Ėksp. Teor. Fiz.*, **20**, 919 (1950).

[4] H. Primakoff: *Bull. Am. Phys. Soc.*, **2**, 63 (1957); N. Bernardes and H. Primakoff: *Phys. Rev. Lett.*, **2**, 290 (1959); **3**, 144 (1959); *Phys. Rev.*, **119**, 968 (1960).

[5] B. Weinstock, B. M. Abraham and D. W. Osborne: *Phys. Rev.*, **82**, 263 (1951); **85**, 158 (1952).

[6] E. F. Hammel: *Progress in Low Temperature Physics*, vol **1** (Amsterdam, 1955).

[7] W. M. Fairbank and G. K. Walters: *Symposium on Liquid and Solid* ^{3}He Columbus, 1957); B. M. Abraham, D. W. Osborne and B. Weinstock: *Symposium on Liquid and Solid* ^{3}He (Columbus, 1957).

[8] J. L. Baum, D. F. Brewer, J. G. Daunt and D. O. Edwards: *Phys. Rev. Lett.*, **3**, 127 (1959); D. O. Edwards, J. L. Baum, D. F. Brewer, J. G. Daunt and A. S. McWilliams: *Helium Three* (Columbus, 1960).

[9] S. G. Sydoriak, R. L. Mills and E. R. Grilly: *Phys. Rev. Lett.*, **4**, 495 (1960).

[10] D. M. Lee, H. A. Fairbank and E. J. Walker: *Bull. Am. Phys. Soc.*, **4**, 239 (1959).

[11] D. F. Brewer and J. G. Daunt: *Phys, Rev.*, **115**, 843 (1959).

[12] This conclusion is supported by specific heat and nuclear resonance experiments on solid ^{3}He: see, for example, *Helium Three* (Columbus, 1960). More recent specific heat experiments have been carried out by A. C. Anderson, G. L. Salinger, W. A. Steyert and J. C. Wheatley and by J. G. Daunt, D. O. Edwards *et al.*: private communication.

[13] E. R. Grilly and R. L. Mills: *Ann. of Phys.*, **8**, 1 (1959); R. H. Sherman and F. J. Edeskuty: *Ann. of Phys.*, **9**, 522 (1960).

[14] A. F. Schuch, E. R. Grilly and R. L. Mills: *Phys. Rev.*, **110**, 775 (1958).

[15] A. F. Schuch and R. L. Mills: *Phys. Rev. Lett.*, **6**, 596 (1961).

[16] E. J. Walker and H. A. Fairbank: *Phys. Rev. Lett.*, **5**, 139 (1960).

[17] L. Goldstein: *Phys. Rev.*, **112**, 1483 (1958); *Ann. of Phys.*, **8**, 390 (1959).

[18] Rives and Meyer: (*Phys. Rev. Lett.*, **7**, 217 (1961); Brewer and Keyston: (*Nature*, **191**, 1261 (1961).

Tipografia Compositori - Bologna - Italy